The Downfall

Emile Zola

TRANSLATED BY E. P. ROBINS

IAP © 2009

Zola, Emile.

The Downfall / Emile Zola – 1st ed.

1. Literature – fiction

CONTENTS

PART FIRST

1

In the middle of the broad, fertile plain that stretches away in the direction of the Rhine, a mile and a quarter from Mulhausen, the camp was pitched. In the fitful light of the overcast August day, beneath the lowering sky that was filled with heavy drifting clouds, the long lines of squat white shelter-tents seemed to cower closer to the ground, and the muskets, stacked at regular intervals along the regimental fronts, made little spots of brightness, while over all the sentries with loaded pieces kept watch and ward, motionless as statues, straining their eyes to pierce the purplish mists that lay on the horizon and showed where the mighty river ran.

It was about five o'clock when they had come in from Belfort; it was now eight, and the men had only just received their rations. There could be no distribution of wood, however, the wagons having gone astray, and it had therefore been impossible for them to make fires and warm their soup. They had consequently been obliged to content themselves as best they might, washing down their dry hard-tack with copious draughts of brandy, a proceeding that was not calculated greatly to help their tired legs after their long march. Near the canteen, however, behind the stacks of muskets, there were two soldiers pertinaciously endeavoring to elicit a blaze from a small pile of green wood, the trunks of some small trees that they had chopped down with their sword-bayonets, and that were obstinately determined not to burn. The cloud of thick, black smoke, rising slowly in the evening air, added to the general cheerlessness of the scene.

There were but twelve thousand men there, all of the 7th corps that the general, Felix Douay, had with him at the time. The 1st division had been ordered to Froeschwiller the day before; the 3d was still at Lyons, and it had been decided to leave Belfort and hurry to the front with the 2d division, the reserve artillery, and an incomplete division of cavalry. Fires had been seen at Lorrach. The sous-prefet at Schelestadt had sent a telegram announcing that the Prussians were preparing to pass the Rhine at Markolsheim. The general did not like his unsupported position on the extreme right, where he was cut off from communication with the other corps, and his movement in the direction of the frontier had been accelerated by the intelligence he had received the day before of the disastrous surprise at Wissembourg. Even if he should not be called on to face the enemy on his own front, he felt that he was likely at any moment to be ordered to march to the relief of the 1st corps. There must be fighting going on, away down the river near Froeschwiller, on that dark and threatening Saturday, that ominous 6th of August; there was premonition of it in the sultry air, and the stray puffs of wind passed shudderingly over the camp as if fraught with tidings of impending evil. And for two days the division had believed that it was marching forth to battle; the men had expected to find the Prussians in their front, at the termination of their forced march from Belfort to Mulhausen.

The day was drawing to an end, and from a remote corner of the camp the rattling drums and the shrill bugles sounded retreat, the sound dying away faintly in the distance on the still air of evening. Jean Macquart, who had been securing the tent and driving the pegs home, rose to his feet. When it began to be rumored that there was to be war he had left Rognes, the scene of the bloody drama in which he had lost his wife, Francoise and the acres that she brought him; he had re-enlisted at the age of thirty-nine, and been assigned to the 106th of the line, of which they were at that time filling up the cadres, with his old rank of corporal, and there were moments when he could not help wondering how it ever came about that he, who after Solferino had been so glad to quit the service and cease endangering his own and other people's lives, was again wearing the *capote* of the infantry man. But what is a man to do, when he has neither trade nor calling, neither wife, house, nor home, and his heart is heavy with mingled rage and sorrow? As well go and have a shot at the enemy, if they come where they are not wanted. And he remembered his old battle cry: Ah! bon sang! if he had no longer heart for honest toil, he would go and defend her, his country, the old land of France!

When Jean was on his legs he cast a look about the camp, where the summons of the drums and bugles, taken up by one command after another, produced a momentary bustle, the conclusion of the business of the day. Some men were running to take their places in the ranks, while others, already half asleep, arose and stretched their stiff limbs with an air of exasperated weariness. He stood waiting patiently for roll-call, with that cheerful imperturbability and determination to make the best of everything that made him the good soldier that he was. His comrades were accustomed to say of him that if he had only had education he would have made his mark. He could just barely read and write, and his aspirations did not rise even so high as to a sergeantcy. Once a peasant, always a peasant.

But he found something to interest him in the fire of green wood that was still smoldering and sending up dense volumes of smoke, and he stepped up to speak to the two men who were busying themselves over it, Loubet and Lapoulle, both members of his squad.

"Quit that! You are stifling the whole camp."

Loubet, a lean, active fellow and something of a wag, replied:

"It will burn, corporal; I assure you it will-why don't you blow, you!"

And by way of encouragement he bestowed a kick on Lapoulle, a colossus of a man, who was on his knees puffing away with might and main, his cheeks distended till they were like wine-skins, his face red and swollen, and his eyes starting from their orbits and streaming with tears. Two other men of the squad, Chouteau and Pache, the former stretched at length upon his back like a man who appreciates the delight of idleness, and the latter engrossed in the occupation of putting a patch on his trousers, laughed long and loud at the ridiculous expression on the face of their comrade, the brutish Lapoulle.

Jean did not interfere to check their merriment. Perhaps the time was at hand when they would not have much occasion for laughter, and he, with all his seriousness and his humdrum, literal way of taking things, did not consider that it was part of his duty to be melancholy, preferring rather to close his eyes or look the other way when his men were enjoying themselves. But his attention was attracted to a second group not far away, another soldier of his squad, Maurice Levasseur, who had been conversing earnestly for near an hour with a civilian, a red-haired gentleman who was apparently about thirty-six years old, with an intelligent, honest face, illuminated by a pair of big protruding blue eyes, evidently the eyes of a near-sighted man. They had been joined by an artilleryman, a quartermaster-sergeant from the reserves, a knowing, self-satisfied-looking person with brown mustache and imperial, and the three stood talking like old friends, unmindful of what was going on about them.

In the kindness of his heart, in order to save them a reprimand, if not something worse, Jean stepped up to them and said:

"You had better be going, sir. It is past retreat, and if the lieutenant should see you-" Maurice did not permit him to conclude his sentence:

"Stay where you are, Weiss," he said, and turning to the corporal, curtly added: "This gentleman is my brother-in-law. He has a pass from the colonel, who is acquainted with him."

What business had he to interfere with other people's affairs, that peasant whose hands were still reeking of the manure-heap? *He* was a lawyer, had been admitted to the bar the preceding autumn, had enlisted as a volunteer and been received into the 106th without the formality of passing through the recruiting station, thanks to the favor of the colonel; it was true that he had condescended to carry a musket, but from the very start he had been conscious of a feeling of aversion and rebellion toward that ignorant clown under whose command he was.

"Very well," Jean tranquilly replied; "don't blame me if your friend finds his way to the guardhouse."

Thereon he turned and went away, assured that Maurice had not been lying, for the colonel, M. de Vineuil, with his commanding, high-bred manner and thick white mustache bisecting his long yellow face, passed by just then and saluted Weiss and the soldier with a smile. The colonel pursued his way at a good round pace toward a farmhouse that was visible off to the right among the plum trees, a few hundred feet away, where the staff had taken up their quarters for the night. No one could say whether the general commanding the 7th corps was there or not; he was in deep affliction on account of the death of his brother, slain in the action at Wissembourg. The brigadier, however, Bourgain-Desfeuilles, in whose command the 106th was, was certain to be there, brawling as loud as ever, and trundling his fat body about on his short, pudgy legs, with his red nose and rubicund face, vouchers for the good dinners he had eaten, and not likely ever to become top-heavy by reason of excessive weight in his upper story. There was a stir and movement about the farmhouse that seemed to be momentarily increasing; couriers and orderlies were arriving and departing every minute; they were awaiting there, with feverish anxiety of impatience, the belated dispatches which should advise them of the result of the battle that everyone, all that long August day, had felt to be imminent. Where had it been fought? what had been the issue? As night closed in and darkness shrouded the scene, a foreboding sense of calamity seemed to settle down upon the orchard, upon the scattered stacks of grain about the stables, and spread, and envelop them in waves of inky blackness. It was said, also, that a Prussian spy had been caught roaming about the camp, and that he had been taken to the house to be examined by the general. Perhaps Colonel de Vineuil had received a telegram of some kind, that he was in such great haste.

Meantime Maurice had resumed his conversation with his brother-in-law Weiss and his cousin Honore Fouchard, the quartermaster-sergeant. Retreat, commencing in the remote distance, then gradually swelling in volume as it drew near with its blare and rattle, reached them, passed them, and died away in the solemn stillness of the twilight; they seemed to be quite unconscious of it. The young man was grandson to a hero of the Grand Army, and had first seen the light at Chene-Populeux, where his father, not caring to tread the path of glory, had held an ill-paid position as collector of taxes. His mother, a peasant, had died in giving him birth, him and his twin sister Henriette, who at an early age had become a second mother to him, and that he was now what he was, a private in the ranks, was owing entirely to his own imprudence, the headlong dissipation of a weak and enthusiastic nature, his money squandered and his substance wasted on women, cards, the thousand follies of the all-devouring minotaur, Paris, when he had concluded his law studies there and his relatives had impoverished themselves to make a gentleman of him. His conduct had brought his father to the grave; his sister, when he had stripped her of her little all, had been so fortunate as to find a husband in that excellent young fellow Weiss, who had long held the position of accountant in the great sugar refinery at Chene-Populeux, and was now foreman for M. Delaherche, one of the chief cloth manufacturers of Sedan. And Maurice, always cheered and encouraged when he saw a prospect of amendment in himself, and equally disheartened when his good resolves failed him and he relapsed, generous and enthusiastic but without steadiness of purpose, a weathercock that shifted with every varying breath of impulse, now believed that experience had done its work and taught him the error of his ways. He was a small, light-complexioned man, with a high, well-developed forehead, small nose, and retreating chin, and a pair of attractive gray eyes in a face that indicated intelligence; there were times when his mind seemed to lack balance.

Weiss, on the eve of the commencement of hostilities, had found that there were family matters that made it necessary for him to visit Mulhausen, and had made a hurried trip to that city. That he had been able to employ the good offices of Colonel de Vineuil to afford him an opportunity of shaking hands with his brother-in-law was owing to the circumstance that that officer was own uncle to young Mme. Delaherche, a pretty young widow whom the cloth manufacturer had married the year previous, and whom Maurice and Henriette, thanks to their being neighbors, had known as a girl. In addition to the colonel, moreover, Maurice had discovered that the captain of his company, Beaudoin, was an acquaintance of Gilberte, Delaherche's young wife; report even had it that she and the captain had been on terms of intimacy in the days when she was Mme. Maginot, living at Meziere, wife of M. Maginot, the timber inspector.

"Give Henriette a good kiss for me, Weiss," said the young man, who loved his sister passionately. "Tell her that she shall have no reason to complain of me, that I wish her to be proud of her brother."

Tears rose to his eyes at the remembrance of his misdeeds. The brother-in-law, who was also deeply affected, ended the painful scene by turning to Honore Fouchard, the artilleryman.

"The first time I am anywhere in the neighborhood," he said, "I will run up to Remilly and tell Uncle Fouchard that I saw you and that you are well."

Uncle Fouchard, a peasant, who owned a bit of land and plied the trade of itinerant butcher, serving his customers from a cart, was a brother of Henriette's and Maurice's mother. He lived at Remilly, in a house perched upon a high hill, about four miles from Sedan.

"Good!" Honore calmly answered; "the father don't worry his head a great deal on my account, but go there all the same if you feel inclined."

At that moment there was a movement over in the direction of the farmhouse, and they beheld the straggler, the man who had been arrested as a spy, come forth, free, accompanied only by a single officer. He had likely had papers to show, or had trumped up a story of some kind, for they were simply expelling him from the camp. In the darkening twilight, and at the distance they were, they could not make him out distinctly, only a big, square-shouldered fellow with a rough shock of reddish hair. And yet Maurice gave vent to an exclamation of surprise.

"Honore! look there. If one wouldn't swear he was the Prussian-you know, Goliah!"

The name made the artilleryman start as if he had been shot; he strained his blazing eyes to follow the receding shape. Goliah Steinberg, the journeyman butcher, the man who had set him and his father by the ears, who had stolen from him his Silvine; the whole base, dirty, miserable story, from which he had not yet ceased to suffer! He would have run after, would have caught him by the throat and strangled him, but the man had already crossed the line of stacked muskets, was moving off and vanishing in the darkness.

"Oh!" he murmured, "Goliah! no, it can't be he. He is down yonder, fighting on the other side. If I ever come across him-"

He shook his fist with an air of menace at the dusky horizon, at the wide empurpled stretch of eastern sky that stood for Prussia in his eyes. No one spoke; they heard the strains of retreat again, but very distant now, away at the extreme end of the camp, blended and lost among the hum of other indistinguishable sounds.

8

"Fichtre!" exclaimed Honore, "I shall have the pleasure of sleeping on the soft side of a plank in the guard-house unless I make haste back to roll-call. Good-night-adieu, everybody!"

And grasping Weiss by both his hands and giving them a hearty squeeze, he strode swiftly away toward the slight elevation where the guns of the reserves were parked, without again mentioning his father's name or sending any word to Silvine, whose name lay at the end of his tongue.

The minutes slipped away, and over toward the left, where the 2d brigade lay, a bugle sounded. Another, near at hand, replied, and then a third, in the remote distance, took up the strain. Presently there was a universal blaring, far and near, throughout the camp, whereon Gaude, the bugler of the company, took up his instrument. He was a tall, lank, beardless, melancholy youth, chary of his words, saving his breath for his calls, which he gave conscientiously, with the vigor of a young hurricane.

Forthwith Sergeant Sapin, a ceremonious little man with large vague eyes, stepped forward and began to call the roll. He rattled off the names in a thin, piping voice, while the men, who had come up and ranged themselves in front of him, responded in accents of varying pitch, from the deep rumble of the violoncello to the shrill note of the piccolo. But there came a hitch in the proceedings.

"Lapoulle!" shouted the sergeant, calling the name a second time with increased emphasis.

There was no response, and Jean rushed off to the place where Private Lapoulle, egged on by his comrades, was industriously trying to fan the refractory fuel into a blaze; flat on his stomach before the pile of blackening, spluttering wood, his face resembling an underdone beefsteak, the warrior was now propelling dense clouds of smoke horizontally along the surface of the plain.

"Thunder and ouns! Quit that, will you!" yelled Jean, "and come and answer to your name."

Lapoulle rose to his feet with a dazed look on his face, then appeared to grasp the situation and yelled: "Present!" in such stentorian tones that Loubet, pretending to be upset by the concussion, sank to the ground in a sitting posture. Pache had finished mending his trousers and answered in a voice that was barely audible, that sounded more like the mumbling of a prayer. Chouteau, not even troubling himself to rise, grunted his answer unconcernedly and turned over on his side.

Lieutenant Rochas, the officer of the guard, was meantime standing a few steps away, motionlessly awaiting the conclusion of the ceremony. When Sergeant Sapin had finished calling the roll and came up to report that all were present, the officer, with a glance at Weiss, who was still conversing with Maurice, growled from under his mustache:

"Yes, and one over. What is that civilian doing here?"

"He has the colonel's pass, Lieutenant," explained Jean, who had heard the question.

Rochas made no reply; he shrugged his shoulders disapprovingly and resumed his round among the company streets while waiting for taps to sound. Jean, stiff and sore after his day's march, went and sat down a little way from Maurice, whose murmured words fell indistinctly upon his unlistening ear, for he, too, had vague, half formed reflections of his own that were stirring sluggishly in the recesses of his muddy, torpid mind.

Maurice was a believer in war in the abstract; he considered it one of the necessary evils, essential to the very existence of nations. This was nothing more than the logical sequence of his course in embracing those theories of evolution which in those days exercised such a potent influence on our young men of intelligence and education. Is not life itself an unending battle? Does not all nature owe its being to a series of relentless conflicts, the survival of the fittest, the maintenance and renewal of force by unceasing activity; is not death a necessary condition to young and vigorous life? And he remembered the sensation of gladness that had filled his heart when first the thought occurred to him that he might expiate his errors by enlisting and defending his country on the frontier. It might be that France of the plebiscite, while giving itself over to the Emperor, had not desired war; he himself, only a week previously, had declared it to be a culpable and idiotic measure. There were long discussions concerning the right of a German prince to occupy the throne of Spain; as the question gradually became more and more intricate and muddled it seemed as if everyone must be wrong, no one right; so that it was impossible to tell from which side the provocation came, and the only part of the entire business that was clear to the eyes of all was the inevitable, the fatal law which at a given moment hurls nation against nation. Then Paris was convulsed from center to circumference; he remembered that burning summer's night, the tossing, struggling human tide that filled the boulevards, the bands of men brandishing torches before the Hotel de Ville, and yelling: "On to Berlin! on to Berlin!" and he seemed to hear the strains of the Marseillaise, sung by a beautiful, stately woman with the face of a queen, wrapped in the folds of a flag, from her elevation on the box of a coach. Was it all a lie, was it true that the heart of Paris had not beaten then? And then, as was always the case with him, that condition of nervous excitation had been succeeded by long hours of doubt and disgust; there were all the small annoyances of the soldier's life; his arrival at the barracks, his examination by the adjutant, the fitting of his uniform by the gruff sergeant, the malodorous bedroom with its fetid air and filthy floor, the horseplay and coarse language of his new comrades, the merciless drill that stiffened his limbs and benumbed his brain. In a week's time, however, he had conquered his first squeamishness, and from that time forth was comparatively contented with his lot; and when the regiment was at last ordered forward to Belfort the fever of enthusiasm had again taken possession of him.

For the first few days after they took the field Maurice was convinced that their success was absolutely certain. The Emperor's plan appeared to him perfectly clear: he would advance four hundred thousand men to the left bank of the Rhine, pass the river before the Prussians had completed their preparations, separate northern and southern Germany by a vigorous inroad, and by means of a brilliant victory or two compel Austria and Italy to join hands immediately with France. Had there not been a short-lived rumor that that 7th corps of which his regiment formed a part was to be embarked at Brest and landed in Denmark, where it would create a diversion that would serve to neutralize one of the Prussian armies? They would be taken by surprise; the arrogant nation would be overrun in every direction and crushed utterly within a few brief weeks. It would be a military picnic, a holiday excursion from Strasbourg to Berlin. While they were lying inactive at Belfort, however, his former doubts and fears returned to him. To the 7th corps had been assigned the duty of guarding the entrance to the Black Forest; it had reached its position in a state of confusion that exceeded imagination, deficient in men, material, everything. The 3d division was in Italy; the 2d cavalry brigade had been halted at Lyons to check a threatened rising among the people there, and three batteries had straggled off in some direction-where, no one could say. Then their destitution in the way of stores and supplies was something wonderful; the depots at Belfort, which were to have furnished everything, were empty; not a sign of a tent, no mess-kettles, no flannel belts, no hospital supplies, no farriers' forges, not even a horse-shackle. The quartermaster's and medical departments were without trained assistants. At the very last moment it was discovered that thirty thousand rifles were practically useless owing to the absence of some small pin or other interchangeable mechanism about the breech-blocks, and the officer who posted off in hot haste to Paris succeeded with the greatest difficulty in securing five thousand of the missing implements. Their inactivity, again, was another matter that kept him on pins and needles; why did they idle away their time for two weeks? why did they not advance? He saw clearly that each day of delay was a mistake that could never be repaired, a chance of victory gone. And if the plan of campaign that he had dreamed of was clear and precise, its manner of execution was most lame and impotent, a fact of which he was to learn a great deal more later on and of which he had then only a faint and glimmering perception: the seven army corps dispersed along the extended frontier line en echelon, from Metz to Bitche and from Bitche to Belfort; the many regiments and squadrons that had been recruited up to only half-strength or less, so that the four hundred and thirty thousand men on paper melted away to two hundred and thirty thousand at the outside; the jealousies among the generals, each of whom thought only of securing for himself a marshal's baton, and gave no care to supporting his neighbor; the frightful lack of foresight, mobilization and concentration being carried on simultaneously in order to gain time, a process that resulted in confusion worse confounded; a system, in a word, of dry rot and slow paralysis, which, commencing with the head, with the Emperor himself, shattered in health and lacking in promptness of decision, could not fail ultimately to communicate itself to the whole army, disorganizing it and annihilating its efficiency, leading it into disaster from which it had not the means of extricating itself. And yet, over and above the dull misery of that period of waiting, in the intuitive, shuddering perception of what must infallibly happen, his certainty that they must be victors in the end remained unimpaired.

On the 3d of August the cheerful news had been given to the public of the victory of Sarrebruck, fought and won the day before. It could scarcely be called a great victory, but the columns of the newspapers teemed with enthusiastic gush; the invasion of Germany was begun, it was the first step in their glorious march to triumph, and the little Prince Imperial, who had coolly stooped and picked up a bullet from the battlefield, then commenced to be celebrated in legend. Two days later, however, when intelligence came of the surprise and defeat at Wissembourg, every mouth was opened to emit a cry of rage and distress. That five thousand men, caught in a trap, had faced thirty-five thousand Prussians all one long summer day, that was not a circumstance to daunt the courage of anyone; it simply called for vengeance. Yes, the leaders had doubtless been culpably lacking in vigilance and were to be censured for their want of foresight, but that would soon be mended; MacMahon had sent for the 1st division of the 7th corps, the 1st corps would be supported by the 5th, and the Prussians must be across the Rhine again by that time, with the bayonets of our infantry at their backs to accelerate their movement. And so, beneath the deep, dim vault of heaven, the thought of the battle that must have raged that day, the feverish impatience with which the tidings were awaited, the horrible feeling of suspense that pervaded the air about them, spread from man to man and became each minute more tense and unendurable.

Maurice was just then saying to Weiss:

"Ah! we have certainly given them a righteous good drubbing today."

Weiss made no reply save to nod his head with an air of anxiety. His gaze was directed toward the Rhine, on that Orient region where now the night had settled down in earnest, like a wall of blackness, concealing strange forms and shapes of mystery. The concluding strains of the bugles for roll-call had been succeeded by a deep silence, which had descended upon the drowsy camp and was only broken now and then by the steps and voices of some wakeful soldiers. A light had been lit-it looked like a twinkling star-in the main room of the farmhouse where the staff, which is supposed never to sleep, was awaiting the telegrams that came in occasionally, though as yet they were undecided. And the green wood fire, now finally left to itself, was still emitting its funereal wreaths of dense black smoke, which drifted in the gentle breeze over the unsleeping farmhouse, obscuring the early stars in the heavens above.

"A drubbing!" Weiss at last replied, "God grant it may be so!"

Jean, still seated a few steps away, pricked up his ears, while Lieutenant Rochas, noticing that the wish was attended by a doubt, stopped to listen.

"What!" Maurice rejoined, "have you not confidence? can you believe that defeat is possible?"

His brother-in-law silenced him with a gesture; his hands were trembling with agitation, his kindly pleasant face was pale and bore an expression of deep distress.

"Defeat, ah! Heaven preserve us from that! You know that I was born in this country; my grandfather and grandmother were murdered by the Cossacks in 1814, and whenever I think of invasion it makes me clench my fist and grit my teeth; I could go through fire and flood, like a trooper, in my shirt sleeves! Defeat-no, no! I cannot, I will not believe it possible."

He became calmer, allowing his arms to fall by his side in discouragement.

"But my mind is not easy, do you see. I know Alsace; I was born there; I am just off a business trip through the country, and we civilians have opportunities of seeing many things that the generals persist in ignoring, although they have them thrust beneath their very eyes. Ah, *we* wanted war with Prussia as badly as anyone; for a long, long time we have been waiting patiently for a chance to pay off old scores, but that did not prevent us from being on neighborly terms with the people in Baden and Bavaria; every one of us, almost, has friends or relatives across the Rhine. It was our belief that they felt like us and would not be sorry to humble the intolerable insolence of the Prussians. And now, after our long period of uncomplaining expectation, for the past two weeks we have seen things going from bad to worse, and it vexes and terrifies us. Since the declaration of war the enemy's horse have been suffered to come among us, terrorizing the villages, reconnoitering the country, cutting the telegraph wires. Baden and Bavaria are rising; immense bodies of troops are being concentrated in the Palatinate; information reaches us from every quarter, from the great fairs and markets, that our frontier is threatened, and when the citizens, the mayors of the communes, take the alarm at last and hurry off to tell your officers what they know, those gentlemen shrug their shoulders and reply: Those things spring from the imagination of cowards; there is no enemy near here. And when there is not an hour to lose, days and days are wasted. What are they waiting for? To give the whole German nation time to concentrate on the other bank of the river?"

His words were uttered in a low, mournful, voice, as if he were reciting to himself a story that had long occupied his thoughts.

"Ah! Germany, I know her too well; and the terrible part of the business is that you soldiers seem to know no more about her than you do about China. You must remember my cousin Gunther, Maurice, the young man, who came to pay me a flying visit at Sedan last spring. His mother is a sister of my mother, and married a Berliner; the young man is a German out and out; he detests everything French. He is a captain in the 5th Prussian corps. I accompanied him to the railway station that night, and he said to me in his sharp, peremptory way: 'If France declares war on us, she will be soundly whipped!' I can hear his words ringing in my ears yet."

Forthwith, Lieutenant Rochas, who had managed to contain himself until then, not without some difficulty, stepped forward in a towering rage. He was a tall, lean individual of about fifty, with a long, weather-beaten, and wrinkled face; his inordinately long nose, curved like the beak of a bird of prey, over a strong but well-shaped mouth, concealed by a thick, bristling mustache that was beginning to be touched with silver. And he shouted in a voice of thunder:

"See here, you, sir! what yarns are those that you are retailing to dishearten my men?"

Jean did not interfere with his opinion, but he thought that the last speaker was right, for he, too, while beginning to be conscious of the protracted delay, and the general confusion in their affairs, had never had the slightest doubt about that terrible thrashing they were certain to give the Prussians. There could be no question about the matter, for was not that the reason of their being there?

"But I am not trying to dishearten anyone, Lieutenant," Weiss answered in astonishment. "Quite the reverse; I am desirous that others should know what I know, because then they will be able to act with their eyes open. Look here! that Germany of which we were speaking-"

And he went on in his clear, demonstrative way to explain the reason of his fears: how Prussia had increased her resources since Sadowa; how the national movement had placed her at the head of the other German states, a mighty empire in process of formation and rejuvenation, with the constant hope and desire for unity as the incentive to their irresistible efforts; the system of compulsory military service, which made them a nation of trained soldiers, provided with the most effective arms of modern invention, with generals who were masters in the art of strategy, proudly mindful still of the crushing defeat they had administered to Austria; the intelligence, the moral force that resided in that army, commanded as it was almost exclusively by young generals, who in turn looked up to a commander-in-chief who seemed destined to revolutionize the art of war, whose prudence and foresight were unparalleled, whose correctness of judgment was a thing to wonder at. And in contrast to that picture of Germany he pointed to France: the Empire sinking into senile decrepitude, sanctioned by the plebiscite, but rotten at its foundation, destroying liberty, and therein stifling every idea of patriotism, ready to give up the ghost as soon as it should cease to satisfy the unworthy appetites to which it had given birth; then there was the army, brave, it was true, as was to be expected from men of their race, and covered with Crimean and Italian laurels, but vitiated by the system that permitted men to purchase substitutes for a money consideration, abandoned to the antiquated methods of African routine, too confident of victory to keep abreast with the more perfect science of modern times; and, finally, the generals, men for the most part not above mediocrity, consumed by petty rivalries, some of them of an ignorance beyond all belief, and at their head the Emperor, an ailing, vacillating man, deceiving himself and everyone with whom he had dealings in that desperate venture on which they were embarking, into which they were all rushing blindfold, with no preparation worthy of the name, with the panic and confusion of a flock of sheep on its way to the shambles.

Rochas stood listening, open-mouthed, and with staring eyes; his terrible nose dilated visibly. Then suddenly his lantern jaws parted to emit an obstreperous, Homeric peal of laughter.

"What are you giving us there, you? what do you mean by all that silly lingo? Why, there is not the first word of sense in your whole harangue-it is too idiotic to deserve an answer. Go and tell those things to the recruits, but don't tell them to me; no! not to me, who have seen twenty-seven years of service."

And he gave himself a thump on the breast with his doubled fist. He was the son of a master mason who had come from Limousin to Paris, where the son, not taking kindly to the paternal handicraft, had enlisted at the age of eighteen. He had been a soldier of fortune and had carried the knapsack, was corporal in Africa, sergeant in the Crimea, and after Solferino had been made lieutenant, having devoted fifteen years of laborious toil and heroic bravery to obtaining that rank, and was so illiterate that he had no chance of ever getting his captaincy.

"You, sir, who think you know everything, let me tell you a thing you don't know. Yes, at Mazagran I was scarce nineteen years old, and there were twenty-three of us, not a living soul more, and for more than four days we held out against twelve thousand Arabs. Yes, indeed! for years and years, if you had only been with us out there in Africa, sir, at Mascara, at Biskra, at Dellys, after that in Grand Kabylia, after that again at Laghouat, you would have seen those dirty niggers run like deer as soon as we showed our faces. And at Sebastopol, sir, fichtre! you

13

wouldn't have said it was the pleasantest place in the world. The wind blew fit to take a man's hair out by the roots, it was cold enough to freeze a brass monkey, and those beggars kept us on a continual dance with their feints and sorties. Never mind; we made them dance in the end; we danced them into the big hot frying pan, and to quick music, too! And Solferino, you were not there, sir! then why do you speak of it? Yes, at Solferino, where it was so hot, although I suppose more rain fell there that day than you have seen in your whole life, at Solferino, where we had our little brush with the Austrians, it would have warmed your heart to see how they vanished before our bayonets, riding one another down in their haste to get away from us, as if their coat tails were on fire!"

He laughed the gay, ringing laugh of the daredevil French soldier; he seemed to expand and dilate with satisfaction. It was the old story: the French trooper going about the world with his girl on his arm and a glass of good wine in his hand; thrones upset and kingdoms conquered in the singing of a merry song. Given a corporal and four men, and great armies would bite the dust. His voice suddenly sank to a low, rumbling bass:

"What! whip France? We, whipped by those Prussian pigs, we!" He came up to Weiss and grasped him violently by the lapel of his coat. His entire long frame, lean as that of the immortal Knight Errant, seemed to breathe defiance and unmitigated contempt for the foe, whoever he might be, regardless of time, place, or any other circumstance. "Listen to what I tell you, sir. If the Prussians dare to show their faces here, we will kick them home again. You hear me? we will kick them from here to Berlin." His bearing and manner were superb; the serene tranquillity of the child, the candid conviction of the innocent who knows nothing and fears nothing. "Parbleu! it is so, because it is so, and that's all there is about it!"

Weiss, stunned and almost convinced, made haste to declare that he wished for nothing better. As for Maurice, who had prudently held his tongue, not venturing to express an opinion in presence of his superior officer, he concluded by joining in the other's merriment; he warmed the cockles of his heart, that devil of a man, whom he nevertheless considered rather stupid. Jean, too, had nodded his approval at every one of the lieutenant's assertions. He had also been at Solferino, where it rained so hard. And that showed what it was to have a tongue in one's head and know how to use it. If all the leaders had talked like that they would not be in such a mess, and there would be camp-kettles and flannel belts in abundance.

It was quite dark by this time, and Rochas continued to gesticulate and brandish his long arms in the obscurity. His historical studies had been confined to a stray volume of Napoleonic memoirs that had found its way to his knapsack from a peddler's wagon. His excitement refused to be pacified and all his book-learning burst from his lips in a torrent of eloquence:

"We flogged the Austrians at Castiglione, at Marengo, at Austerlitz, at Wagram; we flogged the Prussians at Eylau, at Jena, at Lutzen; we flogged the Russians at Friedland, at Smolensk and at the Moskowa; we flogged Spain and England everywhere; all creation flogged, flogged, flogged, up and down, far and near, at home and abroad, and now you tell me that it is we who are to take the flogging! Why, pray tell me? How? Is the world coming to an end?" He drew his tall form up higher still and raised his arm aloft, like the staff of a battle-flag. "Look you, there has been a fight today, down yonder, and we are waiting for the news. Well! I will tell you what the news is-I will tell you, I! We have flogged the Prussians, flogged them until they didn't know whether they were a-foot or a-horseback, flogged them to powder, so that they had to be swept up in small pieces!"

At that moment there passed over the camp, beneath the somber heavens, a loud, wailing cry. Was it the plaint of some nocturnal bird? Or was it a mysterious voice, reaching them from some far-distant field of carnage, ominous of disaster? The whole camp shuddered, lying there in the shadows, and the strained, tense sensation of expectant anxiety that hung,

14

miasma-like, in the air became more strained, more feverish, as they waited for telegrams that seemed as if they would never come. In the distance, at the farmhouse, the candle that lighted the dreary watches of the staff burned up more brightly, with an erect, unflickering flame, as if it had been of wax instead of tallow.

But it was ten o'clock, and Gaude, rising to his feet from the ground where he had been lost in the darkness, sounded taps, the first in all the camp. Other bugles, far and near, took up the strain, and it passed away in the distance with a dying, melancholy wail, as if the angel of slumber had already brushed with his wings the weary men. And Weiss, who had lingered there so late, embraced Maurice affectionately; courage, and hope! he would kiss Henriette for her brother and would have many things to tell uncle Fouchard when they met. Then, just as he was turning to go, a rumor began to circulate, accompanied by the wildest excitement. A great victory had been won by Marshal MacMahon, so the report ran; the Crown Prince of Prussia a prisoner, with twenty-five thousand men, the enemy's army repulsed and utterly destroyed, its guns and baggage abandoned to the victors.

"Didn't I tell you so!" shouted Rochas, in his most thundering voice. Then, running after Weiss, who, light of heart, was hastening to get back to Mulhausen: "To Berlin, sir, and we'll kick them every step of the way!"

A quarter of an hour later came another dispatch, announcing that the army had been compelled to evacuate Woerth and was retreating. Ah, what a night was that! Rochas, overpowered by sleep, wrapped his cloak about him, threw himself down on the bare ground, as he had done many a time before. Maurice and Jean sought the shelter of the tent, into which were crowded, a confused tangle of arms and legs, Loubet, Chouteau, Pache, and Lapoulle, their heads resting on their knapsacks. There was room for six, provided they were careful how they disposed of their legs. Loubet, by way of diverting his comrades and making them forget their hunger, had labored for some time to convince Lapoulle that there was to be a ration of poultry issued the next morning, but they were too sleepy to keep up the joke; they were snoring, and the Prussians might come, it was all one to them. Jean lay for a moment without stirring, pressing close against Maurice; notwithstanding his fatigue he was unable to sleep; he could not help thinking of the things that gentleman had said, how all Germany was up in arms and preparing to pour her devastating hordes across the Rhine; and he felt that his tent-mate was not sleeping, either-was thinking of the same things as he. Then the latter turned over impatiently and moved away, and the other understood that his presence was not agreeable. There was a lack of sympathy between the peasant and the man of culture, an enmity of caste and education that amounted almost to physical aversion. The former, however, experienced a sensation of shame and sadness at this condition of affairs; he shrinkingly drew in his limbs so as to occupy as small a space as possible, endeavoring to escape from the hostile scorn that he was vaguely conscious of in his neighbor. But although the night wind without had blown up chill, the crowded tent was so stifling hot and close that Maurice, in a fever of exasperation, raised the flap, darted out, and went and stretched himself on the ground a few steps away. That made Jean still more unhappy, and in his half-sleeping, half-waking condition he had troubled dreams, made up of a regretful feeling that no one cared for him, and a vague apprehension of impending calamity of which he seemed to hear the steps approaching with measured tread from the shadowy, mysterious depths of the unknown.

Two hours passed, and all the camp lay lifeless, motionless under the oppression of the deep, weird darkness, that was instinct with some dreadful horror as yet without a name. Out of the sea of blackness came stifled sighs and moans; from an invisible tent was heard something that sounded like the groan of a dying man, the fitful dream of some tired soldier. Then there were other sounds that to the strained ear lost their familiarity and became menaces of approaching evil; the neighing of a charger, the clank of a sword, the hurrying steps of some belated prowler. And all at once, off toward the canteens, a great light flamed up. The entire front was brilliantly illuminated; the long, regularly aligned array of stacks stood out against the darkness, and the ruddy blaze, reflected from the burnished barrels of the rifles, assumed the hue of new-shed blood; the erect, stern figures of the sentries became visible in the fiery glow. Could it be the enemy, whose presence the leaders had been talking of for the past two days, and on whose trail they had come out from Belfort to Mulhausen? Then a shower of sparks rose high in the air and the conflagration subsided. It was only the pile of green wood that had been so long the object of Loubet's and Lapoulle's care, and which, after having smoldered for many hours, had at last flashed up like a fire of straw.

Jean, alarmed by the vivid light, hastily left the tent and was near falling over Maurice, who had raised himself on his elbow. The darkness seemed by contrast more opaque than it had been before, and the two men lay stretched on the bare ground, a few paces from each other. All that they could descry before them in the dense shadows of the night was the window of the farm-house, faintly illuminated by the dim candle, which shone with a sinister gleam, as if it were doing duty by the bedside of a corpse. What time was it? two o'clock, or three, perhaps. It was plain that the staff had not made acquaintance with their beds that night. They could hear Bourgain-Desfeuilles' loud, disputatious voice; the general was furious that his rest should be broken thus, and it required many cigars and toddies to pacify him. More telegrams came in; things must be going badly; silhouettes of couriers, faintly drawn against the uncertain sky line, could be descried, galloping madly. There was the sound of scuffling steps, imprecations, a smothered cry as of a man suddenly stricken down, followed by a blood-freezing silence. What could it be? Was it the end? A breath, chill and icy as that from the lips of death, had passed over the camp that lay lost in slumber and agonized expectation.

It was at that moment that Jean and Maurice recognized in the tall, thin, spectral form that passed swiftly by, their colonel, de Vineuil. He was accompanied by the regimental surgeon, Major Bouroche, a large man with a leonine face They were conversing in broken, unfinished sentences, whisperingly, such a conversation as we sometimes hear in dreams.

"It came by the way of Basle. Our 1st division all cut to pieces. The battle lasted twelve hours; the whole army is retreating~"

The colonel's specter halted and called by name another specter, which came lightly forward; it was an elegant ghost, faultless in uniform and equipment.

"Is that you, Beaudoin?"

"Yes, Colonel."

"Ah! bad news, my friend, terrible news! MacMahon beaten at Froeschwiller, Frossard beaten at Spickeren, and between them de Failly, held in check where he could give no assistance. At Froeschwiller it was a single corps against an entire army; they fought like heroes. It was a complete rout, a panic, and now France lies open to their advance~"

His tears choked further utterance, the words came from his lips unintelligible, and the three shadows vanished, swallowed up in the obscurity.

Maurice rose to his feet; a shudder ran through his frame.

"Good God!" he stammeringly exclaimed.

And he could think of nothing else to say, while Jean, in whose bones the very marrow seemed to be congealing, murmured in his resigned manner:

"Ah, worse luck! The gentleman, that relative of yours, was right all the same in saying that they are stronger than we."

16

Maurice was beside himself, could have strangled him. The Prussians stronger than the French! The thought made his blood boil. The peasant calmly and stubbornly added:

"That don't matter, mind you. A man don't give up whipped at the first knock-down he gets. We shall have to keep hammering away at them all the same."

But a tall figure arose before them. They recognized Rochas, still wrapped in his long mantle, whom the fugitive sounds about him, or it may have been the intuition of disaster, had awakened from his uneasy slumber. He questioned them, insisted on knowing all. When he was finally brought, with much difficulty, to see how matters stood, stupor, immense and profound, filled his boyish, inexpressive eyes. More than ten times in succession he repeated:

"Beaten! How beaten? Why beaten?"

And that was the calamity that had lain hidden in the blackness of that night of agony. And now the pale dawn was appearing at the portals of the east, heralding a day heavy with bitterest sorrow and striking white upon the silent tents, in one of which began to be visible the ashy faces of Loubet and Lapoulle, of Chouteau and of Pache, who were snoring still with wide-open mouths. Forth from the thin mists that were slowly creeping upward from the river off yonder in the distance came the new day, bringing with it mourning and affliction.

2

About eight o'clock the sun dispersed the heavy clouds, and the broad, fertile plain about Mulhausen lay basking in the warm, bright light of a perfect August Sunday. From the camp, now awake and bustling with life, could be heard the bells of the neighboring parishes, pealing merrily in the limpid air. The cheerful Sunday following so close on ruin and defeat had its own gayety, its sky was as serene as on a holiday.

Gaude suddenly took his bugle and gave the call that announced the distribution of rations, whereat Loubet appeared astonished. What was it? What did it mean? Were they going to give out chickens, as he had promised Lapoulle the night before? He had been born in the Halles, in the Rue de la Cossonerie, was the unacknowledged son of a small huckster, had enlisted "for the money there was in it," as he said, after having been a sort of Jack-of-all-trades, and was now the gourmand, the epicure of the company, continually nosing after something good to eat. But he went off to see what was going on, while Chouteau, the company artist, house-painter by trade at Belleville, something of a dandy and a revolutionary republican, exasperated against the government for having called him back to the colors after he had served his time, was cruelly chaffing Pache, whom he had discovered on his knees, behind the tent, preparing to say his prayers. There was a pious man for you! Couldn't he oblige him, Chouteau, by interceding with God to give him a hundred thousand francs or some such small trifle? But Pache, an insignificant little fellow with a head running up to a point, who had come to them from some hamlet in the wilds of Picardy, received the other's raillery with the uncomplaining gentleness of a martyr. He was the butt of the squad, he and Lapoulle, the colossal brute who had got his growth in the marshes of the Sologne, so utterly ignorant of everything that on the day of his joining the regiment he had asked his comrades to show him the King. And although the terrible tidings of the disaster at Froeschwiller had been known throughout the camp since early morning, the four men laughed, joked, and went about their usual tasks with the indifference of so many machines.

But there arose a murmur of pleased surprise. It was occasioned by Jean, the corporal, coming back from the commissary's, accompanied by Maurice, with a load of firewood. So, they were giving out wood at last, the lack of which the night before had deprived the men of their soup! Twelve hours behind time, only!

"Hurrah for the commissary!" shouted Chouteau.

"Never mind, so long as it is here," said Loubet. "Ah! won't I make you a bully pot-au-feu!"

He was usually quite willing to take charge of the mess arrangements, and no one was inclined to say him nay, for he cooked like an angel. On those occasions, however, Lapoulle would be given the most extraordinary commissions to execute.

"Go and look after the champagne-Go out and buy some truffles-"

On that morning a queer conceit flashed across his mind, such a conceit as only a Parisian *gamin* contemplating the mystification of a greenhorn is capable of entertaining:

"Look alive there, will you! Come, hand me the chicken."

"The chicken! what chicken, where?"

"Why, there on the ground at your feet, stupid; the chicken that I promised you last night, and that the corporal has just brought in."

He pointed to a large, white, round stone, and Lapoulle, speechless with wonder, finally picked it up and turned it about between his fingers.

18

"A thousand thunders! Will you wash the chicken! More yet; wash its claws, wash its neck! Don't be afraid of the water, lazybones!"

And for no reason at all except the joke of it, because the prospect of the soup made him gay and sportive, he tossed the stone along with the meat into the kettle filled with water.

"That's what will give the bouillon a flavor! Ah, you didn't know that, sacree andouille! You shall have the pope's nose; you'll see how tender it is."

The squad roared with laughter at sight of Lapoulle's face, who swallowed everything and was licking his chops in anticipation of the feast. That funny dog, Loubet, he was the man to cure one of the dumps if anybody could! And when the fire began to crackle in the sunlight, and the kettle commenced to hum and bubble, they ranged themselves reverently about it in a circle with an expression of cheerful satisfaction on their faces, watching the meat as it danced up and down and sniffing the appetizing odor that it exhaled. They were as hungry as a pack of wolves, and the prospect of a square meal made them forgetful of all beside. They had had to take a thrashing, but that was no reason why a man should not fill his stomach. Fires were blazing and pots were boiling from one end of the camp to the other, and amid the silvery peals of the bells that floated from Mulhausen steeples mirth and jollity reigned supreme.

But just as the clocks were on the point of striking nine a commotion arose and spread among the men; officers came running up, and Lieutenant Rochas, to whom Captain Beaudoin had come and communicated an order, passed along in front of the tents of his platoon and gave the command:

"Pack everything! Get yourselves ready to march!"

"But the soup?"

"You will have to wait for your soup until some other day; we are to march at once."

Gaude's bugle rang out in imperious accents. Then everywhere was consternation; dumb, deep rage was depicted on every countenance. What, march on an empty stomach! Could they not wait a little hour until the soup was ready! The squad resolved that their bouillon should not go to waste, but it was only so much hot water, and the uncooked meat was like leather to their teeth. Chouteau growled and grumbled, almost mutinously. Jean had to exert all his authority to make the men hasten their preparations. What was the great urgency that made it necessary for them to hurry off like that? What good was there in hazing people about in that style, without giving them time to regain their strength? And Maurice shrugged his shoulders incredulously when someone said in his hearing that they were about to march against the Prussians and settle old scores with them. In less than fifteen minutes the tents were struck, folded, and strapped upon the knapsacks, the stacks were broken, and all that remained of the camp was the dying embers of the fires on the bare ground.

There were reasons, of importance that had induced General Douay's determination to retreat immediately. The despatch from the sous-prefet at Schelestadt, now three days old, was confirmed; there were telegrams that the fires of the Prussians, threatening Markolsheim, had again been seen, and again, another telegram informed them that one of the enemy's army corps was crossing the Rhine at Huningue: the intelligence was definite and abundant; cavalry and artillery had been sighted in force, infantry had been seen, hastening from every direction to their point of concentration. Should they wait an hour the enemy would surely be in their rear and retreat on Belfort would be impossible. And now, in the shock consequent on defeat, after Wissembourg and Froeschwiller, the general, feeling himself unsupported in his exposed position at the front, had nothing left to do but fall back in haste, and the more so that what news he had received that morning made the situation look even worse than it had appeared the night before.

The staff had gone on ahead at a sharp trot, spurring their horses in the fear lest the Prussians might get into Altkirch before them. General Bourgain-Desfeuilles, aware that he had a hard day's work before him, had prudently taken Mulhausen in his way, where he fortified himself with a copious breakfast, denouncing in language more forcible than elegant such hurried movements. And Mulhausen watched with sorrowful eyes the officers trooping through her streets; as the news of the retreat spread the citizens streamed out of their houses, deploring the sudden departure of the army for whose coming they had prayed so earnestly: they were to be abandoned, then, and all the costly merchandise that was stacked up in the railway station was to become the spoil of the enemy; within a few hours their pretty city was to be in the hands of foreigners? The inhabitants of the villages, too, and of isolated houses, as the staff clattered along the country roads, planted themselves before their doors with wonder and consternation depicted on their faces. What! that army, that a short while before they had seen marching forth to battle, was now retiring without having fired a shot? The leaders were gloomy, urged their chargers forward and refused to answer questions, as if ruin and disaster were galloping at their heels. It was true, then, that the Prussians had annihilated the army and were streaming into France from every direction, like the angry waves of a stream that had burst its barriers? And already to the frightened peasants the air seemed filled with the muttering of distant invasion, rising louder and more threatening at every instant, and already they were beginning to forsake their little homes and huddle their poor belongings into farm-carts; entire families might be seen fleeing in single file along the roads that were choked with the retreating cavalry.

In the hurry and confusion of the movement the 106th was brought to a halt at the very first kilometer of their march, near the bridge over the canal of the Rhone and Rhine. The order of march had been badly planned and still more badly executed, so that the entire 2d division was collected there in a huddle, and the way was so narrow, barely more than sixteen feet in width, that the passage of the troops was obstructed.

Two hours elapsed, and still the 106th stood there watching the seemingly endless column that streamed along before their eyes. In the end the men, standing at rest with ordered arms, began to become impatient. Jean's squad, whose position happened to be opposite a break in the line of poplars where the sun had a fair chance at them, felt themselves particularly aggrieved.

"Guess we must be the rear-guard," Loubet observed with good-natured raillery.

But Chouteau scolded: "They don't value us at a brass farthing, and that's why they let us wait this way. We were here first; why didn't we take the road while it was empty?"

And as they began to discern more clearly beyond the canal, across the wide fertile plain, along the level roads lined with hop-poles and fields of ripening grain, the movement of the troops retiring along the same way by which they had advanced but yesterday, gibes and jeers rose on the air in a storm of angry ridicule.

"Ah, we are taking the back track," Chouteau continued. "I wonder if that is the advance against the enemy that they have been dinning in our ears of late! Strikes me as rather queer! No sooner do we get into camp than we turn tail and make off, never even stopping to taste our soup."

The derisive laughter became louder, and Maurice, who was next to Chouteau in the ranks, took sides with him. Why could they not have been allowed to cook their soup and eat it in peace, since they had done nothing for the last two hours but stand there in the road like so many sticks? Their hunger was making itself felt again; they had a resentful recollection of the savory contents of the kettle dumped out prematurely upon the ground, and they could see no necessity for this headlong retrograde movement, which appeared to them idiotic and cowardly. What chicken-livers they must be, those generals!

But Lieutenant Rochas came along and blew up Sergeant Sapin for not keeping his men in better order, and Captain Beaudoin, very prim and starchy, attracted by the disturbance, appeared upon the scene.

"Silence in the ranks!"

Jean, an old soldier of the army of Italy who knew what discipline was, looked in silent amazement at Maurice, who appeared to be amused by Chouteau's angry sneers; and he wondered how it was that a monsieur, a young man of his acquirements, could listen approvingly to things-they might be true, all the same-but that should not be blurted out in public. The army would never accomplish much, that was certain, if the privates were to take to criticizing the generals and giving their opinions.

At last, after another hour's waiting, the order was given for the 106th to advance, but the bridge was still so encumbered by the rear of the division that the greatest confusion prevailed. Several regiments became inextricably mingled, and whole companies were swept away and compelled to cross whether they would or no, while others, crowded off to the side of the road, had to stand there and mark time; and by way of putting the finishing touch to the muddle; a squadron of cavalry insisted on passing, pressing back into the adjoining fields the stragglers that the infantry had scattered along the roadside. At the end of an hour's march the column had entirely lost its formation and was dragging its slow length along, a mere disorderly rabble.

Thus it happened that Jean found himself away at the rear, lost in a sunken road, together with his squad, whom he had been unwilling to abandon. The 106th had disappeared, nor was there a man or an officer of their company in sight. About them were soldiers, singly or in little groups, from all the regiments, a weary, foot-sore crew, knocked up at the beginning of the retreat, each man straggling on at his own sweet will whithersoever the path that he was on might chance to lead him. The sun beat down fiercely, the heat was stifling, and the knapsack, loaded as it was with the tent and implements of every description, made a terrible burden on the shoulders of the exhausted men. To many of them the experience was an entirely new one, and the heavy great-coats they wore seemed to them like vestments of lead. The first to set an example for the others was a little pale faced soldier with watery eyes; he drew beside the road and let his knapsack slide off into the ditch, heaving a deep sigh as he did so, the long drawn breath of a dying man who feels himself coming back to life.

"There's a man who knows what he is about," muttered Chouteau.

He still continued to plod along, however, his back bending beneath its weary burden, but when he saw two others relieve themselves as the first had done he could stand it no longer. "Ah! zut!" he exclaimed, and with a quick upward jerk of the shoulder sent his kit rolling down an embankment. Fifty pounds at the end of his backbone, he had had enough of it, thank you! He was no beast of burden to lug that load about.

Almost at the same moment Loubet followed his lead and incited Lapoulle to do the same. Pache, who had made the sign of the cross at every stone crucifix they came to, unbuckled the straps and carefully deposited his load at the foot of a low wall, as if fully intending to come back for it at some future time. And when Jean turned his head for a look at his men he saw that every one of them had dropped his burden except Maurice.

"Take up your knapsacks unless you want to have me put under arrest!"

But the men, although they did not mutiny as yet, were silent and looked ugly; they kept advancing along the narrow road, pushing the corporal before them.

"Will you take up your knapsacks! if you don't I will report you."

It was as if Maurice had been lashed with a whip across the face. Report them! that brute of a peasant would report those poor devils for easing their aching shoulders! And looking Jean defiantly in the face, he, too, in an impulse of blind rage, slipped the buckles and let his knapsack fall to the road.

"Very well," said the other in his quiet way, knowing that resistance would be of no avail, "we will settle accounts to-night."

Maurice's feet hurt him abominably; the big, stiff shoes, to which he was not accustomed, had chafed the flesh until the blood came. He was not strong; his spinal column felt as if it were one long raw sore, although the knapsack that had caused the suffering was no longer there, and the weight of his piece, which he kept shifting from one shoulder to the other, seemed as if it would drive all the breath from his body. Great as his physical distress was, however, his moral agony was greater still, for he was in the depths of one of those fits of despair to which he was subject. At Paris the sum of his wrongdoing had been merely the foolish outbreaks of "the other man," as he put it, of his weak, boyish nature, capable of more serious delinquency should he be subjected to temptation, but now, in this retreat that was so like a rout, in which he was dragging himself along with weary steps beneath a blazing sun, he felt all hope and courage vanishing from his heart, he was but a beast in that belated, straggling herd that filled the roads and fields. It was the reaction after the terrible disasters at Wissembourg and Froeschwiller, the echo of the thunder-clap that had burst in the remote distance, leagues and leagues away, rattling at the heels of those panic-stricken men who were flying before they had ever seen an enemy. What was there to hope for now? Was it not all ended? They were beaten; all that was left them was to lie down and die.

"It makes no difference," shouted Loubet, with the *blague* of a child of the Halles, "but this is not the Berlin road we are traveling, all the same."

To Berlin! To Berlin! The cry rang in Maurice's ears, the yell of the swarming mob that filled the boulevards on that midsummer night of frenzied madness when he had determined to enlist. The gentle breeze had become a devastating hurricane; there had been a terrific explosion, and all the sanguine temper of his nation had manifested itself in his absolute, enthusiastic confidence, which had vanished utterly at the very first reverse, before the unreasoning impulse of despair that was sweeping him away among those vagrant soldiers, vanquished and dispersed before they had struck a stroke.

"This confounded blunderbuss must weigh a ton, I think," Loubet went on. "This is fine music to march by!" And alluding to the sum he received as substitute: "I don't care what people say, but fifteen hundred 'balls' for a job like this is downright robbery. Just think of the pipes he'll smoke, sitting by his warm fire, the stingy old miser in whose place I'm going to get my brains knocked out!"

"As for me," growled Chouteau, "I had finished my time. I was going to cut the service, and they keep me for their beastly war. Ah! true as I stand here, I must have been born to bad luck to have got myself into such a mess. And now the officers are going to let the Prussians knock us about as they please, and we're dished and done for." He had been swinging his piece to and fro in his hand; in his discouragement he gave it a toss and landed it on the other side of the hedge. "Eh! get you gone for a dirty bit of old iron!"

The musket made two revolutions in the air and fell into a furrow, where it lay, long and motionless, reminding one somehow of a corpse. Others soon flew to join it, and presently the field was filled with abandoned arms, lying in long winrows, a sorrowful spectacle beneath the blazing sky. It was an epidemic of madness, caused by the hunger that was gnawing at their stomach, the shoes that galled their feet, their weary march, the unexpected defeat that had brought the enemy galloping at their heels. There was nothing more to be accomplished; their leaders were looking out for themselves, the commissariat did not even feed them; nothing but weariness and worriment; better to leave the whole business at once, before it was begun. And what then? why, the musket might go and keep the knapsack company; in view of the work that was before them they might at least as well keep their arms free. And all down the long line of stragglers that stretched almost far as the eye could reach in the smooth and fertile country the muskets flew through the air to the accompaniment of jeers and laughter such as would have befitted the inmates of a lunatic asylum out for a holiday.

Loubet, before parting with his, gave it a twirl as a drum-major does his cane. Lapoulle, observing what all his comrades were doing, must have supposed the performance to be some recent innovation in the manual, and followed suit, while Pache, in the confused idea of duty that he owed to his religious education, refused to do as the rest were doing and was loaded with obloquy by Chouteau, who called him a priest's whelp.

"Look at the sniveling papist! And all because his old peasant of a mother used to make him swallow the holy wafer every Sunday in the village church down there! Be off with you and go serve mass; a man who won't stick with his comrades when they are right is a poor-spirited cur."

Maurice toiled along dejectedly in silence, bowing his head beneath the blazing sun. At every step he took he seemed to be advancing deeper into a horrid, phantom-haunted nightmare; it was as if he saw a yawning, gaping gulf before him toward which he was inevitably tending; it meant that he was suffering himself to be degraded to the level of the miserable beings by whom he was surrounded, that he was prostituting his talents and his position as a man of education.

"Hold!" he said abruptly to Chouteau, "what you say is right; there is truth in it."

And already he had deposited his musket upon a pile of stones, when Jean, who had tried without success to check the shameful proceedings of his men, saw what he was doing and hurried toward him.

"Take up your musket, at once! Do you hear me? take it up at once!"

Jean's face had flushed with sudden anger. Meekest and most pacific of men, always prone to measures of conciliation, his eyes were now blazing with wrath, his voice spoke with the thunders of authority. His men had never before seen him in such a state, and they looked at one another in astonishment.

"Take up your musket at once, or you will have me to deal with!"

Maurice was quivering with anger; he let fall one single word, into which he infused all the insult that he had at command:

"Peasant!"

"Yes, that's just it; I am a peasant, while you, you, are a gentleman! And it is for that reason that you are a pig! Yes! a dirty pig! I make no bones of telling you of it."

Yells and cat-calls arose all around him, but the corporal continued with extraordinary force and dignity:

"When a man has learning he shows it by his actions. If we are brutes and peasants, you owe us the benefit of your example, since you know more than we do. Take up your musket, or Nom de Dieu! I will have you shot the first halt we make."

Maurice was daunted; he stooped and raised the weapon in his hand. Tears of rage stood in his eyes. He reeled like a drunken man as he labored onward, surrounded by his comrades, who now were jeering at him for having yielded. Ah, that Jean! he felt that he should never

23

cease to hate him, cut to the quick as he had been by that bitter lesson, which he could not but acknowledge he had deserved. And when Chouteau, marching at his side, growled: "When corporals are that way, we just wait for a battle and blow a hole in 'em," the landscape seemed red before his eyes, and he had a distinct vision of himself blowing Jean's brains out from behind a wall.

But an incident occurred to divert their thoughts; Loubet noticed that while the dispute was going on Pache had also abandoned his musket, laying it down tenderly at the foot of an embankment. Why? What were the reasons that had made him resist the example of his comrades in the first place, and what were the reasons that influenced him now? He probably could not have told himself, nor did he trouble his head about the matter, chuckling inwardly with silent enjoyment, like a schoolboy who, having long been held up as a model for his mates, commits his first offense. He strode along with a self-contented, rakish air, swinging his arms; and still along the dusty, sunlit roads, between the golden grain and the fields of hops that succeeded one another with tiresome monotony, the human tide kept pouring onward; the stragglers, without arms or knapsacks, were now but a shuffling, vagrant mob, a disorderly array of vagabonds and beggars, at whose approach the frightened villagers barred their doors.

Something that happened just then capped the climax of Maurice's misery. A deep, rumbling noise had for some time been audible in the distance; it was the artillery, that had been the last to leave the camp and whose leading guns now wheeled into sight around a bend in the road, barely giving the footsore infantrymen time to seek safety in the fields. It was an entire regiment of six batteries, and came up in column, in splendid order, at a sharp trot, the colonel riding on the flank at the center of the line, every officer at his post. The guns went rattling, bounding by, accurately maintaining their prescribed distances, each accompanied by its caisson, men and horses, beautiful in the perfect symmetry of its arrangement; and in the 5th battery Maurice recognized his cousin Honore. A very smart and soldierly appearance the quartermaster-sergeant presented on horseback in his position on the left hand of the forward driver, a good-looking light-haired man, Adolphe by name, whose mount was a sturdy chestnut, admirably matched with the mate that trotted at his side, while in his proper place among the six men who were seated on the chests of the gun and its caisson was the gunner, Louis, a small, dark man, Adolphe's comrade; they constituted a team, as it is called, in accordance with the rule of the service that couples a mounted and an unmounted man together. They all appeared bigger and taller to Maurice, somehow, than when he first made their acquaintance at the camp, and the gun, to which four horses were attached, followed by the caisson drawn by six, seemed to him as bright and refulgent as a sun, tended and cherished as it was by its attendants, men and animals, who closed around it protectingly as if it had been a living sentient relative; and then, besides, the contemptuous look that Honore, astounded to behold him among that unarmed rabble, cast on the stragglers, distressed him terribly. And now the tail end of the regiment was passing, the *materiel* of the batteries, prolonges, forges, forage-wagons, succeeded by the rag-tag, the spare men and horses, and then all vanished in a cloud of dust at another turn in the road amid the gradually decreasing clatter of hoofs and wheels.

"Pardi!" exclaimed Loubet, "it's not such a difficult matter to cut a dash when one travels with a coach and four!"

The staff had found Altkirch free from the enemy; not a Prussian had shown his face there yet. It had been the general's wish, not knowing at what moment they might fall upon his rear, that the retreat should be continued to Dannemarie, and it was not until five o'clock that the heads of columns reached that place. Tents were hardly pitched and fires lighted at eight, when night closed in, so great was the confusion of the regiments, depleted by the absence of the stragglers. The men were completely used up, were ready to drop with fatigue and hunger. Up to eight o'clock soldiers, singly and in squads, came trailing in, hunting for their commands; all that long train of the halt, the lame, and the disaffected that we have seen scattered along the roads.

As soon as Jean discovered where his regiment lay he went in quest of Lieutenant Rochas to make his report. He found him, together with Captain Beaudoin, in earnest consultation with the colonel at the door of a small inn, all of them anxiously waiting to see what tidings roll-call would give them as to the whereabouts of their missing men. The moment the corporal opened his mouth to address the lieutenant, Colonel Vineuil, who heard what the subject was, called him up and compelled him to tell the whole story. On his long, yellow face, where the intensely black eyes looked blacker still contrasted with the thick snow-white hair and the long, drooping mustache, there was an expression of patient, silent sorrow, and as the narrative proceeded, how the miserable wretches deserted their colors, threw away arms and knapsacks, and wandered off like vagabonds, grief and shame traced two new furrows on his blanched cheeks.

"Colonel," exclaimed Captain Beaudoin, in his incisive voice, not waiting for his superior to give an opinion, "it will best to shoot half a dozen of those wretches."

And the lieutenant nodded his head approvingly. But the colonel's despondent look expressed his powerlessness.

"There are too many of them. Nearly seven hundred! how are we to go to work, whom are we to select? And then you don't know it, but the general is opposed. He wants to be a father to his men, says he never punished a soldier all the time he was in Africa. No, no; we shall have to overlook it. I can do nothing. It is dreadful."

The captain echoed: "Yes, it is dreadful. It means destruction for us all."

Jean was walking off, having said all he had to say, when he heard Major Bouroche, whom he had not seen where he was standing in the doorway of the inn, growl in a smothered voice: "No more punishment, an end to discipline, the army gone to the dogs! Before a week is over the scoundrels will be ripe for kicking their officers out of camp, while if a few of them had been made an example of on the spot it might have brought the remainder to their senses."

No one was punished. Some officers of the rear-guard that was protecting the trains had been thoughtful enough to collect the muskets and knapsacks scattered along the road. They were almost all recovered, and by daybreak the men were equipped again, the operation being conducted very quietly, as if to hush the matter up as much as possible. Orders were given to break camp at five o'clock, but reveille sounded at four and the retreat to Belfort was hurriedly continued, for everyone was certain that the Prussians were only two or three leagues away. Again there was nothing to eat but dry biscuit, and as a consequence of their brief, disturbed rest and the lack of something to warm their stomachs the men were weak as cats. Any attempt to enforce discipline on the march that morning was again rendered nugatory by the manner of their departure.

The day was worse than its predecessor, inexpressibly gloomy and disheartening. The aspect of the landscape had changed, they were now in a rolling country where the roads they were always alternately climbing and descending were bordered with woods of pine and hemlock, while the narrow gorges were golden with tangled thickets of broom. But panic and terror lay heavy on the fair land that slumbered there beneath the bright sun of August, and had been hourly gathering strength since the preceeding day. A fresh dispatch, bidding the mayors of communes warn the people that they would do well to hide their valuables, had excited universal consternation. The enemy was at hand, then! Would time be given them to make their escape? And to all it seemed that the roar of invasion was ringing in their ears, coming nearer and nearer, the roar of the rushing torrent that, starting from Mulhausen, had grown louder and more ominous as it advanced, and to which every village that it encountered in its course contributed its own alarm amid the sound of wailing and lamentation.

Maurice stumbled along as best he might, like a man walking in a dream; his feet were bleeding, his shoulders sore with the weight of gun and knapsack. He had ceased to think, he advanced automatically into the vision of horrors that lay before his eyes; he had ceased to be conscious even of the shuffling tramp of the comrades around him, and the only thing that was not dim and unreal to his sense was Jean, marching at his side and enduring the same fatigue and horrible distress. It was lamentable to behold the villages they passed through, a sight to make a man's heart bleed with anguish. No sooner did the inhabitants catch sight of the troops retreating in disorderly array, with haggard faces and bloodshot eyes, than they bestirred themselves to hasten their flight. They who had been so confident only a short half month ago, those men and women of Alsace, who smiled when war was mentioned, certain that it would be fought out in Germany! And now France was invaded, and it was among them, above their abodes, in their fields, that the tempest was to burst, like one of those dread cataclysms that lay waste a province in an hour when the lightnings flash and the gates of heaven are opened! Carts were backed up against doors and men tumbled their furniture into them in wild confusion, careless of what they broke. From the upper windows the women threw out a last mattress, or handed down the child's cradle, that they had been near forgetting, whereon baby would be tucked in securely and hoisted to the top of the load, where he reposed serenely among a grove of legs of chairs and upturned tables. At the back of another cart was the decrepit old grandfather tied with cords to a wardrobe, and he was hauled away for all the world as if he had been one of the family chattels. Then there were those who did not own a vehicle, so they piled their household goods haphazard on a wheelbarrow, while others carried an armful of clothing, and others still had thought only of saving the clock, which they went off pressing to their bosom as if it had been a darling child. They found they could not remove everything, and there were chairs and tables, and bundles of linen too heavy to carry, lying abandoned in the gutter, Some before leaving had carefully locked their dwellings, and the houses had a deathlike appearance, with their barred doors and windows, but the greater number, in their haste to get away and with the sorrowful conviction that nothing would escape destruction, had left their poor abodes open, and the yawning apertures displayed the nakedness of the dismantled rooms; and those were the saddest to behold, with the horrible sadness of a city upon which some great dread has fallen, depopulating it, those poor houses opened to the winds of heaven, whence the very cats had fled as if forewarned of the impending doom. At every village the pitiful spectacle became more heartrending, the number of the fugitives was greater, as they clove their way through the ever thickening press, with hands upraised, amid oaths and tears.

But in the open country as they drew near Belfort, Maurice's heart was still more sorely wrung, for there the homeless fugitives were in greater numbers and lined the borders of the road in an unbroken cortege. Ah! the unhappy ones, who had believed that they were to find safety under the walls of the fortifications! The father lashed the poor old nag, the mother followed after, leading her crying children by the hand, and in this way entire families, sinking beneath the weight of their burdens, were strung along the white, blinding road in the fierce sunlight, where the tired little legs of the smaller children were unable to keep up with the headlong flight. Many had taken off their shoes and were going barefoot so as to get over the ground more rapidly, and half-dressed mothers gave the breast to their crying babies as they strode along. Affrighted faces turned for a look backward, trembling hands were raised as if to shut out the horizon from their sight, while the gale of panic tumbled their unkempt locks and sported with their ill-adjusted garments. Others there were, farmers and their men, who pushed straight across the fields, driving before them their flocks and herds, cows, oxen, sheep, horses, that they had driven with sticks and cudgels from their stables; these were seeking the shelter of the inaccessible forests, of the deep valleys and the lofty hill-tops, their course marked by clouds of dust, as in the great migrations of other days, when invaded nations made way before their barbarian conquerors. They were going to live in tents, in some lonely nook among the mountains, where the enemy would never venture to follow them; and the bleating and bellowing of the animals and the trampling of their hoofs upon the rocks grew fainter in the distance, and the golden nimbus that overhung them was lost to sight among the thick pines, while down in the road beneath the tide of vehicles and pedestrians was flowing still as strong as ever, blocking the passage of the troops, and as they drew near Belfort the men had to be brought to a halt again and again, so irresistible was the force of that torrent of humanity.

It was during one of those short halts that Maurice witnessed a scene that was destined to remain indelibly impressed upon his memory.

Standing by the road-side was a lonely house, the abode of some poor peasant, whose lean acres extended up the mountainside in the rear. The man had been unwilling to leave the little field that was his all and had remained, for to go away would have been to him like parting with life. He could be seen within the low-ceiled room, sitting stupidly on a bench, watching with dull, lack-luster eyes the passing of the troops whose retreat would give his ripe grain over to be the spoil of the enemy. Standing beside him was his wife, still a young woman, holding in her arms a child, while another was hanging by her skirts; all three were weeping bitterly. Suddenly the door was thrown open with violence and in its enframement appeared the grandmother, a very old woman, tall and lean of form, with bare, sinewy arms like knotted cords that she raised above her head and shook with frantic gestures. Her gray, scanty locks had escaped from her cap and were floating about her skinny face, and such was her fury that the words she shouted choked her utterance and came from her lips almost unintelligible.

At first the soldiers had laughed. Wasn't she a beauty, the old crazy hag! Then words reached their ears; the old woman was screaming:

"Scum! Robbers! Cowards! Cowards!"

With a voice that rose shriller and more piercing still she kept lashing them with her tongue, expectorating insult on them, and taunting them for dastards with the full force of her lungs. And the laughter ceased, it seemed as if a cold wind had blown over the ranks. The men hung their heads, looked any way save that.

"Cowards! Cowards! Cowards!"

Then all at once her stature seemed to dilate; she drew herself up, tragic in her leanness, in her poor old apology for a gown, and sweeping the heavens with her long arm from west to east, with a gesture so broad that it seemed to fill the dome:

"Cowards, the Rhine is not there! The Rhine lies yonder! Cowards, cowards!"

They got under way again at last, and Maurice, whose look just then encountered Jean's, saw that the latter's eyes were filled with tears, and it did not alleviate his distress to think that those rough soldiers, compelled to swallow an insult that they had done nothing to deserve, were shamed by it. He was conscious of nothing save the intolerable aching in his poor head, and in after days could never remember how the march of that day ended, prostrated as he was by his terrible suffering, mental and physical.

The 7th corps had spent the entire day in getting over the fourteen or fifteen miles between Dannemarie and Belfort, and it was night again before the troops got settled in their bivouacs

27

under the walls of the town, in the very same place whence they had started four days before to march against the enemy. Notwithstanding the lateness of the hour and their spent condition, the men insisted on lighting fires and making soup; it was the first time since their departure that they had had an opportunity to put warm food into their stomachs, and seated about the cheerful blaze in the cool air of evening they were dipping their noses in the porringers and grunting inarticulately in token of satisfaction when news came in that burst upon the camp like a thunderbolt, dumfoundering everyone. Two telegrams had just been received: the Prussians had not crossed the Rhine at Markolsheim, and there was not a single Prussian at Huningue. The passage of the Rhine at Markolsheim and the bridge of boats constructed under the electric light had existed merely in imagination, were an unexplained, inexplicable nightmare of the prefet at Schelestadt; and as for the army corps that had menaced Huningue, that famous corps of the Black Forest, that had made so much talk, it was but an insignificant detachment of Wurtemburgers, a couple of battalions of infantry and a squadron of cavalry, which had maneuvered with such address, marching and countermarching, appearing in one place and then suddenly popping up in another at a distance, as to gain for themselves the reputation of being thirty or forty thousand strong. And to think that that morning they had been near blowing up the viaduct at Dannemarie! Twenty leagues of fertile country had been depopulated by the most idiotic of panics, and at the recollection of what they had seen during their lamentable day's march, the inhabitants flying in consternation to the mountains, driving their cattle before them; the press of vehicles, laden with household effects, streaming cityward and surrounded by bands of weeping women and children, the soldiers waxed wroth and gave way to bitter, sneering denunciation of their leaders.

"Ah! it is too ridiculous too talk about!" sputtered Loubet, not stopping to empty his mouth, brandishing his spoon. "They take us out to fight the enemy, and there's not a soul to fight with! Twelve leagues there and twelve leagues back, and not so much as a mouse in front of us! All that for nothing, just for the fun of being scared to death!"

Chouteau, who was noisily absorbing the last drops in his porringer, bellowed his opinion of the generals, without mentioning names:

"The pigs! what miserable boobies they are, hein! A pretty pack of dunghill-cocks the government has given us as commanders! Wonder what they would do if they had an army actually before them, if they show the white feather this way when there's not a Prussian in sight, hein!-Ah no, not any of it in mine, thank you; soldiers don't obey such pigeon-livered gentlemen."

Someone had thrown another armful of wood on the fire for the pleasurable sensation of comfort there was in the bright, dancing flame, and Lapoulle, who was engaged in the luxurious occupation of toasting his shins, suddenly went off into an imbecile fit of laughter without in the least understanding what it was about, whereon Jean, who had thus far turned a deaf ear to their talk, thought it time to interfere, which he did by saying in a fatherly way:

"You had better hold your tongue, you fellows! It might be the worse for you if anyone should hear you."

He himself, in his untutored, common-sense way of viewing things, was exasperated by the stupid incompetency of their commanders, but then discipline must be maintained, and as Chouteau still kept up a low muttering he cut him short:

"Be silent, I say! Here is the lieutenant: address yourself to him if you have anything to say."

Maurice had listened in silence to the conversation from his place a little to one side. Ah, truly, the end was near! Scarcely had they made a beginning, and all was over. That lack of discipline, that seditious spirit among the men at the very first reverse, had already made the army a demoralized, disintegrated rabble that would melt away at the first indication of catastrophe. There they were, under the walls of Belfort, without having sighted a Prussian, and they were whipped.

The succeeding days were a period of monotony, full of uncertainty and anxious forebodings. To keep his troops occupied General Douay set them to work on the defenses of the place, which were in a state of incompleteness; there was great throwing up of earth and cutting through rock. And not the first item of news! Where was MacMahon's army? What was going on at Metz? The wildest rumors were current, and the Parisian journals, by their system of printing news only to contradict it the next day, kept the country in an agony of suspense. Twice, it was said, the general had written and asked for instructions, and had not even received an answer. On the 12th of August, however, the 7th corps was augmented by the 3d division, which landed from Italy, but there were still only two divisions for duty, for the 1st had participated in the defeat at Froeschwiller, had been swept away in the general rout, and as yet no one had learned where it had been stranded by the current. After a week of this abandonment, of this entire separation from the rest of France, a telegram came bringing them the order to march. The news was well received, for anything was preferable to the prison life they were leading in Belfort. And while they were getting themselves in readiness conjecture and surmise were the order of the day, for no one as yet knew what their destination was to be, some saying that they were to be sent to the defense of Strasbourg, while others spoke with confidence of a bold dash into the Black Forest that was to sever the Prussian line of communication.

Early the next morning the 106th was bundled into cattle-cars and started off among the first. The car that contained Jean's squad was particularly crowded, so much so that Loubet declared there was not even room in it to sneeze. It was a load of humanity, sent off to the war just as a load of sacks would have been dispatched to the mill, crowded in so as to get the greatest number into the smallest space, and as rations had been given out in the usual hurried, slovenly manner and the men had received in brandy what they should have received in food, the consequence was that they were all roaring drunk, with a drunkenness that vented itself in obscene songs, varied by shrieks and yells. The heavy train rolled slowly onward; pipes were alight and men could no longer see one another through the dense clouds of smoke; the heat and odor that emanated from that mass of perspiring human flesh were unendurable, while from the jolting, dingy van came volleys of shouts and laughter that drowned the monotonous rattle of the wheels and were lost amid the silence of the deserted fields. And it was not until they reached Langres that the troops learned that they were being carried back to Paris.

"Ah, nom de Dieu!" exclaimed Chouteau, who already, by virtue of his oratorical ability, was the acknowledged sovereign of his corner, "they will station us at Charentonneau, sure, to keep old Bismarck out of the Tuileries."

The others laughed loud and long, considering the joke a very good one, though no one could say why. The most trivial incidents of the journey, however, served to elicit a storm of yells, cat-calls, and laughter: a group of peasants standing beside the roadway, or the anxious faces of the people who hung about the way-stations in the hope of picking up some bits of news from the passing trains, epitomizing on a small scale the breathless, shuddering alarm that pervaded all France in the presence of invasion. And so it happened that as the train thundered by, a fleeting vision of pandemonium, all that the good burghers obtained in the way of intelligence was the salutations of that cargo of food for powder as it hurried onward to its destination, fast as steam could carry it. At a station where they stopped, however, three well-dressed ladies, wealthy bourgeoises of the town, who distributed cups of bouillon among the men, were received with great respect. Some of the soldiers shed tears, and kissed their hands as they thanked them.

But as soon as they were under way again the filthy songs and the wild shouts began afresh, and so it went on until, a little while after leaving Chaumont, they met another train that was conveying some batteries of artillery to Metz. The locomotives slowed down and the soldiers in the two trains fraternized with a frightful uproar. The artillerymen were also apparently very drunk; they stood up in their seats, and thrusting hands and arms out of the car-windows, gave this cry with a vehemence that silenced every other sound:

"To the slaughter! to the slaughter! to the slaughter!"

It was as if a cold wind, a blast from the charnel-house, had swept through the car. Amid the sudden silence that descended on them Loubet's irreverent voice was heard, shouting:

"Not very cheerful companions, those fellows!"

"But they are right," rejoined Chouteau, as if addressing some pot-house assemblage; "it is a beastly thing to send a lot of brave boys to have their brains blown out for a dirty little quarrel about which they don't know the first word."

And much more in the same strain. He was the type of the Belleville agitator, a lazy, dissipated mechanic, perverting his fellow workmen, constantly spouting the ill-digested odds and ends of political harangues that he had heard, belching forth in the same breath the loftiest sentiments and the most asinine revolutionary clap-trap. He knew it all, and tried to inoculate his comrades with his ideas, especially Lapoulle, of whom he had promised to make a lad of spirit.

"Don't you see, old man, it's all perfectly simple. If Badinguet and Bismarck have a quarrel, let 'em go to work with their fists and fight it out and not involve in their row some hundreds of thousands of men who don't even know one another by sight and have not the slightest desire to fight."

The whole car laughed and applauded, and Lapoulle, who did not know who Badinguet[*] was, and could not have told whether it was a king or an emperor in whose cause he was fighting, repeated like the gigantic baby that he was:

[*] Napoleon III.

"Of course, let 'em fight it out, and take a drink together afterward."

But Chouteau had turned to Pache, whom he now proceeded to take in hand.

"You are in the same boat, you, who pretend to believe in the good God. He has forbidden men to fight, your good God has. Why, then, are you here, you great simpleton?"

"Dame!" Pache doubtfully replied, "it is not for any pleasure of mine that I am here-but the gendarmes-"

"Oh, indeed, the gendarmes! let the gendarmes go milk the ducks!-say, do you know what we would do, all of us, if we had the least bit of spirit? I'll tell you; just the minute that they land us from the cars we'd skip; yes, we'd go straight home, and leave that pig of a Badinguet and his gang of two-for-a-penny generals to settle accounts with their beastly Prussians as best they may!"

There was a storm of bravos; the leaven of perversion was doing its work and it was Chouteau's hour of triumph, airing his muddled theories and ringing the changes on the Republic, the Rights of Man, the rottenness of the Empire, which must be destroyed, and the treason of their commanders, who, as it had been proved, had sold themselves to the enemy at the rate of a million a piece. *He* was a revolutionist, he boldly declared; the others could not even say that they were republicans, did not know what their opinions were, in fact, except Loubet, the concocter of stews and hashes, and *he* had an opinion, for he had been for soup, first, last, and always; but they all, carried away by his eloquence, shouted none the less lustily against the Emperor, their officers, the whole d--d shop, which they would leave the first chance they got, see if they wouldn't! And Chouteau, while fanning the flame of their discontent, kept an eye on Maurice, the fine gentleman, who appeared interested and whom he was proud to have for a companion; so that, by way of inflaming *his* passions also, it occurred to him to make an attack on Jean, who had thus far been tranquilly watching the proceedings out of his half-closed eyes, unmoved among the general uproar. If there was any remnant of resentment in the bosom of the volunteer since the time when the corporal had inflicted such a bitter humiliation on him by forcing him to resume his abandoned musket, now was a fine chance to set the two men by the ears.

"I know some folks who talk of shooting us," Chouteau continued, with an ugly look at Jean; "dirty, miserable skunks, who treat us worse than beasts, and, when a man's back is broken with the weight of his knapsack and Brownbess, aie! aie! object to his planting them in the fields to see if a new crop will grow from them. What do you suppose they would say, comrades, hein! now that we are masters, if we should pitch them all out upon the track, and teach them better manners? That's the way to do, hein! We'll show 'em that we won't be bothered any longer with their mangy wars. Down with Badinguet's bed-bugs! Death to the curs who want to make us fight!"

Jean's face was aflame with the crimson tide that never failed to rush to his cheeks in his infrequent fits of anger. He rose, wedged in though as he was between his neighbors as firmly as in a vise, and his blazing eyes and doubled fists had such a look of business about them that the other quailed.

"Tonnerre de Dieu! will you be silent, pig! For hours I have sat here without saying anything, because we have no longer any leaders, and I could not even send you to the guard-house. Yes, there's no doubt of it, it would be a good thing to shoot such men as you and rid the regiment of the vermin. But see here, as there's no longer any discipline, I will attend to your case myself. There's no corporal here now, but a hard-fisted fellow who is tired of listening to your jaw, and he'll see if he can't make you keep your potato-trap shut. Ah! you d--d coward! You won't fight yourself and you want to keep others from fighting! Repeat your words once and I'll knock your head off!"

By this time the whole car, won over by Jean's manly attitude, had deserted Chouteau, who cowered back in his seat as if not anxious to face his opponent's big fists.

"And I care no more for Badinguet than I do for you, do you understand? I despise politics, whether they are republican or imperial, and now, as in the past, when I used to cultivate my little farm, there is but one thing that I wish for, and that is the happiness of all, peace and good-order, freedom for every man to attend to his affairs. No one denies that war is a terrible business, but that is no reason why a man should not be treated to the sight of a firing-party when he comes trying to dishearten people who already have enough to do to keep their courage up. Good Heavens, friends, how it makes a man's pulses leap to be told that the Prussians are in the land and that he is to go help drive them out!"

Then, with the customary fickleness of a mob, the soldiers applauded the corporal, who again announced his determination to thrash the first man of his squad who should declare non-combatant principles. Bravo, the corporal! they would soon settle old Bismarck's hash! And, in the midst of the wild ovation of which he was the object, Jean, who had recovered his self-control, turned politely to Maurice and addressed him as if he had not been one of his men:

"Monsieur, you cannot have anything in common with those poltroons. Come, we haven't had a chance at them yet; we are the boys who will give them a good basting yet, those Prussians!"

It seemed to Maurice at that moment as if a ray of cheering sunshine had penetrated his heart.

He was humiliated, vexed with himself. What! that man was nothing more than an uneducated rustic! And he remembered the fierce hatred that had burned in his bosom the day he was compelled to pick up the musket that he had thrown away in a moment of madness. But he also remembered his emotion at seeing the two big tears that stood in the corporal's eyes when the old grandmother, her gray hairs streaming in the wind, had so bitterly reproached them and pointed to the Rhine that lay beneath the horizon in the distance. Was it the brotherhood of fatigue and suffering endured in common that had served thus to dissipate his wrathful feelings? He was Bonapartist by birth, and had never thought of the Republic except in a speculative, dreamy way; his feeling toward the Emperor, personally, too, inclined to friendliness, and he was favorable to the war, the very condition of national existence, the great regenerative school of nationalities. Hope, all at once, with one of those fitful impulses of the imagination, that were common in his temperament, revived in him, while the enthusiastic ardor that had impelled him to enlist one night again surged through his veins and swelled his heart with confidence of victory.

"Why, of course, Corporal," he gayly replied, "we shall give them a basting!"

And still the car kept rolling onward with its load of human freight, filled with reeking smoke of pipes and emanations of the crowded men, belching its ribald songs and drunken shouts among the expectant throngs of the stations through which it passed, among the rows of white-faced peasants who lined the iron-way. On the 20th of August they were at the Pantin Station in Paris, and that same evening boarded another train which landed them next day at Rheims *en route* for the camp at Chalons.

3

Maurice was greatly surprised when the 106th, leaving the cars at Rheims, received orders to go into camp there. So they were not to go to Chalons, then, and unite with the army there? And when, two hours later, his regiment had stacked muskets a league or so from the city over in the direction of Courcelles, in the broad plain that lies along the canal between the Aisne and Marne, his astonishment was greater still to learn that the entire army of Chalons had been falling back all that morning and was about to bivouac at that place. From one extremity of the horizon to the other, as far as Saint Thierry and Menvillette, even beyond the Laon road, the tents were going up, and when it should be night the fires of four army-corps would be blazing there. It was evident that the plan now was to go and take a position under the walls of Paris and there await the Prussians; and it was fortunate that that plan had received the approbation of the government, for was it not the wisest thing they could do?

Maurice devoted the afternoon of the 21st to strolling about the camp in search of news. The greatest freedom prevailed; discipline appeared to have been relaxed still further, the men went and came at their own sweet will. He found no obstacle in the way of his return to the city, where he desired to cash a money-order for a hundred francs that his sister Henriette had sent him. While in a cafe he heard a sergeant telling of the disaffection that existed in the eighteen battalions of the garde mobile of the Seine, which had just been sent back to Paris; the 6th battalion had been near killing their officers. Not a day passed at the camp that the generals were not insulted, and since Froeschwiller the soldiers had ceased to give Marshal MacMahon the military salute. The cafe resounded with the sound of voices in excited conversation; a violent dispute arose between two sedate burghers in respect to the number of men that MacMahon would have at his disposal. One of them made the wild assertion that there would be three hundred thousand; the other, who seemed to be more at home upon the subject, stated the strength of the four corps: the 12th, which had just been made complete at the camp with great difficulty with the assistance of provisional regiments and a division of infanterie de marine; the 1st, which had been coming straggling in in fragments ever since the 14th of the month and of which they were doing what they could to perfect the organization; the 5th, defeated before it had ever fought a battle, swept away and broken up in the general panic, and finally, the 7th, then landing from the cars, demoralized like all the rest and minus its 1st division, of which it had just recovered the remains at Rheims; in all, one hundred and twenty thousand at the outside, including the cavalry, Bonnemain's and Margueritte's divisions. When the sergeant took a hand in the quarrel, however, speaking of the army in terms of the utmost contempt, characterizing it as a ruffianly rabble, with no esprit de corps, with nothing to keep it together,-a pack of greenhorns with idiots to conduct them, to the slaughter,-the two bourgeois began to be uneasy, and fearing there might be trouble brewing, made themselves scarce.

When outside upon the street Maurice hailed a newsboy and purchased a copy of every paper he could lay hands on, stuffing some in his pockets and reading others as he walked along under the stately trees that line the pleasant avenues of the old city. Where could the German armies be? It seemed as if obscurity had suddenly swallowed them up. Two were over Metz way, of course: the first, the one commanded by General von Steinmetz, observing the place; the second, that of Prince Frederick Charles, aiming to ascend the right bank of the Moselle in order to cut Bazaine off from Paris. But the third army, that of the Crown Prince of Prussia, the army that had been victorious at Wissembourg and Froeschwiller and had driven our 1st and 5th corps, where was it now, where was it to be located amid the tangled mess of contradictory advices? Was it still in camp at Nancy, or was it true that it had arrived before

Chalons, and was that the reason why we had abandoned our camp there in such hot haste, burning our stores, clothing, forage, provisions, everything-property of which the value to the nation was beyond compute? And when the different plans with which our generals were credited came to be taken into consideration, then there was more confusion, a fresh set of contradictory hypotheses to be encountered. Maurice had until now been cut off in a measure from the outside world, and now for the first time learned what had been the course of events in Paris; the blasting effect of defeat upon a populace that had been confident of victory, the terrible commotions in the streets, the convoking of the Chambers, the fall of the liberal ministry that had effected the plebiscite, the abrogation of the Emperor's rank as General of the Army and the transfer of the supreme command to Marshal Bazaine. The Emperor had been present at the camp of Chalons since the 16th, and all the newspapers were filled with a grand council that had been held on the 17th, at which Prince Napoleon and some of the generals were present, but none of them were agreed upon the decisions that had been arrived at outside of the resultant facts, which were that General Trochu had been appointed governor of Paris and Marshal MacMahon given the command of the army of Chalons, and the inference from this was that the Emperor was to be shorn of all his authority. Consternation, irresolution, conflicting plans that were laid aside and replaced by fresh ones hour by hour; these were the things that everybody felt were in the air. And ever and always the question: Where were the German armies? Who were in the right, those who asserted that Bazaine had no force worth mentioning in front of him and was free to make his retreat through the towns of the north whenever he chose to do so, or those who declared that he was already besieged in Metz? There was a constantly recurring rumor of a series of engagements that had raged during an entire week, from the 14th until the 20th, but it failed to receive confirmation.

Maurice's legs ached with fatigue; he went and sat down upon a bench. Around him the life of the city seemed to be going on as usual; there were nursemaids seated in the shade of the handsome trees watching the sports of their little charges, small property owners strolled leisurely about the walks enjoying their daily constitutional. He had taken up his papers again, when his eyes lighted on an article that had escaped his notice, the "leader" in a rabid republican sheet; then everything was made clear to him. The paper stated that at the council of the 17th at the camp of Chalons the retreat of the army on Paris had been fully decided on, and that General Trochu's appointment to the command of the city had no other object than to facilitate the Emperor's return; but those resolutions, the journal went on to say, were rendered unavailing by the attitude of the Empress-regent and the new ministry. It was the Empress's opinion that the Emperor's return would certainly produce a revolution; she was reported to have said: "He will never reach the Tuileries alive." Starting with these premises she insisted with the utmost urgency that the army should advance, at every risk, whatever might be the cost of human life, and effect a junction with the army of Metz, in which course she was supported moreover by General de Palikao, the Minister of War, who had a plan of his own for reaching Bazaine by a rapid and victorious march. And Maurice, letting his paper fall from his hand, his eyes bent on space, believed that he now had the key to the entire mystery; the two conflicting plans, MacMahon's hesitation to undertake that dangerous flank movement with the unreliable army at his command, the impatient orders that came to him from Paris, each more tart and imperative than its predecessor, urging him on to that mad, desperate enterprise. Then, as the central figure in that tragic conflict, the vision of the Emperor suddenly rose distinctly before his inner eyes, deprived of his imperial authority, which he had committed to the hands of the Empress-regent, stripped of his military command, which he had conferred on Marshal Bazaine; a nullity, the vague and unsubstantial shadow of an emperor, a nameless, cumbersome nonentity whom no one knew what to do with, whom Paris rejected and who had ceased to have a position in the army, for he had pledged himself to issue no further orders.

The next morning, however, after a rainy night through which he slept outside his tent on the bare ground, wrapped in his rubber blanket, Maurice was cheered by the tidings that the retreat on Paris had finally carried the day. Another council had been held during the night, it was said, at which M. Rouher, the former vice-Emperor, had been present; he had been sent by the Empress to accelerate the movement toward Verdun, and it would seem that the marshal had succeeded in convincing him of the rashness of such an undertaking. Were there unfavorable tidings from Bazaine? no one could say for certain. But the absence of news was itself a circumstance of evil omen, and all among the most influential of the generals had cast their vote for the march on Paris, for which they would be the relieving army. And Maurice, happy in the conviction that the retrograde movement would commence not later than the morrow, since the orders for it were said to be already issued, thought he would gratify a boyish longing that had been troubling him for some time past, to give the go-by for one day to soldier's fare, to wit and eat his breakfast off a cloth, with the accompaniment of plate, knife and fork, carafe, and a bottle of good wine, things of which it seemed to him that he had been deprived for months and months. He had money in his pocket, so off he started with quickened pulse, as if going out for a lark, to search for a place of entertainment.

It was just at the entrance of the village of Courcelles, across the canal, that he found the breakfast for which his mouth was watering. He had been told the day before that the Emperor had taken up his quarters in one of the houses of the village, and having gone to stroll there out of curiosity, now remembered to have seen at the junction of the two roads this little inn with its arbor, the trellises of which were loaded with big clusters of ripe, golden, luscious grapes. There was an array of green-painted tables set out in the shade of the luxuriant vine, while through the open door of the vast kitchen he had caught glimpses of the antique clock, the colored prints pasted on the walls, and the comfortable landlady watching the revolving spit. It was cheerful, smiling, hospitable; a regular type of the good old-fashioned French hostelry.

A pretty, white-necked waitress came up and asked him with a great display of flashing teeth:

"Will monsieur have breakfast?"

"Of course I will! Give me some eggs, a cutlet, and cheese. And a bottle of white wine!"

She turned to go; he called her back. "Tell me, is it not in one of those houses that the Emperor has his quarters?"

"There, monsieur, in that one right before you. Only you can't see it, for it is concealed by the high wall with the overhanging trees."

He loosed his belt so as to be more at ease in his capote, and entering the arbor, chose his table, on which the sunlight, finding its way here and there through the green canopy above, danced in little golden spangles. And constantly his thoughts kept returning to that high wall behind which was the Emperor. A most mysterious house it was, indeed, shrinking from the public gaze, even its slated roof invisible. Its entrance was on the other side, upon the village street, a narrow winding street between dead-walls, without a shop, without even a window to enliven it. The small garden in the rear, among the sparse dwellings that environed it, was like an island of dense verdure. And across the road he noticed a spacious courtyard, surrounded by sheds and stables, crowded with a countless train of carriages and baggage-wagons, among which men and horses, coming and going, kept up an unceasing bustle.

"Are those all for the service of the Emperor?" he inquired, meaning to say something humorous to the girl, who was laying a snow-white cloth upon the table.

35

"Yes, for the Emperor himself, and no one else!" she pleasantly replied, glad of a chance to show her white teeth once more; and then she went on to enumerate the suite from information that she had probably received from the stablemen, who had been coming to the inn to drink since the preceding day; there were the staff, comprising twenty-five officers, the sixty cent-gardes and the half-troop of guides for escort duty, the six gendarmes of the provost-guard; then the household, seventy-three persons in all, chamberlains, attendants for the table and the bedroom, cooks and scullions; then four saddle-horses and two carriages for the Emperor's personal use, ten horses for the equerries, eight for the grooms and outriders, not mentioning forty-seven post-horses; then a *char a banc* and twelve baggage wagons, two of which, appropriated to the cooks, had particularly excited her admiration by reason of the number and variety of the utensils they contained, all in the most splendid order.

"Oh, sir, you never saw such stew-pans! they shone like silver. And all sorts of dishes, and jars and jugs, and lots of things of which it would puzzle me to tell the use! And a cellar of wine, claret, burgundy, and champagne-yes! enough to supply a wedding feast."

The unusual luxury of the snowy table-cloth and the white wine sparkling in his glass sharpened Maurice's appetite; he devoured his two poached eggs with a zest that made him fear he was developing epicurean tastes. When he turned to the left and looked out through the entrance of the leafy arbor he had before him the spacious plain, covered with long rows of tents: a busy, populous city that had risen like an exhalation from the stubble-fields between Rheims city and the canal. A few clumps of stunted trees, three wind-mills lifting their skeleton arms in the air, were all there was to relieve the monotony of the gray waste, but above the huddled roofs of Rheims, lost in the sea of foliage of the tall chestnut-trees, the huge bulk of the cathedral with its slender spires was profiled against the blue sky, looming colossal, notwithstanding the distance, beside the modest houses. Memories of school and boyhood's days came over him, the tasks he had learned and recited: all about the *sacre* of our kings, the sainte ampoule, Clovis, Jeanne d'Arc, all the long list of glories of old France.

Then Maurice's thoughts reverted again to that unassuming bourgeoise house, so mysterious in its solitude, and its imperial occupant; and directing his eyes upon the high, yellow wall he was surprised to read, scrawled there in great, awkward letters, the legend: Vive Napoleon! among the meaningless obscenities traced by schoolboys. Winter's storms and summer's sun had half effaced the lettering; evidently the inscription was very ancient. How strange, to see upon that wall that old heroic battle-cry, which probably had been placed there in honor of the uncle, not of the nephew! It brought all his childhood back to him, and Maurice was again a boy, scarcely out of his mother's arms, down there in distant Chene-Populeux, listening to the stories of his grandfather, a veteran of the Grand Army. His mother was dead, his father, in the inglorious days that followed the collapse of the empire, had been compelled to accept a humble position as collector, and there the grandfather lived, with nothing to support him save his scanty pension, in the poor home of the small public functionary, his sole comfort to fight his battles o'er again for the benefit of his two little twin grandchildren, the boy and the girl, a pair of golden-haired youngsters to whom he was in some sense a mother. He would place Maurice on his right knee and Henriette on his left, and then for hours on end the narrative would run on in Homeric strain.

But small attention was paid to dates; his story was of the dire shock of conflicting nations, and was not to be hampered by the minute exactitude of the historian. Successively or together English, Austrians, Prussians, Russians appeared upon the scene, according to the then prevailing condition of the ever-changing alliances, and it was not always an easy matter to tell why one nation received a beating in preference to another, but beaten they all were in the end, inevitably beaten from the very commencement, in a whirlwind of genius and heroic daring that swept great armies like chaff from off the earth. There was Marengo, the classic battle of the plain, with the consummate generalship of its broad plan and the faultless retreat of the battalions by squares, silent and impassive under the enemy's terrible fire; the battle, famous in story, lost at three o'clock and won at six, where the eight hundred grenadiers of the Consular Guard withstood the onset of the entire Austrian cavalry, where Desaix arrived to change impending defeat to glorious victory and die. There was Austerlitz, with its sun of glory shining forth from amid the wintry sky, Austerlitz, commencing with the capture of the plateau of Pratzen and ending with the frightful catastrophe on the frozen lake, where an entire Russian corps, men, guns, horses, went crashing through the ice, while Napoleon, who in his divine omniscience had foreseen it all, of course, directed his artillery to play upon the struggling mass. There was Jena, where so many of Prussia's bravest found a grave; at first the red flames of musketry flashing through the October mists, and Ney's impatience, near spoiling all until Augereau comes wheeling into line and saves him; the fierce charge that tore the enemy's center in twain, and finally panic, the headlong rout of their boasted cavalry, whom our hussars mow down like ripened grain, strewing the romantic glen with a harvest of men and horses. And Eylau, cruel Eylau, bloodiest battle of them all, where the maimed corpses cumbered the earth in piles; Eylau, whose new-fallen snow was stained with blood, the burial-place of heroes; Eylau, in whose name reverberates still the thunder of the charge of Murat's eighty squadrons, piercing the Russian lines in every direction, heaping the ground so thick with dead that Napoleon himself could not refrain from tears. Then Friedland, the trap into which the Russians again allowed themselves to be decoyed like a flock of brainless sparrows, the masterpiece of the Emperor's consummate strategy; our left held back as in a leash, motionless, without a sign of life, while Ney was carrying the city, street by street, and destroying the bridges, then the left hurled like a thunderbolt on the enemy's right, driving it into the river and annihilating it in that cul-de-sac; the slaughter so great that at ten o'clock at night the bloody work was not completed, most wonderful of all the successes of the great imperial epic. And Wagram, where it was the aim of the Austrians to cut us off from the Danube; they keep strengthening their left in order to overwhelm Massena, who is wounded and issues his orders from an open carriage, and Napoleon, like a malicious Titan, lets them go on unchecked; then all at once a hundred guns vomit their terrible fire upon their weakened center, driving it backward more than a league, and their left, terror-stricken to find itself unsupported, gives way before the again victorious Massena, sweeping away before it the remainder of the army, as when a broken dike lets loose its torrents upon the fields. And finally the Moskowa, where the bright sun of Austerlitz shone for the last time; where the contending hosts were mingled in confused *melee* amid deeds of the most desperate daring: mamelons carried under an unceasing fire of musketry, redoubts stormed with the naked steel, every inch of ground fought over again and again; such determined resistance on the part of the Russian Guards that our final victory was only assured by Murat's mad charges, the concentrated fire of our three hundred pieces of artillery, and the valor of Ney, who was the hero of that most obstinate of conflicts. And be the battle what it might, ever our flags floated proudly on the evening air, and as the bivouac fires were lighted on the conquered field out rang the old battle-cry: Vive Napoleon! France, carrying her invincible Eagles from end to end of Europe, seemed everywhere at home, having but to raise her finger to make her will respected by the nations, mistress of a world that in vain conspired to crush her and upon which she set her foot.

Maurice was contentedly finishing his cutlet, cheered not so much by the wine that sparkled in his glass as by the glorious memories that were teeming in his brain, when his glance encountered two ragged, dust-stained soldiers, less like soldiers than weary tramps just off the road; they were asking the attendant for information as to the position of the regiments that were encamped along the canal. He hailed them.

"Hallo there, comrades, this way! You are 7th corps men, aren't you?"

"Right you are, sir; 1st division-at least I am, more by token that I was at Froeschwiller, where it was warm enough, I can tell you. The comrade, here, belongs in the 1st corps; he was at Wissembourg, another beastly hole."

They told their story, how they had been swept away in the general panic, had crawled into a ditch half-dead with fatigue and hunger, each of them slightly wounded, and since then had been dragging themselves along in the rear of the army, compelled to lie over in towns when the fever-fits came on, until at last they had reached the camp and were on the lookout to find their regiments.

Maurice, who had a piece of Gruyere before him, noticed the hungry eyes fixed on his plate.

"Hi there, mademoiselle! bring some more cheese, will you-and bread and wine. You will join me, won't you, comrades? It is my treat. Here's to your good health!"

They drew their chairs up to the table, only too delighted with the invitation. Their entertainer watched them as they attacked the food, and a thrill of pity ran through him as he beheld their sorry plight, dirty, ragged, arms gone, their sole attire a pair of red trousers and the capote, kept in place by bits of twine and so patched and pieced with shreds of vari-colored cloth that one would have taken them for men who had been looting some battle-field and were wearing the spoil they had gathered there.

"Ah! foutre, yes!" continued the taller of the two as he plied his jaws, "it was no laughing matter there! You ought to have seen it, -tell him how it was, Coutard."

And the little man told his story with many gestures, describing figures on the air with his bread.

"I was washing my shirt, you see, while the rest of them were making soup. Just try and picture to yourself a miserable hole, a regular trap, all surrounded by dense woods that gave those Prussian pigs a chance to crawl up to us before we ever suspected they were there. So, then, about seven o'clock the shells begin to come tumbling about our ears. Nom de Dieu! but it was lively work! we jumped for our shooting-irons, and up to eleven o'clock it looked as if we were going to polish 'em off in fine style. But you must know that there were only five thousand of us, and the beggars kept coming, coming as if there was no end to them. I was posted on a little hill, behind a bush, and I could see them debouching in front, to right, to left, like rows of black ants swarming from their hill, and when you thought there were none left there were always plenty more. There's no use mincing matters, we all thought that our leaders must be first-class nincompoops to thrust us into such a hornet's nest, with no support at hand, and leave us to be crushed there without coming to our assistance. And then our General, Douay,[*] poor devil! neither a fool nor a coward, that man,-a bullet comes along and lays him on his back. That ended it; no one left to command us! No matter, though, we kept on fighting all the same; but they were too many for us, we had to fall back at last. We held the railway station for a long time, and then we fought behind a wall, and the uproar was enough to wake the dead. And then, when the city was taken, I don't exactly remember how it came about, but we were upon a mountain, the Geissberg, I think they call it, and there we intrenched ourselves in a sort of castle, and how we did give it to the pigs! they jumped about the rocks like kids, and it was fun to pick 'em off and see 'em tumble on their nose. But what would you have? they kept coming, coming, all the time, ten men to our one, and all the artillery they could wish for. Courage is a very good thing in its place, but sometimes it gets a man into difficulties, and so, at last, when it got too hot to stand it any longer, we cut and run. But regarded as nincompoops, our officers were a decided success; don't you think so, Picot?"

[*] This was Abel Douay-not to be confounded with his brother, Felix, who commanded the

38

7th corps.-TR.

There was a brief interval of silence. Picot tossed off a glass of the white wine and wiped his mouth with the back of his hand.

"Of course," said he. "It was just the same at Froeschwiller; the general who would give battle under such circumstances is a fit subject for a lunatic asylum. That's what my captain said, and he's a little man who knows what he is talking about. The truth of the matter is that no one knew anything; we were only forty thousand strong, and we were surprised by a whole army of those pigs. And no one was expecting to fight that day; battle was joined by degrees, one portion after another of our troops became engaged, against the wishes of our commanders, as it seems. Of course, I didn't see the whole of the affair, but what I do know is that the dance lasted by fits and starts all day long; a body would think it was ended; not a bit of it! away would go the music more furiously than ever. The commencement was at Woerth, a pretty little village with a funny clock-tower that looks like a big stove, owing to the earthenware tiles they have stuck all over it. I'll be hanged if I know why we let go our hold of it that morning, for we broke all our teeth and nails trying to get it back again in the afternoon, without succeeding. Oh, my children, if I were to tell you of the slaughter there, the throats that were cut and the brains knocked out, you would refuse to believe me! The next place where we had trouble was around a village with the jaw-breaking name of Elsasshausen. We got a peppering from a lot of guns that banged away at us at their ease from the top of a blasted hill that we had also abandoned that morning, why, no one has ever been able to tell. And there it was that with these very eyes of mine I saw the famous charge of the cuirassiers. Ah, how gallantly they rode to their death, poor fellows! A shame it was, I say, to let men and horses charge over ground like that, covered with brush and furze, cut up by ditches. And on top of it all, nom de Dieu! what good could they accomplish? But it was very *chic* all the same; it was a beautiful sight to see. The next thing for us to do, shouldn't you suppose so? was to go and sit down somewhere and try to get our wind again. They had set fire to the village and it was burning like tinder, and the whole gang of Bavarian, Wurtemburgian and Prussian pigs, more than a hundred and twenty thousand of them there were, as we found out afterward, had got around into our rear and on our flanks. But there was to be no rest for us then, for just at that time the fiddles began to play again a livelier tune than ever around Froeschwiller. For there's no use talking, fellows, MacMahon may be a blockhead but he is a brave man; you ought to have seen him on his big horse, with the shells bursting all about him! The best thing to do would have been to give leg-bail at the beginning, for it is no disgrace to a general to refuse to fight an army of superior numbers, but he, once we had gone in, was bound to see the thing through to the end. And see it through he did! why, I tell you that the men down in Froeschwiller were no longer human beings; they were ravening wolves devouring one another. For near two hours the gutters ran red with blood. All the same, however, we had to knuckle under in the end. And to think that after it was all over they should come and tell us that we had whipped the Bavarians over on our left! By the piper that played before Moses, if we had only had a hundred and twenty thousand men, if *we* had had guns, and leaders with a little pluck!"

Loud and angry were the denunciations of Coutard and Picot in their ragged, dusty uniforms as they cut themselves huge slices of bread and bolted bits of cheese, evoking their bitter memories there in the shade of the pretty trellis, where the sun played hide and seek among the purple and gold of the clusters of ripening grapes. They had come now to the horrible flight that succeeded the defeat; the broken, demoralized, famishing regiments flying through the fields, the highroads blocked with men, horses, wagons, guns, in inextricable confusion; all the wreck and ruin of a beaten army that pressed on, on, on, with the chill breath of panic on their backs. As they had not had wit enough to fall back while there was time and take post

39

among the passes of the Vosges, where ten thousand men would have sufficed to hold in check a hundred thousand, they should at least have blown up the bridges and destroyed the tunnels; but the generals had lost their heads, and both sides were so dazed, each was so ignorant of the other's movements, that for a time each of them was feeling to ascertain the position of its opponent, MacMahon hurrying off toward Luneville, while the Crown Prince of Prussia was looking for him in the direction of the Vosges. On the 7th the remnant of the 1st corps passed through Saverne, like a swollen stream that carries away upon its muddy bosom all with which it comes in contact. On the 8th, at Sarrebourg, the 5th corps came tumbling in upon the 1st, like one mad mountain torrent pouring its waters into another. The 5th was also flying, defeated without having fought a battle, sweeping away with it its commander, poor General de Failly, almost crazy with the thought that to his inactivity was imputed the responsibility of the defeat, when the fault all rested in the Marshal's having failed to send him orders. The mad flight continued on the 9th and 10th, a stampede in which no one turned to look behind him. On the 11th, in order to turn Nancy, which a mistaken rumor had reported to be occupied by the enemy, they made their way in a pouring rainstorm to Bayon; the 12th they camped at Haroue, the 13th at Vicherey, and on the 14th were at Neufchateau, where at last they struck the railroad, and for three days the work went on of loading the weary men into the cars that were to take them to Chalons. Twenty-four hours after the last train rolled out of the station the Prussians entered the town. "Ah, the cursed luck!" said Picot in conclusion; "how we had to ply our legs! And we who should by rights have been in hospital!"

Coutard emptied what was left in the bottle into his own and his comrade's glass. "Yes, we got on our pins, somehow, and are running yet. Bah! it is the best thing for us, after all, since it gives us a chance to drink the health of those who were not knocked over."

Maurice saw through it all. The sledge hammer blow of Froeschwiller, following so close on the heels of the idiotic surprise at Wissembourg, was the lightning flash whose baleful light disclosed to him the entire naked, terrible truth. We were taken unprepared; we had neither guns, nor men, nor generals, while our despised foe was an innumerable host, provided with all modern appliances and faultless in discipline and leadership. The three German armies had burst apart the weak line of our seven corps, scattered between Metz and Strasbourg, like three powerful wedges. We were doomed to fight our battle out unaided; nothing could be hoped for now from Austria and Italy, for all the Emperor's plans were disconcerted by the tardiness of our operations and the incapacity of the commanders. Fate, even, seemed to be working against us, heaping all sorts of obstacles and ill-timed accidents in our path and favoring the secret plan of the Prussians, which was to divide our armies, throwing one portion back on Metz, where it would be cut off from France, while they, having first destroyed the other fragment, should be marching on Paris. It was as plain now as a problem in mathematics that our defeat would be owing to causes that were patent to everyone; it was bravery without intelligent guidance pitted against numbers and cold science. Men might discuss the question as they would in after days; happen what might, defeat was certain in spite of everything, as certain and inexorable as the laws of nature that rule our planet.

In the midst of his uncheerful revery, Maurice's eyes suddenly lighted on the legend scrawled on the wall before him-Vive Napoleon! and a sensation of intolerable distress seemed to pierce his heart like a red hot iron. Could it be true, then, that France, whose victories were the theme of song and story everywhere, the great nation whose drums had sounded throughout the length and breadth of Europe, had been thrown in the dust at the first onset by an insignificant race, despised of everyone? Fifty years had sufficed to compass it; the world had changed, and defeat most fearful had overtaken those who had been deemed invincible. He remembered the words that had been uttered by Weiss his brother-in-law, during that evening of anxiety when they were at Mulhausen. Yes, he alone of them had been clear of vision, had penetrated the hidden causes that had long been slowly sapping our strength, had felt the freshening gale of youth and progress under the impulse of which Germany was being wafted onward to prosperity and power. Was not the old warlike age dying and a new one coming to the front? Woe to that one among the nations which halted in its onward march! the victory is to those who are with the advance-guard, to those who are clear of head and strong of body, to the most powerful.

But just then there came from the smoke-blackened kitchen, where the walls were bright with the colored prints of Epinal, a sound of voices and the squalling of a girl who submits, not unwillingly, to be tousled. It was Lieutenant Rochas, availing himself of his privilege as a conquering hero, to catch and kiss the pretty waitress. He came out into the arbor, where he ordered a cup of coffee to be served him, and as he had heard the concluding words of Picot's narrative, proceeded to take a hand in the conversation:

"Bah! my children, those things that you are speaking of don't amount to anything. It is only the beginning of the dance; you will see the fun commence in earnest presently. Pardi! up to the present time they have been five to our one, but things are going to take a change now; just put that in your pipe and smoke it. We are three hundred thousand strong here, and every move we make, which nobody can see through, is made with the intention of bringing the Prussians down on us, while Bazaine, who has got his eye on them, will take them in their rear. And then we'll smash 'em, crac! just as I smash this fly!"

Bringing his hands together with a sounding clap he caught and crushed a fly on the wing, and he laughed loud and cheerily, believing with all his simple soul in the feasibility of a plan that seemed so simple, steadfast in his faith in the invincibility of French courage. He good-naturedly informed the two soldiers of the exact position of their regiments, then lit a cigar and seated himself contentedly before his *demitasse*.

"The pleasure was all mine, comrades!" Maurice replied to Coutard and Picot, who, as they were leaving, thanked him for the cheese and wine.

He had also called for a cup of coffee and sat watching the Lieutenant, whose hopefulness had communicated itself to him, a little surprised, however, to hear him enumerate their strength at three hundred thousand men, when it was not more than a hundred thousand, and at his happy-go-lucky way of crushing the Prussians between the two armies of Chalons and Metz. But then he, too, felt such need of some comforting illusion! Why should he not continue to hope when all those glorious memories of the past that he had evoked were still ringing in his ears? The old inn was so bright and cheerful, with its trellis hung with the purple grapes of France, ripening in the golden sunlight! And again his confidence gained a momentary ascendancy over the gloomy despair that the late events had engendered in him.

Maurice's eyes had rested for a moment on an officer of chasseurs d'Afrique who, with his orderly, had disappeared at a sharp trot around the corner of the silent house where the Emperor was quartered, and when the orderly came back alone and stopped with his two horses before the inn door he gave utterance to an exclamation of surprise:

"Prosper! Why, I supposed you were at Metz!"

It was a young man of Remilly, a simple farm-laborer, whom he had known as a boy in the days when he used to go and spend his vacations with his uncle Fouchard. He had been drawn, and when the war broke out had been three years in Africa; he cut quite a dashing figure in his sky-blue jacket, his wide red trousers with blue stripes and red woolen belt, with his sun-dried face and strong, sinewy limbs that indicated great strength and activity.

"Hallo! it's Monsieur Maurice! I'm glad to see you!"

He took things very easily, however, conducting the steaming horses to the stable, and to his own, more particularly, giving a paternal attention. It was no doubt his affection for the noble animal, contracted when he was a boy and rode him to the plow, that had made him select the cavalry arm of the service.

"We've just come in from Monthois, more than ten leagues at a stretch," he said when he came back, "and Poulet will be wanting his breakfast."

Poulet was the horse. He declined to eat anything himself; would only accept a cup of coffee. He had to wait for his officer, who had to wait for the Emperor; he might be five minutes, and then again he might be two hours, so his officer had told him to put the horses in the stable. And as Maurice, whose curiosity was aroused, showed some disposition to pump him, his face became as vacant as a blank page.

"Can't say. An errand of some sort-papers to be delivered."

But Rochas looked at the chasseur with an eye of tenderness, for the uniform awakened old memories of Africa.

"Eh! my lad, where were you stationed out there?"

"At Medeah, Lieutenant."

Ah, Medeah! And drawing their chairs closer together they started a conversation, regardless of difference in rank. The life of the desert had become a second nature, for Prosper, where the trumpet was continually calling them to arms, where a large portion of their time was spent on horseback, riding out to battle as they would to the chase, to some grand battue of Arabs. There was just one soup-basin for every six men, or tribe, as it was called, and each tribe was a family by itself, one of its members attending to the cooking, another washing their linen, the others pitching the tent, caring for the horses, and cleaning the arms. By day they scoured the country beneath a sun like a ball of blazing copper, loaded down with the burden of their arms and utensils; at night they built great fires to drive away the mosquitoes and sat around them, singing the songs of France. Often it happened that in the luminous darkness of the night, thick set with stars, they had to rise and restore peace among their four-footed friends, who, in the balmy softness of the air, had set to biting and kicking one another, uprooting their pickets and neighing and snorting furiously. Then there was the delicious coffee, their greatest, indeed their only, luxury, which they ground by the primitive appliances of a carbine-butt and a porringer, and afterward strained through a red woolen sash. But their life was not one of unalloyed enjoyment; there were dark days, also, when they were far from the abodes of civilized man with the enemy before them. No more fires, then; no singing, no good times. There were times when hunger, thirst and want of sleep caused them horrible suffering, but no matter; they loved that daring, adventurous life, that war of skirmishes, so propitious for the display of personal bravery and as interesting as a fairy tale, enlivened by the razzias, which were only public plundering on a larger scale, and by marauding, or the private peculations of the chicken-thieves, which afforded many an amusing story that made even the generals laugh.

"Ah!" said Prosper, with a more serious face, "it's different here; the fighting is done in quite another way."

And in reply to a question asked by Maurice, he told the story of their landing at Toulon and the long and wearisome march to Luneville. It was there that they first received news of Wissembourg and Froeschwiller. After that his account was less clear, for he got the names of towns mixed, Nancy and Saint-Mihiel, Saint-Mihiel and Metz. There must have been heavy fighting on the 14th, for the sky was all on fire, but all he saw of it was four uhlans behind a hedge. On the 16th there was another engagement; they could hear the artillery going as early as six o'clock in the morning, and he had been told that on the 18th they started the dance again, more lively than ever. But the chasseurs were not in it that time, for at Gravelotte on the 16th, as they were standing drawn up along a road waiting to wheel into column, the Emperor, who passed that way in a victoria, took them to act as his escort to Verdun. And a pretty little jaunt it was, twenty-six miles at a hard gallop, with the fear of being cut off by the Prussians at any moment!

"And what of Bazaine?" asked Rochas.

"Bazaine? they say that he is mightily well pleased that the Emperor lets him alone."

But the Lieutenant wanted to know if Bazaine was coming to join them, whereon Prosper made a gesture expressive of uncertainty; what did any one know? Ever since the 16th their time had been spent in marching and countermarching in the rain, out on reconnoissance and grand-guard duty, and they had not seen a sign of an enemy. Now they were part of the army of Chalons. His regiment, together with two regiments of chasseurs de France and one of hussars, formed one of the divisions of the cavalry of reserve, the first division, commanded by General Margueritte, of whom he spoke with most enthusiastic warmth.

"Ah, the bougre! the enemy will catch a Tartar in him! But what's the good talking? the only use they can find for us is to send us pottering about in the mud."

There was silence for a moment, then Maurice gave some brief news of Remilly and uncle Fouchard, and Prosper expressed his regret that he could not go and shake hands with Honore, the quartermaster-sergeant, whose battery was stationed more than a league away, on the other side of the Laon road. But the chasseur pricked up his ears at hearing the whinnying of a horse and rose and went out to make sure that Poulet was not in want of anything. It was the hour sacred to coffee and pousse-cafe, and it was not long before the little hostelry was full to overflowing with officers and men of every arm of the service. There was not a vacant table, and the bright uniforms shone resplendent against the green background of leaves checkered with spots of sunshine. Major Bouroche had just come in and taken a seat beside Rochas, when Jean presented himself with an order.

"Lieutenant, the captain desires me to say that he wishes to see you at three o'clock on company business."

Rochas signified by a nod of the head that he had heard, and Jean did not go away at once, but stood smiling at Maurice, who was lighting a cigarette. Ever since the occurrence in the railway car there had been a sort of tacit truce between the two men; they seemed to be reciprocally studying each other, with an increasing interest and attraction. But just then Prosper came back, a little out of temper.

"I mean to have something to eat unless my officer comes out of that shanty pretty quick. The Emperor is just as likely as not to stay away until dark, confound it all."

"Tell me," said Maurice, his curiosity again getting the better of him, "isn't it possible that the news you are bringing may be from Bazaine?"

"Perhaps so. There was a good deal of talk about him down there at Monthois."

At that moment there was a stir outside in the street, and Jean, who was standing by one of the doors of the arbor, turned and said:

"The Emperor!"

Immediately everyone was on his feet. Along the broad, white road, with its rows of poplars on either side, came a troop of cent-gardes, spick and span in their brilliant uniforms, their cuirasses blazing in the sunlight, and immediately behind them rode the Emperor, accompanied by his staff, in a wide open space, followed by a second troop of cent-gardes.

There was a general uncovering of heads, and here and there a hurrah was heard; and the

Emperor raised his head as he passed; his face looked drawn, the eyes were dim and watery. He had the dazed appearance of one suddenly aroused from slumber, smiled faintly at sight of the cheerful inn, and saluted. From behind them Maurice and Jean distinctly heard old Bouroche growl, having first surveyed the sovereign with his practiced eye:

"There's no mistake about it, that man is in a bad way." Then he succinctly completed his diagnosis: "His jig is up!"

Jean shook his head and thought in his limited, common sense way: "It is a confounded shame to let a man like that have command of the army!" And ten minutes later, when Maurice, comforted by his good breakfast, shook hands with Prosper and strolled away to smoke more cigarettes, he carried with him the picture of the Emperor, seated on his easy-gaited horse, so pale, so gentle, the man of thought, the dreamer, wanting in energy when the moment for action came. He was reputed to be good-hearted, capable, swayed by generous and noble thoughts, a silent man of strong and tenacious will; he was very brave, too, scorning danger with the scorn of the fatalist for whom destiny has no fears; but in critical moments a fatal lethargy seemed to overcome him; he appeared to become paralyzed in presence of results, and powerless thereafter to struggle against Fortune should she prove adverse. And Maurice asked himself if his were not a special physiological condition, aggravated by suffering; if the indecision and increasing incapacity that the Emperor had displayed ever since the opening of the campaign were not to be attributed to his manifest illness. That would explain everything: a minute bit of foreign substance in a man's system, and empires totter.

The camp that evening was all astir with activity; officers were bustling about with orders and arranging for the start the following morning at five o'clock. Maurice experienced a shock of surprise and alarm to learn that once again all their plans were changed, that they were not to fall back on Paris, but proceed to Verdun and effect a junction with Bazaine. There was a report that dispatches had come in during the day from the marshal announcing that he was retreating, and the young man's thoughts reverted to the officer of chasseurs and his rapid ride from Monthois; perhaps he had been the bearer of a copy of the dispatch. So, then, the opinions of the Empress-regent and the Council of Ministers had prevailed with the vacillating MacMahon, in their dread to see the Emperor return to Paris and their inflexible determination to push the army forward in one supreme attempt to save the dynasty; and the poor Emperor, that wretched man for whom there was no place in all his vast empire, was to be bundled to and fro among the baggage of his army like some worthless, worn-out piece of furniture, condemned to the irony of dragging behind him in his suite his imperial household, cent-gardes, horses, carriages, cooks, silver stew-pans and cases of champagne, trailing his flaunting mantle, embroidered with the Napoleonic bees, through the blood and mire of the highways of his retreat.

At midnight Maurice was not asleep; he was feverishly wakeful, and his gloomy reflections kept him tossing and tumbling on his pallet. He finally arose and went outside, where he found comfort and refreshment in the cool night air. The sky was overspread with clouds, the darkness was intense; along the front of the line the expiring watch-fires gleamed with a red and sullen light at distant intervals, and in the deathlike, boding silence could be heard the long-drawn breathing of the hundred thousand men who slumbered there. Then Maurice became more tranquil, and there descended on him a sentiment of brotherhood, full of compassionate kindness for all those slumbering fellow-creatures, of whom thousands would soon be sleeping the sleep of death. Brave fellows! True, many of them were thieves and drunkards, but think of what they had suffered and the excuse there was for them in the universal demoralization! The glorious veterans of Solferino and Sebastopol were but a handful, incorporated in the ranks of the newly raised troops, too few in number to make their example felt. The four corps that had been got together and equipped so hurriedly, devoid of every element of cohesion, were the forlorn hope, the expiatory band that their rulers were sending to the sacrifice in the endeavor to avert the wrath of destiny. They would bear their cross to the bitter end, atoning with their life's blood for the faults of others, glorious amid disaster and defeat.

And then it was that Maurice, there in the darkness that was instinct with life, became conscious that a great duty lay before him. He ceased to beguile himself with the illusive prospect of great victories to be gained; the march to Verdun was a march to death, and he so accepted it, since it was their lot to die, with brave and cheerful resignation.

4

On Tuesday, the 23d of August, at six o'clock in the morning, camp was broken, and as a stream that has momentarily expanded into a lake resumes its course again, the hundred and odd thousand men of the army of Chalons put themselves in motion and soon were pouring onward in a resistless torrent; and notwithstanding the rumors that had been current since the preceding day, it was a great surprise to most to see that instead of continuing their retrograde movement they were leaving Paris behind them and turning their faces toward the unknown regions of the East.

At five o'clock in the morning the 7th corps was still unsupplied with cartridges. For two days the artillerymen had been working like beavers to unload the materiel, horses, and stores that had been streaming from Metz into the overcrowded station, and it was only at the very last moment that some cars of cartridges were discovered among the tangled trains, and that a detail which included Jean among its numbers was enabled to bring back two hundred and forty thousand on carts that they had hurriedly requisitioned. Jean distributed the regulation number, one hundred cartridges to a man, among his squad, just as Gaude, the company bugler, sounded the order to march.

The 106th was not to pass through Rheims, their orders being to turn the city and debouch into the Chalons road farther on, but on this occasion there was the usual failure to regulate the order and time of marching, so that, the four corps having commenced to move at the same moment, they collided when they came out upon the roads that they were to traverse in common and the result was inextricable confusion. Cavalry and artillery were constantly cutting in among the infantry and bringing them to a halt; whole brigades were compelled to leave the road and stand at ordered arms in the plowed fields for more than an hour, waiting until the way should be cleared. And to make matters worse, they had hardly left the camp when a terrible storm broke over them, the rain pelting down in torrents, drenching the men completely and adding intolerably to the weight of knapsacks and great-coats. Just as the rain began to hold up, however, the 106th saw a chance to go forward, while some zouaves in an adjoining field, who were forced to wait yet for a while, amused themselves by pelting one another with balls of moist earth, and the consequent condition of their uniforms afforded them much merriment.

The sun suddenly came shining out again in the clear sky, the warm, bright sun of an August morning, and with it came returning gayety; the men were steaming like a wash of linen hung out to dry in the open air: the moisture evaporated from their clothing in little more time than it takes to tell it, and when they were warm and dry again, like dogs who shake the water from them when they emerge from a pond, they chaffed one another good-naturedly on their bedraggled appearance and the splashes of mud on their red trousers. Wherever two roads intersected another halt was necessitated; the last one was in a little village just beyond the walls of the city, in front of a small saloon that seemed to be doing a thriving business. Thereon it occurred to Maurice to treat the squad to a drink, by way of wishing them all good luck.

"Corporal, will you allow me-"

Jean, after hesitating a moment, accepted a "pony" of brandy for himself. Loubet and Chouteau were of the party (the latter had been watchful and submissive since that day when the corporal had evinced a disposition to use his heavy fists), and also Pache and Lapoulle, a couple of very decent fellows when there was no one to set them a bad example.

"Your good health, corporal!" said Chouteau in a respectful, whining tone.

"Thank you; here's hoping that you may bring back your head and all your legs and arms!" Jean politely replied, while the others laughed approvingly.

But the column was about to move; Captain Beaudoin came up with a scandalized look on his face and a reproof at the tip of his tongue, while Lieutenant Rochas, more indulgent to the small weaknesses of his men, turned his head so as not to see what was going on. And now they were stepping out at a good round pace along the Chalons road, which stretched before them for many a long league, bordered with trees on either side, undeviatingly straight, like a never-ending ribbon unrolled between the fields of yellow stubble that were dotted here and there with tall stacks and wooden windmills brandishing their lean arms. More to the north were rows of telegraph poles, indicating the position of other roads, on which they could distinguish the black, crawling lines of other marching regiments. In many places the troops had left the highway and were moving in deep columns across the open plain. To the left and front a cavalry brigade was seen, jogging along at an easy trot in a blaze of sunshine. The entire wide horizon, usually so silent and deserted, was alive and populous with those streams of men, pressing onward, onward, in long drawn, black array, like the innumerable throng of insects from some gigantic ant-hill.

About nine o'clock the regiment left the Chalons road and wheeled to the left into another that led to Suippe, which, like the first, extended, straight as an arrow's flight, far as the eye could see. The men marched at the route-step in two straggling files along either side of the road, thus leaving the central space free for the officers, and Maurice could not help noticing their anxious, care-worn air, in striking contrast with the jollity and good-humor of the soldiers, who were happy as children to be on the move once more. As the squad was near the head of the column he could even distinguish the Colonel, M. de Vineuil, in the distance, and was impressed by the grave earnestness of his manner, and his tall, rigid form, swaying in cadence to the motion of his charger. The band had been sent back to the rear, to keep company with the regimental wagons; it played but once during that entire campaign. Then came the ambulances and engineer's train attached to the division, and succeeding that the corps train, an interminable procession of forage wagons, closed vans for stores, carts for baggage, and vehicles of every known description, occupying a space of road nearly four miles in length, and which, at the infrequent curves in the highway, they could see winding behind them like the tail of some great serpent. And last of all, at the extreme rear of the column, came the herds, "rations on the hoof," a surging, bleating, bellowing mass of sheep and oxen, urged on by blows and raising clouds of dust, reminding one of the old warlike peoples of the East and their migrations.

Lapoulle meantime would every now and then give a hitch of his shoulders in an attempt to shift the weight of his knapsack when it began to be too heavy. The others, alleging that he was the strongest, were accustomed to make him carry the various utensils that were common to the squad, including the big kettle and the water-pail; on this occasion they had even saddled him with the company shovel, assuring him that it was a badge of honor. So far was he from complaining that he was now laughing at a song with which Loubet, the tenor of the squad, was trying to beguile the tedium of the way. Loubet had made himself quite famous by reason of his knapsack, in which was to be found a little of everything: linen, an extra pair of shoes, haberdashery, chocolate, brushes, a plate and cup, to say nothing of his regular rations of biscuit and coffee, and although the all-devouring receptacle also contained his cartridges, and his blankets were rolled on top of it, together with the shelter-tent and stakes, the load nevertheless appeared light, such an excellent system he had of packing his trunk, as he himself expressed it.

"It's a beastly country, all the same!" Chouteau kept repeating from time to time, casting a look of intense disgust over the dreary plains of "lousy Champagne."

Broad expanses of chalky ground of a dirty white lay before and around them, and seemed to have no end. Not a farmhouse to be seen anywhere, not a living being; nothing but flocks of crows, forming small spots of blackness on the immensity of the gray waste. On the left, far away in the distance, the low hills that bounded the horizon in that direction were crowned by woods of somber pines, while on the right an unbroken wall of trees indicated the course of the river Vesle. But over there behind the hills they had seen for the last hour a dense smoke was rising, the heavy clouds of which obscured the sky and told of a dreadful conflagration raging at no great distance.

"What is burning over there?" was the question that was on the lips of everyone.

The answer was quickly given and ran through the column from front to rear. The camp of Chalons had been fired, it was said, by order of the Emperor, to keep the immense collection of stores there from falling into the hands of the Prussians, and for the last two days it had been going up in flame and smoke. The cavalry of the rear-guard had been instructed to apply the torch to two immense warehouses, filled with tents, tent-poles, mattresses, clothing, shoes, blankets, mess utensils, supplies of every kind sufficient for the equipment of a hundred thousand men. Stacks of forage also had been lighted, and were blazing like huge beacon-fires, and an oppressive silence settled down upon the army as it pursued its march across the wide, solitary plain at sight of that dusky, eddying column that rose from behind the distant hills, filling the heavens with desolation. All that was to be heard in the bright sunlight was the measured tramp of many feet upon the hollow ground, while involuntarily the eyes of all were turned on that livid cloud whose baleful shadows rested on their march for many a league.

Their spirits rose again when they made their midday halt in a field of stubble, where the men could seat themselves on their unslung knapsacks and refresh themselves with a bite. The large square biscuits could only be eaten by crumbling them in the soup, but the little round ones were quite a delicacy, light and appetizing; the only trouble was that they left an intolerable thirst behind them. Pache sang a hymn, being invited thereto, the squad joining in the chorus. Jean smiled good-naturedly without attempting to check them in their amusement, while Maurice, at sight of the universal cheerfulness and the good order with which their first day's march was conducted, felt a revival of confidence. The remainder of the allotted task of the day was performed with the same light-hearted alacrity, although the last five miles tried their endurance. They had abandoned the high road, leaving the village of Prosnes to their right, in order to avail themselves of a short cut across a sandy heath diversified by an occasional thin pine wood, and the entire division, with its interminable train at its heels, turned and twisted in and out among the trees, sinking ankle deep in the yielding sand at every step. It seemed as if the cheerless waste would never end; all that they met was a flock of very lean sheep, guarded by a big black dog.

It was about four o'clock when at last the 106th halted for the night at Dontrien, a small village on the banks of the Suippe. The little stream winds among some pretty groves of trees; the old church stands in the middle of the graveyard, which is shaded in its entire extent by a magnificent chestnut. The regiment pitched its tents on the left bank, in a meadow that sloped gently down to the margin of the river. The officers said that all the four corps would bivouac that evening on the line of the Suippe between Auberive and Hentregiville, occupying the intervening villages of Dontrien, Betheniville and Pont-Faverger, making a line of battle nearly five leagues long.

Gaude immediately gave the call for "distribution," and Jean had to run for it, for the corporal was steward-in-chief, and it behooved him to be on the lookout to protect his men's interests. He had taken Lapoulle with him, and in a quarter of an hour they returned with some ribs of beef and a bundle of firewood. In the short space of time succeeding their arrival three steers of the herd that followed the column had been knocked in the head under a great oak-tree, skinned, and cut up. Lapoulle had to return for bread, which the villagers of Dontrien had been baking all that afternoon in their ovens. There was really no lack of anything on that first day, setting aside wine and tobacco, with which the troops were to be obliged to dispense during the remainder of the campaign.

Upon Jean's return he found Chouteau engaged in raising the tent, assisted by Pache; he looked at them for a moment with the critical eye of an old soldier who had no great opinion of their abilities.

"It will do very well if the weather is fine to-night," he said at last, "but if it should come on to blow we would like enough wake up and find ourselves in the river. Let me show you."

And he was about to send Maurice with the large pail for water, but the young man had sat down on the ground, taken off his shoe, and was examining his right foot.

"Hallo, there! what's the matter with you?"

"My shoe has chafed my foot and raised a blister. My other shoes were worn out, and when we were at Rheims I bought these, like a big fool, because they were a good fit. I should have selected gunboats."

Jean kneeled and took the foot in his hand, turning it over as carefully as if it had been a little child's, with a disapproving shake of his head.

"You must be careful; it is no laughing matter, a thing like that. A soldier without the use of his feet is of no good to himself or anyone else. When we were in Italy my captain used always to say that it is the men's legs that win battles."

He bade Pache go for the water, no very hard task, as the river was but a few yards away, and Loubet, having in the meantime dug a shallow trench and lit his fire, was enabled to commence operations on his pot-au-feu, which he did by putting on the big kettle full of water and plunging into it the meat that he had previously corded together with a bit of twine, *secundum artem*. Then it was solid comfort for them to watch the boiling of the soup; the whole squad, their chores done up and their day's labor ended, stretched themselves on the grass around the fire in a family group, full of tender anxiety for the simmering meat, while Loubet occasionally stirred the pot with a gravity fitted to the importance of his position. Like children and savages, their sole instinct was to eat and sleep, careless of the morrow, while advancing to face unknown risks and dangers.

But Maurice had unpacked his knapsack and come across a newspaper that he had bought at Rheims, and Chouteau asked:

"Is there anything about the Prussians in it? Read us the news!"

They were a happy family under Jean's mild despotism. Maurice good-naturedly read such news as he thought might interest them, while Pache, the seamstress of the company, mended his greatcoat for him and Lapoulle cleaned his musket. The first item was a splendid victory won by Bazaine, who had driven an entire Prussian corps into the quarries of Jaumont, and the trumped-up tale was told with an abundance of dramatic detail, how men and horses went over the precipice and were crushed on the rocks beneath out of all semblance of humanity, so that there was not one whole corpse found for burial. Then there were minute details of the pitiable condition of the German armies ever since they had invaded France: the ill-fed, poorly equipped soldiers were actually falling from inanition and dying by the roadside of horrible diseases. Another article told how the king of Prussia had the diarrhea, and how Bismarck had broken his leg in jumping from the window of an inn where a party of zouaves had just missed capturing him. Capital news! Lapoulle laughed over it as if he would split his sides, while Chouteau and the others, without expressing the faintest doubt, chuckled at the idea that soon they would be picking up Prussians as boys pick up sparrows in a field after a hail-storm. But they laughed loudest at old Bismarck's accident; oh! the zouaves and the turcos, they were the boys for one's money! It was said that the Germans were in an ecstasy of fear and rage, declaring that it was unworthy of a nation that claimed to be civilized to employ such heathen savages in its armies. Although they had been decimated at Froeschwiller, the foreign troops seemed to have a good deal of life left in them.

It was just striking six from the steeple of the little church of Dontrien when Loubet shouted:

"Come to supper!"

The squad lost no time in seating themselves in a circle. At the very last moment Loubet had succeeded in getting some vegetables from a peasant who lived hard by. That made the crowning glory of the feast: a soup perfumed with carrots and onions, that went down the throat soft as velvet-what could they have desired more? The spoons rattled merrily in the little wooden bowls. Then it devolved on Jean, who always served the portions, to distribute the beef, and it behooved him that day to do it with the strictest impartiality, for hungry eyes were watching him and there would have been a growl had anyone received a larger piece than his neighbors. They concluded by licking the porringers, and were smeared with soup up to their eyes.

"Ah, nom de Dieu!" Chouteau declared when he had finished, throwing himself flat on his back; "I would rather take that than a beating, any day!"

Maurice, too, whose foot pained him less now that he could give it a little rest, was conscious of that sensation of well-being that is the result of a full stomach. He was beginning to take more kindly to his rough companions, and to bring himself down nearer to their level under the pressure of the physical necessities of their life in common. That night he slept the same deep sleep as did his five tent-mates; they all huddled close together, finding the sensation of animal warmth not disagreeable in the heavy dew that fell. It is necessary to state that Lapoulle, at the instigation of Loubet, had gone to a stack not far away and feloniously appropriated a quantity of straw, in which our six gentlemen snored as if it had been a bed of down. And from Auberive to Hentregiville, along the pleasant banks of the Suippe as it meandered sluggishly between its willows, the fires of those hundred thousand sleeping men illuminated the starlit night for fifteen miles, like a long array of twinkling stars.

At sunrise they made coffee, pulverizing the berries in a wooden bowl with a musket-butt, throwing the powder into boiling water, and settling it with a drop of cold water. The luminary rose that morning in a bank of purple and gold, affording a spectacle of royal magnificence, but Maurice had no eye for such displays, and Jean, with the weather-wisdom of a peasant, cast an anxious glance at the red disk, which presaged rain; and it was for that reason that, the surplus of bread baked the day before having been distributed and the squad having received three loaves, he reproved severely Loubet and Pache for making them fast on the outside of their knapsacks; but the tents were folded and the knapsacks packed, and so no one paid any attention to him. Six o'clock was sounding from all the bells of the village when the army put itself in motion and stoutly resumed its advance in the bright hopefulness of the dawn of the new day.

The 106th, in order to reach the road that leads from Rheims to Vouziers, struck into a cross-road, and for more than an hour their way was an ascending one. Below them, toward the north, Betheniville was visible among the trees, where the Emperor was reported to have slept, and when they reached the Vouziers road the level country of the preceding day again presented itself to their gaze and the lean fields of "lousy Champagne" stretched before them in wearisome monotony. They now had the Arne, an insignificant stream, flowing on their left, while to the right the treeless, naked country stretched far as the eye could see in an apparently interminable horizon. They passed through a village or two: Saint-Clement, with its single winding street bordered by a double row of houses, Saint-Pierre, a little town of miserly rich men who had barricaded their doors and windows. The long halt occurred about ten o'clock, near another village, Saint-Etienne, where the men were highly delighted to find tobacco once more. The 7th corps had been cut up into several columns, and the 106th headed one of these columns, having behind it only a battalion of chasseurs and the reserve artillery. Maurice turned his head at every bend in the road to catch a glimpse of the long train that had so excited his interest the day before, but in vain; the herds had gone off in some other direction, and all he could see was the guns, looming inordinately large upon those level plains, like monster insects of somber mien.

After leaving Saint-Etienne, however, there was a change for the worse, and the road from bad became abominable, rising by an easy ascent between great sterile fields in which the only signs of vegetation were the everlasting pine woods with their dark verdure, forming a dismal contrast with the gray-white soil. It was the most forlorn spot they had seen yet. The ill-paved road, washed by the recent rains, was a lake of mud, of tenacious, slippery gray clay, which held the men's feet like so much pitch. It was wearisome work; the troops were exhausted and could not get forward, and as if things were not bad enough already, the rain suddenly began to come down most violently. The guns were mired and had to be left in the road.

Chouteau, who had been given the squad's rice to carry, fatigued and exasperated with his heavy load, watched for an opportunity when no one was looking and dropped the package. But Loubet had seen him.

"See here, that's no way! you ought not to do that. The comrades will be hungry by and by."

"Let be!" replied Chouteau. "There is plenty of rice; they will give us more at the end of the march."

And Loubet, who had the bacon, convinced by such cogent reasoning, dropped his load in turn.

Maurice was suffering more and more with his foot, of which the heel was badly inflamed. He limped along in such a pitiable state that Jean's sympathy was aroused.

"Does it hurt? is it no better, eh?" And as the men were halted just then for a breathing spell, he gave him a bit of good advice. "Take off your shoe and go barefoot; the cool earth will ease the pain."

And in that way Maurice found that he could keep up with his comrades with some degree of comfort; he experienced a sentiment of deep gratitude. It was a piece of great good luck that their squad had a corporal like him, a man who had seen service and knew all the tricks of the trade: he was an uncultivated peasant, of course, but a good fellow all the same.

It was late when they reached their place of bivouac at Contreuve, after marching a long time on the Chalons and Vouziers road and descending by a steep path into the valley of the Semide, up which they came through a stretch of narrow meadows. The landscape had undergone a change; they were now in the Ardennes, and from the lofty hills above the village where the engineers had staked off the ground for the 7th corps' camp, the valley of the Aisne was dimly visible in the distance, veiled in the pale mists of the passing shower.

Six o'clock came and there had been no distribution of rations, whereon Jean, in order to keep occupied, apprehensive also of the consequences that might result from the high wind that was springing up, determined to attend in person to the setting up of the tent. He showed his men how it should be done, selecting a bit of ground that sloped away a little to one side, setting the pegs at the proper angle, and digging a little trench around the whole to carry off the water. Maurice was excused from the usual nightly drudgery on account of his sore foot, and was an interested witness of the intelligence and handiness of the big young fellow whose general appearance was so stolid and ungainly. He was completely knocked up with fatigue, but the confidence that they were now advancing with a definite end in view served to sustain him. They had had a hard time of it since they left Rheims, making nearly forty miles in two days' marching; if they could maintain the pace and if they kept straight on in the direction they were pursuing, there could be no doubt that they would destroy the second German army and effect a junction with Bazaine before the third, the Crown Prince of Prussia's, which was said to be at Vitry-le Francois, could get up to Verdun.

"Oh, come now! I wonder if they are going to let us starve!" was Chouteau's remark when, at seven o'clock, there was still no sign of rations.

By way of taking time by the forelock, Jean had instructed Loubet to light the fire and put on the pot, and, as there was no issue of firewood, he had been compelled to be blind to the slight irregularity of the proceeding when that individual remedied the omission by tearing the palings from an adjacent fence. When he suggested knocking up a dish of bacon and rice, however, the truth had to come out, and he was informed that the rice and bacon were lying in the mud of the Saint-Etienne road. Chouteau lied with the greatest effrontery declaring that the package must have slipped from his shoulders without his noticing it.

"You are a couple of pigs!" Jean shouted angrily, "to throw away good victuals, when there are so many poor devils going with an empty stomach!"

It was the same with the three loaves that had been fastened outside the knapsacks; they had not listened to his warning, and the consequence was that the rain had soaked the bread and reduced it to paste.

"A pretty pickle we are in!" he continued. "We had food in plenty, and now here we are, without a crumb! Ah! you are a pair of dirty pigs!"

At that moment the first sergeant's call was heard, and Sergeant Sapin, returning presently with his usual doleful air, informed the men that it would be impossible to distribute rations that evening, and that they would have to content themselves with what eatables they had on their persons. It was reported that the trains had been delayed by the bad weather, and as to the herds, they must have straggled off as a result of conflicting orders. Subsequently it became known that on that day the 5th and 12th corps had got up to Rethel, where the headquarters of the army were established, and the inhabitants of the neighboring villages, possessed with a mad desire to see the Emperor, had inaugurated a hegira toward that town, taking with them everything in the way of provisions; so that when the 7th corps came up they found themselves in a land of nakedness: no bread, no meat, no people, even. To add to their distress a misconception of orders had caused the supplies of the commissary department to be directed on Chene-Populeux. This was a state of affairs that during the entire campaign formed the despair of the wretched commissaries, who had to endure the abuse and execrations of the whole army, while their sole fault lay in being punctual at rendezvous at which the troops failed to appear.

"It serves you right, you dirty pigs!" continued Jean in his wrath, "and you don't deserve the trouble that I am going to have in finding you something to eat, for I suppose it is my duty not to let you starve, all the same." And he started off to see what he could find, as every good corporal does under such circumstances, taking with him Pache, who was a favorite on account of his quiet manner, although he considered him rather too priest-ridden.

But Loubet's attention had just been attracted to a little farmhouse, one of the last dwellings in Contreuve, some two or three hundred yards away, where there seemed to him to be promise of good results. He called Chouteau and Lapoulle to him and said:

"Come along, and let's see what we can do. I've a notion there's grub to be had over that way."

So Maurice was left to keep up the fire and watch the kettle, in which the water was beginning to boil. He had seated himself on his blanket and taken off his shoe in order to give his blister a chance to heal. It amused him to look about the camp and watch the behavior of the different squads now that there was to be no issue of rations; the deduction that he arrived at was that some of them were in a chronic state of destitution, while others reveled in continual abundance, and that these conditions were ascribable to the greater or less degree of tact and foresight of the corporal and his men. Amid the confusion that reigned about the stacks and tents he remarked some squads who had not been able even to start a fire, others of which the men had abandoned hope and lain themselves resignedly down for the night, while others again were ravenously devouring, no one knew what, something good, no doubt. Another thing that impressed him was the good order that prevailed in the artillery, which had its camp above him, on the hillside. The setting sun peeped out from a rift in the clouds and his rays were reflected from the burnished guns, from which the men had cleansed the coat of mud that they had picked up along the road.

In the meantime General Bourgain-Desfeuilles, commanding the brigade, had found quarters suited to his taste in the little farmhouse toward which the designs of Loubet and his companions were directed. He had discovered something that had the semblance of a bed and was seated at table with a roasted chicken and an omelette before him; consequently he was in the best of humors, and as Colonel de Vineuil happened in just then on regimental business, had invited him to dine. They were enjoying their repast, therefore, waited on by a tall, light-haired individual who had been in the farmer's service only three days and claimed to be an Alsatian, one of those who had been forced to leave their country after the disaster of Froeschwiller. The general did not seem to think it necessary to use any restraint in presence of the man, commenting freely on the movements of the army, and finally, forgetful of the fact that he was not an inhabitant of the country, began to question him about localities and distances. His questions displayed such utter ignorance of the country that the colonel, who had once lived at Mezieres, was astounded; he gave such information as he had at command, which elicited from the chief the exclamation:

"It is just like our idiotic government! How can they expect us to fight in a country of which we know nothing?"

The colonel's face assumed a look of vague consternation. He knew that immediately upon the declaration of war maps of Germany had been distributed among the officers, while it was quite certain that not one of them had a map of France. He was amazed and confounded by what he had seen and heard since the opening of the campaign. His unquestioned bravery was his distinctive trait; he was a somewhat weak and not very brilliant commander, which caused him to be more loved than respected in his regiment.

"It's too bad that a man can't eat his dinner in peace!" the general suddenly blurted out. "What does all that uproar mean? Go and see what the matter is, you Alsatian fellow!"

But the farmer anticipated him by appearing at the door, sobbing and gesticulating like a crazy man. They were robbing him, the zouaves and chasseurs were plundering his house. As he was the only one in the village who had anything to sell he had foolishly allowed himself to be persuaded to open shop. At first he had sold his eggs and chickens, his rabbits, and potatoes, without exacting an extortionate profit, pocketing his money and delivering the merchandise; then the customers had streamed in in a constantly increasing throng, jostling and worrying the old man, finally crowding him aside and taking all he had without pretense of payment. And thus it was throughout the war; if many peasants concealed their property and even denied a drink of water to the thirsty soldier, it was because of their fear of the irresistible inroads of that ocean of men, who swept everything clean before them, thrusting the wretched owners from their houses and beggaring them.

"Eh! will you hold your tongue, old man!" shouted the general in disgust. "Those rascals ought to be shot at the rate of a dozen a day. What is one to do?" And to avoid taking the measures that the case demanded he gave orders to close the door, while the colonel explained to him that there had been no issue of rations and the men were hungry.

While these things were going on within the house Loubet outside had discovered a field of potatoes; he and Lapoulle scaled the fence and were digging the precious tubers with their hands and stuffing their pockets with them when Chouteau, who in the pursuit of knowledge was looking over a low wall, gave a shrill whistle that called them hurriedly to his side. They uttered an exclamation of wonder and delight; there was a flock of geese, ten fat, splendid geese, pompously waddling about a small yard. A council of war was held forthwith, and it was decided that Lapoulle should storm the place and make prisoners of the garrison. The conflict was a bloody one; the venerable gander on which the soldier laid his predaceous hands had nearly deprived him of his nose with its bill, hard and sharp as a tailor's shears. Then he caught it by the neck and tried to choke it, but the bird tore his trousers with its strong claws and pummeled him about the body with its great wings. He finally ended the battle by braining it with his fist, and it had not ceased to struggle when he leaped the wall, hotly pursued by the remainder of the flock, pecking viciously at his legs.

When they got back to camp, with the unfortunate gander and the potatoes hidden in a bag, they found that Jean and Pache had also been successful in their expedition, and had enriched the common larder with four loaves of fresh bread and a cheese that they had purchased from a worthy old woman.

"The water is boiling and we will make some coffee," said the corporal. "Here are bread and cheese; it will be a regular feast!"

He could not help laughing, however, when he looked down and saw the goose lying at his feet. He raised it, examining and hefting it with the judgment of an expert.

"Ah! upon my word, a fine bird! it must weigh twenty pounds."

"We were out walking and met the bird," Loubet explained in an unctuously sanctimonious voice, "and it insisted on making our acquaintance."

Jean made no reply, but his manner showed that he wished to hear nothing more of the matter. Men must live, and then why in the name of common sense should not those poor fellows, who had almost forgotten how poultry tasted, have a treat once in a way!

Loubet had already kindled the fire into a roaring blaze; Pache and Lapoulle set to work to pluck the goose; Chouteau, who had run off to the artillerymen and begged a bit of twine, came back and stretched it between two bayonets; the bird was suspended in front of the hot fire and Maurice was given a cleaning rod and enjoined to keep it turning. The big tin basin was set beneath to catch the gravy. It was a triumph of culinary art; the whole regiment, attracted by the savory odor, came and formed a circle about the fire and licked their chops. And what a feast it was! roast goose, boiled potatoes, bread, cheese, and coffee! When Jean had dissected the bird the squad applied itself vigorously to the task before it; there was no talk of portions, every man ate as much as he was capable of holding. They even sent a plate full over to the artillerymen who had furnished the cord.

The officers of the regiment that evening were a very hungry set of men, for owing to some mistake the canteen wagon was among the missing, gone off to look after the corps train, maybe. If the men were inconvenienced when there was no issue of ration they scarcely ever failed to find something to eat in the end; they helped one another out; the men of the different squads "chipped in" their resources, each contributing his mite, while the officer, with no one to look to save himself, was in a fair way of starving as soon as he had not the canteen to fall back on. So there was a sneer on Chouteau's face, buried in the carcass of the goose, as he saw Captain Beaudoin go by with his prim, supercilious air, for he had heard that officer summoning down imprecations on the driver of the missing wagon; and he gave him an evil look out of the corner of his eye.

"Just look at him! See, his nose twitches like a rabbit's. He would give a dollar for the pope's nose."

They all made merry at the expense of the captain, who was too callow and too harsh to be a favorite with his men; they called him a pete-sec. He seemed on the point of taking the squad in hand for the scandal they were creating with their goose dinner, but thought better of the matter, ashamed, probably, to show his hunger, and walked off, holding his head very erect, as if he had seen nothing.

As for Lieutenant Rochas, who was also conscious of a terribly empty sensation in his epigastric region, he put on a brave face and laughed good-naturedly as he passed the thrice-lucky squad. His men adored him, in the first place because he was at sword's points with the captain, that little whipper-snapper from Saint-Cyr, and also because he had once carried a musket like themselves. He was not always easy to get along with, however, and there were times when they would have given a good deal could they have cuffed him for his brutality.

Jean glanced inquiringly at his comrades, and their mute reply being propitious, arose and beckoned to Rochas to follow him behind the tent.

"See here, Lieutenant, I hope you won't be offended, but if it is agreeable to you~"

And he handed him half a loaf of bread and a wooden bowl in which there were a second joint of the bird and six big mealy potatoes.

That night again the six men required no rocking; they digested their dinner while sleeping the sleep of the just. They had reason to thank the corporal for the scientific way in which he had set up their tent, for they were not even conscious of a small hurricane that blew up about two o'clock, accompanied by a sharp down-pour of rain; some of the tents were blown down, and the men, wakened out of their sound slumber, were drenched and had to scamper in the pitchy darkness, while theirs stood firm and they were warm and dry, thanks to the ingenious device of the trench.

Maurice awoke at daylight, and as they were not to march until eight o'clock it occurred to him to walk out to the artillery camp on the hill and say how do you do to his cousin Honore. His foot was less painful after his good night's rest. His wonder and admiration were again excited by the neatness and perfect order that prevailed throughout the encampment, the six guns of a battery aligned with mathematical precision and accompanied by their caissons, prolonges, forage-wagons, and forges. A short way off, lined up to their rope, stood the horses, whinnying impatiently and turning their muzzles to the rising sun. He had no difficulty in finding Honore's tent, thanks to the regulation which assigns to the men of each piece a separate street, so that a single glance at a camp suffices to show the number of guns.

When Maurice reached his destination the artillerymen were already stirring and about to drink their coffee, and a quarrel had arisen between Adolphe, the forward driver, and Louis, the gunner, his mate. For the entire three years that they had been "married," in accordance with the custom which couples a driver with a gunner, they had lived happily together, with the one exception of meal-times. Louis, an intelligent man and the better informed of the two, did not grumble at the airs of superiority that are affected by every mounted over every unmounted man: he pitched the tent, made the soup, and did the chores, while Adolphe groomed his horses with the pride of a reigning potentate. When the former, a little black, lean man, afflicted with an enormous appetite, rose in arms against the exactions of the latter, a big, burly fellow with huge blonde mustaches, who insisted on being waited on like a lord, then the fun began. The subject matter of the dispute on the present morning was that Louis, who had made the coffee, accused Adolphe of having drunk it all. It required some diplomacy to reconcile them.

Not a morning passed that Honore failed to go and look after his piece, seeing to it that it was carefully dried and cleansed from the night dew, as if it had been a favorite animal that he was fearful might take cold, and there it was that Maurice found him, exercising his paternal supervision in the crisp morning air.

"Ah, it's you! I knew that the 106th was somewhere in the vicinity; I got a letter from Remilly yesterday and was intending to start out and hunt you up. Let's go and have a glass of white wine."

For the sake of privacy he conducted his cousin to the little farmhouse that the soldiers had looted the day before, where the old peasant, undeterred by his losses and allured by the prospect of turning an honest penny, had tapped a cask of wine and set up a kind of public bar. He had extemporized a counter from a board rested on two empty barrels before the door of his house, and over it he dealt out his stock in trade at four sous a glass, assisted by the strapping young Alsatian whom he had taken into his service three days before.

As Honore was touching glasses with Maurice his eyes lighted on this man. He gazed at him a moment as if stupefied, then let slip a terrible oath.

"Tonnerre de Dieu! Goliah!"

And he darted forward and would have caught him by the throat, but the peasant, foreseeing in his action a repetition of his yesterday's experience, jumped quickly within the house and locked the door behind him. For a moment confusion reigned about the premises; soldiers came rushing up to see what was going on, while the quartermaster-sergeant shouted at the top of his voice:

"Open the door, open the door, you confounded idiot! It is a spy, I tell you, a Prussian spy!"

Maurice doubted no longer; there was no room for mistake now; the Alsatian was certainly the man whom he had seen arrested at the camp of Mulhausen and released because there was not evidence enough to hold him, and that man was Goliah, old Fouchard's quondam assistant on his farm at Remilly. When finally the peasant opened his door the house was searched from top to bottom, but to no purpose; the bird had flown, the gawky Alsatian, the tow-headed, simple-faced lout whom General Bourgain-Desfeuilles had questioned the day before at dinner without learning anything and before whom, in the innocence of his heart, he had disclosed things that would have better been kept secret. It was evident enough that the scamp had made his escape by a back window which was found open, but the hunt that was immediately started throughout the village and its environs had no results; the fellow, big as he was, had vanished as utterly as a smoke-wreath dissolves upon the air.

Maurice thought it best to take Honore away, lest in his distracted state he might reveal to the spectators unpleasant family secrets which they had no concern to know.

"Tonnerre de Dieu!" he cried again, "it would have done me such good to strangle him!-The letter that I was speaking of revived all my old hatred for him."

And the two of them sat down upon the ground against a stack of rye a little way from the house, and he handed the letter to his cousin.

It was the old story: the course of Honore Fouchard's and Silvine Morange's love had not run smooth. She, a pretty, meek-eyed, brown-haired girl, had in early childhood lost her mother, an operative in one of the factories of Raucourt, and Doctor Dalichamp, her godfather, a worthy man who was greatly addicted to adopting the wretched little beings whom he ushered into the world, had conceived the idea of placing her in Father Fouchard's family as small maid of all work. True it was that the old boor was a terrible skinflint and a harsh, stern taskmaster; he had gone into the butchering business from sordid love of lucre, and his cart was to be seen daily, rain or shine, on the roads of twenty communes; but if the child was willing to work she would have a home and a protector, perhaps some small prospect in the future. At all events she would be spared the contamination of the factory. And naturally enough it came to pass that in old Fouchard's household the son and heir and the little maid of all work fell in love with each other. Honore was then just turned sixteen and she was twelve, and when she was sixteen and he twenty there was a drawing for the army; Honore, to his great delight, secured a lucky number and determined to marry. Nothing had ever passed between them, thanks to the unusual delicacy that was inherent in the lad's tranquil, thoughtful nature, more than an occasional hug and a furtive kiss in the barn. But when he spoke of the marriage to his father, the old man, who had the stubbornness of the mule, angrily told him that his son might kill him, but never, never would he consent, and continued to keep the girl about the house, not worrying about the matter, expecting it would soon blow over. For two years longer the young folks kept on adoring and desiring each other, and never the least breath of scandal sullied their names. Then one day there was a frightful quarrel between the two men, after which the young man, feeling he could no longer endure his father's tyranny, enlisted and was packed off to Africa, while the butcher still retained the servant-maid, because she was useful to him. Soon after that a terrible thing happened: Silvine, who had sworn that she would be true to her lover and await his return, was detected one day, two short weeks after his departure, in the company of a laborer who had been working on the farm for some months past, that Goliah Steinberg, the Prussian, as he was called; a tall, simple young fellow with short, light hair, wearing a perpetual smile on his broad, pink face, who had made himself Honore's chum. Had Father Fouchard traitorously incited the man to take advantage of the girl? or had Silvine, sick at heart and prostrated by the sorrow of parting with her lover, yielded in a moment of unconsciousness? She could not tell herself; was dazed, and saw herself driven by the necessity of her situation to a marriage with Goliah. He, for his part, always with the everlasting smile on his face, made no objection, only insisted on deferring the ceremony until the child should be born. When that event occurred he suddenly disappeared; it was rumored, subsequently that he had found work on another farm, over Beaumont way. These things had happened three years before the breaking out of the war, and now everyone was convinced that that artless, simple Goliah, who had such a way of ingratiating himself with the girls, was none else than one of those Prussian spies who filled our eastern provinces. When Honore learned the tidings over in Africa he was three months in hospital, as if the fierce sun of that country had smitten him on the neck with one of his fiery javelins, and never thereafter did he apply for leave of absence to return to his country for fear lest he might again set eyes on Silvine and her child.

The artilleryman's hands shook with agitation as Maurice perused the letter. It was from Silvine, the first, the only one that she had ever written him. What had been her guiding impulse, that silent, submissive woman, whose handsome black eyes at times manifested a startling fixedness of purpose in the midst of her never-ending slavery? She simply said that she knew he was with the army, and though she might never see him again, she could not endure the thought that he might die and believe that she had ceased to love him. She loved him still, had never loved another; and this she repeated again and again through four closely written pages, in words of unvarying import, without the slightest word of excuse for herself, without even attempting to explain what had happened. There was no mention of the child, nothing but an infinitely mournful and tender farewell.

The letter produced a profound impression upon Maurice, to whom his cousin had once imparted the whole story. He raised his eyes and saw that Honore was weeping; he embraced him like a brother.

"My poor Honore."

But the sergeant quickly got the better of his emotion. He carefully restored the letter to its place over his heart and rebuttoned his jacket.

"Yes, those are things that a man does not forget. Ah! the scoundrel, if I could but have laid hands on him! But we shall see."

The bugles were sounding the signal to prepare for breaking camp, and each had to hurry away to rejoin his command. The preparations for departure dragged, however, and the troops had to stand waiting in heavy marching order until nearly nine o'clock. A feeling of hesitancy seemed to have taken possession of their leaders; there was not the resolute alacrity of the first two days, when the 7th corps had accomplished forty miles in two marches. Strange and alarming news, moreover, had been circulating through the camp since morning, that the three other corps were marching northward, the 1st at Juniville, the 5th and 12th at Rethel, and this deviation from their route was accounted for on the ground of the necessities of the commissariat. Montmedy had ceased to be their objective, then? why were they thus idling away their time again? What was most alarming of all was that the Prussians could not now be far away, for the officers had cautioned their men not to fall behind the column, as all stragglers were liable to be picked up by the enemy's light cavalry. It was the 25th of August, and Maurice, when he subsequently recalled to mind Goliah's disappearance, was certain that the man had been instrumental in affording the German staff exact information as to the movements of the army of Chalons, and thus producing the change of front of their third army. The succeeding morning the Crown Prince of Prussia left Revigny and the great maneuver was initiated, that gigantic movement by the flank, surrounding and enmeshing us by a series of forced marches conducted in the most admirable order through Champagne and the Ardennes. While the French were stumbling aimlessly about the country, oscillating uncertainly between one place and another, the Prussians were making their twenty miles a day and more, gradually contracting their immense circle of beaters upon the band of men whom they held within their toils, and driving their prey onward toward the forests of the frontier.

A start was finally made, and the result of the day's movement showed that the army was pivoting on its left; the 7th corps only traversed the two short leagues between Contreuve and Vouziers, while the 5th and 12th corps did not stir from Rethel, and the 1st went no farther than Attigny. Between Contreuve and the valley of the Aisne the country became level again and was more bare than ever; as they drew near to Vouziers the road wound among desolate hills and naked gray fields, without a tree, without a house, as gloomy and forbidding as a desert, and the day's march, short as it was, was accomplished with such fatigue and distress that it seemed interminably long. Soon after midday, however, the 1st and 3d divisions had passed through the city and encamped in the meadows on the farther bank of the Aisne, while a brigade of the second, which included the 106th, had remained upon the left bank, bivouacking among the waste lands of which the low foot-hills overlooked the valley, observing from their position the Monthois road, which skirts the stream and by which the enemy was expected to make his appearance.

And Maurice was dumfoundered to behold advancing along that Monthois road Margueritte's entire division, the body of cavalry to which had been assigned the duty of supporting the 7th corps and watching the left flank of the army. The report was that it was on its way to Chene-

Populeux. Why was the left wing, where alone they were threatened by the enemy, stripped in that manner? What sense was there in summoning in upon the center, where they could be of no earthly use, those two thousand horsemen, who should have been dispersed upon our flank, leagues away, as videttes to observe the enemy? And what made matters worse was that they caused the greatest confusion among the columns of the 7th corps, cutting in upon their line of march and producing an inextricable jam of horses, guns, and men. A squadron of chasseurs d'Afrique were halted for near two hours at the gate of Vouziers, and by the merest chance Maurice stumbled on Prosper, who had ridden his horse down to the bank of a neighboring pond to let him drink, and the two men were enabled to exchange a few words. The chasseur appeared stunned, dazed, knew nothing and had seen nothing since they left Rheims; yes, though, he had: he had seen two uhlans more; oh! but they were will o' the wisps, phantoms, they were, that appeared and vanished, and no one could tell whence they came nor whither they went. Their fame had spread, and stories of them were already rife throughout the country, such, for instance, as that of four uhlans galloping into a town with drawn revolvers and taking possession of it, when the corps to which they belonged was a dozen miles away. They were everywhere, preceding the columns like a buzzing, stinging swarm of bees, a living curtain, behind which the infantry could mask their movements and march and countermarch as securely as if they were at home upon parade. And Maurice's heart sank in his bosom as he looked at the road, crowded with chasseurs and hussars which our leaders put to such poor use.

"Well, then, au revoir," said he, shaking Prosper by the hand; "perhaps they will find something for you to do down yonder, after all."

But the chasseur appeared disgusted with the task assigned him. He sadly stroked Poulet's neck and answered:

"Ah, what's the use talking! they kill our horses and let us rot in idleness. It is sickening."

When Maurice took off his shoe that evening to have a look at his foot, which was aching and throbbing feverishly, the skin came with it; the blood spurted forth and he uttered a cry of pain. Jean was standing by, and exhibited much pity and concern.

"Look here, that is becoming serious; you are going to lie right down and not attempt to move. That foot of yours must be attended to. Let me see it."

He knelt down, washed the sore with his own hands and bound it up with some clean linen that he took from his knapsack. He displayed the gentleness of a woman and the deftness of a surgeon, whose big fingers can be so pliant when necessity requires it.

A great wave of tenderness swept over Maurice, his eyes were dimmed with tears, the familiar *thou* rose from his heart to his lips with an irresistible impulse of affection, as if in that peasant whom he once had hated and abhorred, whom only yesterday he had despised, he had discovered a long lost brother.

"Thou art a good fellow, thou! Thanks, good friend."

And Jean, too, looking very happy, dropped into the second person singular, with his tranquil smile.

"Now, my little one, wilt thou have a cigarette? I have some tobacco left."

On the morning of the following day, the 26th, Maurice arose with stiffened limbs and an aching back, the result of his night under the tent. He was not accustomed yet to sleeping on the bare ground; orders had been given before the men turned in that they were not to remove their shoes, and during the night the sergeants had gone the rounds, feeling in the darkness to see if all were properly shod and gaitered, so that his foot was much inflamed and very painful. In addition to his other troubles he had imprudently stretched his legs outside the canvas to relieve their cramped feeling and taken cold in them.

Jean said as soon as he set eyes on him:

"If we are to do any marching today, my lad, you had better see the surgeon and get him to give you a place in one of the wagons."

But no one seemed to know what were the plans for the day, and the most conflicting reports prevailed. It appeared for a moment as if they were about to resume their march; the tents were struck and the entire corps took the road and passed through Vouziers, leaving on the right bank of the Aisne only one brigade of the second division, apparently to continue the observation of the Monthois road; but all at once, as soon as they had put the town behind them and were on the left bank of the stream, they halted and stacked muskets in the fields and meadows that skirt the Grand-Pre road on either hand, and the departure of the 4th hussars, who just then moved off on that road at a sharp trot, afforded fresh food for conjecture.

"If we are to remain here I shall stay with you," declared Maurice, who was not attracted by the prospect of riding in an ambulance.

It soon became known that they were to occupy their present camp until General Douay could obtain definite information as to the movements of the enemy. The general had been harassed by an intense and constantly increasing anxiety since the day before, when he had seen Margueritte's division moving toward Chene, for he knew that his flank was uncovered, that there was not a man to watch the passes of the Argonne, and that he was liable to be attacked at any moment. Therefore he had sent out the 4th hussars to reconnoiter the country as far as the defiles of Grand-Pre and Croix-aux-Bois, with strict orders not to return without intelligence.

There had been an issue of bread, meat, and forage the day before, thanks to the efficient mayor of Vouziers, and about ten o'clock that morning permission had been granted the men to make soup, in the fear that they might not soon again have so good an opportunity, when another movement of troops, the departure of Bordas' brigade over the road taken by the hussars, set all tongues wagging afresh. What! were they going to march again? were they not to be given a chance to eat their breakfast in peace, now that the kettle was on the fire? But the officers explained that Bordas' brigade had only been sent to occupy Buzancy, a few kilometers from there. There were others, indeed, who asserted that the hussars had encountered a strong force of the enemy's cavalry and that the brigade had been dispatched to help them out of their difficulty.

Maurice enjoyed a few hours of delicious repose. He had thrown himself on the ground in a field half way up the hill where the regiment had halted, and in a drowsy state between sleeping and waking was contemplating the verdant valley of the Aisne, the smiling meadows dotted with clumps of trees, among which the little stream wound lazily. Before him and closing the valley in that direction lay Vouziers, an amphitheater of roofs rising one above another and overtopped by the church with its slender spire and dome-crowned tower. Below him, near the bridge, smoke was curling upward from the tall chimneys of the tanneries, while farther away a great mill displayed its flour-whitened buildings among the fresh verdure of the growths that lined the waterside. The little town that lay there, bounding his horizon, hidden among the stately trees, appeared to him to possess a gentle charm; it brought him memories of boyhood, of the journeys that he had made to Vouziers in other days, when he had lived at Chene, the village where he was born. For an hour he was oblivious of the outer world.

The soup had long since been made and eaten and everyone was waiting to see what would happen next, when, about half-past two o'clock, the smoldering excitement began to gain strength, and soon pervaded the entire camp. Hurried orders came to abandon the meadows, and the troops ascended a line of hills between two villages, Chestres and Falaise, some two or three miles apart, and took position there. Already the engineers were at work digging rifle-pits and throwing up epaulments; while over to the left the artillery had occupied the summit of a rounded eminence. The rumor spread that General Bordas had sent in a courier to announce that he had encountered the enemy in force at Grand-Pre and had been compelled to fall back on Buzancy, which gave cause to apprehend that he might soon be cut off from retreat on Vouziers. For these reasons, the commander of the 7th corps, believing an attack to be imminent, had placed his men in position to sustain the first onset until the remainder of the army should have time to come to his assistance, and had started off one of his aides-de-camp with a letter to the marshal, apprising him of the danger, and asking him for re-enforcements. Fearing for the safety of the subsistence train, which had come up with the corps during the night and was again dragging its interminable length in the rear, he summarily sent it to the right about and directed it to make the best of its way to Chagny. Things were beginning to look like fight.

"So, it looks like business this time-eh, Lieutenant?" Maurice ventured to ask Rochas.

"Yes, thank goodness," replied the Lieutenant, his long arms going like windmills. "Wait a little; you'll find it warm enough!"

The soldiers were all delighted; the animation in the camp was still more pronounced. A feverish impatience had taken possession of the men, now that they were actually in line of battle between Chestres and Falaise. At last they were to have a sight of those Prussians who, if the newspapers were to be believed, were knocked up by their long marches, decimated by sickness, starving, and in rags, and every man's heart beat high with the prospect of annihilating them at a single blow.

"We are lucky to come across them again," said Jean. "They've been playing hide-and-seek about long enough since they slipped through our fingers after their battle down yonder on the frontier. But are these the same troops that whipped MacMahon, I wonder?"

Maurice could not answer his question with any degree of certainty. It seemed to him hardly probable, in view of what he had read in the newspapers at Rheims, that the third army, commanded by the Crown Prince of Prussia, could be at Vouziers, when, only two days before, it was just on the point of going into camp at Vitry-le-Francois. There had been some talk of a fourth army, under the Prince of Saxony, which was to operate on the line of the Meuse; this was doubtless the one that was now before them, although their promptitude in occupying Grand-Pre was a matter of surprise, considering the distances. But what put the finishing touch to the confusion of his ideas was his stupefaction to hear General Bourgain-Desfeuilles ask a countryman if the Meuse did not flow past Buzancy, and if the bridges there were strong. The general announced, moreover, in the confidence of his sublime ignorance, that a column of one hundred thousand men was on the way from Grand-Pre to attack them, while another, of sixty thousand, was coming up by the way of Sainte-Menehould.

"How's your foot, Maurice?" asked Jean.

"It don't hurt now," the other laughingly replied. "If there is to be a fight, I think it will be quite well."

It was true; his nervous excitement was so great that he was hardly conscious of the ground on which he trod. To think that in the whole campaign he had not yet burned powder! He had gone forth to the frontier, he had endured the agony of that terrible night of expectation before Mulhausen, and had not seen a Prussian, had not fired a shot; then he had retreated with the rest to Belfort, to Rheims, had now been marching five days trying to find the enemy, and his useless *chassepot* was as clean as the day it left the shop, without the least smell of smoke on it. He felt an aching desire to discharge his piece once, if no more, to relieve the tension of his nerves. Since the day, near six weeks ago, when he had enlisted in a fit of enthusiasm, supposing that he would surely have to face the foe in a day or two, all that he had done had been to tramp up and down the country on his poor, sore feet-the feet of a man who had lived in luxury, far from the battle-field; and so, among all those impatient watchers, there was none who watched more impatiently than he the Grand-Pre road, extending straight away to a seemingly infinite distance between two rows of handsome trees. Beneath him was unrolled the panorama of the valley; the Aisne was, like a silver ribbon, flowing between its willows and poplars, and ever his gaze returned, solicited by an irresistible attraction, to that road down yonder that stretched away, far as the eye could see, to the horizon.

About four o'clock the 4th hussars returned, having made a wide circuit in the country round about, and stories, which grew as they were repeated, began to circulate of conflicts with uhlans, tending to confirm the confident belief which everyone had that an attack was imminent. Two hours later a courier came galloping in, breathless with terror, to announce that General Bordas had positive information that the enemy were on the Vouziers road, and dared not leave Grand-Pre. It was evident that that could not be true, since the courier had just passed over the road unharmed, but no one could tell at what moment it might be the case, and General Dumont, commanding the division, set out at once with his remaining brigade to bring off his other brigade that was in difficulty. The sun went down behind Vouziers and the roofs of the town were sharply profiled in black against a great red cloud. For a long time the brigade was visible as it receded between the double row of trees, until finally it was swallowed up in the gathering darkness.

Colonel de Vineuil came to look after his regiment's position for the night. He was surprised not to find Captain Beaudoin at his post, and as that officer just then chanced to come in from Vouziers, where he alleged in excuse for his absence that he had been breakfasting with the Baronne de Ladicourt, he received a sharp reprimand, which he digested in silence, with the rigid manner of a martinet conscious of being in the wrong.

"My children," said the Colonel, as he passed along the line of men, "we shall probably be attacked to-night, or if not, then by day-break to-morrow morning at the latest. Be prepared, and remember that the 106th has never retreated before the enemy."

The little speech was received with loud hurrahs; everyone, in the prevailing suspense and discouragement, preferred to "take the wipe of the dish-clout" and have done with it. Rifles were examined to see that they were in good order, belts were refilled with cartridges. As they had eaten their soup that morning, the men were obliged to content themselves with biscuits and coffee. An order was promulgated that there was to be no sleeping. The grand-guards were out nearly a mile to the front, and a chain of sentinels at frequent intervals extended down to the Aisne. The officers were seated in little groups about the camp-fires, and beside a low wall at the left of the road the fitful blaze occasionally flared up and rescued from the darkness the gold embroideries and bedizened uniforms of the Commander-in-Chief and his staff, flitting to and fro like phantoms, watching the road and listening for the tramp of horses in the mortal anxiety they were in as to the fate of the third division.

It was about one o'clock in the morning when it came Maurice's turn to take his post as sentry at the edge of an orchard of plum-trees, between the road and the river. The night was black as ink, and as soon as his comrades left him and he found himself alone in the deep silence of the sleeping fields he was conscious of a sensation of fear creeping over him, a feeling of abject terror such as he had never known before and which he trembled with rage and shame at his inability to conquer. He turned his head to cheer himself by a sight of the camp-fires, but they were hidden from him by a wood; there was naught behind him but an unfathomable sea of blackness; all that he could discern was a few distant lights still dimly burning in Vouziers, where the inhabitants, doubtless forewarned and trembling at the thought of the impending combat, were keeping anxious vigil. His terror was increased, if that were possible, on bringing his piece to his shoulder to find that he could not even distinguish the sights on it. Then commenced a period of suspense that tried his nerves most cruelly; every faculty of his being was strained and concentrated in the one sense of hearing; sounds so faint as to be imperceptible reverberated in his ears like the crash of thunder; the plash of a distant waterfall, the rustling of a leaf, the movement of an insect in the grass, were like the booming of artillery. Was that the tramp of cavalry, the deep rumbling of gun-carriages driven at speed, that he heard down there to the right? And there on his left, what was that? was it not the sound of stealthy whispers, stifled voices, a party creeping up to surprise him under cover of the darkness? Three times he was on the point of giving the alarm by firing his piece. The fear that he might be mistaken and incur the ridicule of his comrades served to intensify his distress. He had kneeled upon the ground, supporting his left shoulder against a tree; it seemed to him that he had been occupying that position for hours, that they had forgotten him there, that the army had moved away without him. Then suddenly, at once, his fear left him; upon the road, that he knew was not two hundred yards away, he distinctly heard the cadenced tramp of marching men. Immediately it flashed across his mind as a certainty that they were the troops from Grand-Pre, whose coming had been awaited with such anxiety- General Dumont bringing in Bordas' brigade. At that same moment the corporal of the guard came along with the relief; he had been on post a little less than the customary hour.

He had been right; it was the 3d division returning to camp. Everyone felt a sensation of deep relief. Increased precautions were taken, nevertheless, for what fresh intelligence they received tended to confirm what they supposed they already knew of the enemy's approach. A few uhlans, forbidding looking fellows in their long black cloaks, were brought in as prisoners, but they were uncommunicative, and so daylight came at last, the pale, ghastly light of a rainy morning, bringing with it no alleviation of their terrible suspense. No one had dared to close an eye during that long night. About seven o'clock Lieutenant Rochas affirmed that MacMahon was coming up with the whole army. The truth of the matter was that General Douay, in reply to his dispatch of the preceding day announcing that a battle at Vouziers was inevitable, had received a letter from the marshal enjoining him to hold the position until re-enforcements could reach him; the forward movement had been arrested; the 1st corps was being directed on Terron, the 5th on Buzancy, while the 12th was to remain at Chene and constitute our second line. Then the suspense became more breathless still; it was to be no mere skirmish that the peaceful valley of the Aisne was to witness that day, but a great battle, in which would participate the entire army, that was even now turning its back upon the Meuse and marching southward; and there was no making of soup, the men had to content themselves with coffee and hard-tack, for everyone was saying, without troubling himself to ask why, that the "wipe of the dish-clout" was set down for midday. An aide-de-camp had been dispatched to the marshal to urge him to hurry forward their supports, as intelligence received from every quarter made it more and more certain that the two Prussian armies were close at hand, and three hours later still another officer galloped off like mad toward Chene, where general headquarters were located, with a request for instructions, for consternation had risen to a higher pitch then ever with the receipt of fresh tidings from the *maire* of a country commune, who told of having seen a hundred thousand men at Grand-Pre, while another hundred thousand were advancing by way of Buzancy.

Midday came, and not a sign of the Prussians. At one o'clock, at two, it was the same, and a reaction of lassitude and doubt began to prevail among the troops. Derisive jeers were heard at the expense of the generals: perhaps they had seen their shadow on the wall; they should be presented with a pair of spectacles. A pretty set of humbugs they were, to have caused all that trouble for nothing! A fellow who passed for a wit among his comrades shouted:

"It is like it was down there at Mulhausen, eh?"

The words recalled to Maurice's mind a flood of bitter memories. He thought of that idiotic flight, that panic that had swept away the 7th corps when there was not a German visible, nor within ten leagues of where they were, and now he had a distinct certainty that they were to have a renewal of that experience. It was plain that if twenty-four hours had elapsed since the skirmish at Grand-Pre and they had not been attacked, the reason was that the 4th hussars had merely struck up against a reconnoitering body of cavalry; the main body of the Prussians must be far away, probably a day's march or two. Then the thought suddenly struck him of the time they had wasted, and it terrified him; in three days they had only accomplished the distance from Contreuve to Vouziers, a scant two leagues. On the 25th the other corps, alleging scarcity of supplies, had diverted their course to the north, while now, on the 27th, here they were coming southward again to fight a battle with an invisible enemy. Bordas' brigade had followed the 4th hussars into the abandoned passes of the Argonne, and was supposed to have got itself into trouble; the division had gone to its assistance, and that had been succeeded by the corps, and that by the entire army, and all those movements had amounted to nothing. Maurice trembled as he reflected how pricelessly valuable was every hour, every minute, in that mad project of joining forces with Bazaine, a project that could be carried to a successful issue only by an officer of genius, with seasoned troops under him, who should press forward to his end with the resistless energy of a whirlwind, crushing every obstacle that lay in his path.

"It is all up with us!" said he, as the whole truth flashed through his mind, to Jean, who had given way to despair. Then as the corporal, failing to catch his meaning, looked at him wonderingly, he went on in an undertone, for his friend's ear alone, to speak of their commanders:

"They mean well, but they have no sense, that's certain-and no luck! They know nothing; they foresee nothing; they have neither plans nor ideas, nor happy intuitions. Allons! everything is against us; it is all up!"

And by slow degrees that same feeling of discouragement that Maurice had arrived at by a process of reasoning settled down upon the denser intellects of the troops who lay there inactive, anxiously awaiting to see what the end would be. Distrust, as a result of their truer perception of the position they were in, was obscurely burrowing in those darkened minds, and there was no man so ignorant as not to feel a sense of injury at the ignorance and irresolution of their leaders, although he might not have been able to express in distinct terms the causes of his exasperation. In the name of Heaven, what were they doing there, since the Prussians had not shown themselves? either let them fight and have it over with, or else go off to some place where they could get some sleep; they had had enough of that kind of work. Since the departure of the second aide-de-camp, who had been dispatched in quest of orders, this feeling of unrest had been increasing momentarily; men collected in groups, talking loudly and discussing the situation pro and con, and the general inquietude communicating itself to the officers, they knew not what answer to make to those of their men who ventured to question them. They ought to be marching, it would not answer to dawdle thus; and so, when it became known about five o'clock that the aide-de-camp had returned and that they were to retreat, there was a sigh of relief throughout the camp and every heart was lighter.

It seemed that the wiser counsel was to prevail, then, after all! The Emperor and MacMahon had never looked with favor on the movement toward Montmedy, and now, alarmed to learn that they were again out-marched and out-maneuvered, and that they were to have the army of the Prince of Saxony as well as that of the Crown Prince to contend with, they had renounced the hazardous scheme of uniting their forces with Bazaine, and would retreat through the northern strongholds with a view to falling back ultimately on Paris. The 7th corps' destination would be Chagny, by way of Chene, while the 5th corps would be directed on Poix, and the 1st and 12th on Vendresse. But why, since they were about to fall back, had they advanced to the line of the Aisne? Why all that waste of time and labor, when it would have been so easy and so rational to move straight from Rheims and occupy the strong positions in the valley of the Marne? Was there no guiding mind, no military talent, no common sense? But there should be no more questioning; all should be forgiven, in the universal joy at the adoption of that eminently wise counsel, which was the only means at their command of extricating themselves from the hornets' nest into which they had rushed so imprudently. All, officers and men, felt that they would be the stronger for the retrograde movement, that under the walls of Paris they would be invincible, and that there it was that the Prussians would sustain their inevitable defeat. But Vouziers must be evacuated before daybreak, and they must be well on the road to Chene before the enemy should learn of the movement, and forthwith the camp presented a scene of the greatest animation: trumpets sounding, officers hastening to and fro with orders, while the baggage and quartermaster's trains, in order not to encumber the rear-guard, were sent forward in advance.

Maurice was delighted. As he was endeavoring to explain to Jean the rationale of the impending movement, however, a cry of pain escaped him; his excitement had subsided, and he was again conscious of his foot, aching and burning as if it had been a ball of red-hot metal.

"What's the matter? is it hurting you again?" the corporal asked sympathizingly. And with his calm and sensible resourcefulness he said: "See here, little one, you told me yesterday that you have acquaintances in the town, yonder. You ought to get permission from the major and find some one to drive you over to Chene, where you could have a good night's rest in a comfortable bed. We can pick you up as we go by to-morrow if you are fit to march. What do you say to that, hein?"

In Falaise, the village near which the camp was pitched, Maurice had come across a small farmer, an old friend of his father's, who was about to drive his daughter over to Chene to visit an aunt in that town, and the horse was even then standing waiting, hitched to a light carriole. The prospect was far from encouraging, however, when he broached the subject to Major Bouroche.

"I have a sore foot, monsieur the doctor-"

Bouroche, with a savage shake of his big head with its leonine mane, turned on him with a roar:

"I am not monsieur the doctor; who taught you manners?"

65

And when Maurice, taken all aback, made a stammering attempt to excuse himself, he continued:

"Address me as major, do you hear, you great oaf!"

He must have seen that he had not one of the common herd to deal with and felt a little ashamed of himself; he carried it off with a display of more roughness.

"All a cock-and-bull story, that sore foot of yours!-Yes, yes; you may go. Go in a carriage, go in a balloon, if you choose. We have too many of you malingerers in the army!"

When Jean assisted Maurice into the carriole the latter turned to thank him, whereon the two men fell into each other's arms and embraced as if they were never to meet again. Who could tell, amid the confusion and disorder of the retreat, with those bloody Prussians on their track? Maurice could not tell how it was that there was already such a tender affection between him and the young man, and twice he turned to wave him a farewell. As he left the camp they were preparing to light great fires in order to mislead the enemy when they should steal away, in deepest silence, before the dawn of day.

As they jogged along the farmer bewailed the terrible times through which they were passing. He had lacked the courage to remain at Falaise, and already was regretting that he had left it, declaring that if the Prussians burned his house it would ruin him. His daughter, a tall, pale young woman, wept copiously. But Maurice was like a dead man for want of sleep, and had no ears for the farmer's lamentations; he slumbered peacefully, soothed by the easy motion of the vehicle, which the little horse trundled over the ground at such a good round pace that it took them less than an hour and a half to accomplish the four leagues between Vouziers and Chene. It was not quite seven o'clock and scarcely beginning to be dark when the young man rubbed his eyes and alighted in a rather dazed condition on the public square, near the bridge over the canal, in front of the modest house where he was born and had passed twenty years of his life. He got down there in obedience to an involuntary impulse, although the house had been sold eighteen months before to a veterinary surgeon, and in reply to the farmer's questions said that he knew quite well where he was going, adding that he was a thousand times obliged to him for his kindness.

He continued to stand stock-still, however, beside the well in the middle of the little triangular place; he was as if stunned; his memory was a blank. Where had he intended to go? and suddenly his wits returned to him and he remembered that it was to the notary's, whose house was next door to his father's, and whose mother, Madame Desvallieres, an aged and most excellent lady, had petted him when he was an urchin on account of their being neighbors. But he hardly recognized Chene in the midst of the hurly-burly and confusion into which the little town, ordinarily so dead, was thrown by the presence of an army corps encamped at its gates and filling its quiet streets with officers, couriers, soldiers, and camp-followers and stragglers of every description. The canal was there as of old, passing through the town from end to end and bisecting the market-place in the center into two equal-sized triangles connected by a narrow stone bridge; and there, on the other bank, was the old market with its moss-grown roofs, and the Rue Berond leading away to the left and the Sedan road to the right, but filling the Rue de Vouziers in front of him and extending as far as the Hotel de Ville was such a compact, swarming, buzzing crowd that he was obliged to raise his eyes and take a look over the roof of the notary's house at the slate-covered bell tower in order to assure himself that that was the quiet spot where he had played hop-scotch when he was a youngster. There seemed to be an effort making to clear the square; some men were roughly crowding back the throng of idlers and gazers, and looking more closely he was surprised to see, parked like the guns of a battery, a collection of vans, baggage-wagons, and carriages open and closed; a miscellaneous assortment of traps that he had certainly set eyes on before.

66

It was daylight still; the sun had just sunk in the canal at the point where it vanished in the horizon and the long, straight stretch of water was like a sea of blood, and Maurice was trying to make up his mind what to do when a woman who stood near stared at him a moment and then exclaimed:

"Why goodness gracious, is it possible! Are you the Levasseur boy?"

And thereon he recognized Madame Combette, the wife of the druggist, whose shop was on the market-place. As he was trying to explain to her that he was going to ask good Madame Desvallieres to give him a bed for the night she excitedly hurried him away.

"No, no; come to our house. I will tell you why-" When they were in the shop and she had cautiously closed the door she continued: "You could not know, my dear boy, that the Emperor is at the Desvallieres. His officers took possession of the house in his name and the family are not any too well pleased with the great honor done them, I can tell you. To think that the poor old mother, a woman more than seventy, was compelled to give up her room and go up and occupy a servant's bed in the garret! Look, there, on the place. All that you see there is the Emperor's; those are his trunks, don't you see!"

And then Maurice remembered; they were the imperial carriages and baggage-wagons, the entire magnificent train that he had seen at Rheims.

"Ah! my dear boy, if you could but have seen the stuff they took from them, the silver plate, and the bottles of wine, and the baskets of good things, and the beautiful linen, and everything! I can't help wondering where they find room for such heaps of things, for the house is not a large one. Look, look! see what a fire they have lighted in the kitchen!"

He looked over at the small white, two-storied house that stood at the corner of the market-place and the Rue de Vouziers, a comfortable, unassuming house of bourgeois aspect; how well he remembered it, inside and out, with its central hall and four rooms on each floor; why, it was as if he had just left it! There were lights in the corner room on the first floor overlooking the square; the apothecary's wife informed him that it was the bedroom of the Emperor. But the chief center of activity seemed, as she had said, to be the kitchen, the window of which opened on the Rue de Vouziers. In all their lives the good people of Chene had witnessed no such spectacle, and the street before the house was filled with a gaping crowd, constantly coming and going, who stared with all their eyes at the range on which was cooking the dinner of an Emperor. To obtain a breath of air the cooks had thrown open the window to its full extent. They were three in number, in jackets of resplendent whiteness, superintending the roasting of chickens impaled on a huge spit, stirring the gravies and sauces in copper vessels that shone like gold. And the oldest inhabitant, evoking in memory all the civic banquets that he had beheld at the Silver Lion, could truthfully declare that never at any one time had he seen so much wood burning and so much food cooking.

Combette, a bustling, wizened little man, came in from the street in a great state of excitement from all that he had seen and heard. His position as deputy-mayor gave him facilities for knowing what was going on. It was about half-past three o'clock when MacMahon had telegraphed Bazaine that the Crown Prince of Prussia was approaching Chalons, thus necessitating the withdrawal of the army to the places along the Belgian frontier, and further dispatches were also in preparation for the Minister of War, advising him of the projected movement and explaining the terrible dangers of their position. It was uncertain whether or not the dispatch for Bazaine would get through, for communication with Metz had seemed to be interrupted for the past few days, but the second dispatch was another and more serious matter; and lowering his voice almost to a whisper the apothecary repeated the words that he had heard uttered by an officer of rank: "If they get wind of this in Paris, our goose is cooked!" Everyone was aware of the unrelenting persistency with which the Empress and the Council of Ministers urged the advance of the army. Moreover, the confusion went on increasing from hour to hour, the most conflicting advices were continually coming in as to the whereabouts of the German forces. Could it be possible that the Crown Prince was at Chalons? What, then, were the troops that the 7th corps had encountered among the passes of the Argonne?

"They have no information at staff headquarters," continued the little druggist, raising his arms above his head with a despairing gesture. "Ah, what a mess we are in! But all will be well if the army retreats to-morrow." Then, dropping public for private matters, the kind-hearted man said: "Look here, my young friend, I am going to see what I can do for that foot of yours; then we'll give you some dinner and put you to bed in my apprentice's little room, who has cleared out."

But Maurice was tormented by such an itching desire for further intelligence that he could neither eat nor sleep until he had carried into execution his original design of paying a visit to his old friend, Madame Desvallieres, over the way. He was surprised that he was not halted at the door, which, in the universal confusion, had been left wide open, without so much as a sentry to guard it. People were going out and coming in incessantly, military men and officers of the household, and the roar from the blazing kitchen seemed to rise and pervade the whole house. There was no light in the passage and on the staircase, however, and he had to grope his way up as best he might. On reaching the first floor he paused for a few seconds, his heart beating violently, before the door of the apartment that he knew contained the Emperor, but not a sound was to be heard in the room; the stillness that reigned there was as of death. Mounting the last flight he presented himself at the door of the servant's room to which Madame Desvallieres had been consigned; the old lady was at first terrified at sight of him. When she recognized him presently she said:

"Ah, my poor child, what a sad meeting is this! I would cheerfully have surrendered my house to the Emperor, but the people he has about him have no sense of decency. They lay hands on everything, without so much as saying, 'By your leave,' and I am afraid they will burn the house down with their great fires! He, poor man, looks like a corpse, and such sadness in his face~"

And when the young man took leave of her with a few murmured words of comfort she went with him to the door, and leaning over the banister: "Look!" she softly said, "you can see him from where you are. Ah! we are all undone. Adieu, my child!"

Maurice remained planted like a statue on one of the steps of the dark staircase. Craning his neck and directing his glance through the glazed fanlight over the door of the apartment, he beheld a sight that was never to fade from his memory.

In the bare and cheerless room, the conventional bourgeois "parlor," was the Emperor, seated at a table on which his plate was laid, lighted at either end by wax candles in great silver candelabra. Silent in the background stood two aides-de-camp with folded arms. The wine in the glass was untasted, the bread untouched, a breast of chicken was cooling on the plate. The Emperor did not stir; he sat staring down at the cloth with those dim, lusterless, watery eyes that the young man remembered to have seen before at Rheims; but he appeared more weary than then, and when, evidently at the cost of a great effort, he had raised a couple of mouthfuls to his lips, he impatiently pushed the remainder of the food from him with his hand. That was his dinner. His pale face was blanched with an expression of suffering endured in silence.

As Maurice was passing the dining room on the floor beneath, the door was suddenly thrown open, and through the glow of candles and the steam of smoking joints he caught a glimpse of a table of equerries, chamberlains, and aides-de-camp, engaged in devouring the Emperor's game and poultry and drinking his champagne, amid a great hubbub of conversation. Now that the marshal's dispatch had been sent off, all these people were delighted to know that the retreat was assured. In a week they would be at Paris and could sleep between clean sheets.

Then, for the first time, Maurice suddenly became conscious of the terrible fatigue that was oppressing him like a physical burden; there was no longer room for doubt, the whole army was about to fall back, and the best thing for him to do was to get some sleep while waiting for the 7th corps to pass. He made his way back across the square to the house of his friend Combette, where, like one in a dream, he ate some dinner, after which he was mistily conscious of someone dressing his foot and then conducting him upstairs to a bedroom. And then all was blackness and utter annihilation; he slept a dreamless, unstirring sleep. But after an uncertain length of time-hours, days, centuries, he knew not-he gave a start and sat bolt upright in bed in the surrounding darkness. Where was he? What was that continuous rolling sound, like the rattling of thunder, that had aroused him from his slumber? His recollection suddenly returned to him; he ran to the window to see what was going on. In the obscurity of the street beneath, where the night was usually so peaceful, the artillery was passing, horses, men, and guns, in interminable array, with a roar and clatter that made the lifeless houses quake and tremble. The abrupt vision filled him with unreasoning alarm. What time might it be? The great bell in the Hotel de Ville struck four. He was endeavoring to allay his uneasiness by assuring himself that it was simply the initial movement in the retreat that had been ordered the day previous, when, raising his eyes, he beheld a sight that gave him fresh cause for inquietude: there was a light still in the corner window of the notary's house opposite, and the shadow of the Emperor, drawn in dark profile on the curtain, appeared and disappeared at regularly spaced intervals.

Maurice hastily slipped on his trousers preparatory to going down to the street, but just then Combette appeared at the door with a bed-candle in his hand, gesticulating wildly.

"I saw you from the square as I was coming home from the Mairie, and I came up to tell you the news. They have been keeping me out of my bed all this time; would you believe it, for more than two hours the mayor and I have been busy attending to fresh requisitions. Yes, everything is upset again; there has been another change of plans. Ah! he knew what he was about, that officer did, who wanted to keep the folks in Paris from getting wind of matters!"

He went on for a long time in broken, disjointed phrases, and when he had finished the young man, speechless, brokenhearted, saw it all. About midnight the Emperor had received a dispatch from the Minister of War in reply to the one that had been sent by the marshal. Its exact terms were not known, but an aide-de-camp at the Hotel de Ville had stated openly that the Empress and the Council declared there would be a revolution in Paris should the Emperor retrace his steps and abandon Bazaine. The dispatch, which evinced the utmost ignorance as to the position of the German armies and the resources of the army of Chalons, advised, or rather ordered, an immediate forward movement, regardless of all considerations, in spite of everything, with a heat and fury that seemed incredible.

"The Emperor sent for the marshal," added the apothecary, "and they were closeted together for near an hour; of course I am not in position to say what passed between them, but I am told by all the officers that there is to be no more retreating, and the advance to the Meuse is to be resumed at once. We have been requisitioning all the ovens in the city for the 1st corps, which will come up to-morrow morning and take the place of the 12th, whose artillery you see at this moment starting for la Besace. The matter is decided for good this time; you will smell powder before you are much older."

He ceased. He also was gazing at the lighted window over in the notary's house. Then he went on in a low voice, as if talking to himself, with an expression on his face of reflective curiosity:

"I wonder what they had to say to each other? It strikes one as a rather peculiar proceeding, all the same, to run away from a threatened danger at six in the evening, and at midnight, when nothing has occurred to alter the situation, to rush headlong into the very self-same danger."

Below them in the street Maurice still heard the gun-carriages rumbling and rattling over the stones of the little sleeping city, that ceaseless tramp of horse and man, that uninterrupted tide of humanity, pouring onward toward the Meuse, toward the unknown, terrible fate that the morrow had in store for them. And still upon the mean, cheap curtains of that bourgeois dwelling he beheld the shadow of the Emperor passing and repassing at regular intervals, the restless activity of the sick man, to whom his cares made sleep impossible, whose sole repose was motion, in whose ears was ever ringing that tramp of horses and men whom he was suffering to be sent forward to their death. A few brief hours, then, had sufficed; the slaughter was decided on; it was to be. What, indeed, could they have found to say to each other, that Emperor and that marshal, conscious, both of them, of the inevitable disaster that lay before them? Assured as they were at night of defeat, from their knowledge of the wretched condition the army would be in when the time should come for it to meet the enemy, how, knowing as they did that the peril was hourly becoming greater, could they have changed their mind in the morning? Certain it was that General de Palikao's plan of a swift, bold dash on Montmedy, which seemed hazardous on the 23d and was, perhaps, still not impracticable on the 25th, if conducted with veteran troops and a leader of ability, would on the 27th be an act of sheer madness amid the divided counsels of the chiefs and the increasing demoralization of the troops. This they both well knew; why, then, did they obey those merciless drivers who were flogging them onward in their irresolution? why did they hearken to those furious passions that were spurring them forward? The marshal's, it might be said, was the temperament of the soldier, whose duty is limited to obedience to his instructions, great in its abnegation; while the Emperor, who had ceased entirely to issue orders, was waiting on destiny. They were called on to surrender their lives and the life of the army; they surrendered them. It was the accomplishment of a crime, the black, abominable night that witnessed the murder of a nation, for thenceforth the army rested in the shadow of death; a hundred thousand men and more were sent forward to inevitable destruction.

While pursuing this train of thought Maurice was watching the shadow that still kept appearing and vanishing on the muslin of good Madame Desvallieres' curtain, as if it felt the lash of the pitiless voice that came to it from Paris. Had the Empress that night desired the death of the father in order that the son might reign? March! forward ever! with no look backward, through mud, through rain, to bitter death, that the final game of the agonizing empire may be played out, even to the last card. March! march! die a hero's death on the piled corpses of your people, let the whole world gaze in awe-struck admiration, for the honor and glory of your name! And doubtless the Emperor was marching to his death. Below, the fires in the kitchen flamed and flashed no longer; equerries, aides-de-camp and chamberlains were slumbering, the whole house was wrapped in darkness, while ever the lone shade went and came unceasingly, accepting with resignation the sacrifice that was to be, amid the deafening uproar of the 12th corps, that was defiling still through the black night.

Maurice suddenly reflected that, if the advance was to be resumed, the 7th corps would not pass through Chene, and he beheld himself left behind, separated from his regiment, a deserter from his post. His foot no longer pained him; his friend's dressing and a few hours of complete rest had allayed the inflammation. Combette gave him a pair of easy shoes of his own that were comfortable to his feet, and as soon as he had them on he wanted to be off, hoping that he might yet be able to overtake the 106th somewhere on the road between Chene and Vouziers. The apothecary labored vainly to dissuade him, and had almost made up his mind to put his horse in the gig and drive him over in person, trusting to fortune to befriend him in finding the regiment, when Fernand, the apprentice, appeared, alleging as an excuse for his absence that he had been to see his sister. The youth was a tall, tallow-faced individual, who looked as if he had not the spirit of a mouse; the horse was quickly hitched to the carriage and he drove off with Maurice. It was not yet five o'clock; the rain was pouring in torrents from a sky of inky blackness, and the dim carriage-lamps faintly illuminated the road and cast little fitful gleams of light across the streaming fields on either side, over which came mysterious sounds that made them pull up from time to time in the belief that the army was at hand.

Jean, meantime, down there before Vouziers, had not been slumbering. Maurice had explained to him how the retreat was to be salvation to them all, and he was keeping watch, holding his men together and waiting for the order to move, which might come at any minute. About two o'clock, in the intense darkness that was dotted here and there by the red glow of the watch-fires, a great trampling of horses resounded through the camp; it was the advance-guard of cavalry moving off toward Balay and Quatre-Champs so as to observe the roads from Boult-aux-Bois and Croix-aux-Bois; then an hour later the infantry and artillery also put themselves in motion, abandoning at last the positions of Chestre and Falaise that they had defended so persistently for two long days against an enemy who never showed himself. The sky had become overcast, the darkness was profound, and one by one the regiments marched out in deepest silence, an array of phantoms stealing away into the bosom of the night. Every heart beat joyfully, however, as if they were escaping from some treacherous pitfall; already in imagination the troops beheld themselves under the walls of Paris, where their revenge was awaiting them.

Jean looked out into the thick blackness. The road was bordered with trees on either hand and, as far as he could see, appeared to lie between wide meadows. Presently the country became rougher; there was a succession of sharp rises and descents, and just as they were entering a village which he supposed to be Balay, two straggling rows of houses bordering the road, the dense cloud that had obscured the heavens burst in a deluge of rain. The men had received so many duckings within the past few days that they took this one without a murmur, bowing their heads and plodding patiently onward; but when they had left Balay behind them and were crossing a wide extent of level ground near Quatre-Champs a violent wind began to rise. Beyond Quatre-Champs, when they had fought their way upward to the wide plateau that extends in a dreary stretch of waste land as far as Noirval, the wind increased to a hurricane and the driving rain stung their faces. There it was that the order, proceeding from the head of the column and re-echoed down the line, brought the regiments one after another to a halt, and the entire 7th corps, thirty-odd thousand men, found itself once more reunited in the mud and rain of the gray dawn. What was the matter? Why were they halted there? An uneasy feeling was already beginning to pervade the ranks; it was asserted in some quarters that there had been a change of orders. The men had been brought to ordered arms and forbidden to leave the ranks or sit down. At times the wind swept over the elevated plateau with such violence that they had to press closely to one another to keep from being carried off their feet. The rain blinded them and trickled in ice-cold streams beneath their collars down their backs. And two hours passed, a period of waiting that seemed as if it would never end, for what purpose no one could say, in an agony of expectancy that chilled the hearts of all.

As the daylight increased Jean made an attempt to discern where they were. Someone had shown him where the Chene road lay off to the northwest, passing over a hill beyond Quatre-Champs. Why had they turned to the right instead of to the left? Another object of interest to him was the general and his staff, who had established themselves at the Converserie, a farm on the edge of the plateau. There seemed to be a heated discussion going on; officers were going and coming and the conversation was carried on with much gesticulation. What could they be waiting for? nothing was coming that way. The plateau formed a sort of amphitheater, broad expanses of stubble that were commanded to the north and east by wooded heights; to the south were thick woods, while to the west an opening afforded a glimpse of the valley of the Aisne with the little white houses of Vouziers. Below the Converserie rose the slated steeple of Quatre-Champs church, looming dimly through the furious storm, which seemed as if it would sweep away bodily the few poor moss-grown cottages of the village. As Jean's glance wandered down the ascending road he became conscious of a doctor's gig coming up at a sharp trot along the stony road, that was now the bed of a rapid torrent.

It was Maurice, who, at a turn in the road, from the hill that lay beyond the valley, had finally discerned the 7th corps. For two hours he had been wandering about the country, thanks to the stupidity of a peasant who had misdirected him and the sullen ill-will of his driver, whom fear of the Prussians had almost deprived of his wits. As soon as he reached the farmhouse he leaped from the gig and had no further trouble in finding the regiment.

Jean addressed him in amazement:

"What, is it you? What is the meaning of this? I thought you were to wait until we came along."

Maurice's tone and manner told of his rage and sorrow.

"Ah, yes! we are no longer going in that direction; it is down yonder we are to go, to get ourselves knocked in the head, all of us!"

"Very well," said the other presently, with a very white face. "We will die together, at all events."

The two men met, as they had parted, with an embrace. In the drenching rain that still beat down as pitilessly as ever, the humble private resumed his place in the ranks, while the corporal, in his streaming garments, never murmured as he gave him the example of what a soldier should be.

And now the tidings became more definite and spread among the men; they were no longer retreating on Paris; the advance to the Meuse was again the order of the day. An aide-de-camp had brought to the 7th corps instructions from the marshal to go and encamp at Nonart; the 5th was to take the direction of Beauclair, where it would be the right wing of the army, while the 1st was to move up to Chene and relieve the 12th, then on the march to la Besace on the extreme left. And the reason why more than thirty thousand men had been kept waiting there at ordered arms, for nearly three hours in the midst of a blinding storm, was that General Douay, in the deplorable confusion incident on this new change of front, was alarmed for the safety of the train that had been sent forward the day before toward Chagny; the delay was necessary to give the several divisions time to close up. In the confusion of all these conflicting movements it was said that the 12th corps train had blocked the road at Chene, thus cutting off that of the 7th. On the other hand, an important part of the materiel, all the forges of the artillery, had mistaken their road and strayed off in the direction of Terron; they were now trying to find their way back by the Vouziers road, where they were certain to fall into the hands of the Germans. Never was there such utter confusion, never was anxiety so intense.

A feeling of bitterest discouragement took possession of the troops. Many of them in their despair would have preferred to seat themselves on their knapsacks, in the midst of that sodden, wind-swept plain, and wait for death to come to them. They reviled their leaders and loaded them with insult: ah! famous leaders, they; brainless boobies, undoing at night what they had done in the morning, idling and loafing when there was no enemy in sight, and taking to their heels as soon as he showed his face! Each minute added to the demoralization

72

that was already rife, making of that army a rabble, without faith or hope, without discipline, a herd that their chiefs were conducting to the shambles by ways of which they themselves were ignorant. Down in the direction of Vouziers the sound of musketry was heard; shots were being exchanged between the rear-guard of the 7th corps and the German skirmishers; and now every eye was turned upon the valley of the Aisne, where volumes of dense black smoke were whirling upward toward the sky from which the clouds had suddenly been swept away; they all knew it was the village of Falaise burning, fired by the uhlans. Every man felt his blood boil in his veins; so the Prussians were there at last; they had sat and waited two days for them to come up, and then had turned and fled. The most ignorant among the men had felt their cheeks tingle for very shame as, in their dull way, they recognized the idiocy that had prompted that enormous blunder, that imbecile delay, that trap into which they had walked blindfolded; the light cavalry of the IVth army feinting in front of Bordas' brigade and halting and neutralizing, one by one, the several corps of the army of Chalons, solely to give the Crown Prince time to hasten up with the IIId army. And now, thanks to the marshal's complete and astounding ignorance as to the identity of the troops he had before him, the junction was accomplished, and the 5th and 7th corps were to be roughly handled, with the constant menace of disaster overshadowing them.

Maurice's eyes were bent on the horizon, where it was reddened with the flames of burning Falaise. They had one consolation, however: the train that had been believed to be lost came crawling along out of the Chene road. Without delay the 2d division put itself in motion and struck out across the forest for Boult-aux-Bois; the 3d took post on the heights of Belleville to the left in order to keep an eye to the communications, while the 1st remained at Quatre-Champs to wait for the coming up of the train and guard its countless wagons. Just then the rain began to come down again with increased violence, and as the 106th moved off the plateau, resuming the march that should have never been, toward the Meuse, toward the unknown, Maurice thought he beheld again his vision of the night: the shadow of the Emperor, incessantly appearing and vanishing, so sad, so pitiful a sight, on the white curtain of good old Madame Desvallieres. Ah! that doomed army, that army of despair, that was being driven forward to inevitable destruction for the salvation of a dynasty! March, march, onward ever, with no look behind, through mud, through rain, to the bitter end!

6

"Thunder!" Chouteau ejaculated the following morning when he awoke, chilled and with aching bones, under the tent, "I wouldn't mind having a bouillon with plenty of meat in it."

At Boult-aux-Bois, where they were now encamped, the only ration issued to the men the night before had been an extremely slender one of potatoes; the commissariat was daily more and more distracted and disorganized by the everlasting marches and countermarches, never reaching the designated points of rendezvous in time to meet the troops. As for the herds, no one had the faintest idea where they might be upon the crowded roads, and famine was staring the army in the face.

Loubet stretched himself and plaintively replied:

"Ah, fichtre, yes!-No more roast goose for us now."

The squad was out of sorts and sulky. Men couldn't be expected to be lively on an empty stomach. And then there was the rain that poured down incessantly, and the mud in which they had to make their beds.

Observing Pache make the sign of the cross after mumbling his morning prayer, Chouteau captiously growled:

"Ask that good God of yours, if he is good for anything, to send us down a couple of sausages and a mug of beer apiece."

"Ah, if we only had a good big loaf of bread!" sighed Lapoulle, whose ravenous appetite made hunger a more grievous affliction to him than to the others.

But Lieutenant Rochas, passing by just then, made them be silent. It was scandalous, never to think of anything but their stomachs! When *he* was hungry he tightened up the buckle of his trousers. Now that things were becoming decidedly squally and the popping of rifles was to be heard occasionally in the distance, he had recovered all his old serene confidence: it was all plain enough, now; the Prussians were there-well, all they had to do was, go out and lick 'em. And he gave a significant shrug of the shoulders, standing behind Captain Beaudoin, the *very* young man, as he called him, with his pale face and pursed up lips, whom the loss of his baggage had afflicted so grievously that he had even ceased to fume and scold. A man might get along without eating, at a pinch, but that he could not change his linen was a circumstance productive of sorrow and anger.

Maurice awoke to a sensation of despondency and physical discomfort. Thanks to his easy shoes the inflammation in his foot had gone down, but the drenching he had received the day before, from the effects of which his greatcoat seemed to weigh a ton, had left him with a distinct and separate ache in every bone of his body. When he was sent to the spring to get water for the coffee he took a survey of the plain on the edge of which Boult-aux-Bois is situated: forests rise to the west and north, and there is a hill crowned by the hamlet of Belleville, while, over to the east, Buzancy way, there is a broad, level expanse, stretching far as the eye can see, with an occasional shallow depression concealing a small cluster of cottages. Was it from that direction that they were to expect the enemy? As he was returning from the stream with his bucket filled with water, the father of a family of wretched peasants hailed him from the door of his hovel, and asked him if the soldiers were this time going to stay and defend them. In the confusion of conflicting orders the 5th corps had already traversed the region no less than three times. The sound of cannonading had reached them the day before from the direction of Bar; the Prussians could not be more than a couple of leagues away. And when Maurice made answer to the poor folks that doubtless the 7th corps would also be called away after a time, their tears flowed afresh. Then they were to be abandoned to the enemy, and the soldiers had not come there to fight, whom they saw constantly vanishing and reappearing, always on the run?

"Those who like theirs sweet," observed Loubet, as he poured the coffee, "have only to stick their thumb in it and wait for it to melt."

Not a man of them smiled. It was too bad, all the same, to have to drink their coffee without sugar; and then, too, if they only had some biscuit! Most of them had devoured what eatables they had in their knapsacks, to the very last crumb, to while away their time of waiting, the day before, on the plateau of Quatre-Champs. Among them, however, the members of the squad managed to collect a dozen potatoes, which they shared equally.

Maurice, who began to feel a twinging sensation in his stomach, uttered a regretful cry:

"If I had known of this I would have bought some bread at Chene."

Jean listened in silence. He had had a dispute with Chouteau that morning, who, on being ordered to go for firewood, had insolently refused, alleging that it was not his turn. Now that everything was so rapidly going to the dogs, insubordination among the men had increased to such a point that those in authority no longer ventured to reprimand them, and Jean, with his sober good sense and pacific disposition, saw that if he would preserve his influence with his squad he must keep the corporal in the background as far as possible. For this reason he was hail-fellow-well-met with his men, who could not fail to see what a treasure they had in a man of his experience, for if those committed to his care did not always have all they wanted to eat, they had, at all events, not suffered from hunger, as had been the case with so many others. But he was touched by the sight of Maurice's suffering. He saw that he was losing strength, and looked at him anxiously, asking himself how that delicate young man would ever manage to sustain the privations of that horrible campaign.

When Jean heard Maurice bewail the lack of bread he arose quietly, went to his knapsack, and, returning, slipped a biscuit into the other's hand.

"Here! don't let the others see it; I have not enough to go round."

"But what will you do?" asked the young man, deeply affected.

"Oh, don't be alarmed about me-I have two left."

It was true; he had carefully put aside three biscuits, in case there should be a fight, knowing that men are often hungry on the battlefield. And then, besides, he had just eaten a potato; that would be sufficient for him. Perhaps something would turn up later on.

About ten o'clock the 7th corps made a fresh start. The marshal's first intention had been to direct it by way of Buzancy upon Stenay, where it would have passed the Meuse, but the Prussians, outmarching the army of Chalons, were already in Stenay, and were even reported to be at Buzancy. Crowded back in this manner to the northward, the 7th corps had received orders to move to la Besace, some twelve or fifteen miles from Boult-aux-Bois, whence, on the next day, they would proceed to pass the Meuse at Mouzon. The start was made in a very sulky humor; the men, with empty stomachs and bodies unrefreshed by repose, unnerved, mentally and physically, by the experience of the past few days, vented their dissatisfaction by growling and grumbling, while the officers, without a spark of their usual cheerful gayety, with a vague sense of impending disaster awaiting them at the end of their march, taxed the dilatoriness of their chiefs, and reproached them for not going to the assistance of the 5th corps at Buzancy, where the sound of artillery-firing had been heard. That corps, too, was on the retreat, making its way toward Nonart, while the 12th was even then leaving la Besace for Mouzon and the 1st was directing its course toward Raucourt. It was like nothing so much as the passage of a drove of panic-stricken cattle, with the dogs worrying them and snapping at their heels-a wild stampede toward the Meuse.

When, in the outstreaming torrent of the three divisions that striped the plain with columns of marching men, the 106th left Boult-aux-Bois in the rear of the cavalry and artillery, the sky was again overspread with a pall of dull leaden clouds that further lowered the spirits of the soldiers. Its route was along the Buzancy highway, planted on either side with rows of magnificent poplars. When they reached Germond, a village where there was a steaming manure-heap before every one of the doors that lined the two sides of the straggling street, the sobbing women came to their thresholds with their little children in their arms, and held them out to the passing troops, as if begging the men to take them with them. There was not a mouthful of bread to be had in all the hamlet, nor even a potato, After that, the regiment, instead of keeping straight on toward Buzancy, turned to the left and made for Authe, and when the men turned their eyes across the plain and beheld upon the hilltop Belleville, through which they had passed the day before, the fact that they were retracing their steps was impressed more vividly on their consciousness.

"Heavens and earth!" growled Chouteau, "do they take us for tops?"

And Loubet chimed in:

"Those cheap-John generals of ours are all at sea again! They must think that men's legs are cheap."

The anger and disgust were general. It was not right to make men suffer like that, just for the fun of walking them up and down the country. They were advancing in column across the naked plain in two files occupying the sides of the road, leaving a free central space in which the officers could move to and fro and keep an eye on their men, but it was not the same now as it had been in Champagne after they left Rheims, a march of song and jollity, when they tramped along gayly and the knapsack was like a feather to their shoulders, in the belief that soon they would come up with the Prussians and give them a sound drubbing; now they were dragging themselves wearily forward in angry silence, cursing the musket that galled their shoulder and the equipments that seemed to weigh them to the ground, their faith in their leaders gone, and possessed by such bitterness of despair that they only went forward as does a file of manacled galley-slaves, in terror of the lash. The wretched army had begun to ascend its Calvary.

Maurice, however, within the last few minutes had made a discovery that interested him greatly. To their left was a range of hills that rose one above another as they receded from the road, and from the skirt of a little wood, far up on the mountain-side, he had seen a horseman emerge. Then another appeared, and then still another. There they stood, all three of them, without sign of life, apparently no larger than a man's hand and looking like delicately fashioned toys. He thought they were probably part of a detachment of our hussars out on a reconnoissance, when all at once he was surprised to behold little points of light flashing from their shoulders, doubtless the reflection of the sunlight from epaulets of brass.

"Look there!" he said, nudging Jean, who was marching at his side. "Uhlans!"

The corporal stared with all his eyes. "They, uhlans!"

They were indeed uhlans, the first Prussians that the 106th had set eyes on. They had been in the field nearly six weeks now, and in all that time not only had they never smelt powder, but had never even seen an enemy. The news spread through the ranks, and every head was turned to look at them. Not such bad-looking fellows, those uhlans, after all.

"One of them looks like a jolly little fat fellow," Loubet remarked.

But presently an entire squadron came out and showed itself on a plateau to the left of the little wood, and at sight of the threatening demonstration the column halted. An officer came riding up with orders, and the 106th moved off a little and took position on the bank of a small stream behind a clump of trees. The artillery had come hurrying back from the front on a gallop and taken possession of a low, rounded hill. For near two hours they remained there thus in line of battle without the occurrence of anything further; the body of hostile cavalry remained motionless in the distance, and finally, concluding that they were only wasting time that was valuable, the officers set the column moving again.

"Ah well," Jean murmured regretfully, "we are not booked for it this time."

Maurice, too, had felt his finger-tips tingling with the desire to have just one shot. He kept harping on the theme of the mistake they had made the day before in not going to the support of the 5th corps. If the Prussians had not made their attack yet, it must be because their infantry had not got up in sufficient strength, whence it was evident that their display of cavalry in the distance was made with no other end than to harass us and check the advance of our corps. We had again fallen into the trap set for us, and thenceforth the regiment was constantly greeted with the sight of uhlans popping up on its left flank wherever the ground was favorable for them, tracking it like sleuthhounds, disappearing behind a farmhouse only to reappear at the corner of a wood.

It eventually produced a disheartening effect on the troops to see that cordon closing in on them in the distance and enveloping them as in the meshes of some gigantic, invisible net. Even Pache and Lapoulle had an opinion on the subject.

"It is beginning to be tiresome!" they said. "It would be a comfort to send them our compliments in the shape of a musket-ball!"

But they kept toiling wearily onward on their tired feet, that seemed to them as if they were of lead. In the distress and suffering of that day's march there was ever present to all the undefined sensation of the proximity of the enemy, drawing in on them from every quarter, just as we are conscious of the coming storm before we have seen a cloud on the horizon. Instructions were given the rear-guard to use severe measures, if necessary, to keep the column well closed up; but there was not much straggling, aware as everyone was that the Prussians were close in our rear, and ready to snap up every unfortunate that they could lay hands on. Their infantry was coming up with the rapidity of the whirlwind, making its twenty-five miles a day, while the French regiments, in their demoralized condition, seemed in comparison to be marking time.

At Authe the weather cleared, and Maurice, taking his bearings by the position of the sun, noticed that instead of bearing off toward Chene, which lay three good leagues from where they were, they had turned and were moving directly eastward. It was two o'clock; the men, after shivering in the rain for two days, were now suffering from the intense heat. The road ascended, with long sweeping curves, through a region of utter desolation: not a house, not a living being, the only relief to the dreariness of the waste lands an occasional little somber wood; and the oppressive silence communicated itself to the men, who toiled onward with drooping heads, bathed in perspiration. At last Saint-Pierremont appeared before them, a few empty houses on a small elevation. They did not pass through the village. Maurice observed that here they made a sudden wheel to the left, resuming their northern course, toward la Besace. He now understood the route that had been adopted in their attempt to reach Mouzon ahead of the Prussians; but would they succeed, with such weary, demoralized troops? At Saint-Pierremont the three uhlans had shown themselves again, at a turn in the road leading to Buzancy, and just as the rear-guard was leaving the village a battery was unmasked and a few shells came tumbling among them, without doing any injury, however. No response was attempted, and the march was continued with constantly increasing effort.

From Saint-Pierremont to la Besace the distance is three good leagues, and when Maurice imparted that information to Jean the latter made a gesture of discouragement: the men would never be able to accomplish it; they showed it by their shortness of breath, by their haggard faces. The road continued to ascend, between gently sloping hills on either side that were gradually drawing closer together. The condition of the men necessitated a halt, but the only effect of their brief repose was to increase the stiffness of their benumbed limbs, and when the order was given to march the state of affairs was worse than it had been before; the regiments made no progress, men were everywhere falling in the ranks. Jean, noticing Maurice's pallid face and glassy eyes, infringed on what was his usual custom and conversed, endeavoring by his volubility to divert the other's attention and keep him awake as he moved automatically forward, unconscious of his actions.

"Your sister lives in Sedan, you say; perhaps we shall be there before long."

"What, at Sedan? Never! You must be crazy; it don't lie in our way."

"Is your sister young?"

"Just my age; you know I told you we are twins."

"Is she like you?"

"Yes, she is fair-haired, too; and oh! such pretty curling hair! She is a mite of a woman, with a little thin face, not one of your noisy, flashy hoydens, ah, no!-Dear Henriette!"

"You love her very dearly!"

"Yes, yes-"

There was silence between them after that, and Jean, glancing at Maurice, saw that his eyes were closing and he was about to fall.

"Hallo there, old fellow! Come, confound it all, brace up! Let me take your gun a moment; that will give you a chance to rest. They can't have the cruelty to make us march any further today! we shall leave half our men by the roadside."

At that moment he caught sight of Osches lying straight ahead of them, its few poor hovels climbing in straggling fashion up the hillside, and the yellow church, embowered in trees, looking down on them from its perch upon the summit.

"There's where we shall rest, for certain."

He had guessed aright; General Douay saw the exhausted condition of the troops, and was convinced that it would be useless to attempt to reach la Besace that day. What particularly influenced his determination, however, was the arrival of the train, that ill-starred train that had been trailing in his rear since they left Rheims, and of which the nine long miles of vehicles and animals had so terribly impeded his movements. He had given instructions from Quatre-Champs to direct it straight on Saint-Pierremont, and it was not until Osches that the teams came up with the corps, in such a state of exhaustion that the horses refused to stir. It was now five o'clock; the general, not liking the prospect of attempting the pass of Stonne at that late hour, determined to take the responsibility of abridging the task assigned them by the marshal. The corps was halted and proceeded to encamp; the train below in the meadows, guarded by a division, while the artillery took position on the hills to the rear, and the brigade detailed to act as rear-guard on the morrow rested on a height facing Saint-Pierremont. The other division, which included Bourgain-Desfeuilles' brigade, bivouacked on a wide plateau, bordered by an oak wood, behind the church. There was such confusion in locating the bodies of troops that it was dark before the 106th could move into its position at the edge of the wood.

"Zut!" said Chouteau in a furious rage, "no eating for me; I want to sleep!"

And that was the cry of all; they were overcome with fatigue. Many of them lacked strength and courage to erect their tents, but dropping where they stood, at once fell fast asleep on the bare ground. In order to eat, moreover, rations would have been necessary, and the commissary wagons, which were waiting for the 7th corps to come to them at la Besace, could not well be at Osches at the same time. In the universal relaxation of order and system even the customary corporal's call was omitted: it was everyone for himself. There were to be no more issues of rations from that time forth; the soldiers were to subsist on the provisions they were supposed to carry in their knapsacks, and that evening the sacks were empty; few indeed were those who could muster a crust of bread or some crumbs of the abundance in which they had been living at Vouziers of late. There was coffee, and those who were not too tired made and drank it without sugar.

When Jean thought to make a division of his wealth by eating one of his biscuits himself and giving the other to Maurice, he discovered that the latter was sound asleep. He thought at first he would awake him, but changed his mind and stoically replaced the biscuits in his sack, concealing them with as much caution as if they had been bags of gold; he could get along with coffee, like the rest of the boys. He had insisted on having the tent put up, and they were all stretched on the ground beneath its shelter when Loubet returned from a foraging expedition, bringing in some carrots that he had found in a neighboring field. As there was no fire to cook them by they munched them raw, but the vegetables only served to aggravate their hunger, and they made Pache ill.

"No, no; let him sleep," said Jean to Chouteau, who was shaking Maurice to wake him and give him his share.

"Ah," Lapoulle broke in, "we shall be at Angouleme to-morrow, and then we'll have some bread. I had a cousin in the army once, who was stationed at Angouleme. Nice garrison, that."

They all looked surprised, and Chouteau exclaimed:

"Angouleme-what are you talking about! Just listen to the bloody fool, saying he is at Angouleme!"

It was impossible to extract any explanation from Lapoulle. He had insisted that morning that the uhlans that they sighted were some of Bazaine's troops.

Then darkness descended on the camp, black as ink, silent as death. Notwithstanding the coolness of the night air the men had not been permitted to make fires; the Prussians were known to be only a few miles away, and it would not do to put them on the alert; orders even were transmitted in a hushed voice. The officers had notified their men before retiring that the start would be made at about four in the morning, in order that they might have all the rest possible, and all had hastened to turn in and were sleeping greedily, forgetful of their troubles. Above the scattered camps the deep respiration of all those slumbering crowds, rising upon the stillness of the night, was like the long-drawn breathing of old Mother Earth.

Suddenly a shot rang out in the darkness and aroused the sleepers. It was about three o'clock, and the obscurity was profound. Immediately everyone was on foot, the alarm spread through the camp; it was supposed the Prussians were attacking. It was only Loubet who, unable to sleep longer, had taken it in his head to make a foray into the oak-wood, which he thought gave promise of rabbits: what a jolly good lark it would be if he could bring in a pair of nice rabbits for the comrades' breakfast! But as he was looking about for a favorable place in which to conceal himself, he heard the sound of voices and the snapping of dry branches under heavy footsteps; men were coming toward him; he took alarm and discharged his piece, believing the Prussians were at hand. Maurice, Jean, and others came running up in haste, when a hoarse voice made itself heard:

"For God's sake, don't shoot!"

And there at the edge of the wood stood a tall, lanky man, whose thick, bristling beard they could just distinguish in the darkness. He wore a gray blouse, confined at the waist by a red belt, and carried a musket slung by a strap over his shoulder. He hurriedly explained that he was French, a sergeant of francs-tireurs, and had come with two of his men from the wood of Dieulet, bringing important information for the general.

"Hallo there, Cabasse! Ducat!" he shouted, turning his head, "hallo! you infernal poltroons, come here!"

The men were evidently badly scared, but they came forward. Ducat, short and fat, with a pale face and scanty hair; Cabasse short and lean, with a black face and a long nose not much thicker than a knife-blade.

Meantime Maurice had stepped up and taken a closer look at the sergeant; he finally asked him:

"Tell me, are you not Guillaume Sambuc, of Remilly?"

And when the man hesitatingly answered in the affirmative Maurice recoiled a step or two, for this Sambuc had the reputation of being a particularly hard case, the worthy son of a family of woodcutters who had all gone to the bad, the drunken father being found one night lying by the roadside with his throat cut, the mother and daughter, who lived by begging and stealing, having disappeared, most likely, in the seclusion of some penitentiary. He, Guillaume, did a little in the poaching and smuggling lines, and only one of that litter of wolves' whelps had grown up to be an honest man, and that was Prosper, the hussar, who had gone to work on a farm before he was conscripted, because he hated the life of the forest.

"I saw your brother at Vouziers," Maurice continued; "he is well."

Sambuc made no reply. To end the situation he said:

"Take me to the general. Tell him that the francs-tireurs of the wood of Dieulet have something important to say to him."

On the way back to the camp Maurice reflected on those free companies that had excited such great expectations at the time of their formation, and had since been the object of such bitter denunciation throughout the country. Their professed purpose was to wage a sort of guerilla warfare, lying in ambush behind hedges, harassing the enemy, picking off his sentinels, holding the woods, from which not a Prussian was to emerge alive; while the truth of the matter was that they had made themselves the terror of the peasantry, whom they failed utterly to protect and whose fields they devastated. Every ne'er-do-well who hated the restraints of the regular service made haste to join their ranks, well pleased with the chance that exempted him from discipline and enabled him to lead the life of a tramp, tippling in pothouses and sleeping by the roadside at his own sweet will. Some of the companies were recruited from the very worst material imaginable.

"Hallo there, Cabasse! Ducat!" Sambuc was constantly repeating, turning to his henchmen at every step he took, "Come along, will you, you snails!"

Maurice was as little charmed with the two men as with their leader. Cabasse, the little lean fellow, was a native of Toulon, had served as waiter in a cafe at Marseilles, had failed at Sedan as a broker in southern produce, and finally had brought up in a police-court, where it came near going hard with him, in connection with a robbery of which the details were suppressed. Ducat, the little fat man, quondam *huissier* at Blainville, where he had been forced to sell out his business on account of a malodorous woman scrape, had recently been brought face to face with the court of assizes for an indiscretion of a similar nature at Raucourt, where he was accountant in a factory. The latter quoted Latin in his conversation, while the other could scarcely read, but the two were well mated, as unprepossessing a pair as one could expect to meet in a summer's day.

The camp was already astir; Jean and Maurice took the francs-tireurs to Captain Beaudoin, who conducted them to the quarters of Colonel Vineuil. The colonel attempted to question them, but Sambuc, intrenching himself in his dignity, refused to speak to anyone except the general. Now Bourgain-Desfeuilles had taken up his quarters that night with the cure of Osches, and just then appeared, rubbing his eyes, in the doorway of the parsonage; he was in a horribly bad humor at his slumbers having been thus prematurely cut short, and the prospect that he saw before him of another day of famine and fatigue; hence his reception of the men who were brought before him was not exactly lamblike. Who were they? Whence did they come? What did they want? Ah, some of those francs-tireurs gentlemen-eh! Same thing as skulkers and riff-raff!

"General," Sambuc replied, without allowing himself to be disconcerted, "we and our comrades are stationed in the woods of Dieulet-"

"The woods of Dieulet-where's that?"

"Between Stenay and Mouzon, General."

"What do I know of your Stenay and Mouzon? Do you expect me to be familiar with all these strange names?"

The colonel was distressed by his chief's display of ignorance; he hastily interfered to remind him that Stenay and Mouzon were on the Meuse, and that, as the Germans had occupied the former of those towns, the army was about to attempt the passage of the river at the other, which was situated more to the northward.

"So you see, General," Sambuc continued, "we've come to tell you that the woods of Dieulet are alive with Prussians. There was an engagement yesterday as the 5th corps was leaving Bois-les-Dames, somewhere about Nonart-"

"What, yesterday? There was fighting yesterday?"

"Yes, General, the 5th corps was engaged as it was falling back; it must have been at Beaumont last night. So, while some of us hurried off to report to it the movements of the enemy, we thought it best to come and let you know how matters stood, so that you might go to its assistance, for it will certainly have sixty thousand men to deal with in the morning."

General Bourgain-Desfeuilles gave a contemptuous shrug of his shoulders.

"Sixty thousand men! Why the devil don't you call it a hundred thousand at once? You were dreaming, young man; your fright has made you see double. It is impossible there should be sixty thousand Germans so near us without our knowing it."

And so he went on. It was to no purpose that Sambuc appealed to Ducat and Cabasse to confirm his statement.

"We saw the guns," the Provencal declared; "and those chaps must be crazy to take them through the forest, where the rains of the past few days have left the roads in such a state that they sink in the mud up to the hubs."

"They have someone to guide them, for certain," said the ex-bailiff.

Since leaving Vouziers the general had stoutly refused to attach any further credit to reports of the junction of the two German armies which, as he said, they had been trying to stuff down his throat. He did not even consider it worth his while to send the francs-tireurs before his corps commander, to whom the partisans supposed, all along, that they were talking; if they should attempt to listen to all the yarns that were brought them by tramps and peasants, they would have their hands full and be driven from pillar to post without ever advancing a step. He directed the three men to remain with the column, however, since they were acquainted with the country.

"They are good fellows, all the same," Jean said to Maurice, as they were returning to fold the tent, "to have tramped three leagues across lots to let us know."

The young man agreed with him and commended their action, knowing as he did the country, and deeply alarmed to hear that the Prussians were in Dieulet forest and moving on Sommanthe and Beaumont. He had flung himself down by the roadside, exhausted before the march had commenced, with a sorrowing heart and an empty stomach, at the dawning of that day which he felt was to be so disastrous for them all. Distressed to see him looking so pale, the corporal affectionately asked him:

"Are you feeling so badly still? What is it? Does your foot pain you?"

Maurice shook his head. His foot had ceased to trouble him, thanks to the big shoes.

"Then you are hungry." And Jean, seeing that he did not answer, took from his knapsack one of the two remaining biscuits, and with a falsehood for which he may be forgiven: "Here, take it; I kept your share for you. I ate mine a while ago."

Day was breaking when the 7th corps marched out of Osches en route for Mouzon by way of la Besace, where they should have bivouacked. The train, cause of so many woes, had been sent on ahead, guarded by the first division, and if its own wagons, well horsed as for the most part they were, got over the ground at a satisfactory pace, the requisitioned vehicles, most of them empty, delayed the troops and produced sad confusion among the hills of the defile of Stonne. After leaving the hamlet of la Berliere the road rises more sharply between wooded hills on either side. Finally, about eight o'clock, the two remaining divisions got under way, when Marshal MacMahon came galloping up, vexed to find there those troops that he supposed had left la Besace that morning, with only a short march between them and Mouzon; his comment to General Douay on the subject was expressed in warm language. It was determined that the first division and the train should be allowed to proceed on their way to Mouzon, but that the two other divisions, that they might not be further retarded by this cumbrous advance-guard, should move by the way of Raucourt and Autrecourt so as to pass the Meuse at Villers. The movement to the north was dictated by the marshal's intense anxiety to place the river between his army and the enemy; cost what it might, they must be on the right bank that night. The rear-guard had not yet left Osches when a Prussian battery, recommencing the performance of the previous day, began to play on them from a distant eminence, over in the direction of Saint-Pierremont. They made the mistake of firing a few shots in reply; then the last of the troops filed out of the town.

Until nearly eleven o'clock the 106th slowly pursued its way along the road which zigzags through the pass of Stonne between high hills. On the left hand the precipitous summits rear their heads, devoid of vegetation, while to the right the gentler slopes are clad with woods down to the roadside. The sun had come out again, and the heat was intense down in the inclosed valley, where an oppressive solitude prevailed. After leaving la Berliere, which lies at the foot of a lofty and desolate mountain surmounted by a Calvary, there is not a house to be seen, not a human being, not an animal grazing in the meadows. And the men, the day before so faint with hunger, so spent with fatigue, who since that time had had no food to restore, no slumber, to speak of, to refresh them, were now dragging themselves listlessly along, disheartened, filled with sullen anger.

Soon after that, just as the men had been halted for a short rest along the roadside, the roar of artillery was heard away at their right; judging from the distinctness of the detonations the firing could not be more than two leagues distant. Upon the troops, weary with waiting, tired of retreating, the effect was magical; in the twinkling of an eye everyone was on his feet, eager, in a quiver of excitement, no longer mindful of his hunger and fatigue: why did they not advance? They preferred to fight, to die, rather than keep on flying thus, no one knew why or whither.

General Bourgain-Desfeuilles, accompanied by Colonel de Vineuil, had climbed a hill on the right to reconnoiter the country. They were visible up there in a little clearing between two belts of wood, scanning the surrounding hills with their field-glasses, when all at once they dispatched an aide-de-camp to the column, with instructions to send up to them the francs-tireurs if they were still there. A few men, Jean and Maurice among them, accompanied the latter, in case there should be need of messengers.

"A beastly country this, with its everlasting hills and woods!" the general shouted, as soon as he caught sight of Sambuc. "You hear the music-where is it? where is the fighting going on?"

Sambuc, with Ducat and Cabasse close at his heels, listened a moment before he answered, casting his eye over the wide horizon, and Maurice, standing beside him and gazing out over the panorama of valley and forest that lay beneath him, was struck with admiration. It was like a boundless sea, whose gigantic waves had been arrested by some mighty force. In the foreground the somber verdure of the woods made splashes of sober color on the yellow of the fields, while in the brilliant sunlight the distant hills were bathed in purplish vapors. And while nothing was to be seen, not even the tiniest smoke-wreath floating on the cloudless sky, the cannon were thundering away in the distance, like the muttering of a rising storm.

"Here is Sommanthe, to the right," Sambuc said at last, pointing to a high hill crowned by a wood. "Yoncq lies off yonder to the left. The fighting is at Beaumont, General."

"Either at Varniforet or Beaumont," Ducat observed.

The general muttered below his breath: "Beaumont, Beaumont-a man can never tell where he is in this d--d country." Then raising his voice: "And how far may this Beaumont be from here?"

"A little more than six miles, if you take the road from Chene to Stenay, which runs up the valley yonder."

There was no cessation of the firing, which seemed to be advancing from west to east with a continuous succession of reports like peals of thunder. Sambuc added:

"Bigre! it's getting warm. It is just what I expected; you know what I told you this morning, General; it is certainly the batteries that we saw in the wood of Dieulet. By this time the whole army that came up through Buzancy and Beauclair is at work mauling the 5th corps."

There was silence among them, while the battle raging in the distance growled more furiously than ever, and Maurice had to set tight his teeth to keep himself from speaking his mind aloud. Why did they not hasten whither the guns were calling them, without such waste of words? He had never known what it was to be excited thus; every discharge found an echo in his bosom and inspired him with a fierce longing to be present at the conflict, to put an end to it. Were they to pass by that battle, so near almost that they could stretch forth their arm and touch it with their hand, and never expend a cartridge? It must be to decide a wager that some one had made, that since the beginning of the campaign they were dragged about the country thus, always flying before the enemy! At Vouziers they had heard the musketry of the rear-guard, at Osches the German guns had played a moment on their retreating backs; and now they were to run for it again, they were not to be allowed to advance at double-quick to the succor of comrades in distress! Maurice looked at Jean, who was also very pale, his eyes shining with a bright, feverish light. Every heart leaped in every bosom at the loud summons of the artillery.

While they were waiting a general, attended by his staff, was seen ascending the narrow path that wound up the hill. It was Douay, their corps-commander, who came hastening up, with anxiety depicted on his countenance, and when he had questioned the francs-tireurs he gave utterance to an exclamation of despair. But what could he have done, even had he learned their tidings that morning? The marshal's orders were explicit: they must be across the Meuse that night, cost what it might. And then again, how was he to collect his scattered troops, strung out along the road to Raucourt, and direct then on Beaumont? Could they arrive in time to be of use? The 5th corps must be in full retreat on Mouzon by that time, as was indicated by the sound of the firing, which was receding more and more to the eastward, as a deadly hurricane moves off after having accomplished its disastrous work. With a fierce gesture, expressive of his sense of impotency, General Douay outstretched his arms toward the wide horizon of hill and dale, of woods and fields, and the order went forth to proceed with the march to Raucourt.

Ah, what a march was that through that dismal pass of Stonne, with the lofty summits o'erhanging them on either side, while through the woods on their right came the incessant volleying of the artillery. Colonel de Vineuil rode at the head of his regiment, bracing himself firmly in his saddle, his face set and very pale, his eyes winking like those of one trying not to weep. Captain Beaudoin strode along in silence, gnawing his mustache, while Lieutenant Rochas let slip an occasional imprecation, invoking ruin and destruction on himself and everyone besides. Even the most cowardly among the men, those who had the least stomach for fighting, were shamed and angered by their continuous retreat; they felt the bitter humiliation of turning their backs while those beasts of Prussians were murdering their comrades over yonder.

After emerging from the pass the road, from a tortuous path among the hills, increased in width and led through a broad stretch of level country, dotted here and there with small woods. The 106th was now a portion of the rear-guard, and at every moment since leaving Osches had been expecting to feel the enemy's attack, for the Prussians were following the column step by step, never letting it escape their vigilant eyes, waiting, doubtless, for a favorable opportunity to fall on its rear. Their cavalry were on the alert to take advantage of any bit of ground that promised them an opportunity of getting in on our flank; several squadrons of Prussian Guards were seen advancing from behind a wood, but they gave up their purpose upon a demonstration made by a regiment of our hussars, who came up at a gallop, sweeping the road. Thanks to the breathing-spell afforded them by this circumstance the retreat went on in sufficiently good order, and Raucourt was not far away, when a spectacle greeted their eyes that filled them with consternation and completely demoralized the troops. Upon coming to a cross-road they suddenly caught sight of a hurrying, straggling, flying throng, wounded officers, soldiers without arms and without organization, runaway teams from the train, all-men and animals-mingled in wildest confusion, wild with panic. It was the wreck of one of the brigades of the 1st division, which had been sent that morning to escort the train to Mouzon; there had been an unfortunate misconception of orders, and this brigade and a portion of the wagons had taken a wrong road and reached Varniforet, near Beaumont, at the very time when the 5th corps was being driven back in disorder. Taken unawares, overborne by the flank attack of an enemy superior in numbers, they had fled; and bleeding, with haggard faces, crazed with fear, were now returning to spread consternation among their comrades; it was as if they had been wafted thither on the breath of the battle that had been raging incessantly since noon.

Alarm and anxiety possessed everyone, from highest to lowest, as the column poured through Raucourt in wild stampede. Should they turn to the left, toward Autrecourt, and attempt to pass the Meuse at Villers, as had been previously decided? The general hesitated, fearing to encounter difficulties in crossing there, even if the bridge were not already in possession of the Prussians; he finally decided to keep straight on through the defile of Harancourt and thus reach Remilly before nightfall. First Mouzon, then Villers, and last Remilly; they were still pressing on northward, with the tramp of the uhlans on the road behind them. There remained scant four miles for them to accomplish, but it was five o'clock, and the men were sinking with fatigue. They had been under arms since daybreak, twelve hours had been consumed in advancing three short leagues; they were harassed and fatigued as much by their constant halts and the stress of their emotions as by the actual toil of the march. For the last two nights they had had scarce any sleep; their hunger had been unappeased since they left Vouziers. In Raucourt the distress was terrible; men fell in the ranks from sheer inanition.

The little town is rich, with its numerous factories, its handsome thoroughfare lined with two rows of well-built houses, and its pretty church and mairie; but the night before Marshal MacMahon and the Emperor had passed that way with their respective staffs and all the imperial household, and during the whole of the present morning the entire 1st corps had been streaming like a torrent through the main street. The resources of the place had not been adequate to meet the requirements of these hosts; the shelves of the bakers and grocers were empty, and even the houses of the bourgeois had been swept clean of provisions; there was no bread, no wine, no sugar, nothing capable of allaying hunger or thirst. Ladies had been seen to station themselves before their doors and deal out glasses of wine and cups of bouillon until cask and kettle alike were drained of their last drop. And so there was an end, and when, about three o'clock, the first regiments of the 7th corps began to appear the scene was a pitiful one; the broad street was filled from curb to curb with weary, dust-stained men, dying with hunger, and there was not a mouthful of food to give them. Many of them stopped, knocking at doors and extending their hands beseechingly toward windows, begging for a morsel of bread, and women were seen to cry and sob as they motioned that they could not help them, that they had nothing left.

At the corner of the Rue Dix-Potiers Maurice had an attack of dizziness and reeled as if about to fall. To Jean, who came hastening up, he said:

"No, leave me; it is all up with me. I may as well die here!"

He had sunk down upon a door-step. The corporal spoke in a rough tone of displeasure assumed for the occasion:

"Nom de Dieu! why don't you try to behave like a soldier! Do you want the Prussians to catch you? Come, get up!"

Then, as the young man, lividly pale, his eyes tight-closed, almost unconscious, made no reply, he let slip another oath, but in another key this time, in a tone of infinite gentleness and pity:

"Nom de Dieu! Nom de Dieu!"

And running to a drinking-fountain near by, he filled his basin with water and hurried back to bathe his friend's face. Then, without further attempt at concealment, he took from his sack the last remaining biscuit that he had guarded with such jealous caution, and commenced crumbling it into small bits that he introduced between the other's teeth. The famishing man opened his eyes and ate greedily.

"But you," he asked, suddenly recollecting himself, "how comes it that you did not eat it?"

"Oh, I!" said Jean. "I'm tough, I can wait. A good drink of Adam's ale, and I shall be all right."

He went and filled his basin again at the fountain, emptied it at a single draught, and came back smacking his lips in token of satisfaction with his feast. He, too, was cadaverously pale, and so faint with hunger that his hands were trembling like a leaf.

"Come, get up, and let's be going. We must be getting back to the comrades, little one."

Maurice leaned on his arm and suffered himself to be helped along as if he had been a child;

never had woman's arm about him so warmed his heart. In that extremity of distress, with death staring him in the face, it afforded him a deliciously cheering sense of comfort to know that someone loved and cared for him, and the reflection that that heart, which was so entirely his, was the heart of a simple-minded peasant, whose aspirations scarcely rose above the satisfaction of his daily wants, for whom he had recently experienced a feeling of repugnance, served to add to his gratitude a sensation of ineffable joy. Was it not the brotherhood that had prevailed in the world in its earlier days, the friendship that had existed before caste and culture were; that friendship which unites two men and makes them one in their common need of assistance, in the presence of Nature, the common enemy? He felt the tie of humanity uniting him and Jean, and was proud to know that the latter, his comforter and savior, was stronger than he; while to Jean, who did not analyze his sensations, it afforded unalloyed pleasure to be the instrument of protecting, in his friend, that cultivation and intelligence which, in himself, were only rudimentary. Since the death of his wife, who had been snatched away from him by a frightful catastrophe, he had believed that his heart was dead, he had sworn to have nothing more to do with those creatures, who, even when they are not wicked and depraved, are cause of so much suffering to man. And thus, to both of them their friendship was a comfort and relief. There was no need of any demonstrative display of affection; they understood each other; there was close community of sympathy between them, and, notwithstanding their apparent external dissimilarity, the bond of pity and common suffering made them as one during their terrible march that day to Remilly.

As the French rear-guard left Raucourt by one end of the town the Germans came in at the other, and forthwith two of their batteries commenced firing from the position they had taken on the heights to the left; the 106th, retreating along the road that follows the course of the Emmane, was directly in the line of fire. A shell cut down a poplar on the bank of the stream; another came and buried itself in the soft ground close to Captain Beaudoin, but did not burst. From there on to Harancourt, however, the walls of the pass kept approaching nearer and nearer, and the troops were crowded together in a narrow gorge commanded on either side by hills covered with trees. A handful of Prussians in ambush on those heights might have caused incalculable disaster. With the cannon thundering in their rear and the menace of a possible attack on either flank, the men's uneasiness increased with every step they took, and they were in haste to get out of such a dangerous neighborhood; hence they summoned up their reserved strength, and those soldiers who, but now in Raucourt, had scarce been able to drag themselves along, now, with the peril that lay behind them as an incentive, struck out at a good round pace. The very horses seemed to be conscious that the loss of a minute might cost them dear. And the impetus thus given continued; all was going well, the head of the column must have reached Remilly, when, all at once, their progress was arrested.

"Heavens and earth!" said Chouteau, "are they going to leave us here in the road?"

The regiment had not yet reached Harancourt, and the shells were still tumbling about them; while the men were marking time, awaiting the word to go ahead again, one burst, on the right of the column, without injuring anyone, fortunately. Five minutes passed, that seemed to them long as an eternity, and still they did not move; there was some obstacle on ahead that barred their way as effectually as if a strong wall had been built across the road. The colonel, standing up in his stirrups, peered nervously to the front, for he saw that it would require but little to create a panic among his men.

"We are betrayed; everybody can see it," shouted Chouteau.

Murmurs of reproach arose on every side, the sullen muttering of their discontent exasperated by their fears. Yes, yes! they had been brought there to be sold, to be delivered over to the Prussians. In the baleful fatality that pursued them, and among all the blunders of their leaders, those dense intelligences were unable to account for such an uninterrupted succession of disasters on any other ground than that of treachery.

"We are betrayed! we are betrayed!" the men wildly repeated.

Then Loubet's fertile intellect evolved an idea: "It is like enough that that pig of an Emperor has sat himself down in the road, with his baggage, on purpose to keep us here."

The idle fancy was received as true, and immediately spread up and down the line; everyone declared that the imperial household had blocked the road and was responsible for the stoppage. There was a universal chorus of execration, of opprobrious epithets, an unchaining of the hatred and hostility that were inspired by the insolence of the Emperor's attendants, who took possession of the towns where they stopped at night as if they owned them, unpacking their luxuries, their costly wines and plate of gold and silver, before the eyes of the poor soldiers who were destitute of everything, filling the kitchens with the steam of savory viands while they, poor devils, had nothing for it but to tighten the belt of their trousers. Ah! that wretched Emperor, that miserable man, deposed from his throne and stripped of his command, a stranger in his own empire; whom they were conveying up and down the country along with the other baggage, like some piece of useless furniture, whose doom it was ever to drag behind him the irony of his imperial state: cent-gardes, horses, carriages, cooks, and vans, sweeping, as it were, the blood and mire from the roads of his defeat with the magnificence of his court mantle, embroidered with the heraldic bees!

In rapid succession, one after the other, two more shells fell; Lieutenant Rochas had his *kepi* carried away by a fragment. The men huddled closer together and began to crowd forward, the movement gathering strength as it ran from rear to front. Inarticulate cries were heard, Lapoulle shouted furiously to go ahead. A minute longer and there would have been a horrible catastrophe, and many men must have been crushed to death in the mad struggle to escape from the funnel-like gorge.

The colonel-he was very pale-turned and spoke to the soldiers:

"My children, my children, be a little patient. I have sent to see what is the matter-it will only be a moment-"

But they did not advance, and the seconds seemed like centuries. Jean, quite cool and collected, resumed his hold of Maurice's hand, and whispered to him that, in case their comrades began to shove, they two could leave the road, climb the hill on the left, and make their way to the stream. He looked about to see where the francs-tireurs were, thinking he might gain some information from them regarding the roads, but was told they had vanished while the column was passing through Raucourt. Just then the march was resumed, and almost immediately a bend in the road took them out of range of the German batteries. Later in the day it was ascertained that it was four cuirassier regiments of Bonnemain's division who, in the disorder of that ill-starred retreat, had thus blocked the road of the 7th corps and delayed the march.

It was nearly dark when the 106th passed through Angecourt. The wooded hills continued on the right, but to the left the country was more level, and a valley was visible in the distance, veiled in bluish mists. At last, just as the shades of night were descending, they stood on the heights of Remilly and beheld a ribbon of pale silver unrolling its length upon a broad expanse of verdant plain. It was the Meuse, that Meuse they had so longed to see, and where it seemed as if victory awaited them.

Pointing to some lights in the distance that were beginning to twinkle cheerily among the trees, down in that fertile valley that lay there so peaceful in the mellow twilight, Maurice said to Jean, with the glad content of a man revisiting a country that he knows and loves:

"Look! over that way-that is Sedan!"

7

Remilly is built on a hill that rises from the left bank of the Meuse, presenting the appearance of an amphitheater; the one village street that meanders circuitously down the sharp descent was thronged with men, horses, and vehicles in dire confusion. Half-way up the hill, in front of the church, some drivers had managed to interlock the wheels of their guns, and all the oaths and blows of the artillerymen were unavailing to get them forward. Further down, near the woolen mill, where the Emmane tumbles noisily over the dam, the road was choked with a long line of stranded baggage wagons, while close at hand, at the inn of the Maltese Cross, a constantly increasing crowd of angry soldiers pushed and struggled, and could not obtain so much as a glass of wine.

All this mad hurly-burly was going on at the southern end of the village, which is here separated from the Meuse by a little grove of trees, and where the engineers had that morning stretched a bridge of boats across the river. There was a ferry to the right; the ferryman's house stood by itself, white and staring, amid a rank growth of weeds. Great fires had been built on either bank, which, being replenished from time to time, glared ruddily in the darkness and made the stream and both its shores as light as day. They served to show the immense multitude of men massed there, awaiting a chance to cross, while the footway only permitted the passage of two men abreast, and over the bridge proper the cavalry and artillery were obliged to proceed at a walk, so that the crossing promised to be a protracted operation. It was said that the troops still on the left bank comprised a brigade of the 1st corps, an ammunition train, and the four regiments of cuirassiers belonging to Bonnemain's division, while coming up in hot haste behind them was the 7th corps, over thirty thousand strong, possessed with the belief that the enemy was at their heels and pushing on with feverish eagerness to gain the security of the other shore.

For a while despair reigned. What! they had been marching since morning with nothing to eat, they had summoned up all their energies to escape that deadly trap at Harancourt pass, only in the end to be landed in that slough of despond, with an insurmountable wall staring them in the face! It would be hours, perhaps, before it became the last comer's turn to cross, and everyone knew that even if the Prussians should not be enterprising enough to continue their pursuit in the darkness they would be there with the first glimpse of daylight. Orders came for them to stack muskets, however, and they made their camp on the great range of bare hills which slope downward to the meadows of the Meuse, with the Mouzon road running at their base. To their rear and occupying the level plateau on top of the range the guns of the reserve artillery were arranged in battery, pointed so as to sweep the entrance of the pass should there be necessity for it. And thus commenced another period of agonized, grumbling suspense.

When finally the preparations were all completed the 106th found themselves posted in a field of stubble above the road, in a position that commanded a view of the broad plain. The men had parted regretfully with their arms, casting timorous looks behind them that showed they were apprehensive of a night attack. Their faces were stern and set, and silence reigned, only broken from time to time by some sullen murmur of angry complaint. It was nearly nine o'clock, they had been there two hours, and yet many of them, notwithstanding their terrible fatigue, could not sleep; stretched on the bare ground, they would start and bend their ears to catch the faintest sound that rose in the distance. They had ceased to fight their torturing hunger; they would eat over yonder, on the other bank, when they had passed the river; they would eat grass if nothing else was to be found. The crowd at the bridge, however, seemed to increase rather than diminish; the officers that General Douay had stationed there came back to him every few minutes, always bringing the same unwelcome report, that it would be hours and hours before any relief could be expected. Finally the general determined to go down to the bridge in person, and the men saw him on the bank, bestirring himself and others and hurrying the passage of the troops.

Maurice, seated with Jean against a wall, pointed to the north, as he had done before. "There is Sedan in the distance. And look! Bazeilles is over yonder-and then comes Douzy, and then Carignan, more to the right. We shall concentrate at Carignan, I feel sure we shall. Ah! there is plenty of room, as you would see if it were daylight!"

And his sweeping gesture embraced the entire valley that lay beneath them, enfolded in shadow. There was sufficient light remaining in the sky that they could distinguish the pale gleam of the river where it ran its course among the dusky meadows. The scattered trees made clumps of denser shade, especially a row of poplars to the left, whose tops were profiled on the horizon like the fantastic ornaments on some old castle gateway. And in the background, behind Sedan, dotted with countless little points of brilliant light, the shadows had mustered, denser and darker, as if all the forests of the Ardennes had collected the inky blackness of their secular oaks and cast it there.

Jean's gaze came back to the bridge of boats beneath them.

"Look there! everything is against us. We shall never get across."

The fires upon both banks blazed up more brightly just then, and their light was so intense that the whole fearful scene was pictured on the darkness with vivid distinctness. The boats on which the longitudinal girders rested, owing to the weight of the cavalry and artillery that had been crossing uninterruptedly since morning, had settled to such an extent that the floor of the bridge was covered with water. The cuirassiers were passing at the time, two abreast, in a long unbroken file, emerging from the obscurity of the hither shore to be swallowed up in the shadows of the other, and nothing was to be seen of the bridge; they appeared to be marching on the bosom of the ruddy stream, that flashed and danced in the flickering firelight. The horses snorted and hung back, manifesting every indication of terror as they felt the unstable pathway yielding beneath their feet, and the cuirassiers, standing erect in their stirrups and clutching at the reins, poured onward in a steady, unceasing stream, wrapped in their great white mantles, their helmets flashing in the red light of the flames. One might have taken them for some spectral band of knights, with locks of fire, going forth to do battle with the powers of darkness.

Jean's suffering wrested from him a deep-toned exclamation:

"Oh! I am hungry!"

On every side, meantime, the men, notwithstanding the complainings of their empty stomachs, had thrown themselves down to sleep. Their fatigue was so great that it finally got the better of their fears and struck them down upon the bare earth, where they lay on their back, with open mouth and arms outstretched, like logs beneath the moonless sky. The bustle of the camp was stilled, and all along the naked range, from end to end, there reigned a silence as of death.

"Oh! I am hungry; I am so hungry that I could eat dirt!"

Jean, patient as he was and inured to hardship, could not restrain the cry; he had eaten nothing in thirty-six hours, and it was torn from him by sheer stress of physical suffering. Then Maurice, knowing that two or three hours at all events must elapse before their regiment could move to pass the stream, said:

"See here, I have an uncle not far from here-you know, Uncle Fouchard, of whom you have

90

heard me speak. His house is five or six hundred yards from here; I didn't like the idea, but as you are so hungry- The deuce! the old man can't refuse us bread!"

His comrade made no objection and they went off together. Father Fouchard's little farm was situated just at the mouth of Harancourt pass, near the plateau where the artillery was posted. The house was a low structure, surrounded by quite an imposing cluster of dependencies; a barn, a stable, and cow-sheds, while across the road was a disused carriage-house which the old peasant had converted into an abattoir, where he slaughtered with his own hands the cattle which he afterward carried about the country in his wagon to his customers.

Maurice was surprised as he approached the house to see no light.

"Ah, the old miser! he has locked and barred everything tight and fast. Like as not he won't let us in."

But something that he saw brought him to a standstill. Before the house a dozen soldiers were moving to and fro, hungry plunderers, doubtless, on the prowl in quest of something to eat. First they had called, then had knocked, and now, seeing that the place was dark and deserted, they were hammering at the door with the butts of their muskets in an attempt to force it open. A growling chorus of encouragement greeted them from the outsiders of the circle.

"Nom de Dieu! go ahead! smash it in, since there is no one at home!"

All at once the shutter of a window in the garret was thrown back and a tall old man presented himself, bare-headed, wearing the peasant's blouse, with a candle in one hand and a gun in the other. Beneath the thick shock of bristling white hair was a square face, deeply seamed and wrinkled, with a strong nose, large, pale eyes, and stubborn chin.

"You must be robbers, to smash things as you are doing!" he shouted in an angry tone. "What do you want?"

The soldiers, taken by surprise, drew back a little way.

"We are perishing with hunger; we want something to eat."

"I have nothing, not a crust. Do you suppose that I keep victuals in my house to fill a hundred thousand mouths? Others were here before you; yes, General Ducrot's men were here this morning, I tell you, and they cleaned me out of everything."

The soldiers came forward again, one by one.

"Let us in, all the same; we can rest ourselves, and you can hunt up something-"

And they were commencing to hammer at the door again, when the old fellow, placing his candle on the window-sill, raised his gun to his shoulder.

"As true as that candle stands there, I'll put a hole in the first man that touches that door!"

The prospect looked favorable for a row. Oaths and imprecations resounded, and one of the men was heard to shout that they would settle matters with the pig of a peasant, who was like all the rest of them and would throw his bread in the river rather than give a mouthful to a starving soldier. The light of the candle glinted on the barrels of the chassepots as they were brought to an aim; the angry men were about to shoot him where he stood, while he, headstrong and violent, would not yield an inch.

"Nothing, nothing! Not a crust! I tell you they cleaned me out!"

Maurice rushed in in affright, followed by Jean.

"Comrades, comrades-"

He knocked up the soldiers' guns, and raising his eyes, said entreatingly:

"Come, be reasonable. Don't you know me? It is I."

91

"Who, I?"

"Maurice Levasseur, your nephew."

Father Fouchard took up his candle. He recognized his nephew, beyond a doubt, but was firm in his resolve not to give so much as a glass of water.

"How can I tell whether you are my nephew or not in this infernal darkness? Clear out, everyone of you, or I will fire!"

And amid an uproar of execration, and threats to bring him down and burn the shanty, he still had nothing to say but: "Clear out, or I'll fire!" which he repeated more than twenty times.

Suddenly a loud clear voice was heard rising above the din:

"But not on me, father?"

The others stood aside, and in the flickering light of the candle a man appeared, wearing the chevrons of a quartermaster-sergeant. It was Honore, whose battery was a short two hundred yards from there and who had been struggling for the last two hours against an irresistible longing to come and knock at that door. He had sworn never to set foot in that house again, and in all his four years of army life had not exchanged a single letter with that father whom he now addressed so curtly. The marauders had drawn apart and were conversing excitedly among themselves; what, the old man's son, and a "non-com."! it wouldn't answer; better go and try their luck elsewhere! So they slunk away and vanished in the darkness.

When Fouchard saw that he had nothing more to fear he said in a matter-of-course way, as if he had seen his son only the day before:

"It's you- All right, I'll come down."

His descent was a matter of time. He could be heard inside the house opening locked doors and carefully fastening them again, the maneuvers of a man determined to leave nothing at loose ends. At last the door was opened, but only for a few inches, and the strong grasp that held it would let it go no further.

"Come in, thou! and no one besides!"

He could not turn away his nephew, however, notwithstanding his manifest repugnance.

"Well, thou too!"

He shut the door flat in Jean's face, in spite of Maurice's entreaties. But he was obdurate. No, no! he wouldn't have it; he had no use for strangers and robbers in his house, to smash and destroy his furniture! Finally Honore shoved their comrade inside the door by main strength and the old man had to make the best of it, grumbling and growling vindictively. He had carried his gun with him all this time. When at last he had ushered the three men into the common sitting-room and had stood his gun in a corner and placed the candle on the table, he sank into a mulish silence.

"Say, father, we are perishing with hunger. You will let us have a little bread and cheese, won't you?"

He made a pretense of not hearing and did not answer, turning his head at every instant toward the window as if listening for some other band that might be coming to lay siege to his house.

"Uncle, Jean has been a brother to me; he deprived himself of food to give it to me. And we have seen such suffering together!"

He turned and looked about the room to assure himself that nothing was missing, not giving the three soldiers so much as a glance, and at last, still without a word spoken, appeared to come to a decision. He suddenly arose, took the candle and went out, leaving them in darkness and carefully closing and locking the door behind him in order that no one might follow him. They could hear his footsteps on the stairs that led to the cellar. There was another long period of waiting, and when he returned, again locking and bolting everything after him, he placed upon the table a big loaf of bread and a cheese, amid a silence which, once his anger had blown over, was merely the result of cautious cunning, for no one can ever tell what may come of too much talking. The three men threw themselves ravenously upon the food, and the only sound to be heard in the room was the fierce grinding of their jaws.

Honore rose, and going to the sideboard brought back a pitcher of water.

"I think you might have given us some wine, father."

Whereupon Fouchard, now master of himself and no longer fearing that this anger might lead him into unguarded speech, once more found his tongue.

"Wine! I haven't any, not a drop! The others, those fellows of Ducrot's, ate and drank all I had, robbed me of everything!"

He was lying, and try to conceal it as he might the shifty expression in his great light eyes showed it. For the past two days he had been driving away his cattle, as well those reserved for work on the farm as those he had purchased to slaughter, and hiding them, no one knew where, in the depths of some wood or in some abandoned quarry, and he had devoted hours to burying all his household stores, wine, bread, and things of the least value, even to the flour and salt, so that anyone might have ransacked his cupboards and been none the richer for it. He had refused to sell anything to the first soldiers who came along; no one knew, he might be able to do better later on; and the patient, sly old curmudgeon indulged himself with vague dreams of wealth.

Maurice, who was first to satisfy his appetite, commenced to talk.

"Have you seen my sister Henriette lately?"

The old man was pacing up and down the room, casting an occasional glance at Jean, who was bolting huge mouthfuls of bread; after apparently giving the subject long consideration he deliberately answered:

"Henriette, yes, I saw her last month when I was in Sedan. But I saw Weiss, her husband, this morning. He was with Monsieur Delaherche, his boss, who had come over in his carriage to see the soldiers at Mouzon-which is the same as saying that they were out for a good time."

An expression of intense scorn flitted over the old peasant's impenetrable face.

"Perhaps they saw more of the army than they wanted to, and didn't have such a very good time after all, for ever since three o'clock the roads have been impassable on account of the crowds of flying soldiers."

In the same unmoved voice, as if the matter were one of perfect indifference to him, he gave them some tidings of the defeat of the 5th corps, that had been surprised at Beaumont while the men were making their soup and chased by the Bavarians all the way to Mouzon. Some fugitives who had passed through Remilly, mad with terror, had told him that they had been betrayed once more and that de Failly had sold them to Bismarck. Maurice's thoughts reverted to the aimless, blundering movements of the last two days, to Marshal MacMahon hurrying on their retreat and insisting on getting them across the Meuse at every cost, after wasting so many precious hours in incomprehensible delays. It was too late. Doubtless the marshal, who had stormed so on finding the 7th corps still at Osches when he supposed it to be at la Besace, had felt assured that the 5th corps was safe in camp at Mouzon when, lingering in Beaumont, it had come to grief there. But what could they expect from troops so poorly officered, demoralized by suspense and incessant retreat, dying with hunger and fatigue?

Fouchard had finally come and planted himself behind Jean's chair, watching with astonishment the inroads he was making on the bread and cheese. In a coldly sarcastic tone he asked:

"Are you beginning to feel better, hein?"

The corporal raised his head and replied with the same peasant-like directness:

"Just beginning, thank you!"

Honore, notwithstanding his hunger, had ceased from eating whenever it seemed to him that he heard a noise about the house. If he had struggled long, and finally been false to his oath never to set foot in that house again, the reason was that he could no longer withstand his craving desire to see Silvine. The letter that he had received from her at Rheims lay on his bosom, next his skin, that letter, so tenderly passionate, in which she told him that she loved him still, that she should never love anyone save him, despite the cruel past, despite Goliah and little Charlot, that man's child. He was thinking of naught save her, was wondering why he had not seen her yet, all the time watching himself that he might not let his father see his anxiety. At last his passion became too strong for him, however, and he asked in a tone as natural as he could command:

"Is not Silvine with you any longer?"

Fouchard gave his son a glance out of the corner of his eye, chuckling internally.

"Yes, yes."

Then he expectorated and was silent, so that the artillery man had presently to broach the subject again.

"She has gone to bed, then?"

"No, no."

Finally the old fellow condescended to explain that he, too, had been taking an outing that morning, had driven over to Raucourt market in his wagon and taken his little servant with him. He saw no reason, because a lot of soldiers happened to pass that way, why folks should cease to eat meat or why a man should not attend to his business, so he had taken a sheep and a quarter of beef over there, as it was his custom to do every Tuesday, and had just disposed of the last of his stock-in-trade when up came the 7th corps and he found himself in the middle of a terrible hubbub. Everyone was running, pushing, and crowding. Then he became alarmed lest they should take his horse and wagon from him, and drove off, leaving his servant, who was just then making some purchases in the town.

"Oh, Silvine will come back all right," he concluded in his tranquil voice. "She must have taken shelter with Doctor Dalichamp, her godfather. You would think to look at her that she wouldn't dare to say boo to a goose, but she is a girl of courage, all the same. Yes, yes; she has lots of good qualities, Silvine has."

Was it an attempt on his part to be jocose? or did he wish to explain why it was he kept her in his service, that girl who had caused dissension between father and son, whose child by the Prussian was in the house? He again gave his boy that sidelong look and laughed his voiceless laugh.

"Little Charlot is asleep there in his room; she surely won't be long away, now."

Honore, with quivering lips, looked so intently at his father that the old man began to pace the floor again. Mon Dieu! yes, the child was there; doubtless he would have to look on him. A painful silence filled the room, while he mechanically cut himself more bread and began to eat again. Jean also continued his operations in that line, without finding it necessary to say a word. Maurice contemplated the furniture, the old sideboard, the antique clock, and reflected on the long summer days that he had spent at Remilly in bygone times with his sister Henriette. The minutes slipped away, the clock struck eleven.

"The devil!" he murmured, "it will never do to let the regiment go off without us!"

He stepped to the window and opened it, Fouchard making no objection. Beneath lay the valley, a great bowl filled to the brim with blackness; presently, however, when his eyes became more accustomed to the obscurity, he had no difficulty in distinguishing the bridge, illuminated by the fires on the two banks. The cuirassiers were passing still, like phantoms in their long white cloaks, while their steeds trod upon the bosom of the stream and a chill wind of terror breathed on them from behind; and so the spectral train moved on, apparently interminable, in an endless, slow-moving vision of unsubstantial forms. Toward the right, over the bare hills where the slumbering army lay, there brooded a stillness and repose like death.

"Ah well!" said Maurice with a gesture of disappointment, "'twill be to-morrow morning."

He had left the window open, and Father Fouchard, seizing his gun, straddled the sill and stepped outside, as lightly as a young man. For a time they could hear his tramp upon the road, as regular as that of a sentry pacing his beat, but presently it ceased and the only sound that reached their ears was the distant clamor on the crowded bridge; it must be that he had seated himself by the wayside, where he could watch for approaching danger and at slightest sign leap to defend his property.

Honore's anxiety meantime was momentarily increasing; his eyes were fixed constantly on the clock. It was less than four miles from Raucourt to Remilly, an easy hour's walk for a woman as young and strong as Silvine. Why had she not returned in all that time since the old man lost sight of her in the confusion? He thought of the disorder of a retreating army corps, spreading over the country and blocking the roads; some accident must certainly have happened, and he pictured her in distress, wandering among the lonely fields, trampled under foot by the horsemen.

But suddenly the three men rose to their feet, moved by a common impulse. There was a sound of rapid steps coming up the road and the old man was heard to cock his weapon.

"Who goes there?" he shouted. "Is it you, Silvine?"

There was no reply. He repeated his question, threatening to fire. Then a laboring, breathless voice managed to articulate:

"Yes, yes, Father Fouchard; it is I." And she quickly asked: "And Charlot?"

"He is abed and asleep."

"That is well! Thanks."

There was no longer cause for her to hasten; she gave utterance to a deep-drawn sigh, as if to rid herself of her burden of fatigue and distress.

"Go in by the window," said Fouchard. "There is company in there."

She was greatly agitated when, leaping lightly into the room, she beheld the three men. In the uncertain candle-light she gave the impression of being very dark, with thick black hair and a pair of large, fine, lustrous eyes, the chief adornment of a small oval face, strong by reason of its tranquil resignation. The sudden meeting with Honore had sent all the blood rushing from her heart to her cheeks; and yet she was hardly surprised to find him there; he had been in her thoughts all the way home from Raucourt.

He, trembling with agitation, his heart in his throat, spoke with affected calmness:

"Good-evening, Silvine."

"Good-evening, Honore."

Then, to keep from breaking down and bursting into tears, she turned away, and recognizing Maurice, gave him a smile. Jean's presence was embarrassing to her. She felt as if she were choking somehow, and removed the *foulard* that she wore about her neck.

Honore continued, dropping the friendly *thou* of other days:

"We were anxious about you, Silvine, on account of the Prussians being so near at hand."

All at once her face became very pale and showed great distress; raising her hand to her eyes as if to shut out some atrocious vision, and directing an involuntary glance toward the room where Charlot was slumbering, she murmured:

"The Prussians- Oh! yes, yes, I saw them."

Sinking wearily upon a chair she told how, when the 7th corps came into Raucourt, she had fled for shelter to the house of her godfather, Doctor Dalichamp, hoping that Father Fouchard would think to come and take her up before he left the town. The main street was filled with a surging throng, so dense that not even a dog could have squeezed his way through it, and up to four o'clock she had felt no particular alarm, tranquilly employed in scraping lint in company with some of the ladies of the place; for the doctor, with the thought that they might be called on to care for some of the wounded, should there be a battle over in the direction of Metz and Verdun, had been busying himself for the last two weeks with improvising a hospital in the great hall of the *mairie*. Some people who dropped in remarked that they might find use for their hospital sooner than they expected, and sure enough, a little after midday, the roar of artillery had reached their ears from over Beaumont way. But that was not near enough to cause anxiety and no one was alarmed, when, all at once, just as the last of the French troops were filing out of Raucourt, a shell, with a frightful crash, came tearing through the roof of a neighboring house. Two others followed in quick succession; it was a German battery shelling the rear-guard of the 7th corps. Some of the wounded from Beaumont had already been brought in to the mairie, where it was feared that the enemy's projectiles would finish them as they lay on their mattresses waiting for the doctor to come and operate on them. The men were crazed with fear, and would have risen and gone down into the cellars, notwithstanding their mangled limbs, which extorted from them shrieks of agony.

"And then," continued Silvine, "I don't know how it happened, but all at once the uproar was succeeded by a deathlike stillness. I had gone upstairs and was looking from a window that commanded a view of the street and fields. There was not a soul in sight, not a 'red-leg' to be seen anywhere, when I heard the tramp, tramp of heavy footsteps, and then a voice shouted something that I could not understand and all the muskets came to the ground together with a great crash. And I looked down into the street below, and there was a crowd of small, dirty-looking men in black, with ugly, big faces and wearing helmets like those our firemen wear. Someone told me they were Bavarians. Then I raised my eyes again and saw, oh! thousands and thousands of them, streaming in by the roads, across the fields, through the woods, in serried, never-ending columns. In the twinkling of an eye the ground was black with them, a black swarm, a swarm of black locusts, coming thicker and thicker, so that, in no time at all, the earth was hid from sight."

She shivered and repeated her former gesture, veiling her vision from some atrocious spectacle.

"And the things that occurred afterward would exceed belief. It seems those men had been marching three days, and on top of that had fought at Beaumont like tigers; hence they were perishing with hunger, their eyes were starting from their sockets, they were beside

96

themselves. The officers made no effort to restrain them; they broke into shops and private houses, smashing doors and windows, demolishing furniture, searching for something to eat and drink, no matter what, bolting whatever they could lay their hands on. I saw one in the shop of Monsieur Simonin, the grocer, ladling molasses from a cask with his helmet. Others were chewing strips of raw bacon, others again had filled their mouths with flour. They were told that our troops had been passing through the town for the last two days and there was nothing left, but here and there they found some trifling store that had been hid away, not sufficient to feed so many hungry mouths, and that made them think the folks were lying to them, and they went on to smash things more furiously than ever. In less than an hour, there was not a butcher's, grocer's, or baker's shop in the city left ungutted; even the private houses were entered, their cellars emptied, and their closets pillaged. At the doctor's-did you ever hear of such a thing? I caught one big fellow devouring the soap. But the cellar was the place where they did most mischief; we could hear them from upstairs smashing the bottles and yelling like demons, and they drew the spigots of the casks, so that the place was flooded with wine; when they came out their hands were red with the good wine they had spilled. And to show what happens, men when they make such brutes of themselves: a soldier found a large bottle of laudanum and drank it all down, in spite of Monsieur Dalichamp's efforts to prevent him. The poor wretch was in horrible agony when I came away; he must be dead by this time."

A great shudder ran through her, and she put her hand to her eyes to shut out the horrid sight.

"No, no! I cannot bear it; I saw too much!"

Father Fouchard had crossed the road and stationed himself at the open window where he could hear, and the tale of pillage made him uneasy; he had been told that the Prussians paid for all they took; were they going to start out as robbers at that late day? Maurice and Jean, too, were deeply interested in those details about an enemy whom the girl had seen, and whom they had not succeeded in setting eyes on in their whole month's campaigning, while Honore, pensive and with dry, parched lips, was conscious only of the sound of *her* voice; he could think of nothing save her and the misfortune that had parted them.

Just then the door of the adjoining room was opened, and little Charlot appeared. He had heard his mother's voice, and came trotting into the apartment in his nightgown to give her a kiss. He was a chubby, pink little urchin, large and strong for his age, with a thatch of curling, straw-colored hair and big blue eyes. Silvine shivered at his sudden appearance, as if the sight of him had recalled to her mind the image of someone else that affected her disagreeably. Did she no longer recognize him, then, her darling child, that she looked at him thus, as if he were some evocation of that horrid nightmare! She burst into tears.

"My poor, poor child!" she exclaimed, and clasped him wildly to her breast, while Honore, ghastly pale, noted how strikingly like the little one was to Goliah; the same broad, pink face, the true Teutonic type, in all the health and strength of rosy, smiling childhood. The son of the Prussian, the Prussian, as the pothouse wits of Remilly had styled him! And the French mother, who sat there, pressing him to her bosom, her heart still bleeding with the recollection of the cruel sights she had witnessed that day!

"My poor child, be good; come with me back to bed. Say good-night, my poor child."

She vanished, bearing him away. When she returned from the adjoining room she was no longer weeping; her face wore its customary expression of calm and courageous resignation.

It was Honore who, with a trembling voice, started the conversation again.

"And what did the Prussians do then?"

"Ah, yes; the Prussians. Well, they plundered right and left, destroying everything, eating and drinking all they could lay hands on. They stole linen as well, napkins and sheets, and even curtains, tearing them in strips to make bandages for their feet. I saw some whose feet were one raw lump of flesh, so long and hard had been their march. One little group I saw, seated at the edge of the gutter before the doctor's house, who had taken off their shoes and were bandaging themselves with handsome chemises, trimmed with lace, stolen, doubtless, from pretty Madame Lefevre, the manufacturer's wife. The pillage went on until night. The houses had no doors or windows left, and one passing in the street could look within and see the wrecked furniture, a scene of destruction that would have aroused the anger of a saint. For my part, I was almost wild, and could remain there no longer. They tried in vain to keep me, telling me that the roads were blocked, that I would certainly be killed; I started, and as soon as I was out of Raucourt, struck off to the right and took to the fields. Carts, loaded with wounded French and Prussians, were coming in from Beaumont. Two passed quite close to me in the darkness; I could hear the shrieks and groans, and I ran, oh! how I ran, across fields, through woods, I could not begin to tell you where, except that I made a wide circuit over toward Villers.

"Twice I thought I heard soldiers coming and hid, but the only person I met was another woman, a fugitive like myself. She was from Beaumont, she said, and she told me things too horrible to repeat. After that we ran harder than ever. And at last I am here, so wretched, oh! so wretched with what I have seen!"

Her tears flowed again in such abundance as to choke her utterance. The horrors of the day kept rising to her memory and would not down; she related the story that the woman of Beaumont had told her. That person lived in the main street of the village, where she had witnessed the passage of all the German artillery after nightfall. The column was accompanied on either side of the road by a file of soldiers bearing torches of pitch-pine, which illuminated the scene with the red glare of a great conflagration, and between the flaring, smoking lights the impetuous torrent of horses, guns, and men tore onward at a mad gallop. Their feet were winged with the tireless speed of victory as they rushed on in devilish pursuit of the French, to overtake them in some last ditch and crush them, annihilate them there. They stopped for nothing; on, on they went, heedless of what lay in their way. Horses fell; their traces were immediately cut, and they were left to be ground and torn by the pitiless wheels until they were a shapeless, bleeding mass. Human beings, prisoners and wounded men, who attempted to cross the road, were ruthlessly borne down and shared their fate. Although the men were dying with hunger the fierce hurricane poured on unchecked; was a loaf thrown to the drivers, they caught it flying; the torch-bearers passed slices of meat to them on the end of their bayonets, and then, with the same steel that had served that purpose, goaded their maddened horses on to further effort. And the night grew old, and still the artillery was passing, with the mad roar of a tempest let loose upon the land, amid the frantic cheering of the men.

Maurice's fatigue was too much for him, and notwithstanding the interest with which he listened to Silvine's narrative, after the substantial meal he had eaten he let his head decline upon the table on his crossed arms. Jean's resistance lasted a little longer, but presently he too was overcome and fell dead asleep at the other end of the table. Father Fouchard had gone and taken his position in the road again; Honore was alone with Silvine, who was seated, motionless, before the still open window.

The artilleryman rose, and drawing his chair to the window, stationed himself there beside her. The deep peacefulness of the night was instinct with the breathing of the multitude that lay lost in slumber there, but on it now rose other and louder sounds; the straining and creaking of the bridge, the hollow rumble of wheels; the artillery was crossing on the half-submerged structure. Horses reared and plunged in terror at sight of the swift-running stream, the wheel of a caisson ran over the guard-rail; immediately a hundred strong arms seized the encumbrance and hurled the heavy vehicle to the bottom of the river that it might not obstruct the passage. And as the young man watched the slow, toilsome retreat along the opposite bank, a movement that had commenced the day before and certainly would not be ended by the coming dawn, he could not help thinking of that other artillery that had gone storming through Beaumont, bearing down all before it, crushing men and horses in its path that it might not be delayed the fraction of a second.

Honore drew his chair nearer to Silvine, and in the shuddering darkness, alive with all those sounds of menace, gently whispered:

"You are unhappy?"

"Oh! yes; so unhappy!"

She was conscious of the subject on which he was about to speak, and her head sank sorrowfully on her bosom.

"Tell me, how did it happen? I wish to know."

But she could not find words to answer him.

"Did he take advantage of you, or was it with your consent?"

Then she stammered, in a voice that was barely audible:

"Mon Dieu! I do not know; I swear to you, I do not know, more than a babe unborn. I will not lie to you-I cannot! No, I have no excuse to offer; I cannot say he beat me. You had left me, I was beside myself, and it happened, how, I cannot, no, I cannot tell!"

Sobs choked her utterance, and he, ashy pale and with a great lump rising in his throat, waited silently for a moment. The thought that she was unwilling to tell him a lie, however, was an assuagement to his rage and grief; he went on to question her further, anxious to know the many things, that as yet he had been unable to understand.

"My father has kept you here, it seems?"

She replied with her resigned, courageous air, without raising her eyes:

"I work hard for him, it does not cost much to keep me, and as there is now another mouth to feed he has taken advantage of it to reduce my wages. He knows well enough that now, when he orders, there is nothing left for me but to obey."

"But why do you stay with him?"

The question surprised her so that she looked him in the face.

"Where would you have me go? Here my little one and I have at least a home and enough to keep us from starving."

They were silent again, both intently reading in the other's eyes, while up the shadowy valley the sounds of the sleeping camp came faintly to their ears, and the dull rumble of wheels upon the bridge of boats went on unceasingly. There was a shriek, the loud, despairing cry of man or beast in mortal peril, that passed, unspeakably mournful, through the dark night.

"Listen, Silvine," Honore slowly and feelingly went on; "you sent me a letter that afforded me great pleasure. I should have never come back here, but that letter-I have been reading it again this evening-speaks of things that could not have been expressed more delicately-"

She had turned pale when first she heard the subject mentioned. Perhaps he was angry that she had dared to write to him, like one devoid of shame; then, as his meaning became more clear, her face reddened with delight.

"I know you to be truthful, and knowing it, I believe what you wrote in that letter-yes, I believe it now implicitly. You were right in supposing that, if I were to die in battle without seeing you again, it would be a great sorrow to me to leave this world with the thought that you no longer loved me. And therefore, since you love me still, since I am your first and only love-" His tongue became thick, his emotion was so deep that expression failed him. "Listen, Silvine; if those beasts of Prussians let me live, you shall yet be mine, yes, as soon as I have served my time out we will be married."

She rose and stood erect upon her feet, gave a cry of joy, and threw herself upon the young man's bosom. She could not speak a word; every drop of blood in her veins was in her cheeks. He seated himself upon the chair and drew her down upon his lap.

"I have thought the matter over carefully; it was to say what I have said that I came here this evening. Should my father refuse us his consent, the earth is large; we will go away. And your little one, no one shall harm him, mon Dieu! More will come along, and among them all I shall not know him from the others."

She was forgiven, fully and entirely. Such happiness seemed too great to be true; she resisted, murmuring:

"No, it cannot be; it is too much; perhaps you might repent your generosity some day. But how good it is of you, Honore, and how I love you!"

He silenced her with a kiss upon the lips, and strength was wanting her longer to put aside the great, the unhoped-for good fortune that had come to her; a life of happiness where she had looked forward to one of loneliness and sorrow! With an involuntary, irresistible impulse she threw her arms about him, kissing him again and again, straining him to her bosom with all her woman's strength, as a treasure that was lost and found again, that was hers, hers alone, that thenceforth no one was ever to take from her. He was hers once more, he whom she had lost, and she would die rather than let anyone deprive her of him.

At that moment confused sounds reached their ears; the sleeping camp was awaking amid a tumult that rose and filled the dark vault of heaven. Hoarse voices were shouting orders, bugles were sounding, drums beating, and from the naked fields shadowy forms were seen emerging in indistinguishable masses, a surging, billowing sea whose waves were already streaming downward to the road beneath. The fires on the banks of the stream were dying down; all that could be seen there was masses of men moving confusedly to and fro; it was not even possible to tell if the movement across the river was still in progress. Never had the shades of night veiled such depths of distress, such abject misery of terror.

Father Fouchard came to the window and shouted that the troops were moving. Jean and Maurice awoke, stiff and shivering, and got on their feet. Honore took Silvine's hands in his and gave them a swift parting clasp.

"It is a promise. Wait for me."

She could find no word to say in answer, but all her soul went out to him in one long, last look, as he leaped from the window and hurried away to find his battery.

"Good-by, father!"

"Good-by, my boy!"

And that was all; peasant and soldier parted as they had met, without embracing, like a father and son whose existence was of little import to each other.

Maurice and Jean also left the farmhouse, and descended the steep hill on a run. When they reached the bottom the 106th was nowhere to be found; the regiments had all moved off. They made inquiries, running this way and that, and were directed first one way and then another. At last, when they had near lost their wits in the fearful confusion, they stumbled on their company, under the command of Lieutenant Rochas; as for the regiment and Captain Beaudoin, no one could say where they were. And Maurice was astounded when he noticed for the first time that that mob of men, guns, and horses was leaving Remilly and taking the Sedan road that lay on the left bank. Something was wrong again; the passage of the Meuse was abandoned, they were in full retreat to the north!

An officer of chasseurs, who was standing near, spoke up in a loud voice:

"Nom de Dieu! the time for us to make the movement was the 28th, when we were at Chene!"

Others were more explicit in their information; fresh news had been received. About two o'clock in the morning one of Marshal MacMahon's aides had come riding up to say to General Douay that the whole army was ordered to retreat immediately on Sedan, without loss of a minute's time. The disaster of the 5th corps at Beaumont had involved the three other corps. The general, who was at that time down at the bridge of boats superintending operations, was in despair that only a portion of his 3d division had so far crossed the stream; it would soon be day, and they were liable to be attacked at any moment. He therefore sent instructions to the several organizations of his command to make at once for Sedan, each independently of the others, by the most direct roads, while he himself, leaving orders to burn the bridge of boats, took the road on the left bank with his 2d division and the artillery, and the 3d division pursued that on the right bank; the 1st, that had felt the enemy's claws at Beaumont, was flying in disorder across the country, no one knew where. Of the 7th corps, that had not seen a battle, all that remained were those scattered, incoherent fragments, lost among lanes and by-roads, running away in the darkness.

It was not yet three o'clock, and the night was as black as ever. Maurice, although he knew the country, could not make out where they were in the noisy, surging throng that filled the road from ditch to ditch, pouring onward like a brawling mountain stream. Interspersed among the regiments were many fugitives from the rout at Beaumont, in ragged uniforms, begrimed with blood and dirt, who inoculated the others with their own terror. Down the wide valley, from the wooded hills across the stream, came one universal, all-pervading uproar, the scurrying tramp of other hosts in swift retreat; the 1st corps, coming from Carignan and Douzy, the 12th flying from Mouzon with the shattered remnants of the 5th, moved like puppets and driven onward, all of them, by that one same, inexorable, irresistible pressure that since the 28th had been urging the army northward and driving it into the trap where it was to meet its doom.

Day broke as Maurice's company was passing through Pont Maugis, and then he recognized their locality, the hills of Liry to the left, the Meuse running beside the road on the right. Bazeilles and Balan presented an inexpressibly funereal aspect, looming among the exhalations of the meadows in the chill, wan light of dawn, while against the somber background of her great forests Sedan was profiled in livid outlines, indistinct as the creation of some hideous nightmare. When they had left Wadelincourt behind them and were come at last to the Torcy gate, the governor long refused them admission; he only yielded, after a protracted conference, upon their threat to storm the place. It was five o'clock when at last the 7th corps, weary, cold, and hungry, entered Sedan.

8

In the crush on the Place de Torcy that ensued upon the entrance of the troops into the city Jean became separated from Maurice, and all his attempts to find him again among the surging crowd were fruitless. It was a piece of extreme ill-luck, for he had accepted the young man's invitation to go with him to his sister's, where there would be rest and food for them, and even the luxury of a comfortable bed. The confusion was so great-the regiments disintegrated, no discipline, and no officers to enforce it-that the men were free to do pretty much as they pleased. There was plenty of time to look about them and hunt up their commands; they would have a few hours of sleep first.

Jean in his bewilderment found himself on the viaduct of Torcy, overlooking the broad meadows which, by the governor's orders, had been flooded with water from the river. Then, passing through another archway and crossing the Pont de Meuse, he entered the old, rampart-girt city, where, among the tall and crowded houses and the damp, narrow streets, it seemed to him that night was descending again, notwithstanding the increasing daylight. He could not so much as remember the name of Maurice's brother-in-law; he only knew that his sister's name was Henriette. The outlook was not encouraging; all that kept him awake was the automatic movement of walking; he felt that he should drop were he to stop. The indistinct ringing in his ears was the same that is experienced by one drowning; he was only conscious of the ceaseless onpouring of the stream of men and animals that carried him along with it on its current. He had partaken of food at Remilly, sleep was now his great necessity; and the same was true of the shadowy bands that he saw flitting past him in those strange, fantastic streets. At every moment a man would sink upon the sidewalk or tumble into a doorway, and there would remain, as if struck by death.

Raising his eyes, Jean read upon a signboard: Avenue de la Sous-Prefecture. At the end of the street was a monument standing in a public garden, and at the corner of the avenue he beheld a horseman, a chasseur d'Afrique, whose face seemed familiar to him. Was it not Prosper, the young man from Remilly, whom he had seen in Maurice's company at Vouziers? Perhaps he had been sent in with dispatches. He had dismounted, and his skeleton of a horse, so weak that he could scarcely stand, was trying to satisfy his hunger by gnawing at the tail-board of an army wagon that was drawn up against the curb. There had been no forage for the animals for the last two days, and they were literally dying of starvation. The big strong teeth rasped pitifully on the woodwork of the wagon, while the soldier stood by and wept as he watched the poor brute.

Jean was moving away when it occurred to him that the trooper might be able to give him the address of Maurice's sister. He returned, but the other was gone, and it would have been useless to attempt to find him in that dense throng. He was utterly disheartened, and wandering aimlessly from street to street at last found himself again before the Sous-Prefecture, whence he struggled onward to the Place Turenne. Here he was comforted for an instant by catching sight of Lieutenant Rochas, standing in front of the Hotel de Ville with a few men of his company, at the foot of the statue he had seen before; if he could not find his friend he could at all events rejoin the regiment and have a tent to sleep under. Nothing had been seen of Captain Beaudoin; doubtless he had been swept away in the press and landed in some place far away, while the lieutenant was endeavoring to collect his scattered men and fruitlessly inquiring of everyone he met where division headquarters were. As he advanced into the city, however, his numbers, instead of increasing, dwindled. One man, with the gestures of a lunatic, entered an inn and was seen no more. Three others were halted in front of a grocer's shop by a party of zouaves who had obtained possession of a small cask of brandy; one was already lying senseless in the gutter, while the other two tried to get away, but were too stupid and dazed to move. Loubet and Chouteau had nudged each other with the elbow and disappeared down a blind alley in pursuit of a fat woman with a loaf of bread, so that all who remained with the lieutenant were Pache and Lapoulle, with some ten or a dozen more.

Rochas was standing by the base of the bronze statue of Turenne, making heroic efforts to keep his eyes open. When he recognized Jean he murmured:

"Ah, is it you, corporal? Where are your men?"

Jean, by a gesture expressive in its vagueness, intimated that he did not know, but Pache, pointing to Lapoulle, answered with tears in his eyes:

"Here we are; there are none left but us two. The merciful Lord have pity on our sufferings; it is too hard!"

The other, the colossus with the colossal appetite, looked hungrily at Jean's hands, as if to reproach them for being always empty in those days. Perhaps, in his half-sleeping state, he had dreamed that Jean was away at the commissary's for rations.

"D--n the luck!" he grumbled, "we'll have to tighten up our belts another hole!"

Gaude, the bugler, was leaning against the iron railing, waiting for the lieutenant's order to sound the assembly; sleep came to him so suddenly that he slid from his position and within a second was lying flat on his back, unconscious. One by one they all succumbed to the drowsy influence and snored in concert, except Sergeant Sapin alone, who, with his little pinched nose in his small pale face, stood staring with distended eyes at the horizon of that strange city, as if trying to read his destiny there.

Lieutenant Rochas meantime had yielded to an irresistible impulse and seated himself on the ground. He attempted to give an order.

"Corporal, you will-you will-"

And that was as far as he could proceed, for fatigue sealed his lips, and like the rest he suddenly sank down and was lost in slumber.

Jean, not caring to share his comrades' fate and pillow his head on the hard stones, moved away; he was bent on finding a bed in which to sleep. At a window of the Hotel of the Golden Cross, on the opposite side of the square, he caught a glimpse of General Bourgain-Desfeuilles, already half-undressed and on the point of tasting the luxury of clean white sheets. Why should he be more self-denying than the rest of them? he asked himself; why should he suffer longer? And just then a name came to his recollection that caused him a thrill of delight, the name of the manufacturer in whose employment Maurice's brother-in-law was. M. Delaherche! yes, that was it. He accosted an old man who happened to be passing.

"Can you tell me where M. Delaherche lives?"

"In the Rue Maqua, near the corner of the Rue au Beurre; you can't mistake it; it is a big house, with statues in the garden."

The old man turned away, but presently came running back. "I see you belong to the 106th. If it is your regiment you are looking for, it left the city by the Chateau, down there. I just met the colonel, Monsieur de Vineuil; I used to know him when he lived at Mezieres."

But Jean went his way, with an angry gesture of impatience. No, no! no sleeping on the hard ground for him, now that he was certain of finding Maurice. And yet he could not help feeling a twinge of remorse as he thought of the dignified old colonel, who stood fatigue so manfully in spite of his years, sharing the sufferings of his men, with no more luxurious shelter than his tent. He strode across the Grande Rue with rapid steps and soon was in the midst of the tumult and uproar of the city; there he hailed a small boy, who conducted him to the Rue Maqua.

There it was that in the last century a grand-uncle of the present Delaherche had built the monumental structure that had remained in the family a hundred and sixty years. There is more than one cloth factory in Sedan that dates back to the early years of Louis XV.; enormous piles, they are, covering as much ground as the Louvre, and with stately facades of royal magnificence. The one in the Rue Maqua was three stories high, and its tall windows were adorned with carvings of severe simplicity, while the palatial courtyard in the center was filled with grand old trees, gigantic elms that were coeval with the building itself. In it three generations of Delaherches had amassed comfortable fortunes for themselves. The father of Charles, the proprietor in our time, had inherited the property from a cousin who had died without being blessed with children, so that it was now a younger branch that was in possession. The affairs of the house had prospered under the father's control, but he was something of a blade and a roisterer, and his wife's existence with him was not one of unmixed happiness; the consequence of which was that the lady, when she became a widow, not caring to see a repetition by the son of the performances of the father, made haste to find a wife for him in the person of a simple-minded and exceedingly devout young woman, and subsequently kept him tied to her apron string until he had attained the mature age of fifty and over. But no one in this transitory world can tell what time has in store for him; when the devout young person's time came to leave this life Delaherche, who had known none of the joys of youth, fell head over ears in love with a young widow of Charleville, pretty Madame Maginot, who had been the subject of some gossip in her day, and in the autumn preceding the events recorded in this history had married her, in spite of all his mother's prayers and tears. It is proper to add that Sedan, which is very straitlaced in its notions of propriety, has always been inclined to frown on Charleville, the city of laughter and levity. And then again the marriage would never have been effected but for the fact that Gilberte's uncle was Colonel de Vineuil, who it was supposed would soon be made a general. This relationship and the idea that he had married into army circles was to the cloth manufacturer a source of great delight.

That morning Delaherche, when he learned that the army was to pass through Mouzon, had invited Weiss, his accountant, to accompany him on that carriage ride of which we have heard Father Fouchard speak to Maurice. Tall and stout, with a florid complexion, prominent nose and thick lips, he was of a cheerful, sanguine temperament and had all the French bourgeois' boyish love for a handsome display of troops. Having ascertained from the apothecary at Mouzon that the Emperor was at Baybel, a farm in the vicinity, he had driven up there; had seen the monarch, and even had been near speaking to him, an adventure of such thrilling interest that he had talked of it incessantly ever since his return. But what a terrible return that had been, over roads choked with the panic-stricken fugitives from Beaumont! twenty times their cabriolet was near being overturned into the ditch. Obstacle after obstacle they had encountered, and it was night before the two men reached home. The element of the tragic and unforeseen there was in the whole business, that army that Delaherche had driven out to pass in review and which had brought him home with it, whether he would or no, in the mad gallop of its retreat, made him repeat again and again during their long drive:

"I supposed it was moving on Verdun and would have given anything rather than miss seeing it. Ah well! I have seen it now, and I am afraid we shall see more of it in Sedan than we desire."

The following morning he was awakened at five o'clock by the hubbub, like the roar of water escaping from a broken dam, made by the 7th corps as it streamed through the city; he dressed in haste and went out, and almost the first person he set eyes on in the Place Turenne was Captain Beaudoin. When pretty Madame Maginot was living at Charleville the year before the captain had been one of her best friends, and Gilberte had introduced him to her husband before they were married. Rumor had it that the captain had abdicated his position as first favorite and made way for the cloth merchant from motives of delicacy, not caring to stand in the way of the great good fortune that seemed coming to his fair friend.

"Hallo, is that you?" exclaimed Delaherche. "Good Heavens, what a state you're in!"

It was but too true; the dandified Beaudoin, usually so trim and spruce, presented a sorry spectacle that morning in his soiled uniform and with his grimy face and hands. Greatly to his disgust he had had a party of Turcos for traveling companions, and could not explain how he had become separated from his company. Like all the others he was ready to drop with fatigue and hunger, but that was not what most afflicted him; he had not been able to change his linen since leaving Rheims, and was inconsolable.

"Just think of it!" he wailed, "those idiots, those scoundrels, lost my baggage at Vouziers. If I ever catch them I will break every bone in their body! And now I haven't a thing, not a handkerchief, not a pair of socks! Upon my word, it is enough to make one mad!"

Delaherche was for taking him home to his house forthwith, but he resisted. No, no; he was no longer a human being, he would not frighten people out of their wits. The manufacturer had to make solemn oath that neither his wife nor his mother had risen yet; and besides he should have soap, water, linen, everything he needed.

It was seven o'clock when Captain Beaudoin, having done what he could with the means at his disposal to improve his appearance, and comforted by the sensation of wearing under his uniform a clean shirt of his host's, made his appearance in the spacious, high-ceiled dining room with its somber wainscoting. The elder Madame Delaherche was already there, for she was always on foot at daybreak, notwithstanding she was seventy-eight years old. Her hair was snowy white; in her long, lean face was a nose almost preternaturally thin and sharp and a mouth that had long since forgotten how to laugh. She rose, and with stately politeness invited the captain to be seated before one of the cups of *cafe au lait* that stood on the table.

"But, perhaps, sir, you would prefer meat and wine after the fatigue to which you have been subjected?"

He declined the offer, however. "A thousand thanks, madame; a little milk, with bread and butter, will be best for me."

At that moment a door was smartly opened and Gilberte entered the room with outstretched hand. Delaherche must have told her who was there, for her ordinary hour of rising was ten o'clock. She was tall, lithe of form and well-proportioned, with an abundance of handsome black hair, a pair of handsome black eyes, and a very rosy, wholesome complexion withal; she had a laughing, rather free and easy way with her, and it did not seem possible she could ever look angry. Her peignoir of beige, embroidered with red silk, was evidently of Parisian manufacture.

"Ah, Captain," she rapidly said, shaking hands with the young man, "how nice of you to stop and see us, away up in this out-of-the-world place!" But she was the first to see that she had "put her foot in it" and laugh at her own blunder. "Oh, what a stupid thing I am! I might know you would rather be somewhere else than at Sedan, under the circumstances. But I am very glad to see you once more."

She showed it; her face was bright and animated, while Madame Delaherche, who could not have failed to hear something of the gossip that had been current among the scandalmongers of Charleville, watched the pair closely with her puritanical air. The captain was very reserved in his behavior, however, manifesting nothing more than a pleasant recollection of hospitalities previously received in the house where he was visiting.

They had no more than sat down at table than Delaherche, burning to relieve himself of the subject that filled his mind, commenced to relate his experiences of the day before.

"You know I saw the Emperor at Baybel."

He was fairly started and nothing could stop him. He began by describing the farmhouse, a large structure with an interior court, surrounded by an iron railing, and situated on a gentle eminence overlooking Mouzon, to the left of the Carignan road. Then he came back to the 12th corps, whom he had visited in their camp among the vines on the hillsides; splendid troops they were, with their equipments brightly shining in the sunlight, and the sight of them had caused his heart to beat with patriotic ardor.

"And there I was, sir, when the Emperor, who had alighted to breakfast and rest himself a bit, came out of the farmhouse. He wore a general's uniform and carried an overcoat across his arm, although the sun was very hot. He was followed by a servant bearing a camp stool. He did not look to me like a well man; ah no, far from it; his stooping form, the sallowness of his complexion, the feebleness of his movements, all indicated him to be in a very bad way. I was not surprised, for the druggist at Mouzon, when he recommended me to drive on to Baybel, told me that an aide-de-camp had just been in his shop to get some medicine-you understand what I mean, medicine for-" The presence of his wife and mother prevented him from alluding more explicitly to the nature of the Emperor's complaint, which was an obstinate diarrhea that he had contracted at Chene and which compelled him to make those frequent halts at houses along the road. "Well, then, the attendant opened the camp stool and placed it in the shade of a clump of trees at the edge of a field of wheat, and the Emperor sat down on it. Sitting there in a limp, dejected attitude, perfectly still, he looked for all the world like a small shopkeeper taking a sun bath for his rheumatism. His dull eyes wandered over the wide horizon, the Meuse coursing through the valley at his feet, before him the range of wooded heights whose summits recede and are lost in the distance, on the left the waving tree-tops of Dieulet forest, on the right the verdure-clad eminence of Sommanthe. He was surrounded by his military family, aides and officers of rank, and a colonel of dragoons, who had already applied to me for information about the country, had just motioned me not to go away, when all at once-" Delaherche rose from his chair, for he had reached the point where the dramatic interest of his story culminated and it became necessary to re-enforce words by gestures. "All at once there is a succession of sharp reports and right in front of us, over the wood of Dieulet, shells are seen circling through the air. It produced on me no more effect than a display of fireworks in broad daylight, sir, upon my word it didn't! The people about the Emperor, of course, showed a good deal of agitation and uneasiness. The colonel of dragoons comes running up again to ask if I can give them an idea whence the firing proceeds. I answer him off-hand: 'It is at Beaumont; there is not the slightest doubt about it.' He returns to the Emperor, on whose knees an aide-de-camp was unfolding a map. The Emperor was evidently of opinion that the fighting was not at Beaumont, for he sent the colonel back to me a third time. But I couldn't well do otherwise than stick to what I had said before, could I, now? the more that the shells kept flying through the air, nearer and nearer, following the line of the Mouzon road. And then, sir, as sure as I see you standing there, I saw the Emperor turn his pale face toward me. Yes sir, he looked at me a moment with those dim eyes of his, that were filled with an expression of melancholy and distrust. And then his face declined upon his map again and he made no further movement."

Delaherche, although he was an ardent Bonapartist at the time of the plebiscite, had admitted after our early defeats that the government was responsible for some mistakes, but he stood up

106

for the dynasty, compassionating and excusing Napoleon III., deceived and betrayed as he was by everyone. It was his firm opinion that the men at whose door should be laid the responsibility for all our disasters were none other than those Republican deputies of the opposition who had stood in the way of voting the necessary men and money.

"And did the Emperor return to the farmhouse?" asked Captain Beaudoin.

"That's more than I can say, my dear sir; I left him sitting on his stool. It was midday, the battle was drawing nearer, and it occurred to me that it was time to be thinking of my own return. All that I can tell you besides is that a general to whom I pointed out the position of Carignan in the distance, in the plain to our rear, appeared greatly surprised to learn that the Belgian frontier lay in that direction and was only a few miles away. Ah, that the poor Emperor should have to rely on such servants!"

Gilberte, all smiles, was giving her attention to the captain and keeping him supplied with buttered toast, as much at ease as she had ever been in bygone days when she received him in her salon during her widowhood. She insisted that he should accept a bed with them, but he declined, and it was agreed that he should rest for an hour or two on a sofa in Delaherche's study before going out to find his regiment. As he was taking the sugar bowl from the young woman's hands old Madame Delaherche, who had kept her eye on them, distinctly saw him squeeze her fingers, and the old lady's suspicions were confirmed. At that moment a servant came to the door.

"Monsieur, there is a soldier outside who wants to know the address of Monsieur Weiss."

There was nothing "stuck-up" about Delaherche, people said; he was fond of popularity and was always delighted to have a chat with those of an inferior station.

"He wants Weiss's address! that's odd. Bring the soldier in here."

Jean entered the room in such an exhausted state that he reeled as if he had been drunk. He started at seeing his captain seated at the table with two ladies, and involuntarily withdrew the hand that he had extended toward a chair in order to steady himself; he replied briefly to the questions of the manufacturer, who played his part of the soldier's friend with great cordiality. In a few words he explained his relation toward Maurice and the reason why he was looking for him.

"He is a corporal in my company," the captain finally said by way of cutting short the conversation, and inaugurated a series of questions on his own account to learn what had become of the regiment. As Jean went on to tell that the colonel had been seen crossing the city to reach his camp at the head of what few men were left him, Gilberte again thoughtlessly spoke up, with the vivacity of a woman whose beauty is supposed to atone for her indiscretion:

"Oh! he is my uncle; why does he not come and breakfast with us? We could fix up a room for him here. Can't we send someone for him?"

But the old lady discouraged the project with an authority there was no disputing. The good old bourgeois blood of the frontier towns flowed in her veins; her austerely patriotic sentiments were almost those of a man. She broke the stern silence that she had preserved during the meal by saying:

"Never mind Monsieur de Vineuil; he is doing his duty."

Her short speech was productive of embarrassment among the party. Delaherche conducted the captain to his study, where he saw him safely bestowed upon the sofa; Gilberte moved lightly off about her business, no more disconcerted by her rebuff than is the bird that shakes its wings in gay defiance of the shower; while the handmaid to whom Jean had been intrusted led him by a very labyrinth of passages and staircases through the various departments of the

107

factory.

The Weiss family lived in the Rue des Voyards, but their house, which was Delaherche's property, communicated with the great structure in the Rue Maqua. The Rue des Voyards was at that time one of the most squalid streets in Sedan, being nothing more than a damp, narrow lane, its normal darkness intensified by the proximity of the ramparts, which ran parallel to it. The roofs of the tall houses almost met, the dark passages were like the mouths of caverns, and more particularly so at that end where rose the high college walls. Weiss, however, with free quarters and free fuel on his third floor, found the location a convenient one on account of its nearness to his office, to which he could descend in slippers without having to go around by the street. His life had been a happy one since his marriage with Henriette, so long the object of his hopes and wishes since first he came to know her at Chene, filling her dead mother's place when only six years old and keeping the house for her father, the tax-collector; while he, entering the big refinery almost on the footing of a laborer, was picking up an education as best he could, and fitting himself for the accountant's position which was the reward of his unremitting toil. And even when he had attained to that measure of success his dream was not to be realized; not until the father had been removed by death, not until the brother at Paris had been guilty of those excesses: that brother Maurice to whom his twin sister had in some sort made herself a servant, to whom she had sacrificed her little all to make him a gentleman-not until then was Henriette to be his wife. She had never been aught more than a little drudge at home; she could barely read and write; she had sold house, furniture, all she had, to pay the young man's debts, when good, kind Weiss came to her with the offer of his savings, together with his heart and his two strong arms; and she had accepted him with grateful tears, bringing him in return for his devotion a steadfast, virtuous affection, replete with tender esteem, if not the stormier ardors of a passionate love. Fortune had smiled on them; Delaherche had spoken of giving Weiss an interest in the business, and when children should come to bless their union their felicity would be complete.

"Look out!" the servant said to Jean; "the stairs are steep."

He was stumbling upward as well as the intense darkness of the place would let him, when suddenly a door above was thrown open, a broad belt of light streamed out across the landing, and he heard a soft voice saying:

"It is he."

"Madame Weiss," cried the servant, "here is a soldier who has been inquiring for you."

There came the sound of a low, pleased laugh, and the same soft voice replied:

"Good! good! I know who it is." Then to the corporal, who was hesitating, rather diffidently, on the landing: "Come in, Monsieur Jean. Maurice has been here nearly two hours, and we have been wondering what detained you."

Then, in the pale sunlight that filled the room, he saw how like she was to Maurice, with that wonderful resemblance that often makes twins so like each other as to be indistinguishable. She was smaller and slighter than he, however; more fragile in appearance, with a rather large mouth and delicately molded features, surmounted by an opulence of the most beautiful hair imaginable, of the golden yellow of ripened grain. The feature where she least resembled him was her gray eyes, great calm, brave orbs, instinct with the spirit of the grandfather, the hero of the Grand Army. She used few words, was noiseless in her movements, and was so gentle, so cheerful, so helpfully active that where she passed her presence seemed to linger in the air, like a fragrant caress.

"Come this way, Monsieur Jean," she said. "Everything will soon be ready for you."

He stammered something inarticulately, for his emotion was such that he could find no word of thanks. In addition to that his eyes were closing he beheld her through the irresistible drowsiness that was settling on him as a sea-fog drifts in and settles on the land, in which she seemed floating in a vague, unreal way, as if her feet no longer touched the earth. Could it be that it was all a delightful apparition, that friendly young woman who smiled on him with such sweet simplicity? He fancied for a moment that she had touched his hand and that he had felt the pressure of hers, cool and firm, loyal as the clasp of an old tried friend.

That was the last moment in which Jean was distinctly conscious of what was going on about him. They were in the dining room; bread and meat were set out on the table, but for the life of him he could not have raised a morsel to his lips. A man was there, seated on a chair. Presently he knew it was Weiss, whom he had seen at Mulhausen, but he had no idea what the man was saying with such a sober, sorrowful air, with slow and emphatic gestures. Maurice was already sound asleep, with the tranquillity of death resting on his face, on a bed that had been improvised for him beside the stove, and Henriette was busying herself about a sofa on which a mattress had been thrown; she brought in a bolster, pillow and coverings; with nimble, dexterous hands she spread the white sheets, snowy white, dazzling in their whiteness.

Ah! those clean, white sheets, so long coveted, so ardently desired; Jean had eyes for naught save them. For six weeks he had not had his clothes off, had not slept in a bed. He was as impatient as a child waiting for some promised treat, or a lover expectant of his mistress's coming; the time seemed long, terribly long to him, until he could plunge into those cool, white depths and lose himself there. Quickly, as soon as he was alone, he removed his shoes and tossed his uniform across a chair, then, with a deep sigh of satisfaction, threw himself on the bed. He opened his eyes a little way for a last look about him before his final plunge into unconsciousness, and in the pale morning light that streamed in through the lofty window beheld a repetition of his former pleasant vision, only fainter, more aerial; a vision of Henriette entering the room on tiptoe, and placing on the table at his side a water-jug and glass that had been forgotten before. She seemed to linger there a moment, looking at the sleeping pair, him and her brother, with her tranquil, ineffably tender smile upon her lips, then faded into air, and he, between his white sheets, was as if he were not.

Hours-or was it years? slipped by; Jean and Maurice were like dead men, without a dream, without consciousness of the life that was within them. Whether it was ten years or ten minutes, time had stood still for them; the overtaxed body had risen against its oppressor and annihilated their every faculty. They awoke simultaneously with a great start and looked at each other inquiringly; where were they? what had happened? how long had they slept? The same pale light was entering through the tall window. They felt as if they had been racked; joints stiffer, limbs wearier, mouth more hot and dry than when they had lain down; they could not have slept more than an hour, fortunately. It did not surprise them to see Weiss sitting where they had seen him before, in the same dejected attitude, apparently waiting for them to awake.

"Fichtre!" exclaimed Jean, "we must get up and report ourselves to the first sergeant before noon."

He uttered a smothered cry of pain as he jumped to the floor and began to dress.

"Before noon!" said Weiss. "Are you aware that it is seven o'clock in the evening? You have slept about twelve hours."

Great heavens, seven o'clock! They were thunderstruck. Jean, who by that time was completely dressed, would have run for it, but Maurice, still in bed, found he no longer had control of his legs; how were they ever to find their comrades? would not the army have marched away? They took Weiss to task for having let them sleep so long. But the accountant shook his head sorrowfully and said:

"You have done just as well to remain in bed, for all that has been accomplished."

All that day, from early morning, he had been scouring Sedan and its environs in quest of news, and was just come in, discouraged with the inactivity of the troops and the inexplicable delay that had lost them the whole of that precious day, the 31st. The sole excuse was that the men were worn out and rest was an absolute necessity for them, but granting that, he could not see why the retreat should not have been continued after giving them a few hours of repose.

"I do not pretend to be a judge of such matters," he continued, "but I have a feeling, so strong as to be almost a conviction, that the army is very badly situated at Sedan. The 12th corps is at Bazeilles, where there was a little fighting this morning; the 1st is strung out along the Givonne between la Moncelle and Holly, while the 7th is encamped on the plateau of Floing, and the 5th, what is left of it, is crowded together under the ramparts of the city, on the side of the Chateau. And that is what alarms me, to see them all concentrated thus about the city, waiting for the coming of the Prussians. If I were in command I would retreat on Mezieres, and lose no time about it, either. I know the country; it is the only line of retreat that is open to us, and if we take any other course we shall be driven into Belgium. Come here! let me show you something."

He took Jean by the hand and led him to the window.

"Tell me what you see over yonder on the crest of the hills."

Looking from the window over the ramparts, over the adjacent buildings, their view embraced the valley of the Meuse to the southward of Sedan. There was the river, winding through broad meadows; there, to the left, was Remilly in the background, Pont Maugis and Wadelincourt before them and Frenois to the right; and shutting in the landscape the ranges of verdant hills, Liry first, then la Marfee and la Croix Piau, with their dense forests. A deep tranquillity, a crystalline clearness reigned over the wide prospect that lay there in the mellow light of the declining day.

"Do you see that moving line of black upon the hilltops, that procession of small black ants?"

Jean stared in amazement, while Maurice, kneeling on his bed, craned his neck to see.

"Yes, yes!" they cried. "There is a line, there is another, and another, and another! They are everywhere."

"Well," continued Weiss, "those are Prussians. I have been watching them since morning, and they have been coming, coming, as if there were no end to them! You may be sure of one thing: if our troops are waiting for them, they have no intention of disappointing us. And not I alone, but every soul in the city saw them; it is only the generals who persist in being blind. I was talking with a general officer a little while ago; he shrugged his shoulders and told me that Marshal MacMahon was absolutely certain that he had not over seventy thousand men in his front. God grant he may be right! But look and see for yourselves; the ground is hid by them! they keep coming, ever coming, the black swarm!"

At this juncture Maurice threw himself back in his bed and gave way to a violent fit of sobbing. Henriette came in, a smile on her face. She hastened to him in alarm.

"What is it?"

But he pushed her away. "No, no! leave me, have nothing more to do with me; I have never been anything but a burden to you. When I think that you were making yourself a drudge, a slave, while I was attending college-oh! to what miserable use have I turned that education! And I was near bringing dishonor on our name; I shudder to think where I might be now, had you not beggared yourself to pay for my extravagance and folly."

Her smile came back to her face, together with her serenity.

"Is that all? Your sleep don't seem to have done you good, my poor friend. But since that is all gone and past, forget it! Are you not doing your duty now, like a good Frenchman? I am very proud of you, I assure you, now that you are a soldier."

She had turned toward Jean, as if to ask him to come to her assistance, and he looked at her with some surprise that she appeared to him less beautiful than yesterday; she was paler, thinner, now that the glamour was no longer in his drowsy eyes. The one striking point that remained unchanged was her resemblance to her brother, and yet the difference in their two natures was never more strongly marked than at that moment; he, weak and nervous as a woman, swayed by the impulse of the hour, displaying in his person all the fitful and emotional temperament of his nation, vibrating from one moment to another between the loftiest enthusiasm and the most abject despair; she, the patient, indomitable housewife, such an inconsiderable little creature in her resignation and self-effacement, meeting adversity with a brave face and eyes full of inexpugnable courage and resolution, fashioned from the stuff of which heroes are made.

"Proud of me!" cried Maurice. "Ah! truly, you have great reason to be. For a month and more now we have been flying, like the cowards that we are!"

"What of it? we are not the only ones," said Jean with his practical common sense; "we do what we are told to do."

But the young man broke out more furiously than ever: "I have had enough of it, I tell you! Our imbecile leaders, our continual defeats, our brave soldiers led like sheep to the slaughter- is it not enough, seeing all these things, to make one weep tears of blood? We are here now in Sedan, caught in a trap from which there is no escape; you can see the Prussians closing in on us from every quarter, and certain destruction is staring us in the face; there is no hope, the end is come. No! I shall remain where I am; I may as well be shot as a deserter. Jean, do you go, and leave me here. No! I won't go back there; I will stay here."

He sank upon the pillow in a renewed outpour of tears. It was an utter breakdown of the nervous system, sweeping everything before it, one of those sudden lapses into hopelessness to which he was so subject, in which he despised himself and all the world. His sister, knowing as she did the best way of treating such crises, kept an unruffled face.

"That would not be a nice thing to do, dear Maurice-desert your post in the hour of danger."

He rose impetuously to a sitting posture: "Then give me my musket! I will go and blow my brains out; that will be the shortest way of ending it." Then, pointing with outstretched arm to Weiss, where he sat silent and motionless, he said: "There! that is the only sensible man I have seen; yes, he is the only one who saw things as they were. You remember what he said to me, Jean, at Mulhausen, a month ago?"

"It is true," the corporal assented; "the gentleman said we should be beaten."

And the scene rose again before their mind's eye, that night of anxious vigil, the agonized suspense, the prescience of the disaster at Froeschwiller hanging in the sultry heavy air, while the Alsatian told his prophetic fears; Germany in readiness, with the best of arms and the best of leaders, rising to a man in a grand outburst of patriotism; France dazed, a century behind the age, debauched, and a prey to intestine disorder, having neither commanders, men, nor arms to enable her to cope with her powerful adversary. How quickly the horrible prediction had proved itself true!

Weiss raised his trembling hands. Profound sorrow was depicted on his kind, honest face, with its red hair and beard and its great prominent blue eyes.

111

"Ah!" he murmured, "I take no credit to myself for being right. I don't claim to be wiser than others, but it was all so clear, when one only knew the true condition of affairs! But if we are to be beaten we shall first have the pleasure of killing some of those Prussians of perdition. There is that comfort for us; I believe that many of us are to leave their bones there, and I hope there will be plenty of Prussians to keep them company; I would like to see the ground down there in the valley heaped with dead Prussians!" He arose and pointed down the valley of the Meuse. Fire flashed from his myopic eyes, which had exempted him from service with the army. "A thousand thunders! I would fight, yes, I would, if they would have me. I don't know whether it is seeing them assume the airs of masters in my country-in this country where once the Cossacks did such mischief; but whenever I think of their being here, of their entering our houses, I am seized with an uncontrollable desire to cut a dozen of their throats. Ah! if it were not for my eyes, if they would take me, I would go!" Then, after a moment's silence: "And besides; who can tell?"

It was the hope that sprang eternal, even in the breast of the least confident, of the possibility of victory, and Maurice, ashamed by this time of his tears, listened and caught at the pleasing speculation. Was it not true that only the day before there had been a rumor that Bazaine was at Verdun? Truly, it was time that Fortune should work a miracle for that France whose glories she had so long protected. Henriette, with an imperceptible smile on her lips, silently left the room, and was not the least bit surprised when she returned to find her brother up and dressed, and ready to go back to his duty. She insisted, however, that he and Jean should take some nourishment first. They seated themselves at the table, but the morsels choked them; their stomachs, weakened by their heavy slumber, revolted at the food. Like a prudent old campaigner Jean cut a loaf in two halves and placed one in Maurice's sack, the other in his own. It was growing dark, it behooved them to be going. Henriette, who was standing at the window watching the Prussian troops incessantly defiling on distant la Marfee, the swarming legions of black ants that were gradually being swallowed up in the gathering shadows, involuntarily murmured:

"Oh, war! what a dreadful thing it is!"

Maurice, seeing an opportunity to retort her sermon to him, immediately took her up:

"How is this, little sister? you are anxious to have people fight, and you speak disrespectfully of war!"

She turned and faced him, valiantly as ever: "It is true; I abhor it, because it is an abomination and an injustice. It may be simply because I am a woman, but the thought of such butchery sickens me. Why cannot nations adjust their differences without shedding blood?"

Jean, the good fellow, seconded her with a nod of the head, and nothing to him, too, seemed easier-to him, the unlettered man-than to come together and settle matters after a fair, honest talk; but Maurice, mindful of his scientific theories, reflected on the necessity of war-war, which is itself existence, the universal law. Was it not poor, pitiful man who conceived the idea of justice and peace, while impassive nature revels in continual slaughter?

"That is all very fine!" he cried. "Yes, centuries hence, if it shall come to pass that then all the nations shall be merged in one; centuries hence man may look forward to the coming of that golden age; and even in that case would not the end of war be the end of humanity? I was a fool but now; we must go and fight, since it is nature's law." He smiled and repeated his brother-in-law's expression: "And besides, who can tell?"

He saw things now through the mirage of his vivid self-delusion, they came to his vision distorted through the lens of his diseased nervous sensibility.

"By the way," he continued cheerfully, "what do you hear of our cousin Gunther? You know

112

we have not seen a German yet, so you can't look to me to give you any foreign news."

The question was addressed to his brother-in-law, who had relapsed into a thoughtful silence and answered by a motion of his hand, expressive of his ignorance.

"Cousin Gunther?" said Henriette, "Why, he belongs to the Vth corps and is with the Crown Prince's army; I read it in one of the newspapers, I don't remember which. Is that army in this neighborhood?"

Weiss repeated his gesture, which was imitated by the two soldiers, who could not be supposed to know what enemies were in front of them when their generals did not know. Rising to his feet, the master of the house at last made use of articulate speech.

"Come along; I will go with you. I learned this afternoon where the 106th's camp is situated." He told his wife that she need not expect to see him again that night, as he would sleep at Bazeilles, where they had recently bought and furnished a little place to serve them as a residence during the hot months. It was near a dyehouse that belonged to M. Delaherche. The accountant's mind was ill at ease in relation to certain stores that he had placed in the cellar-a cask of wine and a couple of sacks of potatoes; the house would certainly be visited by marauders if it was left unprotected, he said, while by occupying it that night he would doubtless save it from pillage. His wife watched him closely while he was speaking.

"You need not be alarmed," he added, with a smile; "I harbor no darker design than the protection of our property, and I pledge my word that if the village is attacked, or if there is any appearance of danger, I will come home at once."

"Well, then, go," she said. "But remember, if you are not back in good season you will see me out there looking for you."

Henriette went with them to the door, where she embraced Maurice tenderly and gave Jean a warm clasp of the hand.

"I intrust my brother to your care once more. He has told me of your kindness to him, and I love you for it."

He was too flustered to do more than return the pressure of the small, firm hand. His first impression returned to him again, and he beheld Henriette in the light in which she had first appeared to him, with her bright hair of the hue of ripe golden grain, so alert, so sunny, so unselfish, that her presence seemed to pervade the air like a caress.

Once they were outside they found the same gloomy and forbidding Sedan that had greeted their eyes that morning. Twilight with its shadows had invaded the narrow streets, sidewalk and carriage-way alike were filled with a confused, surging throng. Most of the shops were closed, the houses seemed to be dead or sleeping, while out of doors the crowd was so dense that men trod on one another. With some little difficulty, however, they succeeded in reaching the Place de l'Hotel de Ville, where they encountered M. Delaherche, intent on picking up the latest news and seeing what was to be seen. He at once came up and greeted them, apparently delighted to meet Maurice, to whom he said that he had just returned from accompanying Captain Beaudoin over to Floing, where the regiment was posted, and he became, if that were possible, even more gracious than ever upon learning that Weiss proposed to pass the night at Bazeilles, where he himself, he declared, had just been telling the captain that he intended to take a bed, in order to see how things were looking at the dyehouse.

"We'll go together and be company for each other, Weiss. But first let's go as far as the Sous-Prefecture; we may be able to catch a glimpse of the Emperor."

Ever since he had been so near having the famous conversation with him at Baybel his mind had been full of Napoleon III.; he was not satisfied until he had induced the two soldiers to accompany him. The Place de la Sous-Prefecture was comparatively empty; a few men were standing about in groups, engaged in whispered conversation, while occasionally an officer hurried by, haggard and careworn. The bright hues of the foliage were beginning to fade and grow dim in the melancholy, thick-gathering shades of night; the hoarse murmur of the Meuse was heard as its current poured onward beneath the houses to the right. Among the whisperers it was related how the Emperor-who with the greatest difficulty had been prevailed on to leave Carignan the night before about eleven o'clock-when entreated to push on to Mezieres had refused point-blank to abandon the post of danger and take a step that would prove so demoralizing to the troops. Others asserted that he was no longer in the city, that he had fled, leaving behind him a dummy emperor, one of his officers dressed in his uniform, a man whose startling resemblance to his imperial master had often puzzled the army. Others again declared, and called upon their honor to substantiate their story, that they had seen the army wagons containing the imperial treasure, one hundred millions, all in brand-new twenty-franc pieces, drive into the courtyard of the Prefecture. This convoy was, in fact, neither more nor less than the vehicles for the personal use of the Emperor and his suite, the char a banc, the two caleches, the twelve baggage and supply wagons, which had almost excited a riot in the villages through which they had passed-Courcelles, le Chene, Raucourt; assuming in men's imagination the dimensions of a huge train that had blocked the road and arrested the march of armies, and which now, shorn of their glory, execrated by all, had come in shame and disgrace to hide themselves among the sous-prefect's lilac bushes.

While Delaherche was raising himself on tiptoe and trying to peer through the windows of the rez-de-chaussee, an old woman at his side, some poor day-worker of the neighborhood, with shapeless form and hands calloused and distorted by many years of toil, was mumbling between her teeth:

"An emperor-I should like to see one once-just once-so I could say I had seen him."

Suddenly Delaherche exclaimed, seizing Maurice by the arm:

"See, there he is! at the window, to the left. I had a good view of him yesterday; I can't be mistaken. There, he has just raised the curtain; see, that pale face, close to the glass."

The old woman had overheard him and stood staring with wide-open mouth and eyes, for there, full in the window, was an apparition that resembled a corpse more than a living being; its eyes were lifeless, its features distorted; even the mustache had assumed a ghastly whiteness in that final agony. The old woman was dumfounded; forthwith she turned her back and marched off with a look of supreme contempt.

"That thing an emperor! a likely story."

A zouave was standing near, one of those fugitive soldiers who were in no haste to rejoin their commands. Brandishing his chassepot and expectorating threats and maledictions, he said to his companion:

"Wait! see me put a bullet in his head!"

Delaherche remonstrated angrily, but by that time the Emperor had disappeared. The hoarse murmur of the Meuse continued uninterruptedly; a wailing lament, inexpressibly mournful, seemed to pass above them through the air, where the darkness was gathering intensity. Other sounds rose in the distance, like the hollow muttering of the rising storm; were they the "March! march!" that terrible order from Paris that had driven that ill-starred man onward day by day, dragging behind him along the roads of his defeat the irony of his imperial escort, until now he was brought face to face with the ruin he had foreseen and come forth to meet? What multitudes of brave men were to lay down their lives for his mistakes, and how complete the wreck, in all his being, of that sick man, that sentimental dreamer, awaiting in gloomy silence the fulfillment of his destiny!

Weiss and Delaherche accompanied the two soldiers to the plateau of Floing, where the 7th corps camps were.

"Adieu!" said Maurice as he embraced his brother-in-law.

"No, no; not adieu, the deuce! Au revoir!" the manufacturer gayly cried.

Jean's instinct led him at once to their regiment, the tents of which were pitched behind the cemetery, where the ground of the plateau begins to fall away. It was nearly dark, but there was sufficient light yet remaining in the sky to enable them to distinguish the black huddle of roofs above the city, and further in the distance Balan and Bazeilles, lying in the broad meadows that stretch away to the range of hills between Remilly and Frenois, while to the right was the dusky wood of la Garenne, and to the left the broad bosom of the Meuse had the dull gleam of frosted silver in the dying daylight. Maurice surveyed the broad landscape that was momentarily fading in the descending shadows.

"Ah, here is the corporal!" said Chouteau. "I wonder if he has been looking after our rations!"

The camp was astir with life and bustle. All day the men had been coming in, singly and in little groups, and the crowd and confusion were such that the officers made no pretense of punishing or even reprimanding them; they accepted thankfully those who were so kind as to return and asked no questions. Captain Beaudoin had made his appearance only a short time before, and it was about two o'clock when Lieutenant Rochas had brought in his collection of stragglers, about one-third of the company strength. Now the ranks were nearly full once more. Some of the men were drunk, others had not been able to secure even a morsel of bread and were sinking from inanition; again there had been no distribution of rations. Loubet, however, had discovered some cabbages in a neighboring garden, and cooked them after a fashion, but there was no salt or lard; the empty stomachs continued to assert their claims.

"Come, now, corporal, you are a knowing old file," Chouteau tauntingly continued, "what have you got for us? Oh, it's not for myself I care; Loubet and I had a good breakfast; a lady gave it us. You were not at distribution, then?"

Jean beheld a circle of expectant eyes bent on him; the squad had been waiting for him with anxiety, Pache and Lapoulle in particular, luckless dogs, who had found nothing they could appropriate; they all relied on him, who, as they expressed it, could get bread out of a stone. And the corporal's conscience smote him for having abandoned his men; he took pity on them and divided among them half the bread that he had in his sack.

"Name o' God! Name o' God!" grunted Lapoulle as he contentedly munched the dry bread; it was all he could find to say; while Pache repeated a *Pater* and an *Ave* under his breath to make sure that Heaven should not forget to send him his breakfast in the morning.

Gaude, the bugler, with his darkly mysterious air, as of a man who has had troubles of which he does not care to speak, sounded the call for evening muster with a glorious fanfare; but there was no necessity for sounding taps that night, the camp was immediately enveloped in profound silence. And when he had verified the names and seen that none of his half-section were missing, Sergeant Sapin, with his thin, sickly face and his pinched nose, softly said:

"There will be one less to-morrow night."

Then, as he saw Jean looking at him inquiringly, he added with calm conviction, his eyes bent upon the blackness of the night, as if reading there the destiny that he predicted:

"It will be mine; I shall be killed to-morrow."

It was nine o'clock, with promise of a chilly, uncomfortable night, for a dense mist had risen from the surface of the river, so that the stars were no longer visible. Maurice shivered, where he lay with Jean beneath a hedge, and said they would do better to go and seek the shelter of

115

the tent; the rest they had taken that day had left them wakeful, their joints seemed stiffer and their bones sorer than before; neither could sleep. They envied Lieutenant Rochas, who, stretched on the damp ground and wrapped in his blanket, was snoring like a trooper, not far away. For a long time after that they watched with interest the feeble light of a candle that was burning in a large tent where the colonel and some officers were in consultation. All that evening M. de Vineuil had manifested great uneasiness that he had received no instructions to guide him in the morning. He felt that his regiment was too much "in the air," too much advanced, although it had already fallen back from the exposed position that it had occupied earlier in the day. Nothing had been seen of General Bourgain-Desfeuilles, who was said to be ill in bed at the Hotel of the Golden Cross, and the colonel decided to send one of his officers to advise him of the danger of their new position in the too extended line of the 7th corps, which had to cover the long stretch from the bend in the Meuse to the wood of la Garenne. There could be no doubt that the enemy would attack with the first glimpse of daylight; only for seven or eight hours now would that deep tranquillity remain unbroken. And shortly after the dim light in the colonel's tent was extinguished Maurice was amazed to see Captain Beaudoin glide by, keeping close to the hedge, with furtive steps, and vanish in the direction of Sedan.

The darkness settled down on them, denser and denser; the chill mists rose from the stream and enshrouded everything in a dank, noisome fog.

"Are you asleep, Jean?"

Jean was asleep, and Maurice was alone. He could not endure the thought of going to the tent where Lapoulle and the rest of them were slumbering; he heard their snoring, responsive to Rochas' strains, and envied them. If our great captains sleep soundly the night before a battle, it is like enough for the reason that their fatigue will not let them do otherwise. He was conscious of no sound save the equal, deep-drawn breathing of that slumbering multitude, rising from the darkening camp like the gentle respiration of some huge monster; beyond that all was void. He only knew that the 5th corps was close at hand, encamped beneath the rampart, that the 1st's line extended from the wood of la Garenne to la Moncelle, while the 12th was posted on the other side of the city, at Bazeilles; and all were sleeping; the whole length of that long line, from the nearest tent to the most remote, for miles and miles, that low, faint murmur ascended in rhythmic unison from the dark, mysterious bosom of the night. Then outside this circle lay another region, the realm of the unknown, whence also sounds came intermittently to his ears, so vague, so distant, that he scarcely knew whether they were not the throbbings of his own excited pulses; the indistinct trot of cavalry plashing over the low ground, the dull rumble of gun and caisson along the roads, and, still more marked, the heavy tramp of marching men; the gathering on the heights above of that black swarm, engaged in strengthening the meshes of their net, from which night itself had not served to divert them. And below, there by the river's side, was there not the flash of lights suddenly extinguished, was not that the sound of hoarse voices shouting orders, adding to the dread suspense of that long night of terror while waiting for the coming of the dawn?

Maurice put forth his hand and felt for Jean's; at last he slumbered, comforted by the sense of human companionship. From a steeple in Sedan came the deep tones of a bell, slowly, mournfully, tolling the hour; then all was blank and void.

PART SECOND

1

Weiss, in the obscurity of his little room at Bazeilles, was aroused by a commotion that caused him to leap from his bed. It was the roar of artillery. Groping about in the darkness he found and lit a candle to enable him to consult his watch: it was four o'clock, just beginning to be light. He adjusted his double eyeglass upon his nose and looked out into the main street of the village, the road that leads to Douzy, but it was filled with a thick cloud of something that resembled dust, which made it impossible to distinguish anything. He passed into the other room, the windows of which commanded a view of the Meuse and the intervening meadows, and saw that the cause of his obstructed vision was the morning mist arising from the river. In the distance, behind the veil of fog, the guns were barking more fiercely across the stream. All at once a French battery, close at hand, opened in reply, with such a tremendous crash that the walls of the little house were shaken.

Weiss's house was situated near the middle of the village, on the right of the road and not far from the Place de l'Eglise. Its front, standing back a little from the street, displayed a single story with three windows, surmounted by an attic; in the rear was a garden of some extent that sloped gently downward toward the meadows and commanded a wide panoramic view of the encircling hills, from Remilly to Frenois. Weiss, with the sense of responsibility of his new proprietorship strong upon him, had spent the night in burying his provisions in the cellar and protecting his furniture, as far as possible, against shot and shell by applying mattresses to the windows, so that it was nearly two o'clock before he got to bed. His blood boiled at the idea that the Prussians might come and plunder the house, for which he had toiled so long and which had as yet afforded him so little enjoyment.

He heard a voice summoning him from the street.

"I say, Weiss, are you awake?"

He descended and found it was Delaherche, who had passed the night at his dyehouse, a large brick structure, next door to the accountant's abode. The operatives had all fled, taking to the woods and making for the Belgian frontier, and there was no one left to guard the property but the woman concierge, Francoise Quittard by name, the widow of a mason; and she also, beside herself with terror, would have gone with the others had it not been for her ten-year-old boy Charles, who was so ill with typhoid fever that he could not be moved.

"I say," Delaherche continued, "do you hear that? It is a promising beginning. Our best course is to get back to Sedan as soon as possible."

Weiss's promise to his wife, that he would leave Bazeilles at the first sign of danger, had been given in perfect good faith, and he had fully intended to keep it; but as yet there was only an artillery duel at long range, and the aim could not be accurate enough to do much damage in the uncertain, misty light of early morning.

"Wait a bit, confound it!" he replied. "There is no hurry."

Delaherche, too, was curious to see what would happen; his curiosity made him valiant. He had been so interested in the preparations for defending the place that he had not slept a wink. General Lebrun, commanding the 12th corps, had received notice that he would be attacked at daybreak, and had kept his men occupied during the night in strengthening the defenses of Bazeilles, which he had instructions to hold in spite of everything. Barricades had been thrown up across the Douzy road, and all the smaller streets; small parties of soldiers had been thrown into the houses by way of garrison; every narrow lane, every garden had become a fortress, and since three o'clock the troops, awakened from their slumbers without beat of drum or call of bugle in the inky blackness, had been at their posts, their chassepots freshly greased and cartridge boxes filled with the obligatory ninety rounds of ammunition. It followed that when the enemy opened their fire no one was taken unprepared, and the French batteries, posted to the rear between Balan and Bazeilles, immediately commenced to answer, rather with the idea of showing they were awake than for any other purpose, for in the dense fog that enveloped everything the practice was of the wildest.

"The dyehouse will be well defended," said Delaherche. "I have a whole section in it. Come and see."

It was true; forty and odd men of the infanterie de marine had been posted there under the command of a lieutenant, a tall, light-haired young fellow, scarcely more than a boy, but with an expression of energy and determination on his face. His men had already taken full possession of the building, some of them being engaged in loopholing the shutters of the ground-floor windows that commanded the street, while others, in the courtyard that overlooked the meadows in the rear, were breaching the wall for musketry. It was in this courtyard that Delaherche and Weiss found the young officer, straining his eyes to discover what was hidden behind the impenetrable mist.

"Confound this fog!" he murmured. "We can't fight when we don't know where the enemy is." Presently he asked, with no apparent change of voice or manner: "What day of the week is this?"

"Thursday," Weiss replied.

"Thursday, that's so. Hanged if I don't think the world might come to an end and we not know it!"

But just at that moment the uninterrupted roar of the artillery was diversified by a brisk rattle of musketry proceeding from the edge of the meadows, at a distance of two or three hundred yards. And at the same time there was a transformation, as rapid and startling, almost, as the stage effect in a fairy spectacle: the sun rose, the exhalations of the Meuse were whirled away like bits of finest, filmiest gauze, and the blue sky was revealed, in serene limpidity, undimmed by a single cloud. It was the exquisite morning of a faultless summer day.

"Ah!" exclaimed Delaherche, "they are crossing the railway bridge. See, they are making their way along the track. How stupid of us not to have blown up the bridge!"

The officer's face bore an expression of dumb rage. The mines had been prepared and charged, he averred, but they had fought four hours the day before to regain possession of the bridge and then had forgot to touch them off.

"It is just our luck," he curtly said.

Weiss was silent, watching the course of events and endeavoring to form some idea of the true state of affairs. The position of the French in Bazeilles was a very strong one. The village commanded the meadows, and was bisected by the Douzy road, which, turning sharp to the left, passed under the walls of the Chateau, while another road, the one that led to the railway bridge, bent around to the right and forked at the Place de l'Eglise. There was no cover for any force advancing by these two approaches; the Germans would be obliged to traverse the meadows and the wide, bare level that lay between the outskirts of the village and the Meuse and the railway. Their prudence in avoiding unnecessary risks was notorious, hence it seemed improbable that the real attack would come from that quarter. They kept coming across the bridge, however, in deep masses, and that notwithstanding the slaughter that a battery of mitrailleuses, posted at the edge of the village, effected in their ranks, and all at once those who had crossed rushed forward in open order, under cover of the straggling willows, the columns were re-formed and began to advance. It was from there that the musketry fire, which was growing hotter, had proceeded.

"Oh, those are Bavarians," Weiss remarked. "I recognize them by the braid on their helmets."

But there were other columns, moving to the right and partially concealed by the railway embankment, whose object, it seemed to him, was to gain the cover of some trees in the distance, whence they might descend and take Bazeilles in flank and rear. Should they succeed in effecting a lodgment in the park of Montivilliers, the village might become untenable. This was no more than a vague, half-formed idea, that flitted through his mind for a moment and faded as rapidly as it had come; the attack in front was becoming more determined, and his every faculty was concentrated on the struggle that was assuming, with every moment, larger dimensions.

Suddenly he turned his head and looked away to the north, over the city of Sedan, where the heights of Floing were visible in the distance. A battery had just commenced firing from that quarter; the smoke rose in the bright sunshine in little curls and wreaths, and the reports came to his ears very distinctly. It was in the neighborhood of five o'clock.

"Well, well," he murmured, "they are all going to have a hand in the business, it seems."

The lieutenant of marines, who had turned his eyes in the same direction, spoke up confidently:

"Oh! Bazeilles is the key of the position. This is the spot where the battle will be won or lost."

"Do you think so?" Weiss exclaimed.

"There is not the slightest doubt of it. It is certainly the marshal's opinion, for he was here last night and told us that we must hold the village if it cost the life of every man of us."

Weiss slowly shook his head, and swept the horizon with a glance; then in a low, faltering voice, as if speaking to himself, he said:

"No-no! I am sure that is a mistake. I fear the danger lies in another quarter-where, or what it is, I dare not say-"

He said no more. He simply opened wide his arms, like the jaws of a vise, then, turning to the north, brought his hands together, as if the vise had closed suddenly upon some object there.

This was the fear that had filled his mind for the last twenty-four hours, for he was thoroughly acquainted with the country and had watched narrowly every movement of the troops during the previous day, and now, again, while the broad valley before him lay basking in the radiant sunlight, his gaze reverted to the hills of the left bank, where, for the space of all one day and all one night, his eyes had beheld the black swarm of the Prussian hosts moving steadily onward to some appointed end. A battery had opened fire from Remilly, over to the left, but the one from which the shells were now beginning to reach the French position was posted at Pont-Maugis, on the river bank. He adjusted his binocle by folding the glasses over, the one upon the other, to lengthen its range and enable him to discern what was hidden among the recesses of the wooded slopes, but could distinguish nothing save the white smoke-wreaths that rose momentarily on the tranquil air and floated lazily away over the crests. That human torrent that he had seen so lately streaming over those hills, where was it now-where were massed those innumerable hosts? At last, at the corner of a pine wood, above Noyers and Frenois, he succeeded in making out a little cluster of mounted men in uniform-some general, doubtless, and his staff. And off there to the west the Meuse curved in a great loop, and in that direction lay their sole line of retreat on Mezieres, a narrow road that traversed the pass of Saint-Albert, between that loop and the dark forest of Ardennes. While reconnoitering the day before he had met a general officer who, he afterward learned, was Ducrot, commanding the 1st corps, on a by-road in the valley of Givonne, and had made bold to call his attention to the importance of that, their only line of retreat. If the army did not retire at once by that road while it was still open to them, if it waited until the Prussians should have crossed the Meuse at Donchery and come up in force to occupy the pass, it would be hemmed in and driven back on the Belgian frontier. As early even as the evening of that day the movement would have been too late. It was asserted that the uhlans had possession of the bridge, another bridge that had not been destroyed, for the reason, this time, that some one had neglected to provide the necessary powder. And Weiss sorrowfully acknowledged to himself that the human torrent, the invading horde, could now be nowhere else than on the plain of Donchery, invisible to him, pressing onward to occupy Saint-Albert pass, pushing forward its advanced guards to Saint-Menges and Floing, whither, the day previous, he had conducted Jean and Maurice. In the brilliant sunshine the steeple of Floing church appeared like a slender needle of dazzling whiteness.

And off to the eastward the other arm of the powerful vise was slowly closing in on them. Casting his eyes to the north, where there was a stretch of level ground between the plateaus of Illy and of Floing, he could make out the line of battle of the 7th corps, feebly supported by the 5th, which was posted in reserve under the ramparts of the city; but he could not discern what was occurring to the east, along the valley of the Givonne, where the 1st corps was stationed, its line stretching from the wood of la Garenne to Daigny village. Now, however, the guns were beginning to thunder in that direction also; the conflict seemed to be raging in Chevalier's wood, in front of Daigny. His uneasiness was owing to reports that had been brought in by peasants the day previous, that the Prussian advance had reached Francheval, so that the movement which was being conducted at the west, by way of Donchery, was also in process of execution at the east, by way of Francheval, and the two jaws of the vise would come together up there at the north, near the Calvary of Illy, unless the two-fold flanking movement could be promptly checked. He knew nothing of tactics or strategy, had nothing but his common sense to guide him; but he looked with fear and trembling on that great triangle that had the Meuse for one of its sides, and for the other two the 7th and 1st corps on the north and east respectively, while the extreme angle at the south was occupied by the 12th at Bazeilles-all the three corps facing outward on the periphery of a semicircle, awaiting the appearance of an enemy who was to deliver his attack at some one point, where or when no one could say, but who, instead, fell on them from every direction at once. And at the very center of all, as at the bottom of a pit, lay the city of Sedan, her ramparts furnished with antiquated guns, destitute of ammunition and provisions.

"Understand," said Weiss, with a repetition of his previous gesture, extending his arms and bringing his hands slowly together, "that is how it will be unless your generals keep their eyes open. The movement at Bazeilles is only a feint-"

But his explanation was confused and unintelligible to the lieutenant, who knew nothing of the country, and the young man shrugged his shoulders with an expression of impatience and disdain for the bourgeois in spectacles and frock coat who presumed to set his opinion against the marshal's. Irritated to hear Weiss reiterate his view that the attack on Bazeilles was intended only to mask other and more important movements, he finally shouted:

"Hold your tongue, will you! We shall drive them all into the Meuse, those Bavarian friends of yours, and that is all they will get by their precious feint."

While they were talking the enemy's skirmishers seemed to have come up closer; every now and then their bullets were heard thudding against the dyehouse wall, and our men, kneeling behind the low parapet of the courtyard, were beginning to reply. Every second the report of a chassepot rang out, sharp and clear, upon the air.

"Oh, of course! drive them into the Meuse, by all means," muttered Weiss, "and while we are about it we might as well ride them down and regain possession of the Carignan road." Then addressing himself to Delaherche, who had stationed himself behind the pump where he might be out of the way of the bullets: "All the same, it would have been their wisest course to make tracks last night for Mezieres, and if I were in their place I would much rather be there than here. As it is, however, they have got to show fight, since retreat is out of the question now."

"Are you coming?" asked Delaherche, who, notwithstanding his eager curiosity, was beginning to look pale in the face. "We shall be unable to get into the city if we remain here longer."

"Yes, in one minute I will be with you."

In spite of the danger that attended the movement he raised himself on tiptoe, possessed by an irresistible desire to see how things were shaping. On the right lay the meadows that had been flooded by order of the governor for the protection of the city, now a broad lake stretching from Torcy to Balan, its unruffled bosom glimmering in the morning sunlight with a delicate azure luster. The water did not extend as far as Bazeilles, however, and the Prussians had worked their way forward across the fields, availing themselves of the shelter of every ditch, of every little shrub and tree. They were now distant some five hundred yards, and Weiss was impressed by the caution with which they moved, the dogged resolution and patience with which they advanced, gaining ground inch by inch and exposing themselves as little as possible. They had a powerful artillery fire, moreover, to sustain them; the pure, cool air was vocal with the shrieking of shells. Raising his eyes he saw that the Pont-Maugis battery was not the only one that was playing on Bazeilles; two others, posted half way up the hill of Liry, had opened fire, and their projectiles not only reached the village, but swept the naked plain of la Moncelle beyond, where the reserves of the 12th corps were, and even the wooded slopes of Daigny, held by a division of the 1st corps, were not beyond their range. There was not a summit, moreover, on the left bank of the stream that was not tipped with flame. The guns seemed to spring spontaneously from the soil, like some noxious growth; it was a zone of fire that grew hotter and fiercer every moment; there were batteries at Noyers shelling Balan, batteries at Wadelincourt shelling Sedan, and at Frenois, down under la Marfee, there was a battery whose guns, heavier than the rest, sent their missiles hurtling over the city to burst among the troops of the 7th corps on the plateau of Floing. Those hills that he had always loved so well, that he had supposed were planted there solely to delight the eye, encircling with their verdurous slopes the pretty, peaceful valley that lay beneath, were now become a gigantic, frowning fortress, vomiting ruin and destruction on the feeble defenses of Sedan, and Weiss looked on them with terror and detestation. Why had steps not been taken to defend them the day before, if their leaders had suspected this, or why, rather, had they insisted on holding the position?

A sound of falling plaster caused him to raise his head; a shot had grazed his house, the front of which was visible to him above the party wall. It angered him excessively, and he growled:

"Are they going to knock it about my ears, the brigands!"

Then close behind him there was a little dull, strange sound that he had never heard before, and turning quickly he saw a soldier, shot through the heart, in the act of falling backward. There was a brief convulsive movement of the legs; the youthful, tranquil expression of the face remained, stamped there unalterably by the hand of death. It was the first casualty, and the accountant was startled by the crash of the musket falling and rebounding from the stone pavement of the courtyard.

"Ah, I have seen enough, I am going," stammered Delaherche. "Come, if you are coming; if not, I shall go without you."

The lieutenant, whom their presence made uneasy, spoke up:

"It will certainly be best for you to go, gentlemen. The enemy may attempt to carry the place at any moment."

Then at last, casting a parting glance at the meadows, where the Bavarians were still gaining ground, Weiss gave in and followed Delaherche, but when they had gained the street he insisted upon going to see if the fastening of his door was secure, and when he came back to his companion there was a fresh spectacle, which brought them both to a halt.

At the end of the street, some three hundred yards from where they stood, a strong Bavarian column had debouched from the Douzy road and was charging up the Place de l'Eglise. The square was held by a regiment of sailor-boys, who appeared to slacken their fire for a moment as if with the intention of drawing their assailants on; then, when the close-massed column was directly opposite their front, a most surprising maneuver was swiftly executed: the men abandoned their formation, some of them stepping from the ranks and flattening themselves against the house fronts, others casting themselves prone upon the ground, and down the vacant space thus suddenly formed the mitrailleuses that had been placed in battery at the farther end poured a perfect hailstorm of bullets. The column disappeared as if it had been swept bodily from off the face of the earth. The recumbent men sprang to their feet with a bound and charged the scattered Bavarians with the bayonet, driving them and making the rout complete. Twice the maneuver was repeated, each time with the same success. Two women, unwilling to abandon their home, a small house at the corner of an intersecting lane, were sitting at their window; they laughed approvingly and clapped their hands, apparently glad to have an opportunity to behold such a spectacle.

"There, confound it!" Weiss suddenly said, "I forgot to lock the cellar door! I must go back. Wait for me; I won't be a minute."

There was no indication that the enemy contemplated a renewal of their attack, and Delaherche, whose curiosity was reviving after the shock it had sustained, was less eager to get away. He had halted in front of his dyehouse and was conversing with the concierge, who had come for a moment to the door of the room she occupied in the rez-de-chaussée.

"My poor Francoise, you had better come along with us. A lone woman among such dreadful sights~I can't bear to think of it!"

She raised her trembling hands. "Ah, sir, I would have gone when the others went, indeed I would, if it had not been for my poor sick boy. Come in, sir, and look at him."

He did not enter, but glanced into the apartment from the threshold, and shook his head sorrowfully at sight of the little fellow in his clean, white bed, his face exhibiting the scarlet hue of the disease, and his glassy, burning eyes bent wistfully on his mother.

"But why can't you take him with you?" he urged. "I will find quarters for you in Sedan. Wrap him up warmly in a blanket, and come along with us."

"Oh, no, sir, I cannot. The doctor told me it would kill him. If only his poor father were alive! but we two are all that are left, and we must live for each other. And then, perhaps the Prussians will be merciful; perhaps they won't harm a lone woman and a sick boy."

Just then Weiss reappeared, having secured his premises to his satisfaction. "There, I think it will trouble them some to get in now. Come on! And it is not going to be a very pleasant journey, either; keep close to the houses, unless you want to come to grief."

There were indications, indeed, that the enemy were making ready for another assault. The infantry fire was spluttering away more furiously than ever, and the screaming of the shells was incessant. Two had already fallen in the street a hundred yards away, and a third had imbedded itself, without bursting, in the soft ground of the adjacent garden.

"Ah, here is Francoise," continued the accountant. "I must have a look at your little Charles. Come, come, you have no cause for alarm; he will be all right in a couple of days. Keep your courage up, and the first thing you do go inside, and don't put your nose outside the door." And the two men at last started to go.

"Au revoir, Francoise."

"Au revoir, sirs."

And as they spoke, there came an appalling crash. It was a shell, which, having first wrecked the chimney of Weiss's house, fell upon the sidewalk, where it exploded with such terrific force as to break every window in the vicinity. At first it was impossible to distinguish anything in the dense cloud of dust and smoke that rose in the air, but presently this drifted away, disclosing the ruined facade of the dyehouse, and there, stretched across the threshold, Francoise, a corpse, horribly torn and mangled, her skull crushed in, a fearful spectacle.

Weiss sprang to her side. Language failed him; he could only express his feelings by oaths and imprecations.

"Nom de Dieu! Nom de Dieu!"

Yes, she was dead. He had stooped to feel her pulse, and as he arose he saw before him the scarlet face of little Charles, who had raised himself in bed to look at his mother. He spoke no word, he uttered no cry; he gazed with blazing, tearless eyes, distended as if they would start from their sockets, upon the shapeless mass that was strange, unknown to him; and nothing more.

Weiss found words at last: "Nom de Dieu! they have taken to killing women!"

He had risen to his feet; he shook his fist at the Bavarians, whose braid-trimmed helmets were commencing to appear again in the direction of the church. The chimney, in falling, had crushed a great hole in the roof of his house, and the sight of the havoc made him furious.

"Dirty loafers! You murder women, you have destroyed my house. No, no! I will not go now, I cannot; I shall stay here."

He darted away and came running back with the dead soldier's rifle and ammunition. He was accustomed to carry a pair of spectacles on his person for use on occasions of emergency, when he wished to see with great distinctness, but did not wear them habitually out of respect for the wishes of his young wife. He now impatiently tore off his double eyeglass and substituted the spectacles, and the big, burly bourgeois, his overcoat flapping about his legs, his honest, kindly, round face ablaze with wrath, who would have been ridiculous had he not been so superbly heroic, proceeded to open fire, peppering away at the Bavarians at the bottom of the street. It was in his blood, he said; he had been hankering for something of the kind ever since the days of his boyhood, down there in Alsace, when he had been told all those tales of 1814. "Ah! you dirty loafers! you dirty loafers!" And he kept firing away with such eagerness that, finally, the barrel of his musket became so hot it burned his fingers.

The assault was made with great vigor and determination. There was no longer any sound of musketry in the direction of the meadows. The Bavarians had gained possession of a narrow stream, fringed with willows and poplars, and were making preparations for storming the houses, or rather fortresses, in the Place de l'Eglise. Their skirmishers had fallen back with the same caution that characterized their advance, and the wide grassy plain, dotted here and there with a black form where some poor fellow had laid down his life, lay spread in the mellow, slumbrous sunshine like a great cloth of gold. The lieutenant, knowing that the street

123

was now to be the scene of action, had evacuated the courtyard of the dyehouse, leaving there only one man as guard. He rapidly posted his men along the sidewalk with instructions, should the enemy carry the position, to withdraw into the building, barricade the first floor, and defend themselves there as long as they had a cartridge left. The men fired at will, lying prone upon the ground, and sheltering themselves as best they might behind posts and every little projection of the walls, and the storm of lead, interspersed with tongues of flame and puffs of smoke, that tore through that broad, deserted, sunny avenue was like a downpour of hail beaten level by the fierce blast of winter. A woman was seen to cross the roadway, running with wild, uncertain steps, and she escaped uninjured. Next, an old man, a peasant, in his blouse, who would not be satisfied until he saw his worthless nag stabled, received a bullet square in his forehead, and the violence of the impact was such that it hurled him into the middle of the street. A shell had gone crashing through the roof of the church; two others fell and set fire to houses, which burned with a pale flame in the intense daylight, with a loud snapping and crackling of their timbers. And that poor woman, who lay crushed and bleeding in the doorway of the house where her sick boy was, that old man with a bullet in his brain, all that work of ruin and devastation, maddened the few inhabitants who had chosen to end their days in their native village rather than seek safety in Belgium. Other bourgeois, and workingmen as well, the neatly attired citizen alongside the man in overalls, had possessed themselves of the weapons of dead soldiers, and were in the street defending their firesides or firing vengefully from the windows.

"Ah!" suddenly said Weiss, "the scoundrels have got around to our rear. I saw them sneaking along the railroad track. Hark! don't you hear them off there to the left?"

The heavy fire of musketry that was now audible behind the park of Montivilliers, the trees of which overhung the road, made it evident that something of importance was occurring in that direction. Should the enemy gain possession of the park Bazeilles would be at their mercy, but the briskness of the firing was in itself proof that the general commanding the 12th corps had anticipated the movement and that the position was adequately defended.

"Look out, there, you blockhead!" exclaimed the lieutenant, violently forcing Weiss up against the wall; "do you want to get yourself blown to pieces?"

He could not help laughing a little at the queer figure of the big gentleman in spectacles, but his bravery had inspired him with a very genuine feeling of respect, so, when his practiced ear detected a shell coming their way, he had acted the part of a friend and placed the civilian in a safer position. The missile landed some ten paces from where they were and exploded, covering them both with earth and debris. The citizen kept his feet and received not so much as a scratch, while the officer had both legs broken.

"It is well!" was all he said; "they have sent me my reckoning!"

He caused his men to take him across the sidewalk and place him with his back to the wall, near where the dead woman lay, stretched across her doorstep. His boyish face had lost nothing of its energy and determination.

"It don't matter, my children; listen to what I say. Don't fire too hurriedly; take your time. When the time comes for you to charge, I will tell you."

And he continued to command them still, with head erect, watchful of the movements of the distant enemy. Another house was burning, directly across the street. The crash and rattle of musketry, the roar of bursting shells, rent the air, thick with dust and sulphurous smoke. Men dropped at the corner of every lane and alley; corpses scattered here and there upon the pavement, singly or in little groups, made splotches of dark color, hideously splashed with red. And over the doomed village a frightful uproar rose and swelled, the vindictive shouts of thousands, devoting to destruction a few hundred brave men, resolute to die.

Then Delaherche, who all this time had been frantically shouting to Weiss without intermission, addressed him one last appeal:

"You won't come? Very well! then I shall leave you to your fate. Adieu!"

It was seven o'clock, and he had delayed his departure too long. So long as the houses were there to afford him shelter he took advantage of every doorway, of every bit of projecting wall, shrinking at every volley into cavities that were ridiculously small in comparison with his bulk. He turned and twisted in and out with the sinuous dexterity of the serpent; he would never have supposed that there was so much of his youthful agility left in him. When he reached the end of the village, however, and had to make his way for a space of some three hundred yards along the deserted, empty road, swept by the batteries on Liry hill, although the perspiration was streaming from his face and body, he shivered and his teeth chattered. For a minute or so he advanced cautiously along the bed of a dry ditch, bent almost double, then, suddenly forsaking the protecting shelter, burst into the open and ran for it with might and main, wildly, aimlessly, his ears ringing with detonations that sounded to him like thunder-claps. His eyes burned like coals of fire; it seemed to him that he was wrapt in flame. It was an eternity of torture. Then he suddenly caught sight of a little house to his left, and he rushed for the friendly refuge, gained it, with a sensation as if an immense load had been lifted from his breast. The place was tenanted, there were men and horses there. At first he could distinguish nothing. What he beheld subsequently filled him with amazement.

Was not that the Emperor, attended by his brilliant staff? He hesitated, although for the last two days he had been boasting of his acquaintance with him, then stood staring, open-mouthed. It was indeed Napoleon III.; he appeared larger, somehow, and more imposing on horseback, and his mustache was so stiffly waxed, there was such a brilliant color on his cheeks, that Delaherche saw at once he had been "made up" and painted like an actor. He had had recourse to cosmetics to conceal from his army the ravages that anxiety and illness had wrought in his countenance, the ghastly pallor of his face, his pinched nose, his dull, sunken eyes, and having been notified at five o'clock that there was fighting at Bazeilles, had come forth to see, sadly and silently, like a phantom with rouged cheeks.

There was a brick-kiln near by, behind which there was safety from the rain of bullets that kept pattering incessantly on its other front and the shells that burst at every second on the road. The mounted group had halted.

"Sire," someone murmured, "you are in danger-"

But the Emperor turned and motioned to his staff to take refuge in the narrow road that skirted the kiln, where men and horses would be sheltered from the fire.

"Really, Sire, this is madness. Sire, we entreat you-"

His only answer was to repeat his gesture; probably he thought that the appearance of a group of brilliant uniforms on that deserted road would draw the fire of the batteries on the left bank. Entirely unattended he rode forward into the midst of the storm of shot and shell, calmly, unhurriedly, with his unvarying air of resigned indifference, the air of one who goes to meet his appointed fate. Could it be that he heard behind him the implacable voice that was urging him onward, that voice from Paris: "March! march! die the hero's death on the piled corpses of thy countrymen, let the whole world look on in awe-struck admiration, so that thy son may reign!" -could that be what he heard? He rode forward, controlling his charger to a slow walk. For the space of a hundred yards he thus rode forward, then halted, awaiting the death he had come there to seek. The bullets sang in concert with a music like the fierce autumnal blast; a shell burst in front of him and covered him with earth. He maintained his attitude of patient waiting. His steed, with distended eyes and quivering frame, instinctively recoiled before the grim presence who was so close at hand and yet refused to smite horse or rider. At last the trying experience came to an end, and the Emperor, with his stoic fatalism, understanding that his time was not yet come, tranquilly retraced his steps, as if his only object had been to reconnoiter the position of the German batteries.

"What courage, Sire! We beseech you, do not expose yourself further-"

But, unmindful of their solicitations, he beckoned to his staff to follow him, not offering at present to consult their safety more than he did his own, and turned his horse's head toward la Moncelle, quitting the road and taking the abandoned fields of la Ripaille. A captain was mortally wounded, two horses were killed. As he passed along the line of the 12th corps, appearing and vanishing like a specter, the men eyed him with curiosity, but did not cheer.

To all these events had Delaherche been witness, and now he trembled at the thought that he, too, as soon as he should have left the brick works, would have to run the gauntlet of those terrible projectiles. He lingered, listening to the conversation of some dismounted officers who had remained there.

"I tell you he was killed on the spot; cut in two by a shell."

"You are wrong, I saw him carried off the field. His wound was not severe; a splinter struck him on the hip."

"What time was it?"

"Why, about an hour ago-say half-past six. It was up there around la Moncelle, in a sunken road."

"I know he is dead."

"But I tell you he is not! He even sat his horse for a moment after he was hit, then he fainted and they carried him into a cottage to attend to his wound."

"And then returned to Sedan?"

"Certainly; he is in Sedan now."

Of whom could they be speaking? Delaherche quickly learned that it was of Marshal MacMahon, who had been wounded while paying a visit of inspection to his advanced posts. The marshal wounded! it was "just our luck," as the lieutenant of marines had put it. He was reflecting on what the consequences of the mishap were likely to be when an *estafette* dashed by at top speed, shouting to a comrade, whom he recognized:

"General Ducrot is made commander-in-chief! The army is ordered to concentrate at Illy in order to retreat on Mezieres!"

The courier was already far away, galloping into Bazeilles under the constantly increasing fire, when Delaherche, startled by the strange tidings that came to him in such quick succession and not relishing the prospect of being involved in the confusion of the retreating troops, plucked up courage and started on a run for Balan, whence he regained Sedan without much difficulty.

The *estafette* tore through Bazeilles on a gallop, disseminating the news, hunting up the commanders to give them their instructions, and as he sped swiftly on the intelligence spread among the troops: Marshal MacMahon wounded, General Ducrot in command, the army falling back on Illy!

"What is that they are saying?" cried Weiss, whose face by this time was grimy with powder. "Retreat on Mezieres at this late hour! but it is absurd, they will never get through!"

And his conscience pricked him, he repented bitterly having given that counsel the day before to that very general who was now invested with the supreme command. Yes, certainly, that was yesterday the best, the only plan, to retreat, without loss of a minute's time, by the Saint-

Albert pass, but now the way could be no longer open to them, the black swarms of Prussians had certainly anticipated them and were on the plain of Donchery. There were two courses left for them to pursue, both desperate; and the most promising, as well as the bravest, of them was to drive the Bavarians into the Meuse, and cut their way through and regain possession of the Carignan road.

Weiss, whose spectacles were constantly slipping down upon his nose, adjusted them nervously and proceeded to explain matters to the lieutenant, who was still seated against the wall with his two stumps of legs, very pale and slowly bleeding to death.

"Lieutenant, I assure you I am right. Tell your men to stand their ground. You can see for yourself that we are doing well. One more effort like the last, and we shall drive them into the river."

It was true that the Bavarians' second attack had been repulsed. The mitrailleuses had again swept the Place de l'Eglise, the heaps of corpses in the square resembled barricades, and our troops, emerging from every cross street, had driven the enemy at the point of the bayonet through the meadows toward the river in headlong flight, which might easily have been converted into a general rout had there been fresh troops to support the sailor-boys, who had suffered severely and were by this time much distressed. And in Montivilliers Park, again, the firing did not seem to advance, which was a sign that in that quarter, also, reinforcements, could they have been had, would have cleared the wood.

"Order your men to charge them with the bayonet, lieutenant."

The waxen pallor of death was on the poor boy-officer's face; yet he had strength to murmur in feeble accents:

"You hear, my children; give them the bayonet!"

It was his last utterance; his spirit passed, his ingenuous, resolute face and his wide open eyes still turned on the battle. The flies already were beginning to buzz about Francoise's head and settle there, while lying on his bed little Charles, in an access of delirium, was calling on his mother in pitiful, beseeching tones to give him something to quench his thirst.

"Mother, mother, awake; get up-I am thirsty, I am so thirsty."

But the instructions of the new chief were imperative, and the officers, vexed and grieved to see the successes they had achieved thus rendered nugatory, had nothing for it but to give orders for the retreat. It was plain that the commander-in-chief, possessed by a haunting dread of the enemy's turning movement, was determined to sacrifice everything in order to escape from the toils. The Place de l'Eglise was evacuated, the troops fell back from street to street; soon the broad avenue was emptied of its defenders. Women shrieked and sobbed, men swore and shook their fists at the retiring troops, furious to see themselves abandoned thus. Many shut themselves in their houses, resolved to die in their defense.

"Well, *I* am not going to give up the ship!" shouted Weiss, beside himself with rage. "No! I will leave my skin here first. Let them come on! let them come and smash my furniture and drink my wine!"

Wrath filled his mind to the exclusion of all else, a wild, fierce desire to fight, to kill, at the thought that the hated foreigner should enter his house, sit in his chair, drink from his glass. It wrought a change in all his nature; everything that went to make up his daily life-wife, business, the methodical prudence of the small bourgeois-seemed suddenly to become unstable and drift away from him. And he shut himself up in his house and barricaded it, he paced the empty apartments with the restless impatience of a caged wild beast, going from room to room to make sure that all the doors and windows were securely fastened. He counted his cartridges and found he had forty left, then, as he was about to give a final look to

127

the meadows to see whether any attack was to be apprehended from that quarter, the sight of the hills on the left bank arrested his attention for a moment. The smoke-wreaths indicated distinctly the position of the Prussian batteries, and at the corner of a little wood on la Marfee, over the powerful battery at Frenois, he again beheld the group of uniforms, more numerous than before, and so distinct in the bright sunlight that by supplementing his spectacles with his binocle he could make out the gold of their epaulettes and helmets.

"You dirty scoundrels, you dirty scoundrels!" he twice repeated, extending his clenched fist in impotent menace.

Those who were up there on la Marfee were King William and his staff. As early as seven o'clock he had ridden up from Vendresse, where he had had quarters for the night, and now was up there on the heights, out of reach of danger, while at his feet lay the valley of the Meuse and the vast panorama of the field of battle. Far as the eye could reach, from north to south, the bird's-eye view extended, and standing on the summit of the hill, as from his throne in some colossal opera box, the monarch surveyed the scene.

In the central foreground of the picture, and standing out in bold relief against the venerable forests of the Ardennes, that stretched away on either hand from right to left, filling the northern horizon like a curtain of dark verdure, was the city of Sedan, with the geometrical lines and angles of its fortifications, protected on the south and west by the flooded meadows and the river. In Bazeilles houses were already burning, and the dark cloud of war hung heavy over the pretty village. Turning his eyes eastward he might discover, holding the line between la Moncelle and Givonne, some regiments of the 12th and 1st corps, looking like diminutive insects at that distance and lost to sight at intervals in the dip of the narrow valley in which the hamlets lay concealed; and beyond that valley rose the further slope, an uninhabited, uncultivated heath, of which the pale tints made the dark green of Chevalier's Wood look black by contrast. To the north the 7th corps was more distinctly visible in its position on the plateau of Floing, a broad belt of sere, dun fields, that sloped downward from the little wood of la Garenne to the verdant border of the stream. Further still were Floing, Saint-Menges, Fleigneux, Illy, small villages that lay nestled in the hollows of that billowing region where the landscape was a succession of hill and dale. And there, too, to the left was the great bend of the Meuse, where the sluggish stream, shimmering like molten silver in the bright sunlight, swept lazily in a great horseshoe around the peninsula of Iges and barred the road to Mezieres, leaving between its further bank and the impassable forest but one single gateway, the defile of Saint-Albert.

It was in that triangular space that the hundred thousand men and five hundred guns of the French army had now been crowded and brought to bay, and when His Prussian Majesty condescended to turn his gaze still further to the westward he might perceive another plain, the plain of Donchery, a succession of bare fields stretching away toward Briancourt, Marancourt, and Vrigne-aux-Bois, a desolate expanse of gray waste beneath the clear blue sky; and did he turn him to the east, he again had before his eyes, facing the lines in which the French were so closely hemmed, a vast level stretch of country in which were numerous villages, first Douzy and Carignan, then more to the north Rubecourt, Pourru-aux-Bois, Francheval, Villers-Cernay, and last of all, near the frontier, Chapelle. All about him, far as he could see, the land was his; he could direct the movements of the quarter of a million of men and the eight hundred guns that constituted his army, could master at a glance every detail of the operations of his invading host. Even then the XIth corps was pressing forward toward Saint-Menges, while the Vth was at Vrigne-aux-Bois, and the Wurtemburg division was near Donchery, awaiting orders. This was what he beheld to the west, and if, turning to the east, he found his view obstructed in that quarter by tree-clad hills, he could picture to himself what was passing, for he had seen the XIIth corps entering the wood of Chevalier, he knew that by

128

that time the Guards were at Villers-Cernay. There were the two arms of the gigantic vise, the army of the Crown Prince of Prussia on the left, the Saxon Prince's army on the right, slowly, irresistibly closing on each other, while the two Bavarian corps were hammering away at Bazeilles.

Underneath the King's position the long line of batteries, stretching with hardly an interval from Remilly to Frenois, kept up an unintermittent fire, pouring their shells into Daigny and la Moncelle, sending them hurtling over Sedan city to sweep the northern plateaus. It was barely eight o'clock, and with eyes fixed on the gigantic board he directed the movements of the game, awaiting the inevitable end, calmly controlling the black cloud of men that beneath him swept, an array of pigmies, athwart the smiling landscape.

2

In the dense fog up on the plateau of Floing Gaude, the bugler, sounded reveille at peep of day with all the lung-power he was possessed of, but the inspiring strain died away and was lost in the damp, heavy air, and the men, who had not had courage even to erect their tents and had thrown themselves, wrapped in their blankets, upon the muddy ground, did not awake or stir, but lay like corpses, their ashen features set and rigid in the slumber of utter exhaustion. To arouse them from their trance-like sleep they had to be shaken, one by one, and, with ghastly faces and haggard eyes, they rose to their feet, like beings summoned, against their will, back from another world. It was Jean who awoke Maurice.

"What is it? Where are we!" asked the younger man. He looked affrightedly around him, and beheld only that gray waste, in which were floating the unsubstantial forms of his comrades. Objects twenty yards away were undistinguishable; his knowledge of the country availed him not; he could not even have indicated in which direction lay Sedan. Just then, however, the boom of cannon, somewhere in the distance, fell upon his ear. "Ah! I remember; the battle is for today; they are fighting. So much the better; there will be an end to our suspense!"

He heard other voices around him expressing the same idea. There was a feeling of stern satisfaction that at last their long nightmare was to be dispelled, that at last they were to have a sight of those Prussians whom they had come out to look for, and before whom they had been retreating so many weary days; that they were to be given a chance to try a shot at them, and lighten the load of cartridges that had been tugging at their belts so long, with never an opportunity to burn a single one of them. Everyone felt that, this time, the battle would not, could not be avoided.

But the guns began to thunder more loudly down at Bazeilles, and Jean bent his ear to listen.

"Where is the firing?"

"Faith," replied Maurice, "it seems to me to be over toward the Meuse; but I'll be hanged if I know where we are."

"Look here, youngster," said the corporal, "you are going to stick close by me today, for unless a man has his wits about him, don't you see, he is likely to get in trouble. Now, I have been there before, and can keep an eye out for both of us."

The others of the squad, meantime, were growling angrily because they had nothing with which to warm their stomachs. There was no possibility of kindling fires without dry wood in such weather as prevailed then, and so, at the very moment when they were about to go into battle, the inner man put in his claim for recognition, and would not be denied. Hunger is not conducive to heroism; to those poor fellows eating was the great, the momentous question of life; how lovingly they watched the boiling pot on those red-letter days when the soup was rich and thick; how like children or savages they were in their wrath when rations were not forthcoming!

"No eat, no fight!" declared Chouteau. "I'll be blowed if I am going to risk my skin today!"

The radical was cropping out again in the great hulking house-painter, the orator of Belleville, the pothouse politician, who drowned what few correct ideas he picked up here and there in a nauseous mixture of ineffable folly and falsehood.

"Besides," he went on, "what good was there in making fools of us as they have been doing all along, telling us that the Prussians were dying of hunger and disease, that they had not so much as a shirt to their back, and were tramping along the highways like ragged, filthy paupers!"

Loubet laughed the laugh of the Parisian gamin, who has experienced the various vicissitudes of life in the Halles.

"Oh, that's all in my eye! it is we fellows who have been catching it right along; we are the poor devils whose leaky brogans and tattered toggery would make folks throw us a copper. And then those great victories about which they made such a fuss! What precious liars they must be, to tell us that old Bismarck had been made prisoner and that a German army had been driven over a quarry and dashed to pieces! Oh yes, they fooled us in great shape."

Pache and Lapoulle, who were standing near, shook their heads and clenched their fists ominously. There were others, also, who made no attempt to conceal their anger, for the course of the newspapers in constantly printing bogus news had had most disastrous results; all confidence was destroyed, men had ceased to believe anything or anybody. And so it was that in the soldiers, children of a larger growth, their bright dreams of other days had now been supplanted by exaggerated anticipations of misfortune.

"Pardi!" continued Chouteau, "the thing is accounted for easily enough, since our rulers have been selling us to the enemy right from the beginning. You all know that it is so."

Lapoulle's rustic simplicity revolted at the idea.

"For shame! what wicked people they must be!"

"Yes, sold, as Judas sold his master," murmured Pache, mindful of his studies in sacred history.

It was Chouteau's hour of triumph. "Mon Dieu! it is as plain as the nose on your face. MacMahon got three millions and each of the other generals got a million, as the price of bringing us up here. The bargain was made at Paris last spring, and last night they sent up a rocket as a signal to let Bismarck know that everything was fixed and he might come and take us."

The story was so inanely stupid that Maurice was disgusted. There had been a time when Chouteau, thanks to his facundity of the faubourg, had interested and almost convinced him, but now he had come to detest that apostle of falsehood, that snake in the grass, who calumniated honest effort of every kind in order to sicken others of it.

"Why do you talk such nonsense?" he exclaimed. "You know very well there is no truth in it."

"What, not true? Do you mean to say it is not true that we are betrayed? Ah, come, my aristocratic friend, perhaps you are one of them, perhaps you belong to the d-d band of dirty traitors?" He came forward threateningly. "If you are you have only to say so, my fine gentleman, for we will attend to your case right here, and won't wait for your friend Bismarck, either."

The others were also beginning to growl and show their teeth, and Jean thought it time that he should interfere.

"Silence there! I will report the first man who says another word!"

But Chouteau sneered and jeered at him; what did he care whether he reported him or not! He was not going to fight unless he chose, and they need not try to ride him rough-shod, because he had cartridges in his box for other people beside the Prussians. They were going into action now, and what discipline had been maintained by fear would be at an end: what could they do to him, anyway? he would just skip as soon as he thought he had enough of it. And he was profane and obscene, egging the men on against the corporal, who had been allowing them to starve. Yes, it was his fault that the squad had had nothing to eat in the last three days, while their neighbors had soup and fresh meat in plenty, but "monsieur" had to go off to town with the "aristo" and enjoy himself with the girls. People had spotted 'em, over in Sedan.

"You stole the money belonging to the squad; deny it if you dare, you *bougre* of a belly-god!"

Things were beginning to assume an ugly complexion; Lapoulle was doubling his big fists in a way that looked like business, and Pache, with the pangs of hunger gnawing at his vitals, laid aside his natural douceness and insisted on an explanation. The only reasonable one among them was Loubet, who gave one of his pawky laughs and suggested that, being Frenchmen, they might as well dine off the Prussians as eat one another. For his part, he took no stock in fighting, either with fists or firearms, and alluding to the few hundred francs that he had earned as substitute, added:

"And so, that was all they thought my hide was worth! Well, I am not going to give them more than their money's worth."

Maurice and Jean were in a towering rage at the idotic onslaught, talking loudly and repelling Chouteau's insinuations, when out from the fog came a stentorian voice, bellowing:

"What's this? what's this? Show me the rascals who dare quarrel in the company street!"

And Lieutenant Rochas appeared upon the scene, in his old kepi, whence the rain had washed all the color, and his great coat, minus many of its buttons, evincing in all his lean, shambling person the extreme of poverty and distress. Notwithstanding his forlorn aspect, however, his sparkling eye and bristling mustache showed that his old time confidence had suffered no impairment.

Jean spoke up, scarce able to restrain himself. "Lieutenant, it is these men, who persist in saying that we are betrayed. Yes, they dare to assert that our generals have sold us-"

The idea of treason did not appear so extremely unnatural to Rochas's thick understanding, for it served to explain those reverses that he could not account for otherwise.

"Well, suppose they are sold, is it any of their business? What concern is it of theirs? The Prussians are there all the same, aren't they? and we are going to give them one of the old-fashioned hidings, such as they won't forget in one while." Down below them in the thick sea of fog the guns at Bazeilles were still pounding away, and he extended his arms with a broad, sweeping gesture: "Hein! this is the time that we've got them! We'll see them back home, and kick them every step of the way!"

All the trials and troubles of the past were to him as if they had not been, now that his ears were gladdened by the roar of the guns: the delays and conflicting orders of the chiefs, the demoralization of the troops, the stampede at Beaumont, the distress of the recent forced retreat on Sedan-all were forgotten. Now that they were about to fight at last, was not victory certain? He had learned nothing and forgotten nothing; his blustering, boastful contempt of the enemy, his entire ignorance of the new arts and appliances of war, his rooted conviction that an old soldier of Africa, Italy, and the Crimea could by no possibility be beaten, had suffered no change. It was really a little too comical that a man at his age should take the back track and begin at the beginning again!

All at once his lantern jaws parted and gave utterance to a loud laugh. He was visited by one of those impulses of good-fellowship that made his men swear by him, despite the roughness of the jobations that he frequently bestowed on them.

"Look here, my children, in place of quarreling it will be a great deal better to take a good nip all around. Come, I'm going to treat, and you shall drink my health."

From the capacious pocket of his capote he extracted a bottle of brandy, adding, with his all-conquering air, that it was the gift of a lady. (He had been seen the day before, seated at the table of a tavern in Floing and holding the waitress on his lap, evidently on the best of terms with her.) The soldiers laughed and winked at one another, holding out their porringers, into which he gayly poured the golden liquor.

"Drink to your sweethearts, my children, if you have any and don't forget to drink to the glory of France. Them's my sentiments, so vive la joie!"

"That's right, Lieutenant. Here's to your health, and everybody else's!"

They all drank, and their hearts were warmed and peace reigned once more. The "nip" had much of comfort in it, in the chill morning, just as they were going into action, and Maurice felt it tingling in his veins, giving him cheer and a sort of what is known colloquially as "Dutch courage." Why should they not whip the Prussians? Have not battles their surprises? has not history embalmed many an instance of the fickleness of fortune? That mighty man of war, the lieutenant, added that Bazaine was on the way to join them, would be with them before the day was over: oh, the information was positive; he had it from an aid to one of the generals; and although, in speaking of the route the marshal was to come by, he pointed to the frontier of Belgium, Maurice yielded to one of those spasmodic attacks of hopefulness of his, without which life to him would not have been worth living. Might it not be that the day of reckoning was at hand?

"Why don't we move, Lieutenant?" he made bold to ask. "What are we waiting for?"

Rochas made a gesture, which the other interpreted to mean that no orders had been received. Presently he asked:

"Has anybody seen the captain?"

No one answered. Jean remembered perfectly having seen him making for Sedan the night before, but to the soldier who knows what is good for himself, his officers are always invisible when they are not on duty. He held his tongue, therefore, until happening to turn his head, he caught sight of a shadowy form flitting along the hedge.

"Here he is," said he.

It was Captain Beaudoin in the flesh. They were all surprised by the nattiness of his appearance, his resplendent shoes, his well-brushed uniform, affording such a striking contrast to the lieutenant's pitiful state. And there was a finicking completeness, moreover, about his toilet, greater than the male being is accustomed to bestow upon himself, in his scrupulously white hands and his carefully curled mustache, and a faint perfume of Persian lilac, which had the effect of reminding one in some mysterious way of the dressing room of a young and pretty woman.

"Hallo!" said Loubet, with a sneer, "the captain has recovered his baggage!"

But no one laughed, for they all knew him to be a man with whom it was not well to joke. He was stiff and consequential with his men, and was detested accordingly; a pete sec, to use Rochas's expression. He had seemed to regard the early reverses of the campaign as personal affronts, and the disaster that all had prognosticated was to him an unpardonable crime. He was a strong Bonapartist by conviction; his prospects for promotion were of the brightest; he had several important salons looking after his interests; naturally, he did not take kindly to the changed condition of affairs that promised to make his cake dough. He was said to have a remarkably fine tenor voice, which had helped him no little in his advancement. He was not devoid of intelligence, though perfectly ignorant as regarded everything connected with his profession; eager to please, and very brave, when there was occasion for being so, without superfluous rashness.

"What a nasty fog!" was all he said, pleased to have found his company at last, for which he had been searching for more than half an hour.

At the same time their orders came, and the battalion moved forward. They had to proceed with caution, feeling their way, for the exhalations continued to rise from the stream and were now so dense that they were precipitated in a fine, drizzling rain. A vision rose before

Maurice's eyes that impressed him deeply; it was Colonel de Vineuil, who loomed suddenly from out the mist, sitting his horse, erect and motionless, at the intersection of two roads-the man appearing of preternatural size, and so pale and rigid that he might have served a sculptor as a study for a statue of despair; the steed shivering in the raw, chill air of morning, his dilated nostrils turned in the direction of the distant firing. Some ten paces to their rear were the regimental colors, which the sous-lieutenant whose duty it was to bear them had thus early taken from their case and proudly raised aloft, and as the driving, vaporous rack eddied and swirled about them, they shone like a radiant vision of glory emblazoned on the heavens, soon to fade and vanish from the sight. Water was dripping from the gilded eagle, and the tattered, shot-riddled tri-color, on which were embroidered the names of former victories, was stained and its bright hues dimmed by the smoke of many a battlefield; the sole bit of brilliant color in all the faded splendor was the enameled cross of honor that was attached to the *cravate*.

Another billow of vapor came scurrying up from the river, enshrouding in its fleecy depths colonel, standard, and all, and the battalion passed on, whitherward no one could tell. First their route had conducted them over descending ground, now they were climbing a hill. On reaching the summit the command, halt! started at the front and ran down the column; the men were cautioned not to leave the ranks, arms were ordered, and there they remained, the heavy knapsacks forming a grievous burden to weary shoulders. It was evident that they were on a plateau, but to discern localities was out of the question; twenty paces was the extreme range of vision. It was now seven o'clock; the sound of firing reached them more distinctly, other batteries were apparently opening on Sedan from the opposite bank.

"Oh! I," said Sergeant Sapin with a start, addressing Jean and Maurice, "I shall be killed today."

It was the first time he had opened his lips that morning; an expression of dreamy melancholy had rested on his thin face, with its big, handsome eyes and thin, pinched nose.

"What an idea!" Jean exclaimed; "who can tell what is going to happen him? Every bullet has its billet, they say, but you stand no worse chance than the rest of us."

"Oh, but me-I am as good as dead now. I tell you I shall be killed today."

The near files turned and looked at him curiously, asking him if he had had a dream. No, he had dreamed nothing, but he felt it; it was there.

"And it is a pity, all the same, because I was to be married when I got my discharge."

A vague expression came into his eyes again; his past life rose before him. He was the son of a small retail grocer at Lyons, and had been petted and spoiled by his mother up to the time of her death; then rejecting the proffer of his father, with whom he did not hit it off well, to assist in purchasing his discharge, he had remained with the army, weary and disgusted with life and with his surroundings. Coming home on furlough, however, he fell in love with a cousin and they became engaged; their intention was to open a little shop on the small capital which she would bring him, and then existence once more became desirable. He had received an elementary education; could read, write, and cipher. For the past year he had lived only in anticipation of this happy future.

He shivered, and gave himself a shake to dispel his revery, repeating with his tranquil air:

"Yes, it is too bad; I shall be killed today."

No one spoke; the uncertainty and suspense continued. They knew not whether the enemy was on their front or in their rear. Strange sounds came to their ears from time to time from out the depths of the mysterious fog: the rumble of wheels, the deadened tramp of moving masses, the distant clatter of horses' hoofs; it was the evolutions of troops, hidden from view behind the misty curtain, the batteries, battalions, and squadrons of the 7th corps taking up their positions in line of battle. Now, however, it began to look as if the fog was about to lift; it parted here and there and fragments floated lightly off, like strips of gauze torn from a veil, and bits of sky appeared, not transparently blue, as on a bright summer's day, but opaque and of the hue of burnished steel, like the cheerless bosom of some deep, sullen mountain tarn. It was in one of those brighter moments when the sun was endeavoring to struggle forth that the regiments of chasseurs d'Afrique, constituting part of Margueritte's division, came riding by, giving the impression of a band of spectral horsemen. They sat very stiff and erect in the saddle, with their short cavalry jackets, broad red sashes and smart little kepis, accurate in distance and alignment and managing admirably their lean, wiry mounts, which were almost invisible under the heterogeneous collection of tools and camp equipage that they had to carry. Squadron after squadron they swept by in long array, to be swallowed in the gloom from which they had just emerged, vanishing as if dissolved by the fine rain. The truth was, probably, that they were in the way, and their leaders, not knowing what use to put them to, had packed them off the field, as had often been the case since the opening of the campaign. They had scarcely ever been employed on scouting or reconnoitering duty, and as soon as there was prospect of a fight were trotted about for shelter from valley to valley, useless objects, but too costly to be endangered.

Maurice thought of Prosper as he watched them. "That fellow, yonder, looks like him," he said, under his breath. "I wonder if it is he?"

"Of whom are you speaking?" asked Jean.

"Of that young man of Remilly, whose brother we met at Osches, you remember."

Behind the chasseurs, when they had all passed, came a general officer and his staff dashing down the descending road, and Maurice recognized the general of their brigade, Bourgain-Desfeuilles, shouting and gesticulating wildly. He had torn himself reluctantly from his comfortable quarters at the Hotel of the Golden Cross, and it was evident from the horrible temper he was in that the condition of affairs that morning was not satisfactory to him. In a tone of voice so loud that everyone could hear he roared:

"In the devil's name, what stream is that off yonder, the Meuse or the Moselle?"

The fog dispersed at last, this time in earnest. As at Bazeilles the effect was theatrical; the curtain rolled slowly upward to the flies, disclosing the setting of the stage. From a sky of transparent blue the sun poured down a flood of bright, golden light, and Maurice was no longer at a loss to recognize their position.

"Ah!" he said to Jean, "we are on the plateau de l'Algerie. That village that you see across the valley, directly in our front, is Floing, and that more distant one is Saint-Menges, and that one, more distant still, a little to the right, is Fleigneux. Then those scrubby trees on the horizon, away in the background, are the forest of the Ardennes, and there lies the frontier-"

He went on to explain their position, naming each locality and pointing to it with outstretched hand. The plateau de l'Algerie was a belt of reddish ground, something less than two miles in length, sloping gently downward from the wood of la Garenne toward the Meuse, from which it was separated by the meadows. On it the line of the 7th corps had been established by General Douay, who felt that his numbers were not sufficient to defend so extended a position and properly maintain his touch with the 1st corps, which was posted at right angles with his line, occupying the valley of la Givonne, from the wood of la Garenne to Daigny.

"Oh, isn't it grand, isn't it magnificent!"

And Maurice, revolving on his heel, made with his hand a sweeping gesture that embraced the entire horizon. From their position on the plateau the whole wide field of battle lay stretched before them to the south and west: Sedan, almost at their feet, whose citadel they could see overtopping the roofs, then Balan and Bazeilles, dimly seen through the dun smoke-clouds that hung heavily in the motionless air, and further in the distance the hills of the left bank, Liry, la Marfee, la Croix-Piau. It was away toward the west, however, in the direction of Donchery, that the prospect was most extensive. There the Meuse curved horseshoe-wise, encircling the peninsula of Iges with a ribbon of pale silver, and at the northern extremity of the loop was distinctly visible the narrow road of the Saint-Albert pass, winding between the river bank and a beetling, overhanging hill that was crowned with the little wood of Seugnon, an offshoot of the forest of la Falizette. At the summit of the hill, at the *carrefour* of la Maison-Rouge, the road from Donchery to Vrigne-aux-Bois debouched into the Mezieres pike.

"See, that is the road by which we might retreat on Mezieres."

Even as he spoke the first gun was fired from Saint-Menges. The fog still hung over the bottom-lands in shreds and patches, and through it they dimly descried a shadowy body of men moving through the Saint-Albert defile.

"Ah, they are there," continued Maurice, instinctively lowering his voice. "Too late, too late; they have intercepted us!"

It was not eight o'clock. The guns, which were thundering more fiercely than ever in the direction of Bazeilles, now also began to make themselves heard at the eastward, in the valley of la Givonne, which was hid from view; it was the army of the Crown Prince of Saxony, debouching from the Chevalier wood and attacking the 1st corps, in front of Daigny village; and now that the XIth Prussian corps, moving on Floing, had opened fire on General Douay's troops, the investment was complete at every point of the great periphery of several leagues' extent, and the action was general all along the line.

Maurice suddenly perceived the enormity of their blunder in not retreating on Mezieres during the night; but as yet the consequences were not clear to him; he could not foresee all the disaster that was to result from that fatal error of judgment. Moved by some indefinable instinct of danger, he looked with apprehension on the adjacent heights that commanded the plateau de l'Algerie. If time had not been allowed them to make good their retreat, why had they not backed up against the frontier and occupied those heights of Illy and Saint-Menges, whence, if they could not maintain their position, they would at least have been free to cross over into Belgium? There were two points that appeared to him especially threatening, the *mamelon* of Hattoy, to the north of Floing on the left, and the Calvary of Illy, a stone cross with a linden tree on either side, the highest bit of ground in the surrounding country, to the right. General Douay was keenly alive to the importance of these eminences, and the day before had sent two battalions to occupy Hattoy; but the men, feeling that they were "in the air" and too remote from support, had fallen back early that morning. It was understood that the left wing of the 1st corps was to take care of the Calvary of Illy. The wide expanse of naked country between Sedan and the Ardennes forest was intersected by deep ravines, and the key of the position was manifestly there, in the shadow of that cross and the two lindens, whence their guns might sweep the fields in every direction for a long distance.

Two more cannon shots rang out, quickly succeeded by a salvo; they detected the bluish smoke rising from the underbrush of a low hill to the left of Saint-Menges.

"Our turn is coming now," said Jean.

Nothing more startling occurred just then, however. The men, still preserving their formation and standing at ordered arms, found something to occupy their attention in the fine appearance made by the 2d division, posted in front of Floing, with their left refused and facing the Meuse, so as to guard against a possible attack from that quarter. The ground to the east, as far as the wood of la Garenne, beneath Illy village, was held by the 3d division, while the 1st, which had lost heavily at Beaumont, formed a second line. All night long the engineers had been busy with pick and shovel, and even after the Prussians had opened fire they were still digging away at their shelter trenches and throwing up epaulments.

Then a sharp rattle of musketry, quickly silenced, however, was heard proceeding from a point beneath Floing, and Captain Beaudoin received orders to move his company three hundred yards to the rear. Their new position was in a great field of cabbages, upon reaching which the captain made his men lie down. The sun had not yet drunk up the moisture that had descended on the vegetables in the darkness, and every fold and crease of the thick, golden-green leaves was filled with trembling drops, as pellucid and luminous as brilliants of the fairest water.

"Sight for four hundred yards," the captain ordered.

Maurice rested the barrel of his musket on a cabbage that reared its head conveniently before him, but it was impossible to see anything in his recumbent position: only the blurred surface of the fields traversed by his level glance, diversified by an occasional tree or shrub. Giving Jean, who was beside him, a nudge with his elbow, he asked what they were to do there. The corporal, whose experience in such matters was greater, pointed to an elevation not far away, where a battery was just taking its position; it was evident that they had been placed there to support that battery, should there be need of their services. Maurice, wondering whether Honore and his guns were not of the party, raised his head to look, but the reserve artillery was at the rear, in the shelter of a little grove of trees.

"Nom de Dieu!" yelled Rochas, "will you lie down!"

And Maurice had barely more than complied with this intimation when a shell passed screaming over him. From that time forth there seemed to be no end to them. The enemy's gunners were slow in obtaining the range, their first projectiles passing over and landing well to the rear of the battery, which was now opening in reply. Many of their shells, too, fell upon the soft ground, in which they buried themselves without exploding, and for a time there was a great display of rather heavy wit at the expense of those bloody sauerkraut eaters.

"Well, well!" said Loubet, "their fireworks are a fizzle!"

"They ought to take them in out of the rain," sneered Chouteau.

Even Rochas thought it necessary to say something. "Didn't I tell you that the dunderheads don't know enough even to point a gun?"

But they were less inclined to laugh when a shell burst only ten yards from them and sent a shower of earth flying over the company; Loubet affected to make light of it by ordering his comrades to get out their brushes from the knapsacks, but Chouteau suddenly became very pale and had not a word to say. He had never been under fire, nor had Pache and Lapoulle, nor any member of the squad, in fact, except Jean. Over eyes that had suddenly lost their brightness lids flickered tremulously; voices had an unnatural, muffled sound, as if arrested by some obstruction in the throat. Maurice, who was sufficiently master of himself as yet, endeavored to diagnose his symptoms; he could not be afraid, for he was not conscious that he was in danger; he only felt a slight sensation of discomfort in the epigastric region, and his head seemed strangely light and empty; ideas and images came and went independent of his will. His recollection of the brave show made by the troops of the 2d division made him hopeful, almost to buoyancy; victory appeared certain to him if only they might be allowed to go at the enemy with the bayonet.

"Listen!" he murmured, "how the flies buzz; the place is full of them." Thrice he had heard something that sounded like the humming of a swarm of bees.

"That was not a fly," Jean said, with a laugh. "It was a bullet."

Again and again the hum of those invisible wings made itself heard. The men craned their necks and looked about them with eager interest; their curiosity was uncontrollable-would not allow them to remain quiet.

"See here," Loubet said mysteriously to Lapoulle, with a view to raise a laugh at the expense of his simple-minded comrade, "when you see a bullet coming toward you you must raise your forefinger before your nose-like that; it divides the air, and the bullet will go by to the right or left."

"But I can't see them," said Lapoulle.

A loud guffaw burst from those near.

"Oh, crickey! he says he can't see them! Open your garret windows, stupid! See! there's one-see! there's another. Didn't you see that one? It was of the most beautiful green."

And Lapoulle rolled his eyes and stared, placing his finger before his nose, while Pache fingered the scapular he wore and wished it was large enough to shield his entire person.

Rochas, who had remained on his feet, spoke up and said jocosely:

"Children, there is no objection to your ducking to the shells when you see them coming. As for the bullets, it is useless; they are too numerous!"

At that very instant a soldier in the front rank was struck on the head by a fragment of an exploding shell. There was no outcry; simply a spurt of blood and brain, and all was over.

"Poor devil!" tranquilly said Sergeant Sapin, who was quite cool and exceedingly pale. "Next!"

But the uproar had by this time become so deafening that the men could no longer hear one another's voice; Maurice's nerves, in particular, suffered from the infernal *charivari*. The neighboring battery was banging away as fast as the gunners could load the pieces; the continuous roar seemed to shake the ground, and the mitrailleuses were even more intolerable with their rasping, grating, grunting noise. Were they to remain forever reclining there among the cabbages? There was nothing to be seen, nothing to be learned; no one had any idea how the battle was going. And *was* it a battle, after all-a genuine affair? All that Maurice could make out, projecting his eyes along the level surface of the fields, was the rounded, wood-clad summit of Hattoy in the remote distance, and still unoccupied. Neither was there a Prussian to be seen anywhere on the horizon; the only evidence of life were the faint, blue smoke-wreaths that rose and floated an instant in the sunlight. Chancing to turn his head, he was greatly surprised to behold at the bottom of a deep, sheltered valley, surrounded by precipitous heights, a peasant calmly tilling his little field, driving the plow through the furrow with the assistance of a big white horse. Why should he lose a day? The corn would keep growing, let them fight as they would, and folks must live.

Unable longer to control his impatience, the young man jumped to his feet. He had a fleeting vision of the batteries of Saint-Menges, crowned with tawny vapors and spewing shot and shell upon them; he had also time to see, what he had seen before and had not forgotten, the road from Saint-Albert's pass black with minute moving objects-the swarming hordes of the invader. Then Jean seized him by the legs and pulled him violently to his place again.

"Are you crazy? Do you want to leave your bones here?"

And Rochas chimed in:

"Lie down, will you! What am I to do with such d--d rascals, who get themselves killed without orders!"

"But you don't lie down, lieutenant," said Maurice.

"That's a different thing. I have to know what is going on."

Captain Beaudoin, too, kept his legs like a man, but never opened his lips to say an encouraging word to his men, having nothing in common with them. He appeared nervous and unable to remain long in one place, striding up and down the field, impatiently awaiting orders.

No orders came, nothing occurred to relieve their suspense. Maurice's knapsack was causing him horrible suffering; it seemed to be crushing his back and chest in that recumbent position, so painful when maintained for any length of time. The men had been cautioned against throwing away their sacks unless in case of actual necessity, and he kept turning over, first on his right side, then on the left, to ease himself a moment of his burden by resting it on the ground. The shells continued to fall around them, but the German gunners did not succeed in getting the exact range; no one was killed after the poor fellow who lay there on his stomach with his skull fractured.

"Say, is this thing to last all day?" Maurice finally asked Jean, in sheer desperation.

"Like enough. At Solferino they put us in a field of carrots, and there we stayed five mortal hours with our noses to the ground." Then he added, like the sensible fellow he was: "Why do you grumble? we are not so badly off here. You will have an opportunity to distinguish yourself before the day is over. Let everyone have his chance, don't you see; if we should all be killed at the beginning there would be none left for the end."

"Look," Maurice abruptly broke in, "look at that smoke over Hattoy. They have taken Hattoy; we shall have plenty of music to dance to now!"

For a moment his burning curiosity, which he was conscious was now for the first time beginning to be dashed with personal fear, had sufficient to occupy it; his gaze was riveted on the rounded summit of the mamelon, the only elevation that was within his range of vision, dominating the broad expanse of plain that lay level with his eye. Hattoy was too far distant to permit him to distinguish the gunners of the batteries that the Prussians had posted there; he could see nothing at all, in fact, save the smoke that at each discharge rose above a thin belt of woods that served to mask the guns. The enemy's occupation of the position, of which General Douay had been forced to abandon the defense, was, as Maurice had instinctively felt, an event of the gravest importance and destined to result in the most disastrous consequences; its possessors would have entire command of all the surrounding plateau. This was quickly seen to be the case, for the batteries that opened on the second division of the 7th corps did fearful execution. They had now perfected their range, and the French battery, near which Beaudoin's company was stationed, had two men killed in quick succession. A quartermaster's man in the company had his left heel carried away by a splinter and began to howl most dismally, as if visited by a sudden attack of madness.

"Shut up, you great calf!" said Rochas. "What do you mean by yelling like that for a little scratch!"

The man suddenly ceased his outcries and subsided into a stupid silence, nursing his foot in his hand.

And still the tremendous artillery duel raged, and the death-dealing missiles went screaming over the recumbent ranks of the regiments that lay there on the sullen, sweltering plain, where no thing of life was to be seen beneath the blazing sun. The crashing thunder, the destroying hurricane, were masters in that solitude, and many long hours would pass before the end. But even thus early in the day the Germans had demonstrated the superiority of their artillery; their percussion shells had an enormous range, and exploded, with hardly an exception, on reaching their destination, while the French time-fuse shells, with a much shorter range, burst for the most part in the air and were wasted. And there was nothing left for the poor fellows exposed to that murderous fire save to hug the ground and make themselves as small as possible; they were even denied the privilege of firing in reply, which

139

would have kept their mind occupied and given them a measure of relief; but upon whom or what were they to direct their rifles? since there was not a living soul to be seen upon the entire horizon!

"Are we never to have a shot at them? I would give a dollar for just one chance!" said Maurice, in a frenzy of impatience. "It is disgusting to have them blazing away at us like this and not be allowed to answer."

"Be patient; the time will come," Jean imperturbably replied.

Their attention was attracted by the sound of mounted men approaching on their left, and turning their heads they beheld General Douay, who, accompanied by his staff, had come galloping up to see how his troops were behaving under the terrible fire from Hattoy. He appeared well pleased with what he saw and was in the act of making some suggestions to the officers grouped around him, when, emerging from a sunken road, General Bourgain-Desfeuilles also rode up. This officer, though he owed his advancement to "influence" was wedded to the antiquated African routine and had learned nothing by experience, sat his horse with great composure under the storm of projectiles. He was shouting to the men and gesticulating wildly, after the manner of Rochas: "They are coming, they will be here right away, and then we'll let them have the bayonet!" when he caught sight of General Douay and drew up to his side.

"Is it true that the marshal is wounded, general?" he asked.

"It is but too true, unfortunately. I received a note from Ducrot only a few minutes ago, in which he advises me of the fact, and also notifies me that, by the marshal's appointment, he is in command of the army."

"Ah! so it is Ducrot who is to have his place! And what are the orders now?"

The general shook his head sorrowfully. He had felt that the army was doomed, and for the last twenty-four hours had been strenuously recommending the occupation of Illy and Saint-Menges in order to keep a way of retreat open on Mezieres.

"Ducrot will carry out the plan we talked of yesterday: the whole army is to be concentrated on the plateau of Illy."

And he repeated his previous gesture, as if to say it was too late.

His words were partly inaudible in the roar of the artillery, but Maurice caught their significance clearly enough, and it left him dumfounded by astonishment and alarm. What! Marshal MacMahon wounded since early that morning, General Ducrot commanding in his place for the last two hours, the entire army retreating to the northward of Sedan-and all these important events kept from the poor devils of soldiers who were squandering their life's blood! and all their destinies, dependent on the life of a single man, were to be intrusted to the direction of fresh and untried hands! He had a distinct consciousness of the fate that was in reserve for the army of Chalons, deprived of its commander, destitute of any guiding principle of action, dragged purposelessly in this direction and in that, while the Germans went straight and swift to their preconcerted end with mechanical precision and directness.

Bourgain-Desfeuilles had wheeled his horse and was moving away, when General Douay, to whom a grimy, dust-stained hussar had galloped up with another dispatch, excitedly summoned him back.

"General! General!"

His voice rang out so loud and clear, with such an accent of surprise, that it drowned the uproar of the guns.

"General, Ducrot is no longer in command; de Wimpffen is chief. You know he reached here yesterday, just in the very thick of the disaster at Beaumont, to relieve de Failly at the head of the 5th corps-and he writes me that he has written instructions from the Minister of War assigning him to the command of the army in case the post should become vacant. And there is to be no more retreating; the orders now are to reoccupy our old positions, and defend them to the last."

General Bourgain-Desfeuilles drank in the tidings, his eyes bulging with astonishment. "Nom de Dieu!" he at last succeeded in ejaculating, "one would like to know- But it is no business of mine, anyhow." And off he galloped, not allowing himself to be greatly agitated by this unexpected turn of affairs, for he had gone into the war solely in the hope of seeing his name raised a grade higher in the army list, and it was his great desire to behold the end of the beastly campaign as soon as possible, since it was productive of so little satisfaction to anyone.

Then there was an explosion of derision and contempt among the men of Beaudoin's company. Maurice said nothing, but he shared the opinion of Chouteau and Loubet, who chaffed and blackguarded everyone without mercy. "See-saw, up and down, move as I pull the string! A fine gang they were, those generals! they understood one another; they were not going to pull all the blankets off the bed! What was a poor devil of a soldier to do when he had such leaders put over him? Three commanders in two hours' time, three great numskulls, none of whom knew what was the right thing to do, and all of them giving different orders! Demoralized, were they? Good Heavens, it was enough to demoralize God Almighty himself, and all His angels!" And the inevitable accusation of treason was again made to do duty; Ducrot and de Wimpffen wanted to get three millions apiece out of Bismarck, as MacMahon had done.

Alone in advance of his staff General Douay sat on his horse a long time, his gaze bent on the distant positions of the enemy and in his eyes an expression of infinite melancholy. He made a minute and protracted observation of Hattoy, the shells from which came tumbling almost at his very feet; then, giving a glance at the plateau of Illy, called up an officer to carry an order to the brigade of the 5th corps that he had borrowed the day previous from General de Wimpffen, and which served to connect his right with the left of General Ducrot. He was distinctly heard to say these words:

"If the Prussians should once get possession of the Calvary it would be impossible for us to hold this position an hour; we should be driven into Sedan."

He rode off and was lost to view, together with his escort, at the entrance of the sunken road, and the German fire became hotter than before. They had doubtless observed the presence of the group of mounted officers; but now the shells, which hitherto had come from the front, began to fall upon them laterally, from the left; the batteries at Frenois, together with one which the enemy had carried across the river and posted on the peninsula of Iges, had established, in connection with the guns on Hattoy, an enfilading fire which swept the plateau de l'Algerie in its entire length and breadth. The position of the company now became most lamentable; the men, with death in front of them and on their flank, knew not which way to turn or which of the menacing perils to guard themselves against. In rapid succession three men were killed outright and two severely wounded.

It was then that Sergeant Sapin met the death that he had predicted for himself. He had turned his head, and caught sight of the approaching missile when it was too late for him to avoid it.

"Ah, here it is!" was all he said.

There was no terror in the thin face, with its big handsome eyes; it was only pale; very pale and inexpressibly mournful. The wound was in the abdomen.

"Oh! do not leave me here," he pleaded; "take me to the ambulance, I beseech you. Take me to the rear."

Rochas endeavored to silence him, and it was on his brutal lips to say that it was useless to imperil two comrades' lives for one whose wound was so evidently mortal, when his better nature made its influence felt and he murmured:

"Be patient for a little, my poor boy, and the litter-bearers will come and get you."

But the wretched man, whose tears were now flowing, kept crying, as one distraught that his dream of happiness was vanishing with his trickling life-blood:

"Take me away, take me away~"

Finally Captain Beaudoin, whose already unstrung nerves were further irritated by his pitiful cries, called for two volunteers to carry him to a little piece of woods a short way off where a flying ambulance had been established. Chouteau and Loubet jumped to their feet simultaneously, anticipating the others, seized the sergeant, one of them by the shoulders, the other by the legs, and bore him away on a run. They had gone but a little way, however, when they felt the body becoming rigid in the final convulsion; he was dying.

"I say, he's dead," exclaimed Loubet. "Let's leave him here."

But Chouteau, without relaxing his speed, angrily replied:

"Go ahead, you booby, will you! Do you take me for a fool, to leave him here and have them call us back!"

They pursued their course with the corpse until they came to the little wood, threw it down at the foot of a tree, and went their way. That was the last that was seen of them until nightfall.

The battery beside them had been strengthened by three additional guns; the cannonade on either side went on with increased fury, and in the hideous uproar terror-a wild, unreasoning terror-filled Maurice's soul. It was his first experience of the sensation; he had not until now felt that cold sweat trickling down his back, that terrible sinking at the pit of the stomach, that unconquerable desire to get on his feet and run, yelling and screaming, from the field. It was nothing more than the strain from which his nervous, high-strung temperament was suffering from reflex action; but Jean, who was observing him narrowly, detected the incipient crisis in the wandering, vacant eyes, and seizing him with his strong hand, held him down firmly at his side. The corporal lectured him paternally in a whisper, not mincing his words, but employing good, vigorous language to restore him to a sense of self-respect, for he knew by experience that a man in panic is not to be coaxed out of his cowardice. There were others also who were showing the white feather, among them Pache, who was whimpering involuntarily, in the low, soft voice of a little baby, his eyes suffused with tears. Lapoulle's stomach betrayed him and he was very ill; and there were many others who also found relief in vomiting, amid their comrade's loud jeers and laughter, which helped to restore their courage to them all.

"My God!" ejaculated Maurice, ghastly pale, his teeth chattering. "My God!"

Jean shook him roughly. "You infernal coward, are you going to be sick like those fellows over yonder? Behave yourself, or I'll box your ears."

He was trying to put heart into his friend by gruff but friendly speeches like the above, when they suddenly beheld a dozen dark forms emerging from a little wood upon their front and about four hundred yards away. Their spiked helmets announced them to be Prussians; the first Prussians they had had within reach of their rifles since the opening of the campaign. This first squad was succeeded by others, and in front of their position the little dust clouds that rose where the French shells struck were distinctly visible. It was all very vivid and clear-cut in the transparent air of morning; the Germans, outlined against the dark forest, presented the toy-like appearance of those miniature soldiers of lead that are the delight of children; then, as the enemy's shells began to drop in their vicinity with uncomfortable frequency, they withdrew and were lost to sight within the wood whence they had come.

But Beaudoin's company had seen them there once, and to their eyes they were there still; the chassepots seemed to go off of their own accord. Maurice was the first man to discharge his piece; Jean, Pache, Lapoulle and the others all followed suit. There had been no order given to commence firing, and the captain made an attempt to check it, but desisted upon Rochas's representation that it was absolutely necessary as a measure of relief for the men's pent-up feelings. So, then, they were at liberty to shoot at last, they could use up those cartridges that they had been lugging around with them for the last month, without ever burning a single one! The effect on Maurice in particular was electrical; the noise he made had the effect of dispelling his fear and blunting the keenness of his sensations. The little wood had resumed its former deserted aspect; not a leaf stirred, no more Prussians showed themselves; and still they kept on blazing away as madly as ever at the immovable trees.

Raising his eyes presently Maurice was startled to see Colonel de Vineuil sitting his big horse at no great distance, man and steed impassive and motionless as if carved from stone, patient were they under the leaden hail, with face turned toward the enemy. The entire regiment was now collected in that vicinity, the other companies being posted in the adjacent fields; the musketry fire seemed to be drawing nearer. The young man also beheld the regimental colors a little to the rear, borne aloft by the sturdy arm of the standard-bearer, but it was no longer the phantom flag that he had seen that morning, shrouded in mist and fog; the golden eagle flashed and blazed in the fierce sunlight, and the tri-colored silk, despite the rents and stains of many a battle, flaunted its bright hues defiantly to the breeze. Waving in the breath of the cannon, floating proudly against the blue of heaven, it shone like an emblem of victory.

And why, now that the day of battle had arrived, should not victory perch upon that banner? With that reflection Maurice and his companions kept on industriously wasting their powder on the distant wood, producing havoc there among the leaves and twigs.

3

Sleep did not visit Henriette's eyes that night. She knew her husband to be a prudent man, but the thought that he was in Bazeilles, so near the German lines, was cause to her of deep anxiety. She tried to soothe her apprehensions by reminding herself that she had his solemn promise to return at the first appearance of danger; it availed not, and at every instant she detected herself listening to catch the sound of his footstep on the stair. At ten o'clock, as she was about to go to bed, she opened her window, and resting her elbows on the sill, gazed out into the night.

The darkness was intense; looking downward, she could scarce discern the pavement of the Rue des Voyards, a narrow, obscure passage, overhung by old frowning mansions. Further on, in the direction of the college, a smoky street lamp burned dimly. A nitrous exhalation rose from the street; the squall of a vagrant cat; the heavy step of a belated soldier. From the city at her back came strange and alarming sounds: the patter of hurrying feet, an ominous, incessant rumbling, a muffled murmur without a name that chilled her blood. Her heart beat loudly in her bosom as she bent her ear to listen, and still she heard not the familiar echo of her husband's step at the turning of the street below.

Hours passed, and now distant lights that began to twinkle in the open fields beyond the ramparts excited afresh her apprehensions. It was so dark that it cost her an effort of memory to recall localities. She knew that the broad expanse that lay beneath her, reflecting a dim light, was the flooded meadows, and that flame that blazed up and was suddenly extinguished, surely it must be on la Marfee. But never, to her certain knowledge, had there been farmer's house or peasant's cottage on those heights; what, then, was the meaning of that light? And then on every hand, at Pont-Maugis, Noyers, Frenois, other fires arose, coruscating fitfully for an instant and giving mysterious indication of the presence of the swarming host that lay hidden in the bosom of the night. Yet more: there were strange sounds and voices in the air, subdued murmurings such as she had never heard before, and that made her start in terror; the stifled hum of marching men, the neighing and snorting of steeds, the clash of arms, hoarse words of command, given in guttural accents; an evil dream of a demoniac crew, a witch's sabbat, in the depths of those unholy shades. Suddenly a single cannon-shot rang out, ear-rending, adding fresh terror to the dead silence that succeeded it. It froze her very marrow; what could it mean? A signal, doubtless, telling of the successful completion of some movement, announcing that everything was ready, down there, and that now the sun might rise.

It was about two o'clock when Henriette, forgetting even to close her window, at last threw herself, fully dressed, upon her bed. Her anxiety and fatigue had stupefied her and benumbed her faculties. What could ail her, thus to shiver and burn alternately, she who was always so calm and self-reliant, moving with so light a step that those about her were unconscious of her existence? Finally she sank into a fitful, broken slumber that brought with it no repose, in which was present still that persistent sensation of impending evil that filled the dusky heavens. All at once, arousing her from her unrefreshing stupor, the firing commenced again, faint and muffled in the distance, not a single shot this time, but peal after peal following one another in quick succession. Trembling, she sat upright in bed. The firing continued. Where was she? The place seemed strange to her; she could not distinguish the objects in her chamber, which appeared to be filled with dense clouds of smoke. Then she remembered: the fog must have rolled in from the near-by river and entered the room through the window. Without, the distant firing was growing fiercer. She leaped from her bed and ran to the casement to listen.

Four o'clock was striking from a steeple in Sedan, and day was breaking, tingeing the purplish mists with a sickly, sinister light. It was impossible to discern objects; even the college buildings, distant but a few yards, were undistinguishable. Where could the firing be, mon Dieu! Her first thought was for her brother Maurice; for the reports were so indistinct that they seemed to her to come from the north, above the city; then, listening more attentively, her doubt became certainty; the cannonading was there, before her, and she trembled for her husband. It was surely at Bazeilles. For a little time, however, she suffered herself to be cheered by a ray of hope, for there were moments when the reports seemed to come from the right. Perhaps the fighting was at Donchery, where she knew that the French had not succeeded in blowing up the bridge. Then she lapsed into a condition of most horrible uncertainty; it seemed to be now at Donchery, now at Bazeilles; which, it was impossible to decide, there was such a ringing, buzzing sensation in her head. At last the feeling of suspense became so acute that she felt she could not endure it longer; she *must* know; every nerve in her body was quivering with the ungovernable desire, so she threw a shawl over her shoulders and left the house in quest of news.

When she had descended and was in the street Henriette hesitated a brief moment, for the little light that was in the east had not yet crept downward along the weather-blackened house-fronts to the roadway, and in the old city, shrouded in opaque fog, the darkness still reigned impenetrable. In the tap-room of a low pot-house in the Rue au Beurre, dimly lighted by a tallow candle, she saw two drunken Turcos and a woman. It was not until she turned into the Rue Maqua that she encountered any signs of life: soldiers slinking furtively along the sidewalk and hugging the walls, deserters probably, on the lookout for a place in which to hide; a stalwart trooper with despatches, searching for his captain and knocking thunderously at every door; a group of fat burghers, trembling with fear lest they had tarried there too long, and preparing to crowd themselves into one small carriole if so be they might yet reach Bouillon, in Belgium, whither half the population of Sedan had emigrated within the last two days. She instinctively turned her steps toward the Sous-Prefecture, where she might depend on receiving information, and her desire to avoid meeting acquaintances determined her to take a short cut through lanes and by-ways. On reaching the Rue du Four and the Rue des Laboureurs, however, she found an obstacle in her way; the place had been pre-empted by the ordnance department, and guns, caissons, forges were there in interminable array, having apparently been parked away in that remote corner the day before and then forgotten there. There was not so much as a sentry to guard them. It sent a chill to her heart to see all that artillery lying there silent and ineffective, sleeping its neglected sleep in the concealment of those deserted alleys. She was compelled to retrace her steps, therefore, which she did by passing through the Place du College to the Grande-Rue, where in front of the Hotel de l'Europe she saw a group of orderlies holding the chargers of some general officers, whose high-pitched voices were audible from the brilliantly lighted dining room. On the Place du Rivage and the Place Turenne the crowd was even greater still, composed of anxious groups of citizens, with women and children interspersed among the struggling, terror-stricken throng, hurrying in every direction; and there she saw a general emerge from the Hotel of the Golden Cross, swearing like a pirate, and spur his horse off up the street at a mad gallop, careless whom he might overturn. For a moment she seemed about to enter the Hotel de Ville, then changed her mind, and taking the Rue du Pont-de-Meuse, pushed on to the Sous-Prefecture.

Never had Sedan appeared to her in a light so tragically sinister as now, when she beheld it in the livid, forbidding light of early dawn, enveloped in its shroud of fog. The houses were lifeless and silent as tombs; many of them had been empty and abandoned for the last two days, others the terrified owners had closely locked and barred. Shuddering, the city awoke to the cares and occupations of the new day; the morning was fraught with chill misery in those streets, still half deserted, peopled only by a few frightened pedestrians and those hurrying fugitives, the remnant of the exodus of previous days. Soon the sun would rise and send down its cheerful light upon the scene; soon the city, overwhelmed in the swift-rising tide of disaster, would be crowded as it had never been before. It was half-past five o'clock; the roar of the cannon, caught and deadened among the tall dingy houses, sounded more faintly in her ears.

At the Sous-Prefecture Henriette had some acquaintance with the concierge's daughter, Rose by name, a pretty little blonde of refined appearance who was employed in Delaherche's factory. She made her way at once to the lodge; the mother was not there, but Rose received her with her usual amiability.

"Oh! dear lady, we are so tired we can scarcely stand; mamma has gone to lie down and rest a while. Just think! all night long people have been coming and going, and we have not been able to get a wink of sleep."

And burning to tell all the wonderful sights that she had been witness to since the preceding day, she did not wait to be questioned, but ran on volubly with her narrative.

"As for the marshal, he slept very well, but that poor Emperor! you can't think what suffering he has to endure! Yesterday evening, do you know, I had gone upstairs to help give out the linen, and as I entered the apartment that adjoins his dressing-room I heard groans, oh, *such* groans! just like someone dying. I thought a moment and knew it must be the Emperor, and I was so frightened I couldn't move; I just stood and trembled. It seems he has some terrible complaint that makes him cry out that way. When there are people around he holds in, but as soon as he is alone it is too much for him, and he groans and shrieks in a way to make your hair stand on end."

"Do you know where the fighting is this morning?" asked Henriette, desiring to check her loquacity.

Rose dismissed the question with a wave of her little hand and went on with her narrative.

"That made me curious to know more, you see, and I went upstairs four or five times during the night and listened, and every time it was just the same; I don't believe he was quiet an instant all night long, or got a minute's sleep. Oh! what a terrible thing it is to suffer like that with all he has to worry him! for everything is upside down; it is all a most dreadful mess. Upon my word, I believe those generals are out of their senses; such ghostly faces and frightened eyes! And people coming all the time, and doors banging and some men scolding and others crying, and the whole place like a sailor's boarding-house; officers drinking from bottles and going to bed in their boots! The Emperor is the best of the whole lot, and the one who gives least trouble, in the corner where he conceals himself and his suffering!" Then, in reply to Henriette's reiterated question: "The fighting? there has been fighting at Bazeilles this morning. A mounted officer brought word of it to the marshal, who went immediately to notify the Emperor. The marshal has been gone ten minutes, and I shouldn't wonder if the Emperor intends to follow him, for they are dressing him upstairs. I just now saw them combing him and plastering his face with all sorts of cosmetics."

But Henriette, having finally learned what she desired to know, rose to go.

"Thank you, Rose. I am in somewhat of a hurry this morning."

The young girl went with her to the street door, and took leave of her with a courteous:

"Glad to have been of service to you, Madame Weiss. I know that anything said to you will go no further."

Henriette hurried back to her house in the Rue des Voyards. She felt quite certain that her

146

husband would have returned, and even reflected that he would be alarmed at not finding her there, and hastened her steps in consequence. As she drew near the house she raised her eyes in the expectation of seeing him at the window watching for her, but the window, wide open as she had left it when she went out, was vacant, and when she had run up the stairs and given a rapid glance through her three rooms, it was with a sinking heart that she saw they were untenanted save for the chill fog and continuous roar of the cannonade. The distant firing was still going on. She went and stood for a moment at the window; although the encircling wall of vapor was not less dense than it had been before, she seemed to have a clearer apprehension, now that she had received oral information, of the details of the conflict raging at Bazeilles, the grinding sound of the mitrailleuses, the crashing volleys of the French batteries answering the German batteries in the distance. The reports seemed to be drawing nearer to the city, the battle to be waxing fiercer and fiercer with every moment.

Why did not Weiss return? He had pledged himself so faithfully not to outstay the first attack! And Henriette began to be seriously alarmed, depicting to herself the various obstacles that might have detained him: perhaps he had not been able to leave the village, perhaps the roads were blocked or rendered impassable by the projectiles. It might even be that something had happened him, but she put the thought aside and would not dwell on it, preferring to view things on their brighter side and finding in hope her safest mainstay and reliance. For an instant she harbored the design of starting out and trying to find her husband, but there were considerations that seemed to render that course inadvisable: supposing him to have started on his return, what would become of her should she miss him on the way? and what would be his anxiety should he come in and find her absent? Her guiding principle in all her thoughts and actions was her gentle, affectionate devotedness, and she saw nothing strange or out of the way in a visit to Bazeilles under such extraordinary circumstances, accustomed as she was, like an affectionate little woman, to perform her duty in silence and do the thing that she deemed best for their common interest. Where her husband was, there was her place; that was all there was about it.

She gave a sudden start and left the window, saying:

"Monsieur Delaherche, how could I forget-"

It had just come to her recollection that the cloth manufacturer had also passed the night at Bazeilles, and if he had returned would be able to give her the intelligence she wanted. She ran swiftly down the stairs again. In place of taking the more roundabout way by the Rue des Voyards, she crossed the little courtyard of her house and entered the passage that conducted to the huge structure that fronted on the Rue Maqua. As she came out into the great central garden, paved with flagstones now and retaining of its pristine glories only a few venerable trees, magnificent century-old elms, she was astonished to see a sentry mounting guard at the door of a carriage-house; then it occurred to her that she had been told the day before that the camp chests of the 7th corps had been deposited there for safe keeping, and it produced a strange impression on her mind that all the gold, millions, it was said to amount to, should be lying in that shed while the men for whom it was destined were being killed not far away. As she was about to ascend the private staircase, however, that conducted to the apartment of Gilberte, young Madame Delaherche, she experienced another surprise in an encounter that startled her so that she retraced her steps a little way, doubtful whether it would not be better to abandon her intention, and go home again. An officer, a captain, had crossed her path, as noiselessly as a phantom and vanishing as swiftly, and yet she had had time to recognize him, having seen him in the past at Gilberte's house in Charleville, in the days when she was still Madame Maginot. She stepped back a few steps in the courtyard and raised her eyes to the two tall windows of the bedroom, the blinds of which were closed, then dismissed her scruples and entered.

Upon reaching the first floor, availing herself of that privilege of old acquaintanceship by virtue of which one woman often drops in upon another for an unceremonious early morning chat, she was about to knock at the door of the dressing-room, but apparently someone had left the room hastily and failed to secure the door, so that it was standing ajar, and all she had to do was give it a push to find herself in the dressing room, whence she passed into the bedroom. From the lofty ceiling of the latter apartment depended voluminous curtains of red velvet, protecting the large double bed. The warm, moist air was fragrant with a faint perfume of Persian lilac, and there was no sound to break the silence save a gentle, regular respiration, scarcely audible.

"Gilberte!" said Henriette, very softly.

The young woman was sleeping peacefully, and the dim light that entered the room between the red curtains of the high windows displayed her exquisitely rounded head resting upon a naked arm and her profusion of beautiful hair straying in disorder over the pillow. Her lips were parted in a smile.

"Gilberte!"

She slightly moved and stretched her arms, without opening her eyes.

"Yes, yes; good-by. Oh! please-" Then, raising her head and recognizing Henriette: "What, is it you! How late is it?"

When she learned that it had not yet struck six she seemed disconcerted, assuming a sportive air to hide her embarrassment, saying it was unfair to come waking people up at such an hour. Then, to her friend, questioning her about her husband, she made answer:

"Why, he has not returned; I don't look for him much before nine o'clock. What makes you so eager to see him at this hour of the morning?"

Henriette's voice had a trace of sternness in it as she answered, seeing the other so smiling, so dull of comprehension in her happy waking.

"I tell you there has been fighting all the morning at Bazeilles, and I am anxious about my husband."

"Oh, my dear," exclaimed Gilberte, "I assure you there is not the slightest reason for your feeling so. My husband is so prudent that he would have been home long ago had there been any danger. Until you see him back here you may rest easy, take my word for it."

Henriette was struck by the justness of the argument; Delaherche, it was true, was distinctly not a man to expose himself uselessly. She was reassured, and went and drew the curtains and threw back the blinds; the tawny light from without, where the sun was beginning to pierce the fog with his golden javelins, streamed in a bright flood into the apartment. One of the windows was part way open, and in the soft air of the spacious bedroom, but now so close and stuffy, the two women could hear the sound of the guns. Gilberte, half recumbent, her elbow resting on the pillow, gazed out upon the sky with her lustrous, vacant eyes.

"So, then, they are fighting," she murmured. Her chemise had slipped downward, exposing a rosy, rounded shoulder, half hidden beneath the wandering raven tresses, and her person exhaled a subtle, penetrating odor, the odor of love. "They are fighting, so early in the morning, mon Dieu! It would be ridiculous if it were not for the horror of it."

But Henriette, in looking about the room, had caught sight of a pair of gauntlets, the gloves of a man, lying forgotten on a small table, and she started perceptibly. Gilberte blushed deeply, and extending her arms with a conscious, caressing movement, drew her friend to her and rested her head upon her bosom.

"Yes," she almost whispered, "I saw that you noticed it. Darling, you must not judge me too

148

severely. He is an old friend; I told you all about it at Charleville, long ago, you remember." Her voice sank lower still; there was something that sounded very like a laugh of satisfaction in her tender tones. "He pleaded so with me yesterday that I would see him just once more. Just think, this morning he is in action; he may be dead by this. How could I refuse him?" It was all so heroic and so charming, the contrast was so delicious between war's stern reality and tender sentiment; thoughtless as a linnet, she smiled again, notwithstanding her confusion. Never could she have found it in her heart to drive him from her door, when circumstances all were propitious for the interview. "Do you condemn me?"

Henriette had listened to her confidences with a very grave face. Such things surprised her, for she could not understand them; it must be that she was constituted differently from other women. Her heart that morning was with her husband, her brother, down there where the battle was raging. How was it possible that anyone could sleep so peacefully and be so gay and cheerful when the loved ones were in peril?

"But think of your husband, my dear, and of that poor young man as well. Does not your heart yearn to be with them? You do not reflect that their lifeless forms may be brought in and laid before your eyes at any moment."

Gilberte raised her adorable bare arm before her face to shield her vision from the frightful picture.

"O Heaven! what is that you say? It is cruel of you to destroy all the pleasure of my morning in this way. No, no; I won't think of such things. They are too mournful."

Henriette could not refrain from smiling in spite of her anxiety. She was thinking of the days of their girlhood, and how Gilberte's father, Captain de Vineuil, an old naval officer who had been made collector of customs at Charleville when his wounds had incapacitated him for active service, hearing his daughter cough and fearing for her the fate of his young wife, who had been snatched from his arms by that terrible disease, consumption, had sent her to live at a farm-house near Chene-Populeux. The little maid was not nine years old, and already she was a consummate actress-a perfect type of the village coquette, queening it over her playmates, tricked out in what old finery she could lay hands on, adorning herself with bracelets and tiaras made from the silver paper wrappings of the chocolate. She had not changed a bit when, later, at the age of twenty, she married Maginot, the inspector of woods and forests. Mezieres, a dark, gloomy town, surrounded by ramparts, was not to her taste, and she continued to live at Charleville, where the gay, generous life, enlivened by many festivities, suited her better. Her father was dead, and with a husband whom, by reason of his inferior social position, her friends and acquaintances treated with scant courtesy, she was absolutely mistress of her own actions. She did not escape the censure of the stern moralists who inhabit our provincial cities, and in those days was credited with many lovers; but of the gay throng of officers who, thanks to her father's old connection and her kinship to Colonel de Vineuil, disported themselves in her drawing-room, Captain Beaudoin was the only one who had really produced an impression. She was light and frivolous-nothing more-adoring pleasure and living entirely in the present, without the least trace of perverse inclination; and if she accepted the captain's attentions, it is pretty certain that she did it out of good-nature and love of admiration.

"You did very wrong to see him again," Henriette finally said, in her matter-of-fact way.

"Oh! my dear, since I could not possibly do otherwise, and it was only for just that once. You know very well I would die rather than deceive my new husband."

She spoke with much feeling, and seemed distressed to see her friend shake her head disapprovingly. They dropped the subject, and clasped each other in an affectionate embrace, notwithstanding their diametrically different natures. Each could hear the beating of the other's heart, and they might have understood the tongues those organs spoke-one, the slave of pleasure, wasting and squandering all that was best in herself; the other, with the mute heroism of a lofty soul, devoting herself to a single ennobling affection.

"But hark! how the cannon are roaring," Gilberte presently exclaimed. "I must make haste and dress."

The reports sounded more distinctly in the silent room now that their conversation had ceased. Leaving her bed, the young woman accepted the assistance of her friend, not caring to summon her maid, and rapidly made her toilet for the day, in order that she might be ready to go downstairs should she be needed there. As she was completing the arrangement of her hair there was a knock at the door, and, recognizing the voice of the elder Madame Delaherche, she hastened to admit her.

"Certainly, dear mother, you may come in."

With the thoughtlessness that was part of her nature, she allowed the old lady to enter without having first removed the gauntlets from the table. It was in vain that Henriette darted forward to seize them and throw them behind a chair. Madame Delaherche stood glaring for some seconds at the spot where they had been with an expression on her face as if she were slowly suffocating. Then her glance wandered involuntarily from object to object in the room, stopping finally at the great red-curtained bed, the coverings thrown back in disorder.

"I see that Madame Weiss has disturbed your slumbers. Then you were able to sleep, daughter?"

It was plain that she had had another purpose in coming there than to make that speech. Ah, that marriage that her son had insisted on contracting, contrary to her wish, at the mature age of fifty, after twenty years of joyless married life with a shrewish, bony wife; he, who had always until then deferred so to her will, now swayed only by his passion for this gay young widow, lighter than thistle-down! She had promised herself to keep watch over the present, and there was the past coming back to plague her. But ought she to speak? Her life in the household was one of silent reproach and protest; she kept herself almost constantly imprisoned in her chamber, devoting herself rigidly to the observances of her austere religion. Now, however, the wrong was so flagrant that she resolved to speak to her son.

Gilberte blushingly replied, without an excessive manifestation of embarrassment, however:

"Oh, yes, I had a few hours of refreshing sleep. You know that Jules has not returned-"

Madame Delaherche interrupted her with a grave nod of her head. Ever since the artillery had commenced to roar she had been watching eagerly for her son's return, but she was a Spartan mother, and concealed her gnawing anxiety under a cloak of brave silence. And then she remembered what was the object of her visit there.

"Your uncle, the colonel, has sent the regimental surgeon with a note in pencil, to ask if we will allow them to establish a hospital here. He knows that we have abundance of space in the factory, and I have already authorized the gentlemen to make use of the courtyard and the big drying-room. But you should go down in person-"

"Oh, at once, at once!" exclaimed Henriette, hastening toward the door. "We will do what we can to help."

Gilberte also displayed much enthusiasm for her new occupation as nurse; she barely took the time to throw a lace scarf over her head, and the three women went downstairs. When they reached the bottom and stood in the spacious vestibule, looking out through the main entrance, of which the leaves had been thrown wide back, they beheld a crowd collected in the street before the house. A low-hung carriage was advancing slowly along the roadway, a sort of carriole, drawn by a single horse, which a lieutenant of zouaves was leading by the bridle. They took it to be a wounded man that they were bringing to them, the first of their patients.

"Yes, yes! This is the place; this way!"

But they were quickly undeceived. The sufferer recumbent in the carriole was Marshal MacMahon, severely wounded in the hip, who, his hurt having been provisionally cared for in the cottage of a gardener, was now being taken to the Sous-Prefecture. He was bareheaded and partially divested of his clothing, and the gold embroidery on his uniform was tarnished with dust and blood. He spoke no word, but had raised his head from the pillow where it lay and was looking about him with a sorrowful expression, and perceiving the three women where they stood, wide eyed with horror, their joined hands resting on their bosom, in presence of that great calamity, the whole army stricken in the person of its chief at the very beginning of the conflict, he slightly bowed his head, with a faint, paternal smile. A few of those about him removed their hats; others, who had no time for such idle ceremony, were circulating the report of General Ducrot's appointment to the command of the army. It was half-past seven o'clock.

"And what of the Emperor?" Henriette inquired of a bookseller, who was standing at his door.

"He left the city near an hour ago," replied the neighbor. "I was standing by and saw him pass out at the Balan gate. There is a rumor that his head was taken off by a cannon ball."

But this made the grocer across the street furious. "Hold your tongue," he shouted, "it is an infernal lie! None but the brave will leave their bones there today!"

When near the Place du College the marshal's carriole was lost to sight in the gathering crowd, among whose numbers the most strange and contradictory reports from the field of battle were now beginning to circulate. The fog was clearing; the streets were bright with sunshine.

A hail, in no gentle terms, was heard proceeding from the courtyard: "Now then, ladies, here is where you are wanted, not outside!"

They all three hastened inside and found themselves in presence of Major Bouroche, who had thrown his uniform coat upon the floor, in a corner of the room, and donned a great white apron. Above the broad expanse of, as yet, unspotted white, his blazing, leonine eyes and enormous head, with shock of harsh, bristling hair, seemed to exhale energy and determination. So terrible did he appear to them that the women were his most humble servants from the very start, obedient to his every sign, treading on one another to anticipate his wishes.

"There is nothing here that is needed. Get me some linen; try and see if you can't find some more mattresses; show my men where the pump is-"

And they ran as if their life was at stake to do his bidding; were so active that they seemed to be ubiquitous.

The factory was admirably adapted for a hospital. The drying-room was a particularly noticeable feature, a vast apartment with numerous and lofty windows for light and ventilation, where they could put in a hundred beds and yet have room to spare, and at one side was a shed that seemed to have been built there especially for the convenience of the operators: three long tables had been brought in, the pump was close at hand, and a small grass-plot adjacent might serve as ante-chamber for the patients while awaiting their turn. And the handsome old elms, with their deliciously cool shade, roofed the spot in most agreeably.

Bouroche had considered it would be best to establish himself in Sedan at the commencement, foreseeing the dreadful slaughter and the inevitable panic that would sooner or later drive the troops to the shelter of the ramparts. All that he had deemed it necessary to leave with the regiment was two flying ambulances and some "first aids," that were to send him in the casualties as rapidly as possible after applying the primary dressings. The details of litter-bearers were all out there, whose duty it was to pick up the wounded under fire, and with them were the ambulance wagons and *fourgons* of the medical train. The two assistant-surgeons and three hospital stewards whom he had retained, leaving two assistants on the field, would doubtless be sufficient to perform what operations were necessary. He had also a corps of dressers under him. But he was not gentle in manner and language, for all he did was done impulsively, zealously, with all his heart and soul.

"Tonnerre de Dieu! how do you suppose we are going to distinguish the cases from one another when they begin to come in presently? Take a piece of charcoal and number each bed with a big figure on the wall overhead, and place those mattresses closer together, do you hear? We can strew some straw on the floor in that corner if it becomes necessary."

The guns were barking, preparing his work for him; he knew that at any moment now the first carriage might drive up and discharge its load of maimed and bleeding flesh, and he hastened to get all in readiness in the great, bare room. Outside in the shed the preparations were of another nature: the chests were opened and their contents arranged in order on a table, packages of lint, bandages, compresses, rollers, splints for fractured limbs, while on another table, alongside a great jar of cerate and a bottle of chloroform, were the surgical cases with their blood-curdling array of glittering instruments, probes, forceps, bistouries, scalpels, scissors, saws, an arsenal of implements of every imaginable shape adapted to pierce, cut, slice, rend, crush. But there was a deficient supply of basins.

"You must have pails, pots, jars about the house-something that will hold water. We can't work besmeared with blood all day, that's certain. And sponges, try to get me some sponges."

Madame Delaherche hurried away and returned, followed by three women bearing a supply of the desired vessels. Gilberte, standing by the table where the instruments were laid out, summoned Henriette to her side by a look and pointed to them with a little shudder. They grasped each other's hand and stood for a moment without speaking, but their mute clasp was eloquent of the solemn feeling of terror and pity that filled both their souls. And yet there was a difference, for one retained, even in her distress, the involuntary smile of her bright youth, while in the eyes of the other, pale as death, was the grave earnestness of the heart which, one love lost, can never love again.

"How terrible it must be, dear, to have an arm or leg cut off!"

"Poor fellows!"

Bouroche had just finished placing a mattress on each of the three tables, covering them carefully with oil-cloth, when the sound of horses' hoofs was heard outside and the first ambulance wagon rolled into the court. There were ten men in it, seated on the lateral benches, only slightly wounded; two or three of them carrying their arm in a sling, but the majority hurt about the head. They alighted with but little assistance, and the inspection of their cases commenced forthwith.

One of them, scarcely more than a boy, had been shot through the shoulder, and as Henriette was tenderly assisting him to draw off his greatcoat, an operation that elicited cries of pain, she took notice of the number of his regiment.

"Why, you belong to the 106th! Are you in Captain Beaudoin's company?"

No, he belonged to Captain Bonnaud's company, but for all that he was well acquainted with Corporal Macquart and felt pretty certain that his squad had not been under fire as yet. The tidings, meager as they were, sufficed to remove a great load from the young woman's heart:

her brother was alive and well; if now her husband would only return, as she was expecting every moment he would do, her mind would be quite at rest.

At that moment, just as Henriette raised her head to listen to the cannonade, which was then roaring with increased viciousness, she was thunderstruck to see Delaherche standing only a few steps away in the middle of a group of men, to whom he was telling the story of the frightful dangers he had encountered in getting from Bazeilles to Sedan. How did he happen to be there? She had not seen him come in. She darted toward him.

"Is not my husband with you?"

But Delaherche, who was just then replying to the fond questions of his wife and mother, was in no haste to answer.

"Wait, wait a moment." And resuming his narrative: "Twenty times between Bazeilles and Balan I just missed being killed. It was a storm, a regular hurricane, of shot and shell! And I saw the Emperor, too. Oh! but he is a brave man!-And after leaving Balan I ran-"

Henriette shook him by the arm.

"My husband?"

"Weiss? why, he stayed behind there, Weiss did."

"What do you mean, behind there?"

"Why, yes; he picked up the musket of a dead soldier, and is fighting away with the best of them."

"He is fighting, you say?-and why?"

"He must be out of his head, I think. He would not come with me, and of course I had to leave him."

Henriette gazed at him fixedly, with wide-dilated eyes. For a moment no one spoke; then in a calm voice she declared her resolution.

"It is well; I will go to him."

What, she, go to him? But it was impossible, it was preposterous! Delaherche had more to say of his hurricane of shot and shell. Gilberte seized her by the wrists to detain her, while Madame Delaherche used all her persuasive powers to convince her of the folly of the mad undertaking. In the same gentle, determined tone she repeated:

"It is useless; I will go to him."

She would only wait to adjust upon her head the lace scarf that Gilberte had been wearing and which the latter insisted she should accept. In the hope that his offer might cause her to abandon her resolve Delaherche declared that he would go with her at least as far as the Balan gate, but just then he caught sight of the sentry, who, in all the turmoil and confusion of the time, had been pacing uninterruptedly up and down before the building that contained the treasure chests of the 7th corps, and suddenly he remembered, was alarmed, went to give a look and assure himself that the millions were there still. In the meantime Henriette had reached the portico and was about to pass out into the street.

"Wait for me, won't you? Upon my word, you are as mad as your husband!"

Another ambulance had driven up, moreover, and they had to wait to let it pass in. It was smaller than the other, having but two wheels, and the two men whom it contained, both severely wounded, rested on stretchers placed upon the floor. The first one whom the attendants took out, using the most tender precaution, had one hand broken and his side torn by a splinter of shell; he was a mass of bleeding flesh. The second had his left leg shattered; and Bouroche, giving orders to extend the latter on one of the oil-cloth-covered mattresses, proceeded forthwith to operate on him, surrounded by the staring, pushing crowd of dressers and assistants. Madame Delaherche and Gilberte were seated near the grass-plot, employed in rolling bandages.

In the street outside Delaherche had caught up with Henriette.

"Come, my dear Madame Weiss, abandon this foolhardy undertaking. How can you expect to find Weiss in all that confusion? Most likely he is no longer there by this time; he is probably making his way home through the fields. I assure you that Bazeilles is inaccessible."

But she did not even listen to him, only increasing her speed, and had now entered the Rue de Menil, her shortest way to the Balan gate. It was nearly nine o'clock, and Sedan no longer wore the forbidding, funereal aspect of the morning, when it awoke to grope and shudder amid the despair and gloom of its black fog. The shadows of the houses were sharply defined upon the pavement in the bright sunlight, the streets were filled with an excited, anxious throng, through which orderlies and staff officers were constantly pushing their way at a gallop. The chief centers of attraction were the straggling soldiers who, even at this early hour of the day, had begun to stream into the city, minus arms and equipments, some of them slightly wounded, others in an extreme condition of nervous excitation, shouting and gesticulating like lunatics. And yet the place would have had very much its every-day aspect, had it not been for the tight-closed shutters of the shops, the lifeless house-fronts, where not a blind was open. Then there was the cannonade, that never-ceasing cannonade, beneath which earth and rocks, walls and foundations, even to the very slates upon the roofs, shook and trembled.

What between the damage that his reputation as a man of bravery and politeness would inevitably suffer should he desert Henriette in her time of trouble, and his disinclination to again face the iron hail on the Bazeilles road, Delaherche was certainly in a very unpleasant predicament. Just as they reached the Balan gate a bevy of mounted officers, returning to the city, suddenly came riding up, and they were parted. There was a dense crowd of people around the gate, waiting for news. It was all in vain that he ran this way and that, looking for the young woman in the throng; she must have been beyond the walls by that time, speeding along the road, and pocketing his gallantry for use on some future occasion, he said to himself aloud:

"Very well, so much the worse for her; it was too idiotic."

Then the manufacturer strolled about the city, bourgeois-like desirous to lose no portion of the spectacle, and at the same time tormented by a constantly increasing feeling of anxiety. How was it all to end? and would not the city suffer heavily should the army be defeated? The questions were hard ones to answer; he could not give a satisfactory solution to the conundrum when so much depended on circumstances, but none the less he was beginning to feel very uneasy for his factory and house in the Rue Maqua, whence he had already taken the precaution to remove his securities and valuables and bury them in a place of safety. He dropped in at the Hotel de Ville, found the Municipal Council sitting in permanent session, and loitered away a couple of hours there without hearing any fresh news, unless that affairs outside the walls were beginning to look very threatening. The army, under the pushing and hauling process, pushed back to the rear by General Ducrot during the hour and a half while the command was in his hands, hauled forward to the front again by de Wimpffen, his successor, knew not where to yield obedience, and the entire lack of plan and competent leadership, the incomprehensible vacillation, the abandonment of positions only to retake them again at terrible cost of life, all these things could not fail to end in ruin and disaster.

From there Delaherche pushed forward to the Sous-Prefecture to ascertain whether the Emperor had returned yet from the field of battle. The only tidings he gleaned here were of

Marshal MacMahon, who was said to be resting comfortably, his wound, which was not dangerous, having been dressed by a surgeon. About eleven o'clock, however, as he was again going the rounds, his progress was arrested for a moment in the Grande-Rue, opposite the Hotel de l'Europe, by a sorry cavalcade of dust-stained horsemen, whose jaded nags were moving at a walk, and at their head he recognized the Emperor, who was returning after having spent four hours on the battle-field. It was plain that death would have nothing to do with him. The big drops of anguish had washed the rouge from off those painted cheeks, the waxed mustache had lost its stiffness and drooped over the mouth, and in that ashen face, in those dim eyes, was the stupor of one in his last agony. One of the officers alighted in front of the hotel and proceeded to give some friends, who were collected there, an account of their route, from la Moncelle to Givonne, up the entire length of the little valley among the soldiers of the 1st corps, who had already been pressed back by the Saxons across the little stream to the right bank; and they had returned by the sunken road of the Fond de Givonne, which was even then in such an encumbered condition that had the Emperor desired to make his way to the front again he would have found the greatest difficulty in doing so. Besides, what would it have availed?

As Delaherche was drinking in these particulars with greedy ears a loud explosion shook the quarter. It was a shell, which had demolished a chimney in the Rue Sainte-Barbe, near the citadel. There was a general rush and scramble; men swore and women shrieked. He had flattened himself against the wall, when another explosion broke the windows in a house not far away. The consequences would be dreadful if they should shell Sedan; he made his way back to the Rue Maqua on a keen run, and was seized by such an imperious desire to learn the truth that he did not pause below stairs, but hurried to the roof, where there was a terrace that commanded a view of the city and its environs.

A glance of the situation served to reassure him; the German fire was not directed against the city; the batteries at Frenois and la Marfee were shelling the Plateau de l'Algerie over the roofs of the houses, and now that his alarm had subsided he could even watch with a certain degree of admiration the flight of the projectiles as they sailed over Sedan in a wide, majestic curve, leaving behind them a faint trail of smoke upon the air, like gigantic birds, invisible to mortal eye and to be traced only by the gray plumage shed by their pinions. At first it seemed to him quite evident that what damage had been done so far was the result of random practice by the Prussian gunners: they were not bombarding the city yet; then, upon further consideration, he was of opinion that their firing was intended as a response to the ineffectual fire of the few guns mounted on the fortifications of the place. Turning to the north he looked down from his position upon the extended and complex system of defenses of the citadel, the frowning curtains black with age, the green expanses of the turfed glacis, the stern bastions that reared their heads at geometrically accurate angles, prominent among them the three cyclopean salients, the Ecossais, the Grand Jardin, and la Rochette, while further to the west, in extension of the line, were Fort Nassau and Fort Palatinat, above the faubourg of Menil. The sight produced in him a melancholy impression of immensity and futility. Of what avail were they now against the powerful modern guns with their immense range? Besides, the works were not manned; cannon, ammunition, men were wanting. Some three weeks previously the governor had invited the citizens to organize and form a National Guard, and these volunteers were now doing duty as gunners; and thus it was that there were three guns in service at Palatinat, while at the Porte de Paris there may have been a half dozen. As they had only seven or eight rounds to each gun, however, the men husbanded their ammunition, limiting themselves to a shot every half hour, and that only as a sort of salve to their self-respect, for none of their missiles reached the enemy; all were lost in the meadows opposite them. Hence the enemy's batteries, disdainful of such small game, contemptuously pitched a shell at them

155

from time to time, out of charity, as it were.

Those batteries over across the river were objects of great interest to Delaherche. He was eagerly scanning the heights of la Marfee with his naked eye, when all at once he thought of the spy-glass with which he sometimes amused himself by watching the doings of his neighbors from the terrace. He ran downstairs and got it, returned and placed it in position, and as he was slowly sweeping the horizon and trees, fields, houses came within his range of vision, he lighted on that group of uniforms, at the angle of a pine wood, over the main battery at Frenois, of which Weiss had caught a glimpse from Bazeilles. To him, however, thanks to the excellence of his glass, it would have been no difficult matter to count the number of officers of the staff, so distinctly he made them out. Some of them were reclining carelessly on the grass, others were conversing in little groups, and in front of them all stood a solitary figure, a spare, well-proportioned man to appearances, in an unostentatious uniform, who yet asserted in some indefinable way his masterhood. It was the Prussian King, scarce half finger high, one of those miniature leaden toys that afford children such delight. Although he was not certain of this identity until later on the manufacturer found himself, by reason of some inexplicable attraction, constantly returning to that diminutive puppet, whose face, scarce larger than a pin's head, was but a pale point against the immense blue sky.

It was not midday yet, and since nine o'clock the master had been watching the movements, inexorable as fate, of his armies. Onward, ever onward, they swept, by roads traced for them in advance, completing the circle, slowly but surely closing in and enveloping Sedan in their living wall of men and guns. The army on his left, that had come up across the level plain of Donchery, was debouching still from the pass of Saint-Albert and, leaving Saint-Menges in its rear, was beginning to show its heads of columns at Fleigneux; and, in the rear of the XIth corps, then sharply engaged with General Douay's force, he could discern the Vth corps, availing itself of the shelter of the woods and advancing stealthily on Illy, while battery upon battery came wheeling into position, an ever-lengthening line of thundering guns, until the horizon was an unbroken ring of fire. On the right the army was now in undisputed possession of the valley of the Givonne; the XIIth corps had taken la Moncelle, the Guards had forced the passage of the stream at Daigny, compelling General Ducrot to seek the protection of the wood of la Garenne, and were pushing up the right bank, likewise in full march upon the plateau of Illy. Their task was almost done; one effort more, and up there at the north, among those barren fields, on the very verge of the dark forests of the Ardennes, the Crown Prince of Prussia would join hands with the Crown Prince of Saxony. To the south of Sedan the village of Bazeilles was lost to sight in the dense smoke of its burning houses, in the clouds of dun vapor that rose above the furious conflict.

And tranquilly, ever since the morning, the King had been watching and waiting. An hour yet, two hours, it might be three, it mattered not; it was only a question of time. Wheel and pinion, cog and lever, were working in harmony, the great engine of destruction was in motion, and soon would have run its course. In the center of the immense horizon, beneath the deep vault of sunlit sky, the bounds of the battlefield were ever becoming narrower, the black swarms were converging, closing in on doomed Sedan. There were fiery reflexions in the windows of the city; to the left, in the direction of the Faubourg de la Cassine, it seemed as if a house was burning. And outside the circle of flame and smoke, in the fields no longer trodden by armed men, over by Donchery, over by Carignan, peace, warm and luminous, lay upon the land; the bright waters of the Meuse, the lusty trees rejoicing in their strength, the broad, verdant meadows, the fertile, well-kept farms, all rested peacefully beneath the fervid noonday sun.

Turning to his staff, the King briefly called for information upon some point. It was the royal will to direct each move on the gigantic chessboard; to hold in the hollow of his hand the hosts who looked to him for guidance. At his left, a flock of swallows, affrighted by the noise of the cannonade, rose high in air, wheeled, and vanished in the south.

4

Between the city and Balan, Henriette got over the ground at a good, round pace. It was not yet nine o'clock; the broad footpath, bordered by gardens and pretty cottages, was as yet comparatively free, although as she approached the village it began to be more and more obstructed by flying citizens and moving troops. When she saw a great surge of the human tide advancing on her she hugged the walls and house-fronts, and by dint of address and perseverance slipped through, somehow. The fold of black lace that half concealed her fair hair and small, pale face, the sober gown that enveloped her slight form, made her an inconspicuous object among the throng; she went her way unnoticed by the by-passers, and nothing retarded her light, silent steps.

At Balan, however, she found the road blocked by a regiment of infanterie de marine. It was a compact mass of men, drawn up under the tall trees that concealed them from the enemy's observation, awaiting orders. She raised herself on tiptoe, and could not see the end; still, she made herself as small as she could and attempted to worm her way through. The men shoved her with their elbows, and the butts of their muskets made acquaintance with her ribs; when she had advanced a dozen paces there was a chorus of shouts and angry protests. A captain turned on her and roughly cried:

"Hi, there, you woman! are you crazy? Where are you going?"

"I am going to Bazeilles."

"What, to Bazeilles?"

There was a shout of laughter. The soldiers pointed at her with their fingers; she was the object of their witticisms. The captain, also, greatly amused by the incident, had to have his joke.

"You should take us along with you, my little dear, if you are going to Bazeilles. We were there a short while ago, and I am in hope that we shall go back there, but I can tell you that the temperature of the place is none too cool."

"I am going to Bazeilles to look for my husband," Henriette declared, in her gentle voice, while her blue eyes shone with undiminished resolution.

The laughter ceased; an old sergeant extricated her from the crowd that had collected around her, and forced her to retrace her steps.

"My poor child, you see it is impossible to get through. Bazeilles is no place for you. You will find your husband by and by. Come, listen to reason!"

She had to obey, and stood aside beneath the trees, raising herself on her toes at every moment to peer before her, firm in her resolve to continue her journey as soon as she should be allowed to pass. She learned the condition of affairs from the conversation that went on around her. Some officers were criticising with great acerbity the order for the abandonment of Bazeilles, which had occurred at a quarter-past eight, at the time when General Ducrot, taking over the command from the marshal, had considered it best to concentrate the troops on the plateau of Illy. What made matters worse was, that the valley of the Givonne having fallen into the hands of the Germans through the premature retirement of the 1st corps, the 12th corps, which was even then sustaining a vigorous attack in front, was overlapped on its left flank. Now that General de Wimpffen had relieved General Ducrot, it seemed that the original plan was to be carried out. Orders had been received to retake Bazeilles at every cost, and drive the Bavarians into the Meuse. And so, in the ranks of that regiment that had been halted there in full retreat at the entrance of the village and ordered to resume the offensive, there was much bitter feeling, and angry words were rife. Was ever such stupidity heard of? to make them abandon a position, and immediately tell them to turn round and retake it from the enemy! They were willing enough to risk their life in the cause, but no one cared to throw it away for nothing!

A body of mounted men dashed up the street and General de Wimpffen appeared among them, and raising himself erect on his stirrups, with flashing eyes, he shouted, in ringing tones:

"Friends, we cannot retreat; it would be ruin to us all. And if we do have to retreat, it shall be on Carignan, and not on Mezieres. But we shall be victorious! You beat the enemy this morning; you will beat them again!"

He galloped off on a road that conducted to la Moncelle. It was said that there had been a violent altercation between him and General Ducrot, each upholding his own plan, and decrying the plan of the other-one asserting that retreat by way of Mezieres had been impracticable all that morning; the other predicting that, unless they fell back on Illy, the army would be surrounded before night. And there was a great deal of bitter recrimination, each taxing the other with ignorance of the country and of the situation of the troops. The pity of it was that both were right.

But Henriette, meantime, had made an encounter that caused her to forget her project for a moment. In some poor outcasts; stranded by the wayside, she had recognized a family of honest weavers from Bazeilles, father, mother, and three little girls, of whom the largest was only nine years old. They were utterly disheartened and forlorn, and so weary and footsore that they could go no further, and had thrown themselves down at the foot of a wall.

"Alas! dear lady," the wife and mother said to Henriette, "we have lost our all. Our house-you know where our house stood on the Place de l'Eglise-well, a shell came and burned it. Why we and the children did not stay and share its fate I do not know-"

At these words the three little ones began to cry and sob afresh, while the mother, in distracted language, gave further details of the catastrophe.

"The loom, I saw it burn like seasoned kindling wood, and the bed, the chairs and tables, they blazed like so much straw. And even the clock-yes, the poor old clock that I tried to save and could not."

"My God! my God!" the man exclaimed, his eyes swimming with tears, "what is to become of us?"

Henriette endeavored to comfort them, but it was in a voice that quavered strangely.

"You have been preserved to each other, you are safe and unharmed; your three little girls are left you. What reason have you to complain?"

Then she proceeded to question them to learn how matters stood in Bazeilles, whether they had seen her husband, in what state they had left her house, but in their half-dazed condition they gave conflicting answers. No, they had not seen M. Weiss. One of the little girls, however, declared that she had seen him, and that he was lying on the ground with a great hole in his head, whereon the father gave her a box on the ear, bidding her hold her tongue and not tell such lies to the lady. As for the house, they could say with certainty that it was intact at the time of their flight; they even remembered to have observed, as they passed it, that the doors and windows were tightly secured, as if it was quite deserted. At that time, moreover, the only foothold that the Bavarians had secured for themselves was in the Place de l'Eglise, and to carry the village they would have to fight for it, street by street, house by house. They must have been gaining ground since then, though; all Bazeilles was in flames by that time, like enough, and not a wall left standing, thanks to the fierceness of the assailants and the resolution of the defenders. And so the poor creatures went on, with trembling, affrighted gestures, evoking the horrid sights their eyes had seen and telling their dreadful tale of slaughter and conflagration and corpses lying in heaps upon the ground.

"But my husband?" Henriette asked again.

They made no answer, only continued to cover their face with their hands and sob. Her cruel anxiety, as she stood there erect, with no outward sign of weakness, was only evinced by a slight quivering of the lips. What was she to believe? Vainly she told herself the child was mistaken; her mental vision pictured her husband lying there dead before her in the street with a bullet wound in the head. Again, that house, so securely locked and bolted, was another source of alarm; why was it so? was he no longer in it? The conviction that he was dead sent an icy chill to her heart; but perhaps he was only wounded, perhaps he was breathing still; and so sudden and imperious was the need she felt of flying to his side that she would again have attempted to force her passage through the troops had not the bugles just then sounded the order for them to advance.

The regiment was largely composed of raw, half-drilled recruits from Toulon, Brest, and Rochefort, men who had never fired a shot, but all that morning they had fought with a bravery and firmness that would not have disgraced veteran troops. They had not shown much aptitude for marching on the road from Rheims to Mouzon, weighted as they were with their unaccustomed burdens, but when they came to face the enemy their discipline and sense of duty made themselves felt, and notwithstanding the righteous anger that was in their hearts, the bugle had but to sound and they returned to brave the fire and encounter the foe. Three several times they had been promised a division to support them; it never came. They felt that they were deserted, sacrificed; it was the offering of their life that was demanded of them by those who, having first made them evacuate the place, were now sending them back into the fiery furnace of Bazeilles. And they knew it, and they gave their life, freely, without a murmur, closing up their ranks and leaving the shelter of the trees to meet afresh the storm of shell and bullets.

Henriette gave a deep sigh of relief; at last they were about to move! She followed them, with the hope that she might enter the village unperceived in their rear, prepared to run with them should they take the double-quick. But they had scarcely begun to move when they came to a halt again. The projectiles were now falling thick and fast; to regain possession of Bazeilles it would be necessary to dispute every inch of the road, occupying the cross-streets, the houses and gardens on either side of the way. A brisk fire of musketry proceeded from the head of the column, the advance was irregular, by fits and starts, every petty obstacle entailed a delay of many minutes. She felt that she would never attain her end by remaining there at the rear of the column, waiting for it to fight its way through, and with prompt decision she bent her course to the right and took a path that led downward between two hedges to the meadows.

Henriette's plan now was to reach Bazeilles by those broad levels that border the Meuse. She was not very clear about it in her mind, however, and continued to hasten onward in obedience to that blind instinct which had originally imparted to her its impulse. She had not gone far before she found herself standing and gazing in dismay at a miniature ocean which barred her further progress in that direction. It was the inundated fields, the low-lying lands that a measure of defense had converted into a lake, which had escaped her memory. For a single moment she thought of turning back; then, at the risk of leaving her shoes behind, she pushed on, hugging the bank, through the water that covered the grass and rose above her ankles. For a hundred yards her way, though difficult, was not impracticable; then she encountered a garden-wall directly in her front; the ground fell off sharply, and where the wall terminated the water was six feet deep. Her path was closed effectually; she clenched her

160

little fists and had to summon up all her resolution to keep from bursting into tears. When the first shock of disappointment had passed over she made her way along the enclosure and found a narrow lane that pursued a tortuous course among the scattered houses. She believed that now her troubles were at an end, for she was acquainted with that labyrinth, that tangled maze of passages, which, to one who had the key to them, ended at the village.

But the missiles seemed to be falling there even more thickly than elsewhere. Henriette stopped short in her tracks and all the blood in her body seemed to flow back upon her heart at a frightful detonation, so close that she could feel the wind upon her cheek. A shell had exploded directly before her and only a few yards away. She turned her head and scrutinized for a moment the heights of the left bank, above which the smoke from the German batteries was curling upward; she saw what she must do, and when she started on her way again it was with eyes fixed on the horizon, watching for the shells in order to avoid them. There was method in the rash daring of her proceeding, and all the brave tranquillity that the prudent little housewife had at her command. She was not going to be killed if she could help it; she wished to find her husband and bring him back with her, that they might yet have many days of happy life together. The projectiles still came tumbling frequently as ever; she sped along behind walls, made a cover of boundary stones, availed herself of every slight depression. But presently she came to an open space, a bit of unprotected road where splinters and fragments of exploded shells lay thick, and she was watching behind a shed for a chance to make a dash when she perceived, emerging from a sort of cleft in the ground in front of her, a human head and two bright eyes that peered about inquisitively. It was a little, bare-footed, ten-year-old boy, dressed in a shirt and ragged trousers, an embryonic tramp, who was watching the battle with huge delight. At every report his small black beady eyes would snap and sparkle, and he jubilantly shouted:

"Oh my! aint it bully!-Look out, there comes another one! don't stir! Boom! that was a rouser!-Don't stir! don't stir!"

And each time there came a shell he dived to the bottom of his hole, then reappeared, showing his dirty, elfish face, until it was time to duck again.

Henriette now noticed that the projectiles all came from Liry, while the batteries at Pont-Maugis and Noyers were confining their attention to Balan. At each discharge she could see the smoke distinctly, immediately afterward she heard the scream of the shell, succeeded by the explosion. Just then the gunners afforded them a brief respite; the bluish haze above the heights drifted slowly away upon the wind.

"They've stopped to take a drink, you can go your money on it," said the urchin. "Quick, quick, give me your hand! Now's the time to skip!"

He took her by the hand and dragged her along with him, and in this way they crossed the open together, side by side, running for dear life, with head and shoulders down. When they were safely ensconced behind a stack that opportunely offered its protection at the end of their course and turned to look behind them, they beheld another shell come rushing through the air and alight upon the shed at the very spot they had occupied so lately. The crash was fearful; the shed was knocked to splinters. The little ragamuffin considered that a capital joke, and fairly danced with glee.

"Bravo, hit 'em agin! that's the way to do it!-But it was time for us to skip, though, wasn't it?"

But again Henriette struck up against insurmountable obstacles in the shape of hedges and garden-walls, that offered absolutely no outlet. Her irrepressible companion, still wearing his broad grin and remarking that where there was a will there was a way, climbed to the coping of a wall and assisted her to scale it. On reaching the further side they found themselves in a kitchen garden among beds of peas and string-beans and surrounded by fences on every side;

161

their sole exit was through the little cottage of the gardener. The boy led the way, swinging his arms and whistling unconcernedly, with an expression on his face of most profound indifference. He pushed open a door that admitted him to a bedroom, from which he passed on into another room, where there was an old woman, apparently the only living being upon the premises. She was standing by a table, in a sort of dazed stupor; she looked at the two strangers who thus unceremoniously made a highway of her dwelling, but addressed them no word, nor did they speak a word to her. They vanished as quickly as they had appeared, emerging by the exit opposite their entrance upon an alley that they followed for a moment. After that there were other difficulties to be surmounted, and thus they went on for more than half a mile, scaling walls, struggling through hedges, availing themselves of every short cut that offered, it might be the door of a stable or the window of a cottage, as the exigencies of the case demanded. Dogs howled mournfully; they had a narrow escape from being run down by a cow that was plunging along, wild with terror. It seemed as if they must be approaching the village, however; there was an odor of burning wood in the air, and momentarily volumes of reddish smoke, like veils of finest gauze floating in the wind, passed athwart the sun and obscured his light.

All at once the urchin came to a halt and planted himself in front of Henriette.

"I say, lady, tell us where you're going, will you?"

"You can see very well where I am going; to Bazeilles."

He gave a low whistle of astonishment, following it up with the shrill laugh of the careless vagabond to whom nothing is sacred, who is not particular upon whom or what he launches his irreverent gibes.

"To Bazeilles-oh, no, I guess not; I don't think my business lies that way-I have another engagement. Bye-bye, ta-ta!"

He turned on his heel and was off like a shot, and she was none the wiser as to whence he came or whither he went. She had found him in a hole, she had lost sight of him at the corner of a wall, and never was she to set eyes on him again.

When she was alone again Henriette experienced a strange sensation of fear. He had been no protection to her, that scrubby urchin, but his chatter had been a distraction; he had kept her spirits up by his way of making game of everything, as if it was all one huge raree show. Now she was beginning to tremble, her strength was failing her, she, who by nature was so courageous. The shells no longer fell around her: the Germans had ceased firing on Bazeilles, probably to avoid killing their own men, who were now masters of the village; but within the last few minutes she had heard the whistling of bullets, that peculiar sound like the buzzing of a bluebottle fly, that she recognized by having heard it described. There was such a raging, roaring clamor rising to the heavens in the distance, the confused uproar of other sounds was so violent, that in it she failed to distinguish the report of musketry. As she was turning the corner of a house there was a deadened thud close at her ear, succeeded by the sound of falling plaster, which brought her to a sudden halt; it was a bullet that had struck the facade. She was pale as death, and asked herself if her courage would be sufficient to carry her through to the end; and before she had time to frame an answer, she received what seemed to her a blow from a hammer upon her forehead, and sank, stunned, upon her knees. It was a spent ball that had ricocheted and struck her a little above the left eyebrow with sufficient force to raise an ugly contusion. When she came to, raising her hands to her forehead, she withdrew them covered with blood. But the pressure of her fingers had assured her that the bone beneath was uninjured, and she said aloud, encouraging herself by the sound of her own voice:

"It is nothing, it is nothing. Come, I am not afraid; no, no! I am not afraid."

And it was the truth; she arose, and from that time walked amid the storm of bullets with absolute indifference, like one whose soul is parted from his body, who reasons not, who gives his life. She marched straight onward, with head erect, no longer seeking to shelter herself, and if she struck out at a swifter pace it was only that she might reach her appointed end more quickly. The death-dealing missiles pattered on the road before and behind her; twenty times they were near taking her life; she never noticed them. At last she was at Bazeilles, and struck diagonally across a field of lucerne in order to regain the road, the main street that traversed the village. Just as she turned into it she cast her eyes to the right, and there, some two hundred paces from her, beheld her house in a blaze. The flames were invisible against the bright sunlight; the roof had already fallen in in part, the windows were belching dense clouds of black smoke. She could restrain herself no longer, and ran with all her strength.

Ever since eight o'clock Weiss, abandoned by the retiring troops, had been a self-made prisoner there. His return to Sedan had become an impossibility, for the Bavarians, immediately upon the withdrawal of the French, had swarmed down from the park of Montivilliers and occupied the road. He was alone and defenseless, save for his musket and what few cartridges were left him, when he beheld before his door a little band of soldiers, ten in number, abandoned, like himself, and parted from their comrades, looking about them for a place where they might defend themselves and sell their lives dearly. He ran downstairs to admit them, and thenceforth the house had a garrison, a lieutenant, corporal and eight men, all bitterly inflamed against the enemy, and resolved never to surrender.

"What, Laurent, you here!" he exclaimed, surprised to recognize among the soldiers a tall, lean young man, who held in his hand a musket, doubtless taken from some corpse.

Laurent was dressed in jacket and trousers of blue cloth; he was helper to a gardener of the neighborhood, and had lately lost his mother and his wife, both of whom had been carried off by the same insidious fever.

"And why shouldn't I be?" he replied. "All I have is my skin, and I'm willing to give that. And then I am not such a bad shot, you know, and it will be just fun for me to blaze away at those rascals and knock one of 'em over every time."

The lieutenant and the corporal had already begun to make an inspection of the premises. There was nothing to be done on the ground floor; all they did was to push the furniture against the door and windows in such a way as to form as secure a barricade as possible. After attending to that they proceeded to arrange a plan for the defense of the three small rooms of the first floor and the open attic, making no change, however, in the measures that had been already taken by Weiss, the protection of the windows by mattresses, the loopholes cut here and there in the slats of the blinds. As the lieutenant was leaning from the window to take a survey of their surroundings, he heard the wailing cry of a child.

"What is that?" he asked.

Weiss looked from the window, and, in the adjoining dyehouse, beheld the little sick boy, Charles, his scarlet face resting on the white pillow, imploringly begging his mother to bring him a drink: his mother, who lay dead across the threshold, beyond hearing or answering. With a sorrowful expression he replied:

"It is a poor little child next door, there, crying for his mother, who was killed by a Prussian shell."

"Tonnerre de Dieu!" muttered Laurent, "how are they ever going to pay for all these things!"

As yet only a few random shots had struck the front of the house. Weiss and the lieutenant, accompanied by the corporal and two men, had ascended to the attic, where they were in better position to observe the road, of which they had an oblique view as far as the Place de

163

l'Eglise. The square was now occupied by the Bavarians, but any further advance was attended by difficulties that made them very circumspect. A handful of French soldiers, posted at the mouth of a narrow lane, held them in check for nearly a quarter of an hour, with a fire so rapid and continuous that the dead bodies lay in piles. The next obstacle they encountered was a house on the opposite corner, which also detained them some time before they could get possession of it. At one time a woman, with a musket in her hands, was seen through the smoke, firing from one of the windows. It was the abode of a baker, and a few soldiers were there in addition to the regular occupants; and when the house was finally carried there was a hoarse shout: "No quarter!" a surging, struggling, vociferating throng poured from the door and rolled across the street to the dead-wall opposite, and in the raging torrent were seen the woman's skirt, the jacket of a man, the white hairs of the grandfather; then came the crash of a volley of musketry, and the wall was splashed with blood from base to coping. This was a point on which the Germans were inexorable; everyone caught with arms in his hands and not belonging to some uniformed organization was shot without the formality of a trial, as having violated the law of nations. They were enraged at the obstinate resistance offered them by the village, and the frightful loss they had sustained during the five hours' conflict provoked them to the most atrocious reprisals. The gutters ran red with blood, the piled dead in the streets formed barricades, some of the more open places were charnel-houses, from whose depths rose the death-rattle of men in their last agony. And in every house that they had to carry by assault in this way men were seen distributing wisps of lighted straw, others ran to and fro with blazing torches, others smeared the walls and furniture with petroleum; soon whole streets were burning, Bazeilles was in flames.

And now Weiss's was the only house in the central portion of the village that still continued to hold out, preserving its air of menace, like some stern citadel determined not to yield.

"Look out! here they come!" shouted the lieutenant.

A simultaneous discharge from the attic and the first floor laid low three of the Bavarians, who had come forward hugging the walls. The remainder of the body fell back and posted themselves under cover wherever the street offered facilities, and the siege of the house began; the bullets pelted on the front like rattling hail. For nearly ten minutes the fusillade continued without cessation, damaging the stucco, but not doing much mischief otherwise, until one of the men whom the lieutenant had taken with him to the garret was so imprudent as to show himself at a window, when a bullet struck him square in the forehead, killing him instantly. It was plain that whoever exposed himself would do so at peril of his life.

"Doggone it! there's one gone!" growled the lieutenant. "Be careful, will you; there's not enough of us that we can afford to let ourselves be killed for the fun of it!"

He had taken a musket and was firing away like the rest of them from behind the protection of a shutter, at the same time watching and encouraging his men. It was Laurent, the gardener's helper, however, who more than all the others excited his wonder and admiration. Kneeling on the floor, with his chassepot peering out of the narrow aperture of a loophole, he never fired until absolutely certain of his aim; he even told in advance where he intended hitting his living target.

"That little officer in blue that you see down there, in the heart. -That other fellow, the tall, lean one, between the eyes.-I don't like the looks of that fat man with the red beard; I think I'll let him have it in the stomach."

And each time his man went down as if struck by lightning, hit in the very spot he had mentioned, and he continued to fire at intervals, coolly, without haste, there being no necessity for hurrying himself, as he remarked, since it would require too long a time to kill them all in that way.

"Oh! if I had but my eyes!" Weiss impatiently exclaimed. He had broken his spectacles a while before, to his great sorrow. He had his double eye-glass still, but the perspiration was rolling down his face in such streams that it was impossible to keep it on his nose. His usual calm collectedness was entirely lost in his over-mastering passion; and thus, between his defective vision and his agitated nerves, many of his shots were wasted.

"Don't hurry so, it is only throwing away powder," said Laurent. "Do you see that man who has lost his helmet, over yonder by the grocer's shop? Well, now draw a bead on him,- carefully, don't hurry. That's first-rate! you have broken his paw for him and made him dance a jig in his own blood."

Weiss, rather pale in the face, gave a look at the result of his marksmanship.

"Put him out of his misery," he said.

"What, waste a cartridge! Not, much. Better save it for another of 'em."

The besiegers could not have failed to notice the remarkable practice of the invisible sharpshooter in the attic. Whoever of them showed himself in the open was certain to remain there. They therefore brought up re-enforcements and placed them in position, with instructions to maintain an unremitting fire upon the roof of the building. It was not long before the attic became untenable; the slates were perforated as if they had been tissue paper, the bullets found their way to every nook and corner, buzzing and humming as if the room had been invaded by a swarm of angry bees. Death stared them all in the face if they remained there longer.

"We will go downstairs," said the lieutenant. "We can hold the first floor for awhile yet." But as he was making for the ladder a bullet struck him in the groin and he fell. "Too late, doggone it!"

Weiss and Laurent, aided by the remaining soldiers, carried him below, notwithstanding his vehement protests; he told them not to waste their time on him, his time had come; he might as well die upstairs as down. He was still able to be of service to them, however, when they had laid him on a bed in a room of the first floor, by advising them what was best to do.

"Fire into the mass," he said; "don't stop to take aim. They are too cowardly to risk an advance unless they see your fire begin to slacken."

And so the siege of the little house went on as if it was to last for eternity. Twenty times it seemed as if it must be swept away bodily by the storm of iron that beat upon it, and each time, as the smoke drifted away, it was seen amid the sulphurous blasts, torn, pierced, mangled, but erect and menacing, spitting fire and lead with undiminished venom from each one of its orifices. The assailants, furious that they should be detained for such length of time and lose so many men before such a hovel, yelled and fired wildly in the distance, but had not courage to attempt to carry the lower floor by a rush.

"Look out!" shouted the corporal, "there is a shutter about to fall!"

The concentrated fire had torn one of the inside blinds from its hinges, but Weiss darted forward and pushed a wardrobe before the window, and Laurent was enabled to continue his operations under cover. One of the soldiers was lying at his feet with his jaw broken, losing blood freely. Another received a bullet in his chest, and dragged himself over to the wall, where he lay gasping in protracted agony, while convulsive movements shook his frame at intervals. They were but eight, now, all told, not counting the lieutenant, who, too weak to speak, his back supported by the headboard of the bed, continued to give his directions by signs. As had been the case with the attic, the three rooms of the first floor were beginning to be untenable, for the mangled mattresses no longer afforded protection against the missiles; at every instant the plaster fell in sheets from the walls and ceiling, and the furniture was in process of demolition: the sides of the wardrobe yawned as if they had been cloven by an ax. And worse still, the ammunition was nearly exhausted.

"It's too bad!" grumbled Laurent; "just when everything was going so beautifully!"

But suddenly Weiss was struck with an idea.

"Wait!"

He had thought of the dead soldier up in the garret above, and climbed up the ladder to search for the cartridges he must have about him. A wide space of the roof had been crushed in; he saw the blue sky, a patch of bright, wholesome light that made him start. Not wishing to be killed, he crawled over the floor on his hands and knees, then, when he had the cartridges in his possession, some thirty of them, he made haste down again as fast his legs could carry him.

Downstairs, as he was sharing his newly acquired treasure with the gardener's lad, a soldier uttered a piercing cry and sank to his knees. They were but seven; and presently they were but six, a bullet having entered the corporal's head at the eye and lodged in the brain.

From that time on, Weiss had no distinct consciousness of what was going on around him; he and the five others continued to blaze away like lunatics, expending their cartridges, with not the faintest idea in their heads that there could be such a thing as surrender. In the three small rooms the floor was strewn with fragments of the broken furniture. Ingress and egress were barred by the corpses that lay before the doors; in one corner a wounded man kept up a pitiful wail that was frightful to hear. Every inch of the floor was slippery with blood; a thin stream of blood from the attic was crawling lazily down the stairs. And the air was scarce respirable, an air thick and hot with sulphurous fumes, heavy with smoke, filled with an acrid, nauseating dust; a darkness dense as that of night, through which darted the red flame-tongues of the musketry.

"By God's thunder!" cried Weiss, "they are bringing up artillery!"

It was true. Despairing of ever reducing that handful of madmen, who had consumed so much of their time, the Bavarians had run up a gun to the corner of the Place de l'Eglise, and were putting it into position; perhaps they would be allowed to pass when they should have knocked the house to pieces with their solid shot. And the honor there was to them in the proceeding, the gun trained on them down there in the square, excited the bitter merriment of the besieged; the utmost intensity of scorn was in their gibes. Ah! the cowardly bougres, with their artillery! Kneeling in his old place still, Laurent carefully adjusted his aim and each time picked off a gunner, so that the service of the piece became impossible, and it was five or six minutes before they fired their first shot. It ranged high, moreover, and only clipped away a bit of the roof.

But the end was now at hand. It was all in vain that they searched the dead men's belts; there was not a single cartridge left. With vacillating steps and haggard faces the six groped around the room, seeking what heavy objects they might find to hurl from the windows upon their enemies. One of them showed himself at the casement, vociferating insults, and shaking his fist; instantly he was pierced by a dozen bullets; and there remained but five. What were they to do? go down and endeavor to make their escape by way of the garden and the meadows? The question was never answered, for at that moment a tumult arose below, a furious mob came tumbling up the stairs: it was the Bavarians, who had at last thought of turning the position by breaking down the back door and entering the house by that way. For a brief moment a terrible hand-to-hand conflict raged in the small rooms among the dead bodies and the debris of the furniture. One of the soldiers had his chest transfixed by a bayonet thrust, the two others were made prisoners, while the attitude of the lieutenant, who had given up the ghost, was that of one about to give an order, his mouth open, his arm raised aloft.

While these things were occurring an officer, a big, flaxen-haired man, carrying a revolver in his hand, whose bloodshot eyes seemed bursting from their sockets, had caught sight of Weiss and Laurent, both in their civilian attire; he roared at them in French:

"Who are you, you fellows? and what are you doing here?"

Then, glancing at their faces, black with powder-stains, he saw how matters stood, he heaped insult and abuse on them in guttural German, in a voice that shook with anger. Already he had raised his revolver and was about to send a bullet into their heads, when the soldiers of his command rushed in, seized Laurent and Weiss, and hustled them out to the staircase. The two men were borne along like straws upon a mill-race amidst that seething human torrent, under whose pressure they were hurled from out the door and sent staggering, stumbling across the street to the opposite wall amid a chorus of execration that drowned the sound of their officers' voices. Then, for a space of two or three minutes, while the big fair-haired officer was endeavoring to extricate them in order to proceed with their execution, an opportunity was afforded them to raise themselves erect and look about them.

Other houses had taken fire; Bazeilles was now a roaring, blazing furnace. Flames had begun to appear at the tall windows of the church and were creeping upward toward the roof. Some soldiers who were driving a venerable lady from her home had compelled her to furnish the matches with which to fire her own beds and curtains. Lighted by blazing brands and fed by petroleum in floods, fires were rising and spreading in every quarter; it was no longer civilized warfare, but a conflict of savages, maddened by the long protracted strife, wreaking vengeance for their dead, their heaps of dead, upon whom they trod at every step they took. Yelling, shouting bands traversed the streets amid the scurrying smoke and falling cinders, swelling the hideous uproar into which entered sounds of every kind: shrieks, groans, the rattle of musketry, the crash of falling walls. Men could scarce see one another; great livid clouds drifted athwart the sun and obscured his light, bearing with them an intolerable stench of soot and blood, heavy with the abominations of the slaughter. In every quarter the work of death and destruction still went on: the human brute unchained, the imbecile wrath, the mad fury, of man devouring his brother man.

And Weiss beheld his house burn before his eyes. Some soldiers had applied the torch, others fed the flame by throwing upon it the fragments of the wrecked furniture. The rez-de-chaussee was quickly in a blaze, the smoke poured in dense black volumes from the wounds in the front and roof. But now the dyehouse adjoining was also on fire, and horrible to relate, the voice of little Charles, lying on his bed delirious with fever, could be heard through the crackling of the flames, beseeching his mother to bring him a draught of water, while the skirts of the wretched woman who, with her disfigured face, lay across the door-sill, were even then beginning to kindle.

"Mamma, mamma, I am thirsty! Mamma, bring me a drink of water-"

The weak, faint voice was drowned in the roar of the conflagration; the cheering of the victors rose on the air in the distance.

But rising above all other sounds, dominating the universal clamor, a terrible cry was heard. It was Henriette, who had reached the place at last, and now beheld her husband, backed up against the wall, facing a platoon of men who were loading their muskets.

She flew to him and threw her arms about his neck.

"My God! what is it! They cannot be going to kill you!"

Weiss looked at her with stupid, unseeing eyes. She! his wife, so long the object of his desire, so fondly idolized! A great shudder passed through his frame and he awoke to consciousness of his situation. What had he done? why had he remained there, firing at the enemy, instead of returning to her side, as he had promised he would do? It all flashed upon him now, as the darkness is illuminated by the lightning's glare: he had wrecked their happiness, they were to be parted, forever parted. Then he noticed the blood upon her forehead.

"Are you hurt?" he asked. "You were mad to come-"

She interrupted him with an impatient gesture.

"Never mind me; it is a mere scratch. But you, you! why are you here? They shall not kill you; I will not suffer it!"

The officer, who was endeavoring to clear the road in order to give the firing party the requisite room, came up on hearing the sound of voices, and beholding a woman with her arms about the neck of one of his prisoners, exclaimed loudly in French:

"Come, come, none of this nonsense here! Whence come you? What is your business here?"

"Give me my husband."

"What, is he your husband, that man? His sentence is pronounced; the law must take its course."

"Give me my husband."

"Come, be rational. Stand aside; we do not wish to harm you."

"Give me my husband."

Perceiving the futility of arguing with her, the officer was about to give orders to remove her forcibly from the doomed man's arms when Laurent, who until then had maintained an impassive silence, ventured to interfere.

"See here, Captain, I am the man who killed so many of your men; go ahead and shoot me-that will be all right, especially as I have neither chick nor child in all the world. But this gentleman's case is different; he is a married man, don't you see. Come, now, let him go; then you can settle my business as soon as you choose."

Beside himself with anger, the captain screamed:

"What is all this lingo? Are you trying to make game of me? Come, step out here, some one of you fellows, and take away this woman!"

He had to repeat his order in German, whereon a soldier came forward from the ranks, a short stocky Bavarian, with an enormous head surrounded by a bristling forest of red hair and beard, beneath which all that was to be seen were a pair of big blue eyes and a massive nose. He was besmeared with blood, a hideous spectacle, like nothing so much as some fierce, hairy denizen of the woods, emerging from his cavern and licking his chops, still red with the gore of the victims whose bones he has been crunching.

With a heart-rending cry Henriette repeated:

"Give me my husband, or let me die with him."

This seemed to cause the cup of the officer's exasperation to overrun; he thumped himself violently on the chest, declaring that he was no executioner, that he would rather die than harm a hair of an innocent head. There was nothing against her; he would cut off his right hand rather than do her an injury. And then he repeated his order that she be taken away.

As the Bavarian came up to carry out his instructions Henriette tightened her clasp on Weiss's neck, throwing all her strength into her frantic embrace.

"Oh, my love! Keep me with you, I beseech you; let me die with you-"

Big tears were rolling down his cheeks as, without answering, he endeavored to loosen the convulsive clasp of the fingers of the poor creature he loved so dearly.

"You love me no longer, then, that you wish to die without me. Hold me, keep me, do not let them take me. They will weary at last, and will kill us together."

He had loosened one of the little hands, and carried it to his lips and kissed it, working all the while to make the other release its hold.

"No, no, it shall not be! I will not leave thy bosom; they shall pierce my heart before reaching thine. I will not survive-"

But at last, after a long struggle, he held both the hands in his. Then he broke the silence that he had maintained until then, uttering one single word:

"Farewell, dear wife."

And with his own hands he placed her in the arms of the Bavarian, who carried her away. She shrieked and struggled, while the soldier, probably with intent to soothe her, kept pouring in her ear an uninterrupted stream of words in unmelodious German. And, having freed her head, looking over the shoulder of the man, she beheld the end.

It lasted not five seconds. Weiss, whose eye-glass had slipped from its position in the agitation of their parting, quickly replaced it upon his nose, as if desirous to look death in the face. He stepped back and placed himself against the wall, and the face of the self-contained, strong young man, as he stood there in his tattered coat, was sublimely beautiful in its expression of tranquil courage. Laurent, who stood beside him, had thrust his hands deep down into his pockets. The cold cruelty of the proceeding disgusted him; it seemed to him that they could not be far removed from savagery who could thus slaughter men before the eyes of their wives. He drew himself up, looked them square in the face, and in a tone of deepest contempt expectorated:

"Dirty pigs!"

The officer raised his sword; the signal was succeeded by a crashing volley, and the two men sank to the ground, an inert mass, the gardener's lad upon his face, the other, the accountant, upon his side, lengthwise of the wall. The frame of the latter, before he expired, contracted in a supreme convulsion, the eyelids quivered, the mouth opened as if he was about to speak. The officer came up and stirred him with his foot, to make sure that he was really dead.

Henriette had seen the whole: the fading eyes that sought her in death, the last struggle of the strong man in agony, the brutal boot spurning the corpse. And while the Bavarian still held her in his arms, conveying her further and further from the object of her love, she uttered no cry; she set her teeth, in silent fury, into what was nearest: a human hand, it chanced to be. The soldier gave vent to a howl of anguish and dashed her to the ground; raising his uninjured fist above her head he was on the point of braining her. And for a moment their faces were in contact; she experienced a feeling of intensest loathing for the monster, and that blood-stained hair and beard, those blue eyes, dilated and brimming with hate and rage, were destined to remain forever indelibly imprinted on her memory.

In after days Henriette could never account distinctly to herself for the time immediately succeeding these events. She had but one desire: to return to the spot where her loved one had died, take possession of his remains, and watch and weep over them; but, as in an evil dream, obstacles of every sort arose before her and barred the way. First a heavy infantry fire broke out afresh, and there was great activity among the German troops who were holding Bazeilles; it was due to the arrival of the infanterie de marine and other regiments that had been despatched from Balan to regain possession of the village, and the battle commenced to rage again with the utmost fury. The young woman, in company with a band of terrified citizens, was swept away to the left into a dark alley. The result of the conflict could not remain long doubtful, however; it was too late to reconquer the abandoned positions. For near half an hour the infantry struggled against superior numbers and faced death with splendid bravery, but the enemy's strength was constantly increasing, their re-enforcements were pouring in from every direction, the roads, the meadows, the park of Montivilliers; no force at our command could have dislodged them from the position, so dearly bought, where they had left thousands of their bravest. Destruction and devastation now had done their work; the place was a shambles, disgraceful to humanity, where mangled forms lay scattered among smoking ruins, and poor Bazeilles, having drained the bitter cup, went up at last in smoke and flame.

Henriette turned and gave one last look at her little house, whose floors fell in even as she gazed, sending myriads of little sparks whirling gayly upward on the air. And there, before her, prone at the wall's foot, she saw her husband's corpse, and in her despair and grief would fain have returned to him, but just then another crowd came up and surged around her, the bugles were sounding the signal to retire, she was borne away, she knew not how, among the retreating troops. Her faculty of self-guidance left her; she was as a bit of flotsam swept onward by the eddying human tide that streamed along the way. And that was all she could remember until she became herself again and found she was at Balan, among strangers, her head reclined upon a table in a kitchen, weeping.

5

It was nearly ten o'clock up on the Plateau de l'Algerie, and still the men of Beaudoin's company were resting supine, among the cabbages, in the field whence they had not budged since early morning. The cross fire from the batteries on Hattoy and the peninsula of Iges was hotter than ever; it had just killed two more of their number, and there were no orders for them to advance. Were they to stay there and be shelled all day, without a chance to see anything of the fighting?

They were even denied the relief of discharging their chassepots. Captain Beaudoin had at last put his foot down and stopped the firing, that senseless fusillade against the little wood in front of them, which seemed entirely deserted by the Prussians. The heat was stifling; it seemed to them that they should roast, stretched there on the ground under the blazing sky.

Jean was alarmed, on turning to look at Maurice, to see that he had declined his head and was lying, with closed eyes, apparently inanimate, his cheek against the bare earth. He was very pale, there was no sign of life in his face.

"Hallo there! what's the matter?"

But Maurice was only sleeping. The mental strain, conjointly with his fatigue, had been too much for him, in spite of the dangers that menaced them at every moment. He awoke with a start and stared about him, and the peace that slumber had left in his wide-dilated eyes was immediately supplanted by a look of startled affright as it dawned on him where he was. He had not the remotest idea how long he had slept; all he knew was that the state from which he had been recalled to the horrors of the battlefield was one of blessed oblivion and tranquillity.

"Hallo! that's funny; I must have been asleep!" he murmured. "Ah! it has done me good."

It was true that he suffered less from that pressure about his temples and at his heart, that horrible constriction that seems as if it would crush one's bones. He chaffed Lapoulle, who had manifested much uneasiness since the disappearance of Chouteau and Loubet and spoke of going to look for them. A capital idea! so he might get away and hide behind a tree, and smoke a pipe! Pache thought that the surgeons had detained them at the ambulance, where there was a scarcity of sick-bearers. That was a job that he had no great fancy for, to go around under fire and collect the wounded! And haunted by a lingering superstition of the country where he was born, he added that it was unlucky to touch a corpse; it brought death.

"Shut up, confound you!" roared Lieutenant Rochas. "Who is going to die?"

Colonel de Vineuil, sitting his tall horse, turned his head and gave a smile, the first that had been seen on his face that morning. Then he resumed his statue-like attitude, waiting for orders as impassively as ever under the tumbling shells.

Maurice's attention was attracted to the sick-bearers, whose movements he watched with interest as they searched for wounded men among the depressions of the ground. At the end of a sunken road, and protected by a low ridge not far from their position, a flying ambulance of first aid had been established, and its emissaries had begun to explore the plateau. A tent was quickly erected, while from the hospital van the attendants extracted the necessary supplies; compresses, bandages, linen, and the few indispensable instruments required for the hasty dressings they gave before dispatching the patients to Sedan, which they did as rapidly as they could secure wagons, the supply of which was limited. There was an assistant surgeon in charge, with two subordinates of inferior rank under him. In all the army none showed more gallantry and received less acknowledgment than the litter-bearers. They could be seen all over the field in their gray uniform, with the distinctive red badge on their cap and on their arm, courageously risking their lives and unhurriedly pushing forward through the thickest of the fire to the spots where men had been seen to fall. At times they would creep on hands and knees: would always take advantage of a hedge or ditch, or any shelter that was afforded by the conformation of the ground, never exposing themselves unnecessarily out of bravado. When at last they reached the fallen men their painful task commenced, which was made more difficult and protracted by the fact that many of the subjects had fainted, and it was hard to tell whether they were alive or dead. Some lay face downward with their mouths in a pool of blood, in danger of suffocating, others had bitten the ground until their throats were choked with dry earth, others, where a shell had fallen among a group, were a confused, intertwined heap of mangled limbs and crushed trunks. With infinite care and patience the bearers would go through the tangled mass, separating the living from the dead, arranging their limbs and raising the head to give them air, cleansing the face as well as they could with the means at their command. Each of them carried a bucket of cool water, which he had to use very savingly. And Maurice could see them thus engaged, often for minutes at a time, kneeling by some man whom they were trying to resuscitate, waiting for him to show some sign of life.

He watched one of them, some fifty yards away to the left, working over the wound of a little soldier from the sleeve of whose tunic a thin stream of blood was trickling, drop by drop. The man of the red cross discovered the source of the hemorrhage and finally checked it by compressing the artery. In urgent cases, like that of the little soldier, they rendered these partial attentions, locating fractures, bandaging and immobilizing the limbs so as to reduce the danger of transportation. And the transportation, even, was an affair that called for a great deal of judgment and ingenuity; they assisted those who could walk, and carried others, either in their arms, like little children, or pickaback when the nature of the hurt allowed it; at other times they united in groups of two, three, or four, according to the requirements of the case, and made a chair by joining their hands, or carried the patient off by his legs and shoulders in a recumbent posture. In addition to the stretchers provided by the medical department there were all sorts of temporary makeshifts, such as the stretchers improvised from knapsack straps and a couple of muskets. And in every direction on the unsheltered, shell-swept plain they could be seen, singly or in groups, hastening with their dismal loads to the rear, their heads bowed and picking their steps, an admirable spectacle of prudent heroism.

Maurice saw a pair on his right, a thin, puny little fellow lugging a burly sergeant, with both legs broken, suspended from his neck; the sight reminded the young man of an ant, toiling under a burden many times larger than itself; and even as he watched them a shell burst directly in their path and they were lost to view. When the smoke cleared away the sergeant was seen lying on his back, having received no further injury, while the bearer lay beside him, disemboweled. And another came up, another toiling ant, who, when he had turned his dead comrade on his back and examined him, took the sergeant up and made off with his load.

It gave Maurice a chance to read Lapoulle a lesson.

"I say, if you like the business, why don't you go and give that man a lift!"

For some little time the batteries at Saint-Menges had been thundering as if determined to surpass all previous efforts, and Captain Beaudoin, who was still tramping nervously up and down before his company line, at last stepped up to the colonel. It was a pity, he said, to waste the men's morale in that way and keep their minds on the stretch for hours and hours.

"I can't help it; I have no orders," the colonel stoically replied.

They had another glimpse of General Douay as he flew by at a gallop, followed by his staff. He had just had an interview with General de Wimpffen, who had ridden up to entreat him to hold his ground, which he thought he could promise to do, but only so long as the Calvary of

Illy, on his right, held out; Illy once taken, he would be responsible for nothing; their defeat would be inevitable. General de Wimpffen averred that the 1st corps would look out for the position at Illy, and indeed a regiment of zouaves was presently seen to occupy the Calvary, so that General Douay, his anxiety being relieved on that score, sent Dumont's division to the assistance of the 12th corps, which was then being hard pushed. Scarcely fifteen minutes later, however, as he was returning from the left, whither he had ridden to see how affairs were looking, he was surprised, raising his eyes to the Calvary, to see it was unoccupied; there was not a zouave to be seen there, they had abandoned the plateau that was no longer tenable by reason of the terrific fire from the batteries at Fleigneux. With a despairing presentiment of impending disaster he was spurring as fast as he could to the right, when he encountered Dumont's division, flying in disorder, broken and tangled in inextricable confusion with the debris of the 1st corps. The latter, which, after its retrograde movement, had never been able to regain possession of the posts it had occupied in the morning, leaving Daigny in the hands of the XIIth Saxon corps and Givonne to the Prussian Guards, had been compelled to retreat in a northerly direction across the wood of Garenne, harassed by the batteries that the enemy had posted on every summit from one end of the valley to the other. The terrible circle of fire and flame was contracting; a portion of the Guards had continued their march on Illy, moving from east to west and turning the eminences, while from west to east, in the rear of the XIth corps, now masters of Saint-Menges, the Vth, moving steadily onward, had passed Fleigneux and with insolent temerity was constantly pushing its batteries more and more to the front, and so contemptuous were they of the ignorance and impotence of the French that they did not even wait for the infantry to come up to support their guns. It was midday; the entire horizon was aflame, concentrating its destructive fire on the 7th and 1st corps.

Then General Douay, while the German artillery was thus preparing the way for the decisive movement that should make them masters of the Calvary, resolved to make one last desperate attempt to regain possession of the hill. He dispatched his orders, and throwing himself in person among the fugitives of Dumont's division, succeeded in forming a column which he sent forward to the plateau. It held its ground for a few minutes, but the bullets whistled so thick, the naked, treeless fields were swept by such a tornado of shot and shell, that it was not long before the panic broke out afresh, sweeping the men adown the slopes, rolling them up as straws are whirled before the wind. And the general, unwilling to abandon his project, ordered up other regiments.

A staff officer galloped by, shouting to Colonel de Vineuil as he passed an order that was lost in the universal uproar. Hearing, the colonel was erect in his stirrups in an instant, his face aglow with the gladness of battle, and pointing to the Calvary with a grand movement of his sword:

"Our turn has come at last, boys!" he shouted. "Forward!"

A thrill of enthusiasm ran through the ranks at the brief address, and the regiment put itself in motion. Beaudoin's company was among the first to get on its feet, which it did to the accompaniment of much good-natured chaff, the men declaring they were so rusty they could not move; the gravel must have penetrated their joints. The fire was so hot, however, that by the time they had advanced a few feet they were glad to avail themselves of the protection of a shelter trench that lay in their path, along which they crept in an undignified posture, bent almost double.

"Now, young fellow, look out for yourself!" Jean said to Maurice; "we're in for it. Don't let 'em see so much as the end of your nose, for if you do they will surely snip it off, and keep a sharp lookout for your legs and arms unless you have more than you care to keep. Those who come out of this with a whole skin will be lucky."

Maurice did not hear him very distinctly; the words were lost in the all-pervading clamor that buzzed and hummed in the young man's ears. He could not have told now whether he was afraid or not; he went forward because the others did, borne along with them in their headlong rush, without distinct volition of his own; his sole desire was to have the affair ended as soon as possible. So true was it that he was a mere drop in the on-pouring torrent that when the leading files came to the end of the trench and began to waver at the prospect of climbing the exposed slope that lay before them, he immediately felt himself seized by a sensation of panic, and was ready to turn and fly. It was simply an uncontrollable instinct, a revolt of the muscles, obedient to every passing breath.

Some of the men had already faced about when the colonel came hurrying up.

"Steady there, my children. You won't cause me this great sorrow; you won't behave like cowards. Remember, the 106th has never turned its back upon the enemy; will you be the first to disgrace our flag?"

And he spurred his charger across the path of the fugitives, addressing them individually, speaking to them, of their country, in a voice that trembled with emotion.

Lieutenant Rochas was so moved by his words that he gave way to an ungovernable fit of anger, raising his sword and belaboring the men with the flat as if it had been a club.

"You dirty loafers, I'll see whether you will go up there or not! I'll kick you up! About face! and I'll break the jaw of the first man that refuses to obey!"

But such an extreme measure as kicking a regiment into action was repugnant to the colonel.

"No, no, lieutenant; they will follow me. Won't you, my children? You won't let your old colonel fight it out alone with the Prussians! Up there lies the way; forward!"

He turned his horse and left the trench, and they did all follow, to a man, for he would have been considered the lowest of the low who could have abandoned their leader after that brave, kind speech. He was the only one, however, who, while crossing the open fields, erect on his tall horse, was cool and unconcerned; the men scattered, advancing in open order and availing themselves of every shelter afforded by the ground. The land sloped upward; there were fully five hundred yards of stubble and beet fields between them and the Calvary, and in place of the correctly aligned columns that the spectator sees advancing when a charge is ordered in field maneuvers, all that was to be seen was a loose array of men with rounded backs, singly or in small groups, hugging the ground, now crawling warily a little way on hands and knees, now dashing forward for the next cover, like huge insects fighting their way upward to the crest by dint of agility and address. The enemy's batteries seemed to have become aware of the movement; their fire was so rapid that the reports of the guns were blended in one continuous roar. Five men were killed, a lieutenant was cut in two.

Maurice and Jean had considered themselves fortunate that their way led along a hedge behind which they could push forward unseen, but the man immediately in front of them was shot through the temples and fell back dead in their arms; they had to cast him down at one side. By this time, however, the casualties had ceased to excite attention; they were too numerous. A man went by, uttering frightful shrieks and pressing his hands upon his protruding entrails; they beheld a horse dragging himself along with both thighs broken, and these anguishing sights, these horrors of the battlefield, affected them no longer. They were suffering from the intolerable heat, the noonday sun that beat upon their backs and burned like hot coals.

"How thirsty I am!" Maurice murmured. "My throat is like an ash barrel. Don't you notice that smell of something scorching, a smell like burning woolen?"

Jean nodded. "It was just the same at Solferino; perhaps it is the smell that always goes with

174

war. But hold, I have a little brandy left; we'll have a sup."

And they paused behind the hedge a moment and raised the flask to their lips, but the brandy, instead of relieving their thirst, burned their stomach. It irritated them, that nasty taste of burnt rags in their mouths. Moreover they perceived that their strength was commencing to fail for want of sustenance and would have liked to take a bite from the half loaf that Maurice had in his knapsack, but it would not do to stop and breakfast there under fire, and then they had to keep up with their comrades. There was a steady stream of men coming up behind them along the hedge who pressed them forward, and so, doggedly bending their backs to the task before them, they resumed their course. Presently they made their final rush and reached the crest. They were on the plateau, at the very foot of the Calvary, the old weather-beaten cross that stood between two stunted lindens.

"Good for our side!" exclaimed Jean; "here we are! But the next thing is to remain here!"

He was right; it was not the pleasantest place in the world to be in, as Lapoulle remarked in a doleful tone that excited the laughter of the company. They all lay down again, in a field of stubble, and for all that three men were killed in quick succession. It was pandemonium let loose up there on the heights; the projectiles from Saint-Menges, Fleigneux, and Givonne fell in such numbers that the ground fairly seemed to smoke, as it does at times under a heavy shower of rain. It was clear that the position could not be maintained unless artillery was dispatched at once to the support of the troops who had been sent on such a hopeless undertaking. General Douay, it was said, had given instructions to bring up two batteries of the reserve artillery, and the men were every moment turning their heads, watching anxiously for the guns that did not come.

"It is absurd, ridiculous!" declared Beaudoin, who was again fidgeting up and down before the company. "Who ever heard of placing a regiment in the air like this and giving it no support!" Then, observing a slight depression on their left, he turned to Rochas: "Don't you think, Lieutenant, that the company would be safer there?"

Rochas stood stock still and shrugged his shoulders. "It is six of one and half a dozen of the other, Captain. My opinion is that we will do better to stay where we are."

Then the captain, whose principles were opposed to swearing, forgot himself.

"But, good God! there won't a man of us escape! We can't allow the men to be murdered like this!"

And he determined to investigate for himself the advantages of the position he had mentioned, but had scarcely taken ten steps when he was lost to sight in the smoke of an exploding shell; a splinter of the projectile had fractured his right leg. He fell upon his back, emitting a shrill cry of alarm, like a woman's.

"He might have known as much," Rochas muttered. "There's no use his making such a fuss over it; when the dose is fixed for one, he has to take it."

Some members of the company had risen to their feet on seeing their captain fall, and as he continued to call lustily for assistance, Jean finally ran to him, immediately followed by Maurice.

"Friends, friends, for Heaven's sake do not leave me here; carry me to the ambulance!"

"Dame, Captain, I don't know that we shall be able to get so far, but we can try."

As they were discussing how they could best take hold to raise him they perceived, behind the hedge that had sheltered them on their way up, two stretcher-bearers who seemed to be waiting for something to do, and finally, after protracted signaling, induced them to draw near. All would be well if they could only get the wounded man to the ambulance without accident, but the way was long and the iron hail more pitiless than ever.

The bearers had tightly bandaged the injured limb in order to keep the bones in position and were about to bear the captain off the field on what children call a "chair," formed by joining their hands and slipping an arm of the patient over each of their necks, when Colonel de Vineuil, who had heard of the accident, came up, spurring his horse. He manifested much emotion, for he had known the young man ever since his graduation from Saint-Cyr.

"Cheer up, my poor boy; have courage. You are in no danger; the doctors will save your leg."

The captain's face wore an expression of resignation, as if he had summoned up all his courage to bear his misfortune manfully.

"No, my dear Colonel; I feel it is all up with me, and I would rather have it so. The only thing that distresses me is the waiting for the inevitable end."

The bearers carried him away, and were fortunate enough to reach the hedge in safety, behind which they trotted swiftly away with their burden. The colonel's eyes followed them anxiously, and when he saw them reach the clump of trees where the ambulance was stationed a look of deep relief rose to his face.

"But you, Colonel," Maurice suddenly exclaimed, "you are wounded too!"

He had perceived blood dripping from the colonel's left boot. A projectile of some description had carried away the heel of the foot-covering and forced the steel shank into the flesh.

M. de Vineuil bent over his saddle and glanced unconcernedly at the member, in which the sensation at that time must have been far from pleasurable.

"Yes, yes," he replied, "it is a little remembrance that I received a while ago. A mere scratch, that don't prevent me from sitting my horse-" And he added, as he turned to resume his position to the rear of his regiment: "As long as a man can stick on his horse he's all right."

At last the two batteries of reserve artillery came up. Their arrival was an immense relief to the anxiously expectant men, as if the guns were to be a rampart of protection to them and at the same time demolish the hostile batteries that were thundering against them from every side. And then, too, it was in itself an exhilarating spectacle to see the magnificent order they preserved as they came dashing up, each gun followed by its caisson, the drivers seated on the near horse and holding the off horse by the bridle, the cannoneers bolt upright on the chests, the chiefs of detachment riding in their proper position on the flank. Distances were preserved as accurately as if they were on parade, and all the time they were tearing across the fields at headlong speed, with the roar and crash of a hurricane.

Maurice, who had lain down again, arose and said to Jean in great excitement:

"Look! over there on the left, that is Honore's battery. I can recognize the men."

Jean gave him a back-handed blow that brought him down to his recumbent position.

"Lie down, will you! and make believe dead!"

But they were both deeply interested in watching the maneuvers of the battery, and never once removed their eyes from it; it cheered their heart to witness the cool and intrepid activity of those men, who, they hoped, might yet bring victory to them.

The battery had wheeled into position on a bare summit to the left, where it brought up all standing; then, quick as a flash, the cannoneers leaped from the chests and unhooked the limbers, and the drivers, leaving the gun in position, drove fifteen yards to the rear, where they wheeled again so as to bring team and limber face to the enemy and there remained, motionless as statues. In less time than it takes to tell it the guns were in place, with the proper intervals between them, distributed into three sections of two guns each, each section commanded by a lieutenant, and over the whole a captain, a long maypole of a man, who made a terribly conspicuous landmark on the plateau. And this captain, having first made a brief calculation, was heard to shout:

"Sight for sixteen hundred yards!"

Their fire was to be directed upon a Prussian battery, screened by some bushes, to the left of Fleigneux, the shells from which were rendering the position of the Calvary untenable.

"Honore's piece, you see," Maurice began again, whose excitement was such that he could not keep still, "Honore's piece is in the center section. There he is now, bending over to speak to the gunner; you remember Louis, the gunner, don't you? the little fellow with whom we had a drink at Vouziers? And that fellow in the rear, who sits so straight on his handsome chestnut, is Adolphe, the driver-"

First came the gun with its chief and six cannoneers, then the limber with its four horses ridden by two men, beyond that the caisson with its six horses and three drivers, still further to the rear were the prolonge, forge, and battery wagon; and this array of men, horses and *materiel* extended to the rear in a straight unbroken line of more than a hundred yards in length; to say nothing of the spare caisson and the men and beasts who were to fill the places of those removed by casualties, who were stationed at one side, as much as possible out of the enemy's line of fire.

And now Honore was attending to the loading of his gun. The two men whose duty it was to fetch the cartridge and the projectile returned from the caisson, where the corporal and the artificer were stationed; two other cannoneers, standing at the muzzle of the piece, slipped into the bore the cartridge, a charge of powder in an envelope of serge, and gently drove it home with the rammer, then in like manner introduced the shell, the studs of which creaked faintly in the spirals of the rifling. When the primer was inserted in the vent and all was in readiness, Honore thought he would like to point the gun himself for the first shot, and throwing himself in a semi-recumbent posture on the trail, working with one hand the screw that regulated the elevation, with the other he signaled continually to the gunner, who, standing behind him, moved the piece by imperceptible degrees to right or left with the assistance of the lever.

"That ought to be about right," he said as he arose.

The captain came up, and stooping until his long body was bent almost double, verified the elevation. At each gun stood the assistant gunner, waiting to pull the lanyard that should ignite the fulminate by means of a serrated wire. And the orders were given in succession, deliberately, by number:

"Number one, Fire! Number two, Fire!"

Six reports were heard, the guns recoiled, and while they were being brought back to position the chiefs of detachment observed the effect of the shots and found that the range was short. They made the necessary correction and the evolution was repeated, in exactly the same manner as before; and it was that cool precision, that mechanical routine of duty, without agitation and without haste, that did so much to maintain the *morale* of the men. They were a little family, united by the tie of a common occupation, grouped around the gun, which they loved and reverenced as if it had been a living thing; it was the object of all their care and attention, to it all else was subservient, men, horses, caisson, everything. Thence also arose the spirit of unity and cohesion that animated the battery at large, making all its members work together for the common glory and the common good, like a well-regulated household.

The 106th had cheered lustily at the completion of the first round; they were going to make those bloody Prussian guns shut their mouths at last! but their elation was succeeded by dismay when it was seen that the projectiles fell short, many of them bursting in the air and never reaching the bushes that served to mask the enemy's artillery.

"Honore," Maurice continued, "says that all the other pieces are popguns and that his old girl is the only one that is good for anything. Ah, his old girl! He talks as if she were his wife and there were not another like her in the world! Just notice how jealously he watches her and makes the men clean her off! I suppose he is afraid she will overheat herself and take cold!"

He continued rattling on in this pleasant vein to Jean, both of them cheered and encouraged by the cool bravery with which the artillerymen served their guns; but the Prussian batteries, after firing three rounds, had now got the range, which, too long at the beginning, they had at last ciphered down to such a fine point that their shells were landed invariably among the French pieces, while the latter, notwithstanding the efforts that were made to increase their range, still continued to place their projectiles short of the enemy's position. One of Honore's cannoneers was killed while loading the piece; the others pushed the body out of their way, and the service went on with the same methodical precision, with neither more nor less haste. In the midst of the projectiles that fell and burst continually the same unvarying rhythmical movements went on uninterruptedly about the gun; the cartridge and shell were introduced, the gun was pointed, the lanyard pulled, the carriage brought back to place; and all with such undeviating regularity that the men might have been taken for automatons, devoid of sight and hearing.

What impressed Maurice, however, more than anything else, was the attitude of the drivers, sitting straight and stiff in their saddles fifteen yards to the rear, face to the enemy. There was Adolphe, the broad-chested, with his big blond mustache across his rubicund face; and who shall tell the amount of courage a man must have to enable him to sit without winking and watch the shells coming toward him, and he not allowed even to twirl his thumbs by way of diversion! The men who served the guns had something to occupy their minds, while the drivers, condemned to immobility, had death constantly before their eyes, and plenty of leisure to speculate on probabilities. They were made to face the battlefield because, had they turned their backs to it, the coward that so often lurks at the bottom of man's nature might have got the better of them and swept away man and beast. It is the unseen danger that makes dastards of us; that which we can see we brave. The army has no more gallant set of men in its ranks than the drivers in their obscure position.

Another man had been killed, two horses of a caisson had been disemboweled, and the enemy kept up such a murderous fire that there was a prospect of the entire battery being knocked to pieces should they persist in holding that position longer. It was time to take some step to baffle that tremendous fire, notwithstanding the danger there was in moving, and the captain unhesitatingly gave orders to bring up the limbers.

The risky maneuver was executed with lightning speed; the drivers came up at a gallop, wheeled their limber into position in rear of the gun, when the cannoneers raised the trail of the piece and hooked on. The movement, however, collecting as it did, momentarily, men and horses on the battery front in something of a huddle, created a certain degree of confusion, of which the enemy took advantage by increasing the rapidity of their fire; three more men dropped. The teams darted away at breakneck speed, describing an arc of a circle among the fields, and the battery took up its new position some fifty or sixty yards more to the right, on a gentle eminence that was situated on the other flank of the 106th. The pieces were unlimbered, the drivers resumed their station at the rear, face to the enemy, and the firing was reopened; and so little time was lost between leaving their old post and taking up the new that the earth had barely ceased to tremble under the concussion.

Maurice uttered a cry of dismay, when, after three attempts, the Prussians had again got their range; the first shell landed squarely on Honore's gun. The artilleryman rushed forward, and

178

with a trembling hand felt to ascertain what damage had been done his pet; a great wedge had been chipped from the bronze muzzle. But it was not disabled, and the work went on as before, after they had removed from beneath the wheels the body of another cannoneer, with whose blood the entire carriage was besplashed.

"It was not little Louis; I am glad of that," said Maurice, continuing to think aloud. "There he is now, pointing his gun; he must be wounded, though, for he is only using his left arm. Ah, he is a brave lad, is little Louis; and how well he and Adolphe get on together, in spite of their little tiffs, only provided the gunner, the man who serves on foot, shows a proper amount of respect for the driver, the man who rides a horse, notwithstanding that the latter is by far the more ignorant of the two. Now that they are under fire, though, Louis is as good a man as Adolphe-"

Jean, who had been watching events in silence, gave utterance to a distressful cry:

"They will have to give it up! No troops in the world could stand such a fire."

Within the space of five minutes the second position had become as untenable as was the first; the projectiles kept falling with the same persistency, the same deadly precision. A shell dismounted a gun, fracturing the chase, killing a lieutenant and two men. Not one of the enemy's shots failed to reach, and at each discharge they secured a still greater accuracy of range, so that if the battery should remain there another five minutes they would not have a gun or a man left. The crushing fire threatened to wipe them all out of existence.

Again the captain's ringing voice was heard ordering up the limbers. The drivers dashed up at a gallop and wheeled their teams into place to allow the cannoneers to hook on the guns, but before Adolphe had time to get up Louis was struck by a fragment of shell that tore open his throat and broke his jaw; he fell across the trail of the carriage just as he was on the point of raising it. Adolphe was there instantly, and beholding his prostrate comrade weltering in his blood, jumped from his horse and was about to raise him to his saddle and bear him away. And at that moment, just as the battery was exposed flank to the enemy in the act of wheeling, offering a fair target, a crashing discharge came, and Adolphe reeled and fell to the ground, his chest crushed in, with arms wide extended. In his supreme convulsion he seized his comrade about the body, and thus they lay, locked in each other's arms in a last embrace, "married" even in death.

Notwithstanding the slaughtered horses and the confusion that that death-dealing discharge had caused among the men, the battery had rattled up the slope of a hillock and taken post a few yards from the spot where Jean and Maurice were lying. For the third time the guns were unlimbered, the drivers retired to the rear and faced the enemy, and the cannoneers, with a gallantry that nothing could daunt, at once reopened fire.

"It is as if the end of all things were at hand!" said Maurice, the sound of whose voice was lost in the uproar.

It seemed indeed as if heaven and earth were confounded in that hideous din. Great rocks were cleft asunder, the sun was hid from sight at times in clouds of sulphurous vapor. When the cataclysm was at its height the horses stood with drooping heads, trembling, dazed with terror. The captain's tall form was everywhere upon the eminence; suddenly he was seen no more; a shell had cut him clean in two, and he sank, as a ship's mast that is snapped off at the base.

But it was about Honore's gun, even more than the others, that the conflict raged, with cool efficiency and obstinate determination. The non-commissioned officer found it necessary to forget his chevrons for the time being and lend a hand in working the piece, for he had now but three cannoneers left; he pointed the gun and pulled the lanyard, while the others brought

179

ammunition from the caisson, loaded, and handled the rammer and the sponge. He had sent for men and horses from the battery reserves that were kept to supply the places of those removed by casualties, but they were slow in coming, and in the meantime the survivors must do the work of the dead. It was a great discouragement to all that their projectiles ranged short and burst almost without exception in the air, inflicting no injury on the powerful batteries of the foe, the fire of which was so efficient. And suddenly Honore let slip an oath that was heard above the thunder of the battle; ill-luck, ill-luck, nothing but ill-luck! the right wheel of his piece was smashed! Tonnerre de Dieu! what a state she was in, the poor darling! stretched on her side with a broken paw, her nose buried in the ground, crippled and good for nothing! The sight brought big tears to his eyes, he laid his trembling hand upon the breech, as if the ardor of his love might avail to warm his dear mistress back to life. And the best gun of them all, the only one that had been able to drop a few shells among the enemy! Then suddenly he conceived a daring project, nothing less than to repair the injury there and then, under that terrible fire. Assisted by one of his men he ran back to the caisson and secured the spare wheel that was attached to the rear axle, and then commenced the most dangerous operation that can be executed on a battlefield. Fortunately the extra men and horses that he had sent for came up just then, and he had two cannoneers to lend him a hand.

For the third time, however, the strength of the battery was so reduced as practically to disable it. To push their heroic daring further would be madness; the order was given to abandon the position definitely.

"Make haste, comrades!" Honore exclaimed. "Even if she is fit for no further service we'll carry her off; those fellows shan't have her!"

To save the gun, even as men risk their life to save the flag; that was his idea. And he had not ceased to speak when he was stricken down as by a thunderbolt, his right arm torn from its socket, his left flank laid open. He had fallen upon his gun he loved so well, and lay there as if stretched on a bed of honor, with head erect, his unmutilated face turned toward the enemy, and bearing an expression of proud defiance that made him beautiful in death. From his torn jacket a letter had fallen to the ground and lay in the pool of blood that dribbled slowly from above.

The only lieutenant left alive shouted the order: "Bring up the limbers!"

A caisson had exploded with a roar that rent the skies. They were obliged to take the horses from another caisson in order to save a gun of which the team had been killed. And when, for the last time, the drivers had brought up their smoking horses and the guns had been limbered up, the whole battery flew away at a gallop and never stopped until they reached the edge of the wood of la Garenne, nearly twelve hundred yards away.

Maurice had seen the whole. He shivered with horror, and murmured mechanically, in a faint voice:

"Oh! poor fellow, poor fellow!"

In addition to this feeling of mental distress he had a horrible sensation of physical suffering, as if something was gnawing at his vitals. It was the animal portion of his nature asserting itself; he was at the end of his endurance, was ready to sink with hunger. His perceptions were dimmed, he was not even conscious of the dangerous position the regiment was in now it no longer was protected by the battery. It was more than likely that the enemy would not long delay to attack the plateau in force.

"Look here," he said to Jean, "I *must* eat-if I am to be killed for it the next minute, I must eat."

He opened his knapsack and, taking out the bread with shaking hands, set his teeth in it voraciously. The bullets were whistling above their heads, two shells exploded only a few yards away, but all was as naught to him in comparison with his craving hunger.

"Will you have some, Jean?"

The corporal was watching him with hungry eyes and a stupid expression on his face; his stomach was also twinging him.

"Yes, I don't care if I do; this suffering is more than I can stand."

They divided the loaf between them and each devoured his portion gluttonously, unmindful of what was going on about them so long as a crumb remained. And it was at that time that they saw their colonel for the last time, sitting his big horse, with his blood-stained boot. The regiment was surrounded on every side; already some of the companies had left the field. Then, unable longer to restrain their flight, with tears standing in his eyes and raising his sword above his head:

"My children," cried M. de Vineuil, "I commend you to the protection of God, who thus far has spared us all!"

He rode off down the hill, surrounded by a swarm of fugitives, and vanished from their sight.

Then, they knew not how, Maurice and Jean found themselves once more behind the hedge, with the remnant of their company. Some forty men at the outside were all that remained, with Lieutenant Rochas as their commander, and the regimental standard was with them; the subaltern who carried it had furled the silk about the staff in order to try to save it. They made their way along the hedge, as far as it extended, to a cluster of small trees upon a hillside, where Rochas made them halt and reopen fire. The men, dispersed in skirmishing order and sufficiently protected, could hold their ground, the more that an important calvary movement was in preparation on their right and regiments of infantry were being brought up to support it.

It was at that moment that Maurice comprehended the full scope of that mighty, irresistible turning movement that was now drawing near completion. That morning he had watched the Prussians debouching by the Saint-Albert pass and had seen their advanced guard pushed forward, first to Saint-Menges, then to Fleigneux, and now, behind the wood of la Garenne, he could hear the thunder of the artillery of the Guard, could behold other German uniforms arriving on the scene over the hills of Givonne. Yet a few moments, it might be, and the circle would be complete; the Guard would join hands with the Vth corps, surrounding the French army with a living wall, girdling them about with a belt of flaming artillery. It was with the resolve to make one supreme, desperate effort, to try to hew a passage through that advancing wall, that General Margueritte's division of the reserve cavalry was massing behind a protecting crest preparatory to charging. They were about to charge into the jaws of death, with no possibility of achieving any useful result, solely for the glory of France and the French army. And Maurice, whose thoughts turned to Prosper, was a witness of the terrible spectacle.

What between the messages that were given him to carry and their answers, Prosper had been kept busy since daybreak spurring up and down the plateau of Illy. The cavalrymen had been awakened at peep of dawn, man by man, without sound of trumpet, and to make their morning coffee had devised the ingenious expedient of screening their fires with a greatcoat so as not to attract the attention of the enemy. Then there came a period when they were left entirely to themselves, with nothing to occupy them; they seemed to be forgotten by their commanders. They could hear the sound of the cannonading, could descry the puffs of smoke, could see the distant movements of the infantry, but were utterly ignorant of the battle, its importance, and its results. Prosper, as far as he was concerned, was suffering from want of sleep. The cumulative fatigue induced by many nights of broken rest, the invincible somnolency caused by the easy gait of his mount, made life a burden. He dreamed dreams and

181

saw visions; now he was sleeping comfortably in a bed between clean sheets, now snoring on the bare ground among sharpened flints. For minutes at a time he would actually be sound asleep in his saddle, a lifeless clod, his steed's intelligence answering for both. Under such circumstances comrades had often tumbled from their seats upon the road. They were so fagged that when they slept the trumpets no longer awakened them; the only way to rouse them from their lethargy and get them on their feet was to kick them soundly.

"But what are they going to do, what are they going to do with us?" Prosper kept saying to himself. It was the only thing he could think of to keep himself awake.

For six hours the cannon had been thundering. As they climbed a hill two comrades, riding at his side, had been struck down by a shell, and as they rode onward seven or eight others had bit the dust, pierced by rifle-balls that came no one could say whence. It was becoming tiresome, that slow parade, as useless as it was dangerous, up and down the battlefield. At last- it was about one o'clock-he learned that it had been decided they were to be killed off in a somewhat more decent manner. Margueritte's entire division, comprising three regiments of chasseurs d'Afrique, one of chasseurs de France, and one of hussars, had been drawn in and posted in a shallow valley a little to the south of the Calvary of Illy. The trumpets had sounded: "Dismount!" and then the officers' command ran down the line to tighten girths and look to packs.

Prosper alighted, stretched his cramped limbs, and gave Zephyr a friendly pat upon the neck. Poor Zephyr! he felt the degradation of the ignominious, heartbreaking service they were subjected to almost as keenly as his master; and not only that, but he had to carry a small arsenal of stores and implements of various kinds: the holsters stuffed with his master's linen and underclothing and the greatcoat rolled above, the stable suit, blouse, and overalls, and the sack containing brushes, currycomb, and other articles of equine toilet behind the saddle, the haversack with rations slung at his side, to say nothing of such trifles as side-lines and picket-pins, the watering bucket and the wooden basin. The cavalryman's tender heart was stirred by a feeling of compassion, as he tightened up the girth and looked to see that everything was secure in its place.

It was a trying moment. Prosper was no more a coward than the next man, but his mouth was intolerably dry and hot; he lit a cigarette in the hope that it would relieve the unpleasant sensation. When about to charge no man can assert with any degree of certainty that he will ride back again. The suspense lasted some five or six minutes; it was said that General Margueritte had ridden forward to reconnoiter the ground over which they were to charge; they were awaiting his return. The five regiments had been formed in three columns, each column having a depth of seven squadrons; enough to afford an ample meal to the hostile guns.

Presently the trumpets rang out: "To horse!" and this was succeeded almost immediately by the shrill summons: "Draw sabers!"

The colonel of each regiment had previously ridden out and taken his proper position, twenty-five yards to the front, the captains were all at their posts at the head of their squadrons. Then there was another period of anxious waiting, amid a silence heavy as that of death. Not a sound, not a breath, there, beneath the blazing sun; nothing, save the beating of those brave hearts. One order more, the supreme, the decisive one, and that mass, now so inert and motionless, would become a resistless tornado, sweeping all before it.

At that juncture, however, an officer appeared coming over the crest of the hill in front, wounded, and preserving his seat in the saddle only by the assistance of a man on either side. No one recognized him at first, but presently a deep, ominous murmur began to run from squadron to squadron, which quickly swelled into a furious uproar. It was General

Margueritte, who had received a wound from which he died a few days later; a musket-ball had passed through both cheeks, carrying away a portion of the tongue and palate. He was incapable of speech, but waved his arm in the direction of the enemy. The fury of his men knew no bounds; their cries rose louder still upon the air.

"It is our general! Avenge him, avenge him!"

Then the colonel of the first regiment, raising aloft his saber, shouted in a voice of thunder:

"Charge!"

The trumpets sounded, the column broke into a trot and was away. Prosper was in the leading squadron, but almost at the extreme right of the right wing, a position of less danger than the center, upon which the enemy always naturally concentrate their hottest fire. When they had topped the summit of the Calvary and began to descend the slope beyond that led downward into the broad plain he had a distinct view, some two-thirds of a mile away, of the Prussian squares that were to be the object of their attack. Beside that vision all the rest was dim and confused before his eyes; he moved onward as one in a dream, with a strange ringing in his ears, a sensation of voidness in his mind that left him incapable of framing an idea. He was a part of the great engine that tore along, controlled by a superior will. The command ran along the line: "Keep touch of knees! Keep touch of knees!" in order to keep the men closed up and give their ranks the resistance and rigidity of a wall of granite, and as their trot became swifter and swifter and finally broke into a mad gallop, the chasseurs d'Afrique gave their wild Arab cry that excited their wiry steeds to the verge of frenzy. Onward they tore, faster and faster still, until their gallop was a race of unchained demons, their shouts the shrieks of souls in mortal agony; onward they plunged amid a storm of bullets that rattled on casque and breastplate, on buckle and scabbard, with a sound like hail; into the bosom of that hailstorm flashed that thunderbolt beneath which the earth shook and trembled, leaving behind it, as it passed, an odor of burned woolen and the exhalations of wild beasts.

At five hundred yards the line wavered an instant, then swirled and broke in a frightful eddy that brought Prosper to the ground. He clutched Zephyr by the mane and succeeded in recovering his seat. The center had given way, riddled, almost annihilated as it was by the musketry fire, while the two wings had wheeled and ridden back a little way to renew their formation. It was the foreseen, foredoomed destruction of the leading squadron. Disabled horses covered the ground, some quiet in death, but many struggling violently in their strong agony; and everywhere dismounted riders could be seen, running as fast as their short legs would let them, to capture themselves another mount. Many horses that had lost their master came galloping back to the squadron and took their place in line of their own accord, to rush with their comrades back into the fire again, as if there was some strange attraction for them in the smell of gunpowder. The charge was resumed; the second squadron went forward, like the first, at a constantly accelerated rate of speed, the men bending upon their horses' neck, holding the saber along the thigh, ready for use upon the enemy. Two hundred yards more were gained this time, amid the thunderous, deafening uproar, but again the center broke under the storm of bullets; men and horses went down in heaps, and the piled corpses made an insurmountable barrier for those who followed. Thus was the second squadron in its turn mown down, annihilated, leaving its task to be accomplished by those who came after.

When for the third time the men were called upon to charge and responded with invincible heroism, Prosper found that his companions were principally hussars and chasseurs de France. Regiments and squadrons, as organizations, had ceased to exist; their constituent elements were drops in the mighty wave that alternately broke and reared its crest again, to swallow up all that lay in its destructive path. He had long since lost distinct consciousness of what was going on around him, and suffered his movements to be guided by his mount, faithful Zephyr, who had received a wound in the ear that seemed to madden him. He was now in the center, where all about him horses were rearing, pawing the air, and falling backward; men were dismounted as if torn from their saddle by the blast of a tornado, while others, shot through some vital part, retained their seat and rode onward in the ranks with vacant, sightless eyes. And looking back over the additional two hundred yards that this effort had won for them, they could see the field of yellow stubble strewn thick with dead and dying. Some there were who had fallen headlong from their saddle and buried their face in the soft earth. Others had alighted on their back and were staring up into the sun with terror-stricken eyes that seemed bursting from their sockets. There was a handsome black horse, an officer's charger, that had been disemboweled, and was making frantic efforts to rise, his fore feet entangled in his entrails. Beneath the fire, that became constantly more murderous as they drew nearer, the survivors in the wings wheeled their horses and fell back to concentrate their strength for a fresh onset.

Finally it was the fourth squadron, which, on the fourth attempt, reached the Prussian lines. Prosper made play with his saber, hacking away at helmets and dark uniforms as well as he could distinguish them, for all was dim before him, as in a dense mist. Blood flowed in torrents; Zephyr's mouth was smeared with it, and to account for it he said to himself that the good horse must have been using his teeth on the Prussians. The clamor around him became so great that he could not hear his own voice, although his throat seemed splitting from the yells that issued from it. But behind the first Prussian line there was another, and then another, and then another still. Their gallant efforts went for nothing; those dense masses of men were like a tangled jungle that closed around the horses and riders who entered it and buried them in its rank growths. They might hew down those who were within reach of their sabers; others stood ready to take their place, the last squadrons were lost and swallowed up in their vast numbers. The firing, at point-blank range, was so furious that the men's clothing was ignited. Nothing could stand before it, all went down; and the work that it left unfinished was completed by bayonet and musket butt. Of the brave men who rode into action that day two-thirds remained upon the battlefield, and the sole end achieved by that mad charge was to add another glorious page to history. And then Zephyr, struck by a musket-ball full in the chest, dropped in a heap, crushing beneath him Prosper's right thigh; and the pain was so acute that the young man fainted.

Maurice and Jean, who had watched the gallant effort with burning interest, uttered an exclamation of rage.

"Tonnerre de Dieu! what bravery wasted!"

And they resumed their firing from among the trees of the low hill where they were deployed in skirmishing order. Rochas himself had picked up an abandoned musket and was blazing away with the rest. But the plateau of Illy was lost to them by this time beyond hope of recovery; the Prussians were pouring in upon it from every quarter. It was somewhere in the neighborhood of two o'clock, and their great movement was accomplished; the Vth corps and the Guards had effected their junction, the investment of the French army was complete.

Jean was suddenly brought to the ground.

"I am done for," he murmured.

He had received what seemed to him like a smart blow of a hammer on the crown of his head, and his *kepi* lay behind him with a great furrow plowed through its top. At first he thought that the bullet had certainly penetrated the skull and laid bare the brain; his dread of finding a yawning orifice there was so great that for some seconds he dared not raise his hand to ascertain the truth. When finally he ventured, his fingers, on withdrawing them, were red with an abundant flow of blood, and the pain was so intense that he fainted.

184

Just then Rochas gave the order to fall back. The Prussians had crept up on them and were only two or three hundred yards away; they were in danger of being captured.

"Be cool, don't hurry; face about and give 'em another shot. Rally behind that low wall that you see down there."

Maurice was in despair; he knew not what to do.

"We are not going to leave our corporal behind, are we, lieutenant?"

"What are we to do? he has turned up his toes."

"No, no! he is breathing still. Take him along!"

Rochas shrugged his shoulders as if to say they could not bother themselves for every man that dropped. A wounded man is esteemed of little value on the battlefield. Then Maurice addressed his supplications to Lapoulle and Pache.

"Come, give me a helping hand. I am not strong enough to carry him unassisted."

They were deaf to his entreaties; all they could hear was the voice that urged them to seek safety for themselves. The Prussians were now not more than a hundred yards from them; already they were on their hands and knees, crawling as fast as they could go toward the wall.

And Maurice, weeping tears of rage, thus left alone with his unconscious companion, raised him in his arms and endeavored to lug him away, but he found his puny strength unequal to the task, exhausted as he was by fatigue and the emotions of the day. At the first step he took he reeled and fell with his burden. If only he could catch sight of a stretcher-bearer! He strained his eyes, thought he had discovered one among the crowd of fugitives, and made frantic gestures of appeal; no one came, they were left behind, alone. Summoning up his strength with a determined effort of the will he seized Jean once more and succeeded in advancing some thirty paces, when a shell burst near them and he thought that all was ended, that he, too, was to die on the body of his comrade.

Slowly, cautiously, Maurice picked himself up. He felt his body, arms, and legs; nothing, not a scratch. Why should he not look out for himself and fly, alone? There was time left still; a few bounds would take him to the wall and he would be saved. His horrible sensation of fear returned and made him frantic. He was collecting his energies to break away and run, when a feeling stronger than death intervened and vanquished the base impulse. What, abandon Jean! he could not do it. It would be like mutilating his own being; the brotherly affection that had bourgeoned and grown between him and that rustic had struck its roots down into his life, too deep to be slain like that. The feeling went back to the earliest days, was perhaps as old as the world itself; it was as if there were but they two upon earth, of whom one could not forsake the other without forsaking himself, and being doomed thenceforth to an eternity of solitude. Molded of the same clay, quickened by the same spirit, duty imperiously commanded to save himself in saving his brother.

Had it not been for the crust of bread he ate an hour before under the Prussian shells Maurice could never have done what he did; *how* he did it he could never in subsequent days remember. He must have hoisted Jean upon his shoulders and crawled through the brush and brambles, falling a dozen times only to pick himself up and go on again, stumbling at every rut, at every pebble. His indomitable will sustained him, his dogged resolution would have enabled him to bear a mountain on his back. Behind the low wall he found Rochas and the few men that were left of the squad, firing away as stoutly as ever and defending the flag, which the subaltern held beneath his arm. It had not occurred to anyone to designate lines of retreat for the several army corps in case the day should go against them; owing to this want of foresight every general was at liberty to act as seemed to him best, and at this stage of the conflict they all found themselves being crowded back upon Sedan under the steady,

185

unrelaxing pressure of the German armies. The second division of the 7th corps fell back in comparatively good order, while the remnants of the other divisions, mingled with the debris of the 1st corps, were already streaming into the city in terrible disorder, a roaring torrent of rage and fright that bore all, men and beasts, before it.

But to Maurice, at that moment, was granted the satisfaction of seeing Jean unclose his eyes, and as he was running to a stream that flowed near by, for water with which to bathe his friend's face, he was surprised, looking down on his right into a sheltered valley that lay between rugged slopes, to behold the same peasant whom he had seen that morning, still leisurely driving the plow through the furrow with the assistance of his big white horse. Why should he lose a day? Men might fight, but none the less the corn would keep on growing; and folks must live.

6

Up on his lofty terrace, whither he had betaken himself to watch how affairs were shaping, Delaherche at last became impatient and was seized with an uncontrollable desire for news. He could see that the enemy's shells were passing over the city and that the few projectiles which had fallen on the houses in the vicinity were only responses, made at long intervals, to the irregular and harmless fire from Fort Palatinat, but he could discern nothing of the battle, and his agitation was rising to fever heat; he experienced an imperious longing for intelligence, which was constantly stimulated by the reflection that his life and fortune would be in danger should the army be defeated. He found it impossible to remain there longer, and went downstairs, leaving behind him the telescope on its tripod, turned on the German batteries.

When he had descended, however, he lingered a moment, detained by the aspect of the central garden of the factory. It was near one o'clock, and the ambulance was crowded with wounded men; the wagons kept driving up to the entrance in an unbroken stream. The regular ambulance wagons of the medical department, two-wheeled and four-wheeled, were too few in number to meet the demand, and vehicles of every description from the artillery and other trains, prolonges, provision vans, everything on wheels that could be picked up on the battlefield, came rolling up with their ghastly loads; and later in the day even carrioles and market-gardeners' carts were pressed into the service and harnessed to horses that were found straying along the roads. Into these motley conveyances were huddled the men collected from the flying ambulances, where their hurts had received such hasty attention as could be afforded. It was a sight to move the most callous to behold the unloading of those poor wretches, some with a greenish pallor on their face, others suffused with the purple hue that denotes congestion; many were in a state of coma, others uttered piercing cries of anguish; some there were who, in their semi-conscious condition, yielded themselves to the arms of the attendants with a look of deepest terror in their eyes, while a few, the minute a hand was laid on them, died of the consequent shock. They continued to arrive in such numbers that soon every bed in the vast apartment would have its occupant, and Major Bouroche had given orders to make use of the straw that had been spread thickly upon the floor at one end. He and his assistants had thus far been able to attend to all the cases with reasonable promptness; he had requested Mme. Delaherche to furnish him with another table, with mattress and oilcloth cover, for the shed where he had established his operating room. The assistant would thrust a napkin saturated with chloroform to the patient's nostrils, the keen knife flashed in the air, there was the faint rasping of the saw, barely audible, the blood spurted in short, sharp jets that were checked immediately. As soon as one subject had been operated on another was brought in, and they followed one another in such quick succession that there was barely time to pass a sponge over the protecting oilcloth. At the extremity of the grass plot, screened from sight by a clump of lilac bushes, they had set up a kind of morgue whither they carried the bodies of the dead, which were removed from the beds without a moment's delay in order to make room for the living, and this receptacle also served to receive the amputated legs, and arms, whatever debris of flesh and bone remained upon the table.

Mme. Delaherche and Gilberte, seated at the foot of one of the great trees, found it hard work to keep pace with the demand for bandages. Bouroche, who happened to be passing, his face very red, his apron white no longer, threw a bundle of linen to Delaherche and shouted:

"Here! be doing something; make yourself useful!"

But the manufacturer objected. "Oh! excuse me; I must go and try to pick up some news. One can't tell whether his neck is safe or not." Then, touching his lips to his wife's hair: "My poor Gilberte, to think that a shell may burn us out of house and home at any moment! It is horrible."

She was very pale; she raised her head and glanced about her, shuddering as she did so. Then, involuntarily, her unextinguishable smile returned to her lips.

"Oh, horrible, indeed! and all those poor men that they are cutting and carving. I don't see how it is that I stay here without fainting."

Mme. Delaherche had watched her son as he kissed the young woman's hair. She made a movement as if to part them, thinking of that other man who must have kissed those tresses so short a time ago; then her old hands trembled, she murmured beneath her breath:

"What suffering all about us, mon Dieu! It makes one forget his own."

Delaherche left them, with the assurance that he would be away no longer than was necessary to ascertain the true condition of affairs. In the Rue Maqua he was surprised to observe the crowds of soldiers that were streaming into the city, without arms and in torn, dust-stained uniforms. It was in vain, however, that he endeavored to slake his thirst for news by questioning them; some answered with vacant, stupid looks that they knew nothing, while others told long rambling stories, with the maniacal gestures and whirling words of one bereft of reason. He therefore mechanically turned his steps again toward the Sous Prefecture as the likeliest quarter in which to look for information. As he was passing along the Place du College two guns, probably all that remained of some battery, came dashing up to the curb on a gallop, and were abandoned there. When at last he turned into the Grande Rue he had further evidence that the advanced guards of the fugitives were beginning to take possession, of the city; three dismounted hussars had seated themselves in a doorway and were sharing a loaf of bread; two others were walking their mounts up and down, leading them by the bridle, not knowing where to look for stabling for them; officers were hurrying to and fro distractedly, seemingly without any distinct purpose. On the Place Turenne a lieutenant counseled him not to loiter unnecessarily, for the shells had an unpleasant way of dropping there every now and then; indeed, a splinter had just demolished the railing about the statue of the great commander who overran the Palatinate. And as if to emphasize the officer's advice, while he was making fast time down the Rue de la Sous Prefecture he saw two projectiles explode, with a terrible crash, on the Pont de Meuse.

He was standing in front of the janitor's lodge, debating with himself whether it would be best to send in his card and try to interview one of the aides-de-camp, when he heard a girlish voice calling him by name.

"M. Delaherche! Come in here, quick; it is not safe out there."

It was Rose, his little operative, whose existence he had quite forgotten. She might be a useful ally in assisting him to gain access to headquarters; he entered the lodge and accepted her invitation to be seated.

"Just think, mamma is down sick with the worry and confusion; she can't leave her bed, so, you see, I have to attend to everything, for papa is with the National Guards up in the citadel. A little while ago the Emperor left the building-I suppose he wanted to let people see he is not a coward-and succeeded in getting as far as the bridge down at the end of the street. A shell alighted right in front of him; one of his equerries had his horse killed under him. And then he came back-he couldn't do anything else, could he, now?"

"You must have heard some talk of how the battle is going. What do they say, those gentlemen upstairs?"

She looked at him in surprise. Her pretty face was bright and smiling, with its fluffy golden hair and the clear, childish eyes of one who bestirred herself among her multifarious duties, in the midst of all those horrors, which she did not well understand.

"No, I know nothing. About midday I sent up a letter for Marshal MacMahon, but it could not be given him right away, because the Emperor was in the room. They were together nearly an hour, the Marshal lying on his bed, the Emperor close beside him seated on a chair. That much I know for certain, because I saw them when the door was opened."

"And then, what did they say to each other?"

She looked at him again, and could not help laughing.

"Why, I don't know; how could you expect me to? There's not a living soul knows what they said to each other."

She was right; he made an apologetic gesture in recognition of the stupidity of his question. But the thought of that fateful conversation haunted him; the interest there was in it for him who could have heard it! What decision had they arrived at?

"And now," Rose added, "the Emperor is back in his cabinet again, where he is having a conference with two generals who have just come in from the battlefield." She checked herself, casting a glance at the main entrance of the building. "See! there is one of them, now- and there comes the other."

He hurried from the room, and in the two generals recognized Ducrot and Douay, whose horses were standing before the door. He watched them climb into their saddles and gallop away. They had hastened into the city, each independently of the other, after the plateau of Illy had been captured by the enemy, to notify the Emperor that the battle was lost. They placed the entire situation distinctly before him; the army and Sedan were even then surrounded on every side; the result could not help but be disastrous.

For some minutes the Emperor continued silently to pace the floor of his cabinet, with the feeble, uncertain step of an invalid. There was none with him save an aide-de-camp, who stood by the door, erect and mute. And ever, to and fro, from the window to the fireplace, from the fireplace to the window, the sovereign tramped wearily, the inscrutable face now drawn and twitching spasmodically with a nervous tic. The back was bent, the shoulders bowed, as if the weight of his falling empire pressed on them more heavily, and the lifeless eyes, veiled by their heavy lids, told of the anguish of the fatalist who has played his last card against destiny and lost. Each time, however, that his walk brought him to the half-open window he gave a start and lingered there a second. And during one of those brief stoppages he faltered with trembling lips:

"Oh! those guns, those guns, that have been going since the morning!"

The thunder of the batteries on la Marfee and at Frenois seemed, indeed, to resound with more terrific violence there than elsewhere. It was one continuous, uninterrupted crash, that shook the windows, nay, the very walls themselves; an incessant uproar that exasperated the nerves by its persistency. And he could not banish the reflection from his mind that, as the struggle was now hopeless, further resistance would be criminal. What would avail more bloodshed, more maiming and mangling; why add more corpses to the dead that were already piled high upon that bloody field? They were vanquished, it was all ended; then why not stop the slaughter? The abomination of desolation raised its voice to heaven: let it cease.

The Emperor, again before the window, trembled and raised his hands to his ears, as if to shut out those reproachful voices.

"Oh, those guns, those guns! Will they never be silent!"

Perhaps the dreadful thought of his responsibilities arose before him, with the vision of all those thousands of bleeding forms with which his errors had cumbered the earth; perhaps, again, it was but the compassionate impulse of the tender-hearted dreamer, of the well-meaning man whose mind was stocked with humanitarian theories. At the moment when he beheld utter ruin staring him in the face, in that frightful whirlwind of destruction that broke him like a reed and scattered his fortunes in the dust, he could yet find tears for others. Almost crazed at the thought of the slaughter that was mercilessly going on so near him, he felt he had not strength to endure it longer; each report of that accursed cannonade seemed to pierce his heart and intensified a thousandfold his own private suffering.

"Oh, those guns, those guns! they must be silenced at once, at once!"

And that monarch who no longer had a throne, for he had delegated all his functions to the Empress regent, that chief without an army, since he had turned over the supreme command to Marshal Bazaine, now felt that he must once more take the reins in his hand and be the master. Since they left Chalons he had kept himself in the background, had issued no orders, content to be a nameless nullity without recognized position, a cumbrous burden carried about from place to place among the baggage of his troops, and it was only in their hour of defeat that the Emperor reasserted itself in him; the one order that he was yet to give, out of the pity of his sorrowing heart, was to raise the white flag on the citadel to request an armistice.

"Those guns, oh! those guns! Take a sheet, someone, a tablecloth, it matters not what! only hasten, hasten, and see that it is done!"

The aide-de-camp hurried from the room, and with unsteady steps the Emperor continued to pace his beat, back and forth, between the window and the fireplace, while still the batteries kept thundering, shaking the house from garret to foundation.

Delaherche was still chatting with Rose in the room below when a non-commissioned officer of the guard came running in and interrupted them.

"Mademoiselle, the house is in confusion, I cannot find a servant. Can you let me have something from your linen closet, a white cloth of some kind?"

"Will a napkin answer?"

"No, no, it would not be large enough. Half of a sheet, say."

Rose, eager to oblige, was already fumbling in her closet.

"I don't think I have any half-sheets. No, I don't see anything that looks as if it would serve your purpose. Oh, here is something; could you use a tablecloth?"

"A tablecloth! just the thing. Nothing could be better." And he added as he left the room: "It is to be used as a flag of truce, and hoisted on the citadel to let the enemy know we want to stop the fighting. Much obliged, mademoiselle."

Delaherche gave a little involuntary start of delight; they were to have a respite at last, then! Then he thought it might be unpatriotic to be joyful at such a time, and put on a long face again; but none the less his heart was very glad and he contemplated with much interest a colonel and captain, followed by the sergeant, as they hurriedly left the Sous-Prefecture. The colonel had the tablecloth, rolled in a bundle, beneath his arm. He thought he should like to follow them, and took leave of Rose, who was very proud that her napery was to be put to such use. It was then just striking two o'clock.

In front of the Hotel de Ville Delaherche was jostled by a disorderly mob of half-crazed soldiers who were pushing their way down from the Faubourg de la Cassine; he lost sight of the colonel, and abandoned his design of going to witness the raising of the white flag. He certainly would not be allowed to enter the citadel, and then again he had heard it reported that shells were falling on the college, and a new terror filled his mind; his factory might have been burned since he left it. All his feverish agitation returned to him and he started off on a run; the rapid motion was a relief to him. But the streets were blocked by groups of men, at every crossing he was delayed by some new obstacle. It was only when he reached the Rue Maqua and beheld the monumental facade of his house intact, no smoke or sign of fire about it, that his anxiety was allayed, and he heaved a deep sigh of satisfaction. He entered, and from the doorway shouted to his mother and wife:

"It is all right! they are hoisting the white flag; the cannonade won't last much longer."

He said nothing more, for the appearance presented by the ambulance was truly horrifying.

In the vast drying-room, the wide door of which was standing open, not only was every bed occupied, but there was no more room upon the litter that had been shaken down on the floor at the end of the apartment. They were commencing to strew straw in the spaces between the beds, the wounded were crowded together so closely that they were in contact. Already there were more than two hundred patients there, and more were arriving constantly; through the lofty windows the pitiless white daylight streamed in upon that aggregation of suffering humanity. Now and then an unguarded movement elicited an involuntary cry of anguish. The death-rattle rose on the warm, damp air. Down the room a low, mournful wail, almost a lullaby, went on and ceased not. And all about was silence, intense, profound, the stolid resignation of despair, the solemn stillness of the death-chamber, broken only by the tread and whispers of the attendants. Rents in tattered, shell-torn uniforms disclosed gaping wounds, some of which had received a hasty dressing on the battlefield, while others were still raw and bleeding. There were feet, still incased in their coarse shoes, crushed into a mass like jelly; from knees and elbows, that were as if they had been smashed by a hammer, depended inert limbs. There were broken hands, and fingers almost severed, ready to drop, retained only by a strip of skin. Most numerous among the casualties were the fractures; the poor arms and legs, red and swollen, throbbed intolerably and were heavy as lead. But the most dangerous hurts were those in the abdomen, chest, and head. There were yawning fissures that laid open the entire flank, the knotted viscera were drawn into great hard lumps beneath the tight-drawn skin, while as the effect of certain wounds the patient frothed at the mouth and writhed like an epileptic. Here and there were cases where the lungs had been penetrated, the puncture now so minute as to permit no escape of blood, again a wide, deep orifice through which the red tide of life escaped in torrents; and the internal hemorrhages, those that were hid from sight, were the most terrible in their effects, prostrating their victim like a flash, making him black in the face and delirious. And finally the head, more than any other portion of the frame, gave evidence of hard treatment; a broken jaw, the mouth a pulp of teeth and bleeding tongue, an eye torn from its socket and exposed upon the cheek, a cloven skull that showed the palpitating brain beneath. Those in whose case the bullet had touched the brain or spinal marrow were already as dead men, sunk in the lethargy of coma, while the fractures and other less serious cases tossed restlessly on their pallets and beseechingly called for water to quench their thirst.

Leaving the large room and passing out into the courtyard, the shed where the operations were going on presented another scene of horror. In the rush and hurry that had continued unabated since morning it was impossible to operate on every case that was brought in, so their attention had been confined to those urgent cases that imperatively demanded it. Whenever Bouroche's rapid judgment told him that amputation was necessary, he proceeded at once to perform it. In the same way he lost not a moment's time in probing the wound and extracting the projectile whenever it had lodged in some locality where it might do further mischief, as in the muscles of the neck, the region of the arm pit, the thigh joint, the ligaments of the knee and elbow. Severed arteries, too, had to be tied without delay. Other wounds were

191

merely dressed by one of the hospital stewards under his direction and left to await developments. He had already with his own hand performed four amputations, the only rest that he allowed himself being to attend to some minor cases in the intervals between them, and was beginning to feel fatigue. There were but two tables, his own and another, presided over by one of his assistants; a sheet had been hung between them, to isolate the patients from each other. Although the sponge was kept constantly at work the tables were always red, and the buckets that were emptied over a bed of daisies a few steps away, the clear water in which a single tumbler of blood sufficed to redden, seemed to be buckets of unmixed blood, torrents of blood, inundating the gentle flowers of the parterre. Although the room was thoroughly ventilated a nauseating smell arose from the tables and their horrid burdens, mingled with the sweetly insipid odor of chloroform.

Delaherche, naturally a soft-hearted man, was in a quiver of compassionate emotion at the spectacle that lay before his eyes, when his attention was attracted by a landau that drove up to the door. It was a private carriage, but doubtless the ambulance attendants had found none other ready to their hand and had crowded their patients into it. There were eight of them, sitting on one another's knees, and as the last man alighted the manufacturer recognized Captain Beaudoin, and gave utterance to a cry of terror and surprise.

"Ah, my poor friend! Wait, I will call my mother and my wife."

They came running up, leaving the bandages to be rolled by servants. The attendants had already raised the captain and brought him into the room, and were about to lay him down upon a pile of straw when Delaherche noticed, lying on a bed, a soldier whose ashy face and staring eyes exhibited no sign of life.

"Look, is he not dead, that man?"

"That's so!" replied the attendant. "He may as well make room for someone else!"

He and one of his mates took the body by the arms and legs and carried it off to the morgue that had been extemporized behind the lilac bushes. A dozen corpses were already there in a row, stiff and stark, some drawn out to their full length as if in an attempt to rid themselves of the agony that racked them, others curled and twisted in every attitude of suffering. Some seemed to have left the world with a sneer on their faces, their eyes retroverted till naught was visible but the whites, the grinning lips parted over the glistening teeth, while in others, with faces unspeakably sorrowful, big tears still stood on the cheeks. One, a mere boy, short and slight, half whose face had been shot away by a cannon-ball, had his two hands clasped convulsively above his heart, and in them a woman's photograph, one of those pale, blurred pictures that are made in the quarters of the poor, bedabbled with his blood. And at the feet of the dead had been thrown in a promiscuous pile the amputated arms and legs, the refuse of the knife and saw of the operating table, just as the butcher sweeps into a corner of his shop the offal, the worthless odds and ends of flesh and bone.

Gilberte shuddered as she looked on Captain Beaudoin. Good God! how pale he was, stretched out on his mattress, his face so white beneath the encrusting grime! And the thought that but a few short hours before he had held her in his arms, radiant in all his manly strength and beauty, sent a chill of terror to her heart. She kneeled beside him.

"What a terrible misfortune, my friend! But it won't amount to anything, will it?" And she drew her handkerchief from her pocket and began mechanically to wipe his face, for she could not bear to look at it thus soiled with powder, sweat, and clay. It seemed to her, too, that she would be helping him by cleansing him a little. "Will it? it is only your leg that is hurt; it won't amount to anything."

The captain made an effort to rouse himself from his semi-conscious state, and opened his eyes. He recognized his friends and greeted them with a faint smile.

"Yes, it is only the leg. I was not even aware of being hit; I thought I had made a misstep and fallen~" He spoke with great difficulty. "Oh! I am so thirsty!"

Mme. Delaherche, who was standing at the other side of the mattress, looking down compassionately on the young man, hastily left the room. She returned with a glass and a carafe of water into which a little cognac had been poured, and when the captain had greedily swallowed the contents of the glass, she distributed what remained in the carafe among the occupants of the adjacent beds, who begged with trembling outstretched hands and tearful voices for a drop. A zouave, for whom there was none left, sobbed like a child in his disappointment.

Delaherche was meantime trying to gain the major's ear to see if he could not prevail on him to take up the captain's case out of its regular turn. Bouroche came into the room just then, with his blood-stained apron and lion's mane hanging in confusion about his perspiring face, and the men raised their heads as he passed and endeavored to stop him, all clamoring at once for recognition and immediate attention: "This way, major! It's my turn, major!" Faltering words of entreaty went up to him, trembling hands clutched at his garments, but he, wrapped up in the work that lay before him and puffing with his laborious exertions, continued to plan and calculate and listened to none of them. He communed with himself aloud, counting them over with his finger and classifying them, assigning them their numbers; this one first, then that one, then that other fellow; one, two, three; the jaw, the arm, then the thigh; while the assistant who accompanied him on his round made himself all ears in his effort to memorize his directions.

"Major," said Delaherche, plucking him by the sleeve, "there is an officer over here, Captain Beaudoin…"

Bouroche interrupted him. "What, Beaudoin here! Ah, the poor devil!" And he crossed over at once to the side of the wounded man. A single glance, however, must have sufficed to show him that the case was a bad one, for he added in the same breath, without even stooping to examine the injured member: "Good! I will have them bring him to me at once, just as soon as I am through with the operation that is now in hand."

And he went back to the shed, followed by Delaherche, who would not lose sight of him for fear lest he might forget his promise.

The business that lay before him now was the rescision of a shoulder-joint in accordance with Lisfranc's method, which surgeons never fail to speak of as a "very pretty" operation, something neat and expeditious, barely occupying forty seconds in the performance. The patient was subjected to the influence of chloroform, while an assistant grasped the shoulder with both hands, the fingers under the armpit, the thumbs on top. Bouroche, brandishing the long, keen knife, cried: "Raise him!" seized the deltoid with his left hand and with a swift movement of the right cut through the flesh of the arm and severed the muscle; then, with a deft rearward cut, he disarticulated the joint at a single stroke, and presto! the arm fell on the table, taken off in three motions. The assistant slipped his thumbs over the brachial artery in such manner as to close it. "Let him down!" Bouroche could not restrain a little pleased laugh as he proceeded to secure the artery, for he had done it in thirty-five seconds. All that was left to do now was to bring a flap of skin down over the wound and stitch it, in appearance something like a flat epaulette. It was not only "pretty," but exciting, on account of the danger, for a man will pump all the blood out of his body in two minutes through the brachial, to say nothing of the risk there is in bringing a patient to a sitting posture when under the influence of anaesthetics.

Delaherche was white as a ghost; a thrill of horror ran down his back. He would have turned

193

and fled, but time was not given him; the arm was already off. The soldier was a new recruit, a sturdy peasant lad; on emerging from his state of coma he beheld a hospital attendant carrying away the amputated limb to conceal it behind the lilacs. Giving a quick downward glance at his shoulder, he saw the bleeding stump and knew what had been done, whereon he became furiously angry.

"Ah, nom de Dieu! what have you been doing to me? It is a shame!"

Bouroche was too done up to make him an immediate answer, but presently, in his fatherly way:

"I acted for the best; I didn't want to see you kick the bucket, my boy. Besides, I asked you, and you told me to go ahead."

"I told you to go ahead! I did? How could I know what I was saying!" His anger subsided and he began to weep scalding tears. "What is going to become of me now?"

They carried him away and laid him on the straw, and gave the table and its covering a thorough cleansing; and the buckets of blood-red water that they threw out across the grass plot gave to the pale daisies a still deeper hue of crimson.

When Delaherche had in some degree recovered his equanimity he was astonished to notice that the bombardment was still going on. Why had it not been silenced? Rose's tablecloth must have been hoisted over the citadel by that time, and yet it seemed as if the fire of the Prussian batteries was more rapid and furious than ever. The uproar was such that one could not hear his own voice; the sustained vibration tried the stoutest nerves. On both operators and patients the effect could not but be most unfavorable of those incessant detonations that seemed to penetrate the inmost recesses of one's being. The entire hospital was in a state of feverish alarm and apprehension.

"I supposed it was all over; what can they mean by keeping it up?" exclaimed Delaherche, who was nervously listening, expecting each shot would be the last.

Returning to Bouroche to remind him of his promise and conduct him to the captain, he was astonished to find him seated on a bundle of straw before two pails of iced water, into which he had plunged both his arms, bared to the shoulder. The major, weary and disheartened, overwhelmed by a sensation of deepest melancholy and dejection, had reached one of those terrible moments when the practitioner becomes conscious of his own impotency; he had exhausted his strength, physical and moral, and taken this means to restore it. And yet he was not a weakling; he was steady of hand and firm of heart; but the inexorable question had presented itself to him: "What is the use?" The feeling that he could accomplish so little, that so much must be left undone, had suddenly paralyzed him. What was the use? since Death, in spite of his utmost effort, would always be victorious. Two attendants came in, bearing Captain Beaudoin on a stretcher.

"Major," Delaherche ventured to say, "here is the captain."

Bouroche opened his eyes, withdrew his arms from their cold bath, shook and dried them on the straw. Then, rising to his feet:

"Ah, yes; the next one- Well, well, the day's work is not yet done." And he shook the tawny locks upon his lion's head, rejuvenated and refreshed, restored to himself once more by the invincible habit of duty and the stern discipline of his profession.

"Good! just above the right ankle," said Bouroche, with unusual garrulity, intended to quiet the nerves of the patient. "You displayed wisdom in selecting the location of your wound; one is not much the worse for a hurt in that quarter. Now we'll just take a little look at it."

But Beaudoin's persistently lethargic condition evidently alarmed him. He inspected the

194

contrivance that had been applied by the field attendant to check the flow of blood, which was simply a cord passed around the leg outside the trousers and twisted tight with the assistance of a bayonet sheath, with a growling request to be informed what infernal ignoramus had done that. Then suddenly he saw how matters were and was silent; while they were bringing him in from the field in the overcrowded landau the improvised tourniquet had become loosened and slipped down, thus giving rise to an extensive hemorrhage. He relieved his feelings by storming at the hospital steward who was assisting him.

"You confounded snail, cut! Are you going to keep me here all day?"

The attendant cut away the trousers and drawers, then the shoe and sock, disclosing to view the leg and foot in their pale nudity, stained with blood. Just over the ankle was a frightful laceration, into which the splinter of the bursting shell had driven a piece of the red cloth of the trousers. The muscle protruded from the lips of the gaping orifice, a roll of whitish, mangled tissue.

Gilberte had to support herself against one of the uprights of the shed. Ah! that flesh, that poor flesh that was so white; now all torn and maimed and bleeding! Despite the horror and terror of the sight she could not turn away her eyes.

"Confound it!" Bouroche exclaimed, "they have made a nice mess here!"

He felt the foot and found it cold; the pulse, if any, was so feeble as to be undistinguishable. His face was very grave, and he pursed his lips in a way that was habitual with him when he had a more than usually serious case to deal with.

"Confound it," he repeated, "I don't like the looks of that foot!"

The captain, whom his anxiety had finally aroused from his semi-somnolent state, asked:

"What were you saying, major?"

Bouroche's tactics, whenever an amputation became necessary, were never to appeal directly to the patient for the customary authorization. He preferred to have the patient accede to it voluntarily.

"I was saying that I don't like the looks of that foot," he murmured, as if thinking aloud. "I am afraid we shan't be able to save it."

In a tone of alarm Beaudoin rejoined: "Come, major, there is no use beating about the bush. What is your opinion?"

"My opinion is that you are a brave man, captain, and that you are going to let me do what the necessity of the case demands."

To Captain Beaudoin it seemed as if a sort of reddish vapor arose before his eyes through which he saw things obscurely. He understood. But notwithstanding the intolerable fear that appeared to be clutching at his throat, he replied, unaffectedly and bravely:

"Do as you think best, major."

The preparations did not consume much time. The assistant had saturated a cloth with chloroform and was holding it in readiness; it was at once applied to the patient's nostrils. Then, just at the moment that the brief struggle set in that precedes anaesthesia, two attendants raised the captain and placed him on the mattress upon his back, in such a position that the legs should be free; one of them retained his grasp on the left limb, holding it flexed, while an assistant, seizing the right, clasped it tightly with both his hands in the region of the groin in order to compress the arteries.

Gilberte, when she saw Bouroche approach the victim with the glittering steel, could endure no more.

"Oh, don't! oh, don't! it is too horrible!"

And she would have fallen had it not been that Mme. Delaherche put forth her arm to sustain her.

"But why do you stay here?"

Both the women remained, however. They averted their eyes, not wishing to see the rest; motionless and trembling they stood locked in each other's arms, notwithstanding the little love there was between them.

At no time during the day had the artillery thundered more loudly than now. It was three o'clock, and Delaherche declared angrily that he gave it up-he could not understand it. There could be no doubt about it now, the Prussian batteries, instead of slackening their fire, were extending it. Why? What had happened? It was as if all the forces of the nether regions had been unchained; the earth shook, the heavens were on fire. The ring of flame-belching mouths of bronze that encircled Sedan, the eight hundred guns of the German armies, that were served with such activity and raised such an uproar, were expending their thunders on the adjacent fields; had that concentric fire been focused upon the city, had the batteries on those commanding heights once begun to play upon Sedan, it would have been reduced to ashes and pulverized into dust in less than fifteen minutes. But now the projectiles were again commencing to fall upon the houses, the crash that told of ruin and destruction was heard more frequently. One exploded in the Rue des Voyards, another grazed the tall chimney of the factory, and the bricks and mortar came tumbling to the ground directly in front of the shed where the surgeons were at work. Bouroche looked up and grumbled:

"Are they trying to finish our wounded for us? Really, this racket is intolerable."

In the meantime an attendant had seized the captain's leg, and the major, with a swift circular motion of his hand, made an incision in the skin below the knee and some two inches below the spot where he intended to saw the bone; then, still employing the same thin-bladed knife, that he did not change in order to get on more rapidly, he loosened the skin on the superior side of the incision and turned it back, much as one would peel an orange. But just as he was on the point of dividing the muscles a hospital steward came up and whispered in his ear:

"Number two has just slipped his cable."

The major did not hear, owing to the fearful uproar.

"Speak up, can't you! My ear drums are broken with their d---d cannon."

"Number two has just slipped his cable."

"Who is that, number two?"

"The arm, you know."

"Ah, very good! Well, then, you can bring me number three, the jaw."

And with wonderful dexterity, never changing his position, he cut through the muscles clean down to the bone with a single motion of his wrist. He laid bare the tibia and fibula, introduced between them an implement to keep them in position, drew the saw across them once, and they were sundered. And the foot remained in the hands of the attendant who was holding it.

The flow of blood had been small, thanks to the pressure maintained by the assistant higher up the leg, at the thigh. The ligature of the three arteries was quickly accomplished, but the major shook his head, and when the assistant had removed his fingers he examined the stump, murmuring, certain that the patient could not hear as yet:

"It looks bad; there's no blood coming from the arterioles."

196

And he completed his diagnosis of the case by an expressive gesture: Another poor fellow who was soon to answer the great roll-call! while on his perspiring face was again seen that expression of weariness and utter dejection, that hopeless, unanswerable: "What is the use?" since out of every ten cases that they assumed the terrible responsibility of operating on they did not succeed in saving four. He wiped his forehead, and set to work to draw down the flap of skin and put in the three sutures that were to hold it in place.

Delaherche having told Gilberte that the operation was completed, she turned her gaze once more upon the table; she caught a glimpse of the captain's foot, however, as the attendant was carrying it away to the place behind the lilacs. The charnel house there continued to receive fresh occupants; two more corpses had recently been brought in and added to the ghastly array, one with blackened lips still parted wide as if rending the air with shrieks of anguish, the other, his form so contorted and contracted in the convulsions of the last agony that he was like a stunted, malformed boy. Unfortunately, there was beginning to be a scarcity of room in the little secluded corner, and the human debris had commenced to overflow and invade the adjacent alley. The attendant hesitated a moment, in doubt what to do with the captain's foot, then finally concluded to throw it on the general pile.

"Well, captain, that's over with," the major said to Beaudoin when he regained consciousness. "You'll be all right now."

But the captain did not show the cheeriness that follows a successful operation. He opened his eyes and made an attempt to raise himself, then fell back on his pillow, murmuring wearily, in a faint voice:

"Thanks, major. I'm glad it's over."

He was conscious of the pain, however, when the alcohol of the dressing touched the raw flesh. He flinched a little, complaining that they were burning him. And just as they were bringing up the stretcher preparatory to carrying him back into the other room the factory was shaken to its foundations by a most terrific explosion; a shell had burst directly in the rear of the shed, in the small courtyard where the pump was situated. The glass in the windows was shattered into fragments, and a dense cloud of smoke came pouring into the ambulance. The wounded men, stricken with panic terror, arose from their bed of straw; all were clamoring with affright; all wished to fly at once.

Delaherche rushed from the building in consternation to see what damage had been done. Did they mean to burn his house down over his head? What did it all mean? Why did they open fire again when the Emperor had ordered that it should cease?

"Thunder and lightning! Stir yourselves, will you!" Bouroche shouted to his staff, who were standing about with pallid faces, transfixed by terror. "Wash off the table; go and bring me in number three!"

They cleansed the table; and once more the crimson contents of the buckets were hurled across the grass plot upon the bed of daisies, which was now a sodden, blood-soaked mat of flowers and verdure. And Bouroche, to relieve the tedium until the attendants should bring him "number three," applied himself to probing for a musket-ball, which, having first broken the patient's lower jaw, had lodged in the root of the tongue. The blood flowed freely and collected on his fingers in glutinous masses.

Captain Beaudoin was again resting on his mattress in the large room. Gilberte and Mme. Delaherche had followed the stretcher when he was carried from the operating table, and even Delaherche, notwithstanding his anxiety, came in for a moment's chat.

"Lie here and rest a few minutes, Captain. We will have a room prepared for you, and you shall be our guest."

But the wounded man shook off his lethargy and for a moment had command of his faculties.

"No, it is not worth while; I feel that I am going to die."

And he looked at them with wide eyes, filled with the horror of death.

"Oh, Captain! why do you talk like that?" murmured Gilberte, with a shiver, while she forced a smile to her lips. "You will be quite well a month hence."

He shook his head mournfully, and in the room was conscious of no presence save hers; on all his face was expressed his unutterable yearning for life, his bitter, almost craven regret that he was to be snatched away so young, leaving so many joys behind untasted.

"I am going to die, I am going to die. Oh! 'tis horrible~"

Then suddenly he became conscious of his torn, soiled uniform and the grime upon his hands, and it made him feel uncomfortable to be in the company of women in such a state. It shamed him to show such weakness, and his desire to look and be the gentleman to the last restored to him his manhood. When he spoke again it was in a tone almost of cheerfulness.

"If I have got to die, though, I would rather it should be with clean hands. I should count it a great kindness, madame, if you would moisten a napkin and let me have it."

Gilberte sped away and quickly returned with the napkin, with which she herself cleansed the hands of the dying man. Thenceforth, desirous of quitting the scene with dignity, he displayed much firmness. Delaherche did what he could to cheer him, and assisted his wife in the small attentions she offered for his comfort. Old Mme. Delaherche, too, in presence of the man whose hours were numbered, felt her enmity subsiding. She would be silent, she who knew all and had sworn to impart her knowledge to her son. What would it avail to excite discord in the household, since death would soon obliterate all trace of the wrong?

The end came very soon. Captain Beaudoin, whose strength was ebbing rapidly, relapsed into his comatose condition, and a cold sweat broke out and stood in beads upon his neck and forehead. He opened his eyes again, and began to feebly grope about him with his stiffening fingers, as if feeling for a covering that was not there, pulling at it with a gentle, continuous movement, as if to draw it up around his shoulders.

"It is cold~ Oh! it is so cold."

And so he passed from life, peacefully, without a struggle; and on his wasted, tranquil face rested an expression of unspeakable melancholy.

Delaherche saw to it that the remains, instead of being borne away and placed among the common dead, were deposited in one of the outbuildings of the factory. He endeavored to prevail on Gilberte, who was tearful and disconsolate, to retire to her apartment, but she declared that to be alone now would be more than her nerves could stand, and begged to be allowed to remain with her mother-in-law in the ambulance, where the noise and movement would be a distraction to her. She was seen presently running to carry a drink of water to a chasseur d'Afrique whom his fever had made delirious, and she assisted a hospital steward to dress the hand of a little recruit, a lad of twenty, who had had his thumb shot away and come in on foot from the battlefield; and as he was jolly and amusing, treating his wound with all the levity and nonchalance of the Parisian rollicker, she was soon laughing and joking as merrily as he.

While the captain lay dying the cannonade seemed, if that were possible, to have increased in violence; another shell had landed in the garden, shattering one of the old elms. Terror-stricken men came running in to say that all Sedan was in danger of destruction; a great fire had broken out in the Faubourg de la Cassine. If the bombardment should continue with such fury for any length of time there would be nothing left of the city.

"It can't be; I am going to see about it!" Delaherche exclaimed, violently excited.

"Where are you going, pray?" asked Bouroche.

"Why, to the Sous-Prefecture, to see what the Emperor means by fooling us in this way, with his talk of hoisting the white flag."

For some few seconds the major stood as if petrified at the idea of defeat and capitulation, which presented itself to him then for the first time in the midst of his impotent efforts to save the lives of the poor maimed creatures they were bringing in to him from the field. Rage and grief were in his voice as he shouted:

"Go to the devil, if you will! All you can do won't keep us from being soundly whipped!"

On leaving the factory Delaherche found it no easy task to squeeze his way through the throng; at every instant the crowd of straggling soldiers that filled the streets received fresh accessions. He questioned several of the officers whom he encountered; not one of them had seen the white flag on the citadel. Finally he met a colonel, who declared that he had caught a momentary glimpse of it: that it had been run up and then immediately hauled down. That explained matters; either the Germans had not seen it, or seeing it appear and disappear so quickly, had inferred the distressed condition of the French and redoubled their fire in consequence. There was a story in circulation how a general officer, enraged beyond control at the sight of the flag, had wrested it from its bearer, broken the staff, and trampled it in the mud. And still the Prussian batteries continued to play upon the city, shells were falling upon the roofs and in the streets, houses were in flames; a woman had just been killed at the corner of the Rue Pont de Meuse and the Place Turenne.

At the Sous-Prefecture Delaherche failed to find Rose at her usual station in the janitor's lodge. Everywhere were evidences of disorder; all the doors were standing open; the reign of terror had commenced. As there was no sentry or anyone to prevent, he went upstairs, encountering on the way only a few scared-looking men, none of whom made any offer to stop him. He had reached the first story and was hesitating what to do next when he saw the young girl approaching him.

"Oh, M. Delaherche! isn't this dreadful! Here, quick! this way, if you would like to see the Emperor."

On the left of the corridor a door stood ajar, and through the narrow opening a glimpse could be had of the sovereign, who had resumed his weary, anguished tramp between the fireplace and the window. Back and forth he shuffled with heavy, dragging steps, and ceased not, despite his unendurable suffering. An aide-de-camp had just entered the room ~it was he who had failed to close the door behind him-and Delaherche heard the Emperor ask him in a sorrowfully reproachful voice:

"What is the reason of this continued firing, sir, after I gave orders to hoist the white flag?"

The torture to him had become greater than he could bear, that never-ceasing cannonade, that seemed to grow more furious with every minute. Every time he approached the window it pierced him to the heart. More spilling of blood, more useless squandering of human life! At every moment the piles of corpses were rising higher on the battlefield, and his was the responsibility. The compassionate instincts that entered so largely into his nature revolted at it, and more than ten times already he had asked that question of those who approached him.

"I gave orders to raise the white flag; tell me, why do they continue firing?"

The aide-de-camp made answer in a voice so low that Delaherche failed to catch its purport. The Emperor, moreover, seemed not to pause to listen, drawn by some irresistible attraction to that window at which, each time he approached it, he was greeted by that terrible salvo of artillery that rent and tore his being. His pallor was greater even than it had been before; his poor, pinched, wan face, on which were still visible traces of the rouge that had been applied that morning, bore witness to his anguish.

At that moment a short, quick-motioned man in dust-soiled uniform, whom Delaherche recognized as General Lebrun, hurriedly crossed the corridor and pushed open the door, without waiting to be announced. And scarcely was he in the room when again was heard the Emperor's so oft repeated question.

"Why do they continue to fire, General, when I have given orders to hoist the white flag?"

The aide-de-camp left the apartment, shutting the door behind him, and Delaherche never knew what was the general's answer. The vision had faded from his sight.

"Ah!" said Rose, "things are going badly; I can see that clearly enough by all those gentlemen's faces. It is bad for my tablecloth, too; I am afraid I shall never see it again; somebody told me it had been torn in pieces. But it is for the Emperor that I feel most sorry in all this business, for he is in a great deal worse condition than the marshal; he would be much better off in his bed than in that room, where he is wearing himself out with his everlasting walking."

She spoke with much feeling, and on her pretty pink and white face there was an expression of sincere pity, but Delaherche, whose Bonapartist ardor had somehow cooled considerably during the last two days, said to himself that she was a little fool. He nevertheless remained chatting with her a moment in the hall below while waiting for General Lebrun to take his departure, and when that officer appeared and left the building he followed him.

General Lebrun had explained to the Emperor that if it was thought best to apply for an armistice, etiquette demanded that a letter to that effect, signed by the commander-in-chief of the French forces, should be dispatched to the German commander-in-chief. He had also offered to write the letter, go in search of General de Wimpffen, and obtain his signature to it. He left the Sous-Prefecture with the letter in his pocket, but apprehensive he might not succeed in finding de Wimpffen, entirely ignorant as he was of the general's whereabouts on the field of battle. Within the ramparts of Sedan, moreover, the crowd was so dense that he was compelled to walk his horse, which enabled Delaherche to keep him in sight until he reached the Minil gate.

Once outside upon the road, however, General Lebrun struck into a gallop, and when near Balan had the good fortune to fall in with the chief. Only a few minutes previous to this the latter had written to the Emperor: "Sire, come and put yourself at the head of your troops; they will force a passage through the enemy's lines for you, or perish in the attempt;" therefore he flew into a furious passion at the mere mention of the word armistice. No, no! he would sign nothing, he would fight it out! This was about half-past three o'clock, and it was shortly afterward that occurred the gallant, but mad attempt, the last serious effort of the day, to pierce the Bavarian lines and regain possession of Bazeilles. In order to put heart into the troops a ruse was resorted to: in the streets of Sedan and in the fields outside the walls the shout was raised: "Bazaine is coming up! Bazaine is at hand!" Ever since morning many had allowed themselves to be deluded by that hope; each time that the Germans opened fire with a fresh battery it was confidently asserted to be the guns of the army of Metz. In the neighborhood of twelve hundred men were collected, soldiers of all arms, from every corps, and the little column bravely advanced into the storm of missiles that swept the road, at double time. It was a splendid spectacle of heroism and endurance while it lasted; the numerous casualties did not check the ardor of the survivors, nearly five hundred yards were traversed with a courage and nerve that seemed almost like madness; but soon there were great gaps in the ranks, the bravest began to fall back. What could they do against overwhelming numbers? It was a mad attempt, anyway; the desperate effort of a commander who could not bring himself to acknowledge that he was defeated. And it ended by General de

Wimpffen finding himself and General Lebrun alone together on the Bazeilles road, which they had to make up their mind to abandon to the enemy, for good and all. All that remained for them to do was to retreat and seek security under the walls of Sedan.

Upon losing sight of the general at the Minil gate Delaherche had hurried back to the factory at the best speed he was capable of, impelled by an irresistible longing to have another look from his observatory at what was going on in the distance. Just as he reached his door, however, his progress was arrested a moment by encountering Colonel de Vineuil, who, with his blood-stained boot, was being brought in for treatment in a condition of semi-consciousness, upon a bed of straw that had been prepared for him on the floor of a market-gardener's wagon. The colonel had persisted in his efforts to collect the scattered fragments of his regiment until he dropped from his horse. He was immediately carried upstairs and put to bed in a room on the first floor, and Bouroche, who was summoned at once, finding the injury not of a serious character, had only to apply a dressing to the wound, from which he first extracted some bits of the leather of the boot. The worthy doctor was wrought up to a high pitch of excitement; he exclaimed, as he went downstairs, that he would rather cut off one of his own legs than continue working in that unsatisfactory, slovenly way, without a tithe of either the assistants or the appliances that he ought to have. Below in the ambulance, indeed, they no longer knew where to bestow the cases that were brought them, and had been obliged to have recourse to the lawn, where they laid them on the grass. There were already two long rows of them, exposed beneath the shrieking shells, filling the air with their dismal plaints while waiting for his ministrations. The number of cases brought in since noon exceeded four hundred, and in response to Bouroche's repeated appeals for assistance he had been sent one young doctor from the city. Good as was his will, he was unequal to the task; he probed, sliced, sawed, sewed like a man frantic, and was reduced to despair to see his work continually accumulating before him. Gilberte, satiated with sights of horror, unable longer to endure the sad spectacle of blood and tears, remained upstairs with her uncle, the colonel, leaving to Mme. Delaherche the care of moistening fevered lips and wiping the cold sweat from the brow of the dying.

Rapidly climbing the stairs to his terrace, Delaherche endeavored to form some idea for himself of how matters stood. The city had suffered less injury than was generally supposed; there was one great conflagration, however, over in the Faubourg de la Cassine, from which dense volumes of smoke were rising. Fort Palatinat had discontinued its fire, doubtless because the ammunition was all expended; the guns mounted on the Porte de Paris alone continued to make themselves heard at infrequent intervals. But something that he beheld presently had greater interest for his eyes than all beside; they had run up the white flag on the citadel again, but it must be that it was invisible from the battlefield, for there was no perceptible slackening of the fire. The Balan road was concealed from his vision by the neighboring roofs; he was unable to make out what the troops were doing in that direction. Applying his eye to the telescope, however, which remained as he had left it, directed on la Marfee, he again beheld the cluster of officers that he had seen in that same place about midday. The master of them all, that miniature toy-soldier in lead, half finger high, in whom he had thought to recognize the King of Prussia, was there still, erect in his plain, dark uniform before the other officers, who, in their showy trappings, were for the most part reclining carelessly on the grass. Among them were officers from foreign lands, aides-de-camp, generals, high officials, princes; all of them with field glasses in their hands, with which, since early morning, they had been watching every phase of the death-struggle of the army of Chalons, as if they were at the play. And the direful drama was drawing to its end.

From among the trees that clothed the summit of la Marfee King William had just witnessed the junction of his armies. It was an accomplished fact; the third army, under the leadership of his son, the Crown Prince, advancing by the way of Saint-Menges and Fleigneux, had secured possession of the plateau of Illy, while the fourth, commanded by the Crown Prince of Saxony, turning the wood of la Garenne and, coming up through Givonne and Daigny, had also reached its appointed rendezvous. There, too, the XIth and Vth corps had joined hands with the XIIth corps and the Guards. The gallant but ineffectual charge of Margueritte's division in its supreme effort to break through the hostile lines at the very moment when the circle was being rounded out had elicited from the king the exclamation: "Ah, the brave fellows!" Now the great movement, inexorable as fate, the details of which had been arranged with such mathematical precision, was complete, the jaws of the vise had closed, and stretching on his either hand far in the distance, a mighty wall of adamant surrounding the army of the French, were the countless men and guns that called him master. At the north the contracting lines maintained a constantly increasing pressure on the vanquished, forcing them back upon Sedan under the merciless fire of the batteries that lined the horizon in an array without a break. Toward the south, at Bazeilles, where the conflict had ceased to rage and the scene was one of mournful desolation, great clouds of smoke were rising from the ruins of what had once been happy homes, while the Bavarians, now masters of Balan, had advanced their batteries to within three hundred yards of the city gates. And the other batteries, those posted on the left bank at Pont Maugis, Noyers, Frenois, Wadelincourt, completing the impenetrable rampart of flame and bringing it around to the sovereign's feet on his right, that had been spouting fire uninterruptedly for nearly twelve hours, now thundered more loudly still.

But King William, to give his tired eyes a moment's rest, dropped his glass to his side and continued his observations with unassisted vision. The sun was slanting downward to the woods on his left, about to set in a sky where there was not a cloud, and the golden light that lay upon the landscape was so transcendently clear and limpid that the most insignificant objects stood out with startling distinctness. He could almost count the houses in Sedan, whose windows flashed back the level rays of the departing day-star, and the ramparts and fortifications, outlined in black against the eastern sky, had an unwonted aspect of frowning massiveness. Then, scattered among the fields to right and left, were the pretty, smiling villages, reminding one of the toy villages that come packed in boxes for the little ones; to the west Donchery, seated at the border of her broad plain; Douzy and Carignan to the east, among the meadows. Shutting in the picture to the north was the forest of the Ardennes, an ocean of sunlit verdure, while the Meuse, loitering with sluggish current through the plain with many a bend and curve, was like a stream of purest molten gold in that caressing light. And seen from that height, with the sun's parting kiss resting on it, the horrible battlefield, with its blood and smoke, became an exquisite and highly finished miniature; the dead horsemen and disemboweled steeds on the plateau of Floing were so many splashes of bright color; on the right, in the direction of Givonne, those minute black specks that whirled and eddied with such apparent lack of aim, like motes dancing in the sunshine, were the retreating fragments of the beaten army; while on the left a Bavarian battery on the peninsula of Iges, its guns the size of matches, might have been taken for some mechanical toy as it performed its evolutions with clockwork regularity. The victory was crushing, exceeding all that the victor could have desired or hoped, and the King felt no remorse in presence of all those corpses, of those thousands of men that were as the dust upon the roads of that broad valley where, notwithstanding the burning of Bazeilles, the slaughter of Illy, the anguish of Sedan, impassive nature yet could don her gayest robe and put on her brightest smile as the perfect day faded into the tranquil evening.

But suddenly Delaherche descried a French officer climbing the steep path up the flank of la Marfee; he was a general, wearing a blue tunic, mounted on a black horse, and preceded by a hussar bearing a white flag. It was General Reille, whom the Emperor had entrusted with this communication for the King of Prussia: "My brother, as it has been denied me to die at the head of my army, all that is left me is to surrender my sword to Your Majesty. I am Your Majesty's affectionate brother, Napoleon." Desiring to arrest the butchery and being no longer master, the Emperor yielded himself a prisoner, in the hope to placate the conqueror by the sacrifice. And Delaherche saw General Reille rein up his charger and dismount at ten paces from the King, then advance and deliver his letter; he was unarmed and merely carried a riding whip. The sun was setting in a flood of rosy light; the King seated himself on a chair in the midst of a grassy open space, and resting his hand on the back of another chair that was held in place by a secretary, replied that he accepted the sword and would await the appearance of an officer empowered to settle the terms of the capitulation.

7

As when the ice breaks up and the great cakes come crashing, grinding down upon the bosom of the swollen stream, carrying away all before them, so now, from every position about Sedan that had been wrested from the French, from Floing and the plateau of Illy, from the wood of la Garenne, the valley of la Givonne and the Bazeilles road, the stampede commenced; a mad torrent of horses, guns, and affrighted men came pouring toward the city. It was a most unfortunate inspiration that brought the army under the walls of that fortified place. There was too much in the way of temptation there; the shelter that it afforded the skulker and the deserter, the assurance of safety that even the bravest beheld behind its ramparts, entailed widespread panic and demoralization. Down there behind those protecting walls, so everyone imagined, was safety from that terrible artillery that had been blazing without intermission for near twelve hours; duty, manhood, reason were all lost sight of; the man disappeared and was succeeded by the brute, and their fierce instinct sent them racing wildly for shelter, seeking a place where they might hide their head and lie down and sleep.

When Maurice, bathing Jean's face with cool water behind the shelter of their bit of wall, saw his friend open his eyes once more, he uttered an exclamation of delight.

"Ah, poor old chap, I was beginning to fear you were done for! And don't think I say it to find fault, but really you are not so light as you were when you were a boy."

It seemed to Jean, in his still dazed condition, that he was awaking from some unpleasant dream. Then his recollection returned to him slowly, and two big tears rolled down his cheeks. To think that little Maurice, so frail and slender, whom he had loved and petted like a child, should have found strength to lug him all that distance!

"Let's see what damage your knowledge-box has sustained."

The wound was not serious; the bullet had plowed its way through the scalp and considerable blood had flowed. The hair, which was now matted with the coagulated gore, had served to stanch the current, therefore Maurice refrained from applying water to the hurt, so as not to cause it to bleed afresh.

"There, you look a little more like a civilized being, now that you have a clean face on you. Let's see if I can find something for you to wear on your head." And picking up the *kepi* of a soldier who lay dead not far away, he tenderly adjusted it on his comrade. "It fits you to a T. Now if you can only walk everyone will say we are a very good-looking couple."

Jean got on his legs and gave his head a shake to assure himself it was secure. It seemed a little heavier than usual, that was all; he thought he should get along well enough. A great wave of tenderness swept through his simple soul; he caught Maurice in his arms and hugged him to his bosom, while all he could find to say was:

"Ah! dear boy, dear boy!"

But the Prussians were drawing near: it would not answer to loiter behind the wall. Already Lieutenant Rochas, with what few men were left him, was retreating, guarding the flag, which the sous-lieutenant still carried under his arm, rolled around the staff. Lapoulle's great height enabled him to fire an occasional shot at the advancing enemy over the coping of the wall, while Pache had slung his chassepot across his shoulder by the strap, doubtless considering that he had done a fair day's work and it was time to eat and sleep. Maurice and Jean, stooping until they were bent almost double, hastened to rejoin them. There was no scarcity of muskets and ammunition; all they had to do was stoop and pick them up. They equipped themselves afresh, having left everything behind, knapsacks included, when one lugged the other out of danger on his shoulders. The wall extended to the wood of la Garenne, and the little band, believing that now their safety was assured, made a rush for the protection afforded by some farm buildings, whence they readily gained the shelter of the trees.

"Ah!" said Rochas, drawing a long breath, "we will remain here a moment and get our wind before we resume the offensive." No adversity could shake his unwavering faith.

They had not advanced many steps before all felt that they were entering the valley of death, but it was useless to think of retracing their steps; their only line of retreat lay through the wood, and cross it they must, at every hazard. At that time, instead of la Garenne, its more fitting name would have been the wood of despair and death; the Prussians, knowing that the French troops were retiring in that direction, were riddling it with artillery and musketry. Its shattered branches tossed and groaned as if enduring the scourging of a mighty tempest. The shells hewed down the stalwart trees, the bullets brought the leaves fluttering to the earth in showers; wailing voices seemed to issue from the cleft trunks, sobs accompanied the little twigs as they fell bleeding from the parent stem. It might have been taken for the agony of some vast multitude, held there in chains and unable to flee under the pelting of that pitiless iron hail; the shrieks, the terror of thousands of creatures rooted to the ground. Never was anguish so poignant as of that bombarded forest.

Maurice and Jean, who by this time had caught up with their companions, were greatly alarmed. The wood where they then were was a growth of large trees, and there was no obstacle to their running, but the bullets came whistling about their ears from every direction, making it impossible for them to avail themselves of the shelter of the trunks. Two men were killed, one of them struck in the back, the other in front. A venerable oak, directly in Maurice's path, had its trunk shattered by a shell, and sank, with the stately grace of a mailed paladin, carrying down all before it, and even as the young man was leaping back the top of a gigantic ash on his left, struck by another shell, came crashing to the ground like some tall cathedral spire. Where could they fly? whither bend their steps? Everywhere the branches were falling; it was as one who should endeavor to fly from some vast edifice menaced with destruction, only to find himself in each room he enters in succession confronted with crumbling walls and ceilings. And when, in order to escape being crushed by the big trees, they took refuge in a thicket of bushes, Jean came near being killed by a projectile, only it fortunately failed to explode. They could no longer make any progress now on account of the dense growth of the shrubbery; the supple branches caught them around the shoulders, the rank, tough grass held them by the ankles, impenetrable walls of brambles rose before them and blocked their way, while all the time the foliage was fluttering down about them, clipped by the gigantic scythe that was mowing down the wood. Another man was struck dead beside them by a bullet in the forehead, and he retained his erect position, caught in some vines between two small birch trees. Twenty times, while they were prisoners in that thicket, did they feel death hovering over them.

"Holy Virgin!" said Maurice, "we shall never get out of this alive."

His face was ashy pale, he was shivering again with terror; and Jean, always so brave, who had cheered and comforted him that morning, he, also, was very white and felt a strange, chill sensation creeping down his spine. It was fear, horrible, contagious, irresistible fear. Again they were conscious of a consuming thirst, an intolerable dryness of the mouth, a contraction of the throat, painful as if someone were choking them. These symptoms were accompanied by nausea and qualms at the pit of the stomach, while maleficent goblins kept puncturing their anguish, trembling legs with needles. Another of the physical effects of their fear was that in the congested condition of the blood vessels of the retina they beheld thousands upon thousands of small black specks flitting past them, as if it had been possible to distinguish the flying bullets.

"Confound the luck!" Jean stammered. "It is not worth speaking of, but it's vexatious all the same, to be here getting one's head broken for other folks, when those other folks are at home, smoking their pipe in comfort."

"Yes, that's so," Maurice replied, with a wild look. "Why should it be I rather than someone else?"

It was the revolt of the individual Ego, the unaltruistic refusal of the one to make himself a sacrifice for the benefit of the species.

"And then again," Jean continued, "if a fellow could but know the rights of the matter; if he could be sure that any good was to come from it all." Then turning his head and glancing at the western sky: "Anyway, I wish that blamed sun would hurry up and go to roost. Perhaps they'll stop fighting when it's dark."

With no distinct idea of what o'clock it was and no means of measuring the flight of time, he had long been watching the tardy declination of the fiery disk, which seemed to him to have ceased to move, hanging there in the heavens over the woods of the left bank. And this was not owing to any lack of courage on his part; it was simply the overmastering, ever increasing desire, amounting to an imperious necessity, to be relieved from the screaming and whistling of those projectiles, to run away somewhere and find a hole where he might hide his head and lose himself in oblivion. Were it not for the feeling of shame that is implanted in men's breasts and keeps them from showing the white feather before their comrades, every one of them would lose his head and run, in spite of himself, like the veriest poltroon.

Maurice and Jean, meanwhile, were becoming somewhat more accustomed to their surroundings, and even when their terror was at its highest there came to them a sort of exalted self-unconsciousness that had in it something of bravery. They finally reached a point when they did not even hasten their steps as they made their way through the accursed wood. The horror of the bombardment was even greater than it had been previously among that race of sylvan denizens, killed at their post, struck down on every hand, like gigantic, faithful sentries. In the delicious twilight that reigned, golden-green, beneath their umbrageous branches, among the mysterious recesses of romantic, moss-carpeted retreats, Death showed his ill-favored, grinning face. The solitary fountains were contaminated; men fell dead in distant nooks whose depths had hitherto been trod by none save wandering lovers. A bullet pierced a man's chest; he had time to utter the one word: "hit!" and fell forward on his face, stone dead. Upon the lips of another, who had both legs broken by a shell, the gay laugh remained; unconscious of his hurt, he supposed he had tripped over a root. Others, injured mortally, would run on for some yards, jesting and conversing, until suddenly they went down like a log in the supreme convulsion. The severest wounds were hardly felt at the moment they were received; it was only at a later period that the terrible suffering commenced, venting itself in shrieks and hot tears.

Ah, that accursed wood, that wood of slaughter and despair, where, amid the sobbing of the expiring trees, arose by degrees and swelled the agonized clamor of wounded men. Maurice and Jean saw a zouave, nearly disemboweled, propped against the trunk of an oak, who kept up a most terrific howling, without a moment's intermission. A little way beyond another man was actually being slowly roasted; his clothing had taken fire and the flames had run up and caught his beard, while he, paralyzed by a shot that had broken his back, was silently weeping scalding tears. Then there was a captain, who, one arm torn from its socket and his flank laid open to the thigh, was writhing on the ground in agony unspeakable, beseeching, in heartrending accents, the by-passers to end his suffering. There were others, and others, and others still, whose torments may not be described, strewing the grass-grown paths in such

206

numbers that the utmost caution was required to avoid treading them under foot. But the dead and wounded had ceased to count; the comrade who fell by the way was abandoned to his fate, forgotten as if he had never been. No one turned to look behind. It was his destiny, poor devil! Next it would be someone else, themselves, perhaps.

They were approaching the edge of the wood when a cry of distress was heard behind them.

"Help! help!"

It was the subaltern standard-bearer, who had been shot through the left lung. He had fallen, the blood pouring in a stream from his mouth, and as no one heeded his appeal he collected his fast ebbing strength for another effort:

"To the colors!"

Rochas turned and in a single bound was at his side. He took the flag, the staff of which had been broken in the fall, while the young officer murmured in words that were choked by the bubbling tide of blood and froth:

"Never mind me; I am a goner. Save the flag!"

And they left him to himself in that charming woodland glade to writhe in protracted agony upon the ground, tearing up the grass with his stiffening fingers and praying for death, which would be hours yet ere it came to end his misery.

At last they had left the wood and its horrors behind them. Beside Maurice and Jean all that were left of the little band were Lieutenant Rochas, Lapoulle and Pache. Gaude, who had strayed away from his companions, presently came running from a thicket to rejoin them, his bugle hanging from his neck and thumping against his back with every step he took. It was a great comfort to them all to find themselves once again in the open country, where they could draw their breath; and then, too, there were no longer any whistling bullets and crashing shells to harass them; the firing had ceased on this side of the valley.

The first object they set eyes on was an officer who had reined in his smoking, steaming charger before a farm-yard gate and was venting his towering rage in a volley of Billingsgate. It was General Bourgain-Desfeuilles, the commander of their brigade, covered with dust and looking as if he was about to tumble from his horse with fatigue. The chagrin on his gross, high-colored, animal face told how deeply he took to heart the disaster that he regarded in the light of a personal misfortune. His command had seen nothing of him since morning. Doubtless he was somewhere on the battlefield, striving to rally the remnants of his brigade, for he was not the man to look closely to his own safety in his rage against those Prussian batteries that had at the same time destroyed the empire and the fortunes of a rising officer, the favorite of the Tuileries.

"Tonnerre de Dieu!" he shouted, "is there no one of whom one can ask a question in this d---d country?"

The farmer's people had apparently taken to the woods. At last a very old woman appeared at the door, some servant who had been forgotten, or whose feeble legs had compelled her to remain behind.

"Hallo, old lady, come here! Which way from here is Belgium?"

She looked at him stupidly, as one who failed to catch his meaning. Then he lost all control of himself and effervesced, forgetful that the woman was only a poor peasant, bellowing that he had no idea of going back to Sedan to be caught like a rat in a trap; not he! he was going to make tracks for foreign parts, he was, and d---d quick, too! Some soldiers had come up and stood listening.

"But you won't get through, General," spoke up a sergeant; "the Prussians are everywhere. This morning was the time for you to cut stick."

There were stories even then in circulation of companies that had become separated from their regiments and crossed the frontier without any intention of doing so, and of others that, later in the day, had succeeded in breaking through the enemy's lines before the armies had effected their final junction.

The general shrugged his shoulders impatiently. "What, with a few daring fellows of your stripe, do you mean to say we couldn't go where we please? I think I can find fifty daredevils to risk their skin in the attempt." Then, turning again to the old peasant: "Eh! you old mummy, answer, will you, in the devil's name! where is the frontier?"

She understood him this time. She extended her skinny arm in the direction of the forest.

"That way, that way!"

"Eh? What's that you say? Those houses that we see down there, at the end of the field?"

"Oh! farther, much farther. Down yonder, away down yonder!"

The general seemed as if his anger must suffocate him. "It is too disgusting, an infernal country like this! one can make neither top nor tail of it. There was Belgium, right under our nose; we were all afraid we should put our foot in it without knowing it; and now that one wants to go there it is somewhere else. No, no! it is too much; I've had enough of it; let them take me prisoner if they will, let them do what they choose with me; I am going to bed!" And clapping spurs to his horse, bobbing up and down on his saddle like an inflated wine skin, he galloped off toward Sedan.

A winding path conducted the party down into the Fond de Givonne, an outskirt of the city lying between two hills, where the single village street, running north and south and sloping gently upward toward the forest, was lined with gardens and modest houses. This street was just then so obstructed by flying soldiers that Lieutenant Rochas, with Pache, Lapoulle, and Gaude, found himself caught in the throng and unable for the moment to move in either direction. Maurice and Jean had some difficulty in rejoining them; and all were surprised to hear themselves hailed by a husky, drunken voice, proceeding from the tavern on the corner, near which they were blockaded.

"My stars, if here ain't the gang! Hallo, boys, how are you? My stars, I'm glad to see you!"

They turned, and recognized Chouteau, leaning from a window of the ground floor of the inn. He seemed to be very drunk, and went on, interspersing his speech with hiccoughs:

"Say, fellows, don't stand on ceremony if you're thirsty. There's enough left for the comrades." He turned unsteadily and called to someone who was invisible within the room: "Come here, you lazybones. Give these gentlemen something to drink-"

Loubet appeared in turn, advancing with a flourish and holding aloft in either hand a full bottle, which he waved above his head triumphantly. He was not so far gone as his companion; with his Parisian blague, imitating the nasal drawl of the coco-venders of the boulevards on a public holiday, he cried:

"Here you are, nice and cool, nice and cool! Who'll have a drink?"

Nothing had been seen of the precious pair since they had vanished under pretense of taking Sergeant Sapin into the ambulance. It was sufficiently evident that since then they had been strolling and seeing the sights, taking care to keep out of the way of the shells, until finally they had brought up at this inn that was given over to pillage.

Lieutenant Rochas was very angry. "Wait a bit, you scoundrels, just wait, and I'll attend to your case! deserting and getting drunk while the rest of your company were under fire!"

But Chouteau would have none of his reprimand. "See here, you old lunatic, I want you to understand that the grade of lieutenant is abolished; we are all free and equal now. Aren't you satisfied with the basting the Prussians gave you today, or do you want some more?"

The others had to restrain the lieutenant to keep him from assaulting the socialist. Loubet himself, dandling his bottles affectionately in his arms, did what he could to pour oil upon the troubled waters.

"Quit that, now! what's the use quarreling, when all men are brothers!" And catching sight of Lapoulle and Pache, his companions in the squad: "Don't stand there like great gawks, you fellows! Come in here and take something to wash the dust out of your throats."

Lapoulle hesitated a moment, dimly conscious of the impropriety there was in the indulgence when so many poor devils were in such sore distress, but he was so knocked up with fatigue, so terribly hungry and thirsty! He said not a word, but suddenly making up his mind, gave one bound and landed in the room, pushing before him Pache, who, equally silent, yielded to the temptation he had not strength to resist. And they were seen no more.

"The infernal scoundrels!" muttered Rochas. "They deserve to be shot, every mother's son of them!"

He had now remaining with him of his party only Jean, Maurice, and Gaude, and all four of them, notwithstanding their resistance, were gradually involved and swallowed up in the torrent of stragglers and fugitives that streamed along the road, filling its whole width from ditch to ditch. Soon they were at a distance from the inn. It was the routed army rolling down upon the ramparts of Sedan, a roily, roaring flood, such as the disintegrated mass of earth and boulders that the storm, scouring the mountainside, sweeps down into the valley. From all the surrounding plateaus, down every slope, up every narrow gorge, by the Floing road, by Pierremont, by the cemetery, by the Champ de Mars, as well as through the Fond de Givonne, the same sorry rabble was streaming cityward in panic haste, and every instant brought fresh accessions to its numbers. And who could reproach those wretched men, who, for twelve long, mortal hours, had stood in motionless array under the murderous artillery of an invisible enemy, against whom they could do nothing? The batteries now were playing on them from front, flank, and rear; as they drew nearer the city they presented a fairer mark for the convergent fire; the guns dealt death and destruction out by wholesale on that dense, struggling mass of men in that accursed hole, where there was no escape from the bursting shells. Some regiments of the 7th corps, more particularly those that had been stationed about Floing, had left the field in tolerably good order, but in the Fond de Givonne there was no longer either organization or command; the troops were a pushing, struggling mob, composed of debris from regiments of every description, zouaves, turcos, chasseurs, infantry of the line, most of them without arms, their uniforms soiled and torn, with grimy hands, blackened faces, bloodshot eyes starting from their sockets and lips swollen and distorted from their yells of fear or rage. At times a riderless horse would dash through the throng, overturning those who were in his path and leaving behind him a long wake of consternation. Then some guns went thundering by at breakneck speed, a retreating battery abandoned by its officers, and the drivers, as if drunk, rode down everything and everyone, giving no word of warning. And still the shuffling tramp of many feet along the dusty road went on and ceased not, the close-compacted column pressed on, breast to back, side to side; a retreat en masse, where vacancies in the ranks were filled as soon as made, all moved by one common impulse, to reach the shelter that lay before them and be behind a wall.

Again Jean raised his head and gave an anxious glance toward the west; through the dense clouds of dust raised by the tramp of that great multitude the luminary still poured his scorching rays down upon the exhausted men. The sunset was magnificent, the heavens transparently, beautifully blue.

"It's a nuisance, all the same," he muttered, "that plaguey sun that stays up there and won't go to roost!"

Suddenly Maurice became aware of the presence of a young woman whom the movement of the resistless throng had jammed against a wall and who was in danger of being injured, and on looking more attentively was astounded to recognize in her his sister Henriette. For near a minute he stood gazing at her in open-mouthed amazement, and finally it was she who spoke, without any appearance of surprise, as if she found the meeting entirely natural.

"They shot him at Bazeilles-and I was there. Then, in the hope that they might at least let me have his body, I had an idea-"

She did not mention either Weiss or the Prussians by name; it seemed to her that everyone must understand. Maurice did understand. It made his heart bleed; he gave a great sob.

"My poor darling!"

When, about two o'clock, Henriette recovered consciousness, she found herself at Balan, in the kitchen of some people who were strangers to her, her head resting on a table, weeping. Almost immediately, however, she dried her tears; already the heroic element was reasserting itself in that silent woman, so frail, so gentle, yet of a spirit so indomitable that she could suffer martyrdom for the faith, or the love, that was in her. She knew not fear; her quiet, undemonstrative courage was lofty and invincible. When her distress was deepest she had summoned up her resolution, devoting her reflections to how she might recover her husband's body, so as to give it decent burial. Her first project was neither more nor less than to make her way back to Bazeilles, but everyone advised her against this course, assuring her that it would be absolutely impossible to get through the German lines. She therefore abandoned the idea, and tried to think of someone among her acquaintance who would afford her the protection of his company, or at least assist her in the necessary preliminaries. The person to whom she determined she would apply was a M. Dubreuil, a cousin of hers, who had been assistant superintendent of the refinery at Chene at the time her husband was employed there; Weiss had been a favorite of his; he would not refuse her his assistance. Since the time, now two years ago, when his wife had inherited a handsome fortune, he had been occupying a pretty villa, called the Hermitage, the terraces of which could be seen skirting the hillside of a suburb of Sedan, on the further side of the Fond de Givonne. And thus it was toward the Hermitage that she was now bending her steps, compelled at every moment to pause before some fresh obstacle, continually menaced with being knocked down and trampled to death.

Maurice, to whom she briefly explained her project, gave it his approval.

"Cousin Dubreuil has always been a good friend to us. He will be of service to you."

Then an idea of another nature occurred to him. Lieutenant Rochas was greatly embarrassed as to what disposition he should make of the flag. They all were firmly resolved to save it-to do anything rather than allow it to fall into the hands of the Prussians. It had been suggested to cut it into pieces, of which each should carry one off under his shirt, or else to bury it at the foot of a tree, so noting the locality in memory that they might be able to come and disinter it at some future day; but the idea of mutilating the flag, or burying it like a corpse, affected them too painfully, and they were considering if they might not preserve it in some other manner. When Maurice, therefore, proposed to entrust the standard to a reliable person who would conceal it and, in case of necessity, defend it, until such day as he should restore it to them intact, they all gave their assent.

"Come," said the young man, addressing his sister, "we will go with you to the Hermitage and see if Dubreuil is there. Besides, I do not wish to leave you without protection."

It was no easy matter to extricate themselves from the press, but they succeeded finally and entered a path that led upward on their left. They soon found themselves in a region intersected by a perfect labyrinth of lanes and narrow passages, a district where truck farms and gardens predominated, interspersed with an occasional villa and small holdings of extremely irregular outline, and these lanes and passages wound circuitously between blank walls, turning sharp corners at every few steps and bringing up abruptly in the cul-de-sac of some courtyard, affording admirable facilities for carrying on a guerila warfare; there were spots where ten men might defend themselves for hours against a regiment. Desultory firing was already beginning to be heard, for the suburb commanded Balan, and the Bavarians were already coming up on the other side of the valley.

When Maurice and Henriette, who were in the rear of the others, had turned once to the left, then to the right and then to the left again, following the course of two interminable walls, they suddenly came out before the Hermitage, the door of which stood wide open. The grounds, at the top of which was a small park, were terraced off in three broad terraces, on one of which stood the residence, a roomy, rectangular structure, approached by an avenue of venerable elms. Facing it, and separated from it by the deep, narrow valley, with its steeply sloping banks, were other similar country seats, backed by a wood.

Henriette's anxiety was aroused at sight of the open door, "They are not at home," she said; "they must have gone away."

The truth was that Dubreuil had decided the day before to take his wife and children to Bouillon, where they would be in safety from the disaster he felt was impending. And yet the house was not unoccupied; even at a distance and through the intervening trees the approaching party were conscious of movements going on within its walls. As the young woman advanced into the avenue she recoiled before the dead body of a Prussian soldier.

"The devil!" exclaimed Rochas; "so they have already been exchanging civilities in this quarter!"

Then all hands, desiring to ascertain what was going on, hurried forward to the house, and there their curiosity was quickly gratified; the doors and windows of the rez-de-chaussee had been smashed in with musket-butts and the yawning apertures disclosed the destruction that the marauders had wrought in the rooms within, while on the graveled terrace lay various articles of furniture that had been hurled from the stoop. Particularly noticeable was a drawing-room suite in sky-blue satin, its sofa and twelve fauteuils piled in dire confusion, helter-skelter, on and around a great center table, the marble top of which was broken in twain. And there were zouaves, chasseurs, liners, and men of the infanterie de marine running to and fro excitedly behind the buildings and in the alleys, discharging their pieces into the little wood that faced them across the valley.

"Lieutenant," a zouave said to Rochas, by way of explanation, "we found a pack of those dirty Prussian hounds here, smashing things and raising Cain generally. We settled their hash for them, as you can see for yourself; only they will be coming back here presently, ten to our one, and that won't be so pleasant."

Three other corpses of Prussian soldiers were stretched upon the terrace. As Henriette was looking at them absently, her thoughts doubtless far away with her husband, who, amid the blood and ashes of Bazeilles, was also sleeping his last sleep, a bullet whistled close to her head and struck a tree that stood behind her. Jean sprang forward.

"Madame, don't stay there. Go inside the house, quick, quick!"

His heart overflowed with pity as he beheld the change her terrible affliction had wrought in her, and he recalled her image as she had appeared to him only the day before, her face bright with the kindly smile of the happy, loving wife. At first he had found no word to say to her, hardly knowing even if she would recognize him. He felt that he could gladly give his life, if that would serve to restore her peace of mind.

"Go inside, and don't come out. At the first sign of danger we will come for you, and we will all escape together by way of the wood up yonder."

But she apathetically replied:

"Ah, M. Jean, what is the use?"

Her brother, however, was also urging her, and finally she ascended the stoop and took her position within the vestibule, whence her vision commanded a view of the avenue in its entire length. She was a spectator of the ensuing combat.

Maurice and Jean had posted themselves behind one of the elms near the house. The gigantic trunks of the centenarian monarchs were amply sufficient to afford shelter to two men. A little way from them Gaude, the bugler, had joined forces with Lieutenant Rochas, who, unwilling to confide the flag to other hands, had rested it against the tree at his side while he handled his musket. And every trunk had its defenders; from end to end the avenue was lined with men covered, Indian fashion, by the trees, who only exposed their head when ready to fire.

In the wood across the valley the Prussians appeared to be receiving re-enforcements, for their fire gradually grew warmer. There was no one to be seen; at most, the swiftly vanishing form now and then of a man changing his position. A villa, with green shutters, was occupied by their sharpshooters, who fired from the half-open windows of the rez-de-chaussee. It was about four o'clock, and the noise of the cannonade in the distance was diminishing, the guns were being silenced one by one; and there they were, French and Prussians, in that out-of-the-way-corner whence they could not see the white flag floating over the citadel, still engaged in the work of mutual slaughter, as if their quarrel had been a personal one. Notwithstanding the armistice there were many such points where the battle continued to rage until it was too dark to see; the rattle of musketry was heard in the faubourg of the Fond de Givonne and in the gardens of Petit-Pont long after it had ceased elsewhere.

For a quarter of an hour the bullets flew thick and fast from one side of the valley to the other. Now and again someone who was so incautious as to expose himself went down with a ball in his head or chest. There were three men lying dead in the avenue. The rattling in the throat of another man who had fallen prone upon his face was something horrible to listen to, and no one thought to go and turn him on his back to ease his dying agony. Jean, who happened to look around just at that moment, beheld Henriette glide tranquilly down the steps, approach the wounded man and turn him over, then slip a knapsack beneath his head by way of pillow. He ran and seized her and forcibly brought her back behind the tree where he and Maurice were posted.

"Do you wish to be killed?"

She appeared to be entirely unconscious of the danger to which she had exposed herself.

"Why, no-but I am afraid to remain in that house, all alone. I would rather be outside."

And so she stayed with them. They seated her on the ground at their feet, against the trunk of the tree, and went on expending the few cartridges that were left them, blazing away to right and left, with such fury that they quite forgot their sensations of fear and fatigue. They were utterly unconscious of what was going on around them, acting mechanically, with but one end in view; even the instinct of self-preservation had deserted them.

"Look, Maurice," suddenly said Henriette; "that dead soldier there before us, does he not belong to the Prussian Guard?"

She had been eying attentively for the past minute or two one of the dead bodies that the enemy had left behind them when they retreated, a short, thick-set young man, with big mustaches, lying upon his side on the gravel of the terrace.

The chin-strap had broken, releasing the spiked helmet, which had rolled away a few steps. And it was indisputable that the body was attired in the uniform of the Guard; the dark gray trousers, the blue tunic with white facings, the greatcoat rolled and worn, belt-wise, across the shoulder.

"It is the Guard uniform," she said; "I am quite certain of it. It is exactly like the colored plate I have at home, and then the photograph that Cousin Gunther sent us-" She stopped suddenly, and with her unconcerned, fearless air, before anyone could make a motion to detain her, walked up to the corpse, bent down and read the number of the regiment. "Ah, the Forty-third!" she exclaimed. "I knew it."

And she returned to her position, while a storm of bullets whistled around her ears. "Yes, the Forty-third; Cousin Gunther's regiment -something told me it must be so. Ah! if my poor husband were only here!"

After that all Jean's and Maurice's entreaties were ineffectual to make her keep quiet. She was feverishly restless, constantly protruding her head to peer into the opposite wood, evidently harassed by some anxiety that preyed upon her mind. Her companions continued to load and fire with the same blind fury, pushing her back with their knee whenever she exposed herself too rashly. It looked as if the Prussians were beginning to consider that their numbers would warrant them in attacking, for they showed themselves more frequently and there were evidences of preparations going on behind the trees. They were suffering severely, however, from the fire of the French, whose bullets at that short range rarely failed to bring down their man.

"That may be your cousin," said Jean. "Look, that officer over there, who has just come out of the house with the green shutters."

He was a captain, as could be seen by the gold braid on the collar of his tunic and the golden eagle on his helmet that flashed back the level ray of the setting sun. He had discarded his epaulettes, and carrying his saber in his right hand, was shouting an order in a sharp, imperative voice; and the distance between them was so small, a scant two hundred yards, that every detail of his trim, slender figure was plainly discernible, as well as the pinkish, stern face and slight blond mustache.

Henriette scrutinized him with attentive eyes. "It is he," she replied, apparently unsurprised. "I recognize him perfectly."

With a look of concentrated rage Maurice drew his piece to his shoulder and covered him. "The cousin- Ah! sure as there is a God in heaven he shall pay for Weiss."

But, quivering with excitement, she jumped to her feet and knocked up the weapon, whose charge was wasted on the air.

"Stop, stop! we must not kill acquaintances, relatives! It is too barbarous."

And, all her womanly instincts coming back to her, she sank down behind the tree and gave way to a fit of violent weeping. The horror of it all was too much for her; in her great dread and sorrow she was forgetful of all beside.

Rochas, meantime, was in his element. He had excited the few zouaves and other troops around him to such a pitch of frenzy, their fire had become so murderously effective at sight of the Prussians, that the latter first wavered and then retreated to the shelter of their wood.

"Stand your ground, my boys! don't give way an inch! Aha, see 'em run, the cowards! we'll fix their flint for 'em!"

213

He was in high spirits and seemed to have recovered all his unbounded confidence, certain that victory was yet to crown their efforts. There had been no defeat. The handful of men before him stood in his eyes for the united armies of Germany, and he was going to destroy them at his leisure. All his long, lean form, all his thin, bony face, where the huge nose curved down upon the self-willed, sensual mouth, exhaled a laughing, vain-glorious satisfaction, the joy of the conquering trooper who goes through the world with his sweetheart on his arm and a bottle of good wine in his hand.

"Parbleu, my children, what are we here for, I'd like to know, if not to lick 'em out of their boots? and that's the way this affair is going to end, just mark my words. We shouldn't know ourselves any longer if we should let ourselves be beaten. Beaten! come, come, that is too good! When the neighbors tread on our toes, or when we feel we are beginning to grow rusty for want of something to do, we just turn to and give 'em a thrashing; that's all there is to it. Come, boys, let 'em have it once more, and you'll see 'em run like so many jackrabbits!"

He bellowed and gesticulated like a lunatic, and was such a good fellow withal in the comforting illusion of his ignorance that the men were inoculated with his confidence. He suddenly broke out again:

"And we'll kick 'em, we'll kick 'em, we'll kick 'em to the frontier! Victory, victory!"

But at that juncture, just as the enemy across the valley seemed really to be falling back, a hot fire of musketry came pouring in on them from the left. It was a repetition of the everlasting flanking movement that had done the Prussians such good service; a strong detachment of the Guards had crept around toward the French rear through the Fond de Givonne. It was useless to think of holding the position longer; the little band of men who were defending the terraces were caught between two fires and menaced with being cut off from Sedan. Men fell on every side, and for a moment the confusion was extreme; the Prussians were already scaling the wall of the park, and advancing along the pathways. Some zouaves rushed forward to repel them, and there was a fierce hand-to-hand struggle with the bayonet. There was one zouave, a big, handsome, brown-bearded man, bare-headed and with his jacket hanging in tatters from his shoulders, who did his work with appalling thoroughness, driving his reeking bayonet home through splintering bones and yielding tissues, cleansing it of the gore that it had contracted from one man by plunging it into the flesh of another; and when it broke he laid about him, smashing many a skull, with the butt of his musket; and when finally he made a misstep and lost his weapon he sprung, bare-handed, for the throat of a burly Prussian, with such tigerish fierceness that both men rolled over and over on the gravel to the shattered kitchen door, clasped in a mortal embrace. The trees of the park looked down on many such scenes of slaughter, and the green lawn was piled with corpses. But it was before the stoop, around the sky-blue sofa and fauteuils, that the conflict raged with greatest fury; a maddened mob of savages, firing at one another at point-blank range, so that hair and beards were set on fire, tearing one another with teeth and nails when a knife was wanting to slash the adversary's throat.

Then Gaude, with his sorrowful face, the face of a man who has had his troubles of which he does not care to speak, was seized with a sort of sudden heroic madness. At that moment of irretrievable defeat, when he must have known that the company was annihilated and that there was not a man left to answer his summons, he grasped his bugle, carried it to his lips and sounded the general, in so tempestuous, ear-splitting strains that one would have said he wished to wake the dead. Nearer and nearer came the Prussians, but he never stirred, only sounding the call the louder, with all the strength of his lungs. He fell, pierced with many bullets, and his spirit passed in one long-drawn, parting wail that died away and was lost upon the shuddering air.

Rochas made no attempt to fly; he seemed unable to comprehend. Even more erect than usual, he waited the end, stammering:

"Well, what's the matter? what's the matter?"

Such a possibility had never entered his head as that they could be defeated. They were changing everything in these degenerate days, even to the manner of fighting; had not those fellows a right to remain on their own side of the valley and wait for the French to go and attack them? There was no use killing them; as fast as they were killed more kept popping up. What kind of a d---d war was it, anyway, where they were able to collect ten men against their opponent's one, where they never showed their face until evening, after blazing away at you all day with their artillery until you didn't know on which end you were standing? Aghast and confounded, having failed so far to acquire the first idea of the rationale of the campaign, he was dimly conscious of the existence of some mysterious, superior method which he could not comprehend, against which he ceased to struggle, although in his dogged stubbornness he kept repeating mechanically:

"Courage, my children! victory is before us!"

Meanwhile he had stooped and clutched the flag. That was his last, his only thought, to save the flag, retreating again, if necessary, so that it might not be defiled by contact with Prussian hands. But the staff, although it was broken, became entangled in his legs; he narrowly escaped falling. The bullets whistled past him, he felt that death was near; he stripped the silk from the staff and tore it into shreds, striving to destroy it utterly. And then it was that, stricken at once in the neck, chest, and legs, he sank to earth amid the bright tri-colored rags, as if they had been his pall. He survived a moment yet, gazing before him with fixed, dilated eyes, reading, perhaps, in the vision he beheld on the horizon the stern lesson that War conveys, the cruel, vital struggle that is to be accepted not otherwise than gravely, reverently, as immutable law. Then a slight tremor ran through his frame, and darkness succeeded to his infantine bewilderment; he passed away, like some poor dumb, lowly creature of a day, a joyous insect that mighty, impassive Nature, in her relentless fatality, has caught and crushed. In him died all a legend.

When the Prussians began to draw near Jean and Maurice had retreated, retiring from tree to tree, face to the enemy, and always, as far as possible, keeping Henriette behind them. They did not give over firing, discharging their pieces and then falling back to seek a fresh cover. Maurice knew where there was a little wicket in the wall at the upper part of the park, and they were so fortunate as to find it unfastened. With lighter hearts when they had left it behind them, they found themselves in a narrow by-road that wound between two high walls, but after following it for some distance the sound of firing in front caused them to turn into a path on their left. As luck would have it, it ended in an impasse; they had to retrace their steps, running the gauntlet of the bullets, and take the turning to the right. When they came to exchange reminiscences in later days they could never agree on which road they had taken. In that tangled network of suburban lanes and passages there was firing still going on from every corner that afforded a shelter, protracted battles raged at the gates of farmyards, everything that could be converted into a barricade had its defenders, from whom the assailants tried to wrest it; all with the utmost fury and vindictiveness. And all at once they came out upon the Fond de Givonne road, not far from Sedan.

For the third time Jean raised his eyes toward the western sky, that was all aflame with a bright, rosy light; and he heaved a sigh of unspeakable relief.

"Ah, that pig of a sun! at last he is going to bed!"

And they ran with might and main, all three of them, never once stopping to draw breath. About them, filling the road in all its breadth, was the rear-guard of fugitives from the

215

battlefield, still flowing onward with the irresistible momentum of an unchained mountain torrent. When they came to the Balan gate they had a long period of waiting in the midst of the impatient, ungovernable throng. The chains of the drawbridge had given way, and the only path across the fosse was by the foot-bridge, so that the guns and horses had to turn back and seek admission by the bridge of the chateau, where the jam was said to be even still more fearful. At the gate of la Cassine, too, people were trampled to death in their eagerness to gain admittance. From all the adjacent heights the terror-stricken fragments of the army came tumbling into the city, as into a cesspool, with the hollow roar of pent-up water that has burst its dam. The fatal attraction of those walls had ended by making cowards of the bravest; men trod one another down in their blind haste to be under cover.

Maurice had caught Henriette in his arms, and in a voice that trembled with suspense:

"It cannot be," he said, "that they will have the cruelty to close the gate and shut us out."

That was what the crowd feared would be done. To right and left, however, upon the glacis soldiers were already arranging their bivouacs, while entire batteries, guns, caissons, and horses, in confusion worse confounded, had thrown themselves pell-mell into the fosse for safety.

But now shrill, impatient bugle calls rose on the evening air, followed soon by the long-drawn strains of retreat. They were summoning the belated soldiers back to their comrades, who came running in, singly and in groups. A dropping fire of musketry still continued in the faubourgs, but it was gradually dying out. Heavy guards were stationed on the banquette behind the parapet to protect the approaches, and at last the gate was closed. The Prussians were within a hundred yards of the sally-port; they could be seen moving on the Balan road, tranquilly establishing themselves in the houses and gardens.

Maurice and Jean, pushing Henriette before them to protect her from the jostling of the throng, were among the last to enter Sedan. Six o'clock was striking. The artillery fire had ceased nearly an hour ago. Soon the distant musketry fire, too, was silenced. Then, to the deafening uproar, to the vengeful thunder that had been roaring since morning, there succeeded a stillness as of death. Night came, and with it came a boding silence, fraught with terror.

8

At half-past five o'clock, after the closing of the gates, Delaherche, in his eager thirst for news, now that he knew the battle lost, had again returned to the Sous-Prefecture. He hung persistently about the approaches of the janitor's lodge, tramping up and down the paved courtyard with feverish impatience, for more than three hours, watching for every officer who came up and interviewing him, and thus it was that he had become acquainted, piecemeal, with the rapid series of events; how General de Wimpffen had tendered his resignation and then withdrawn it upon the peremptory refusal of Generals Ducrot and Douay to append their names to the articles of capitulation, how the Emperor had thereupon invested the General with full authority to proceed to the Prussian headquarters and treat for the surrender of the vanquished army on the most advantageous terms obtainable; how, finally, a council of war had been convened with the object of deciding what possibilities there were of further protracting the struggle successfully by the defense of the fortress. During the deliberations of this council, which consisted of some twenty officers of the highest rank and seemed to him as if it would never end, the cloth manufacturer climbed the steps of the huge public building at least twenty times, and at last his curiosity was gratified by beholding General de Wimpffen emerge, very red in the face and his eyelids puffed and swollen with tears, behind whom came two other generals and a colonel. They leaped into the saddle and rode away over the Pont de Meuse. The bells had struck eight some time before; the inevitable capitulation was now to be accomplished, from which there was no escape.

Delaherche, somewhat relieved in mind by what he had heard and seen, remembered that it was a long time since he had tasted food and resolved to turn his steps homeward, but the terrific crowd that had collected since he first came made him pause in dismay. It is no exaggeration to say that the streets and squares were so congested, so thronged, so densely packed with horses, men, and guns, that one would have declared the closely compacted mass could only have been squeezed and wedged in there thus by the effort of some gigantic mechanism. While the ramparts were occupied by the bivouacs of such regiments as had fallen back in good order, the city had been invaded and submerged by an angry, surging, desperate flood, the broken remnants of the various corps, stragglers and fugitives from all arms of the service, and the dammed-up tide made it impossible for one to stir foot or hand. The wheels of the guns, of the caissons, and the innumerable vehicles of every description, had interlocked and were tangled in confusion worse confounded, while the poor horses, flogged unmercifully by their drivers and pulled, now in this direction, now in that, could only dance in their bewilderment, unable to move a step either forward or back. And the men, deaf to reproaches and threats alike, forced their way into the houses, devoured whatever they could lay hands on, flung themselves down to sleep wherever they could find a vacant space, it might be in the best bedroom or in the cellar. Many of them had fallen in doorways, where they blocked the vestibule; others, without strength to go farther, lay extended on the sidewalks and slept the sleep of death, not even rising when some by-passer trod on them and bruised an arm or leg, preferring the risk of death to the fatigue of changing their location.

These things all helped to make Delaherche still more keenly conscious of the necessity of immediate capitulation. There were some quarters in which numerous caissons were packed so close together that they were in contact, and a single Prussian shell alighting on one of them must inevitably have exploded them all, entailing the immediate destruction of the city by conflagration. Then, too, what could be accomplished with such an assemblage of miserable wretches, deprived of all their powers, mental and physical, by reason of their long-endured privations, and destitute of either ammunition or subsistence? Merely to clear the

217

streets and reduce them to a condition of something like order would require a whole day. The place was entirely incapable of defense, having neither guns nor provisions.

These were the considerations that had prevailed at the council among those more reasonable officers who, in the midst of their grief and sorrow for their country and the army, had retained a clear and undistorted view of the situation as it was; and the more hot-headed among them, those who cried with emotion that it was impossible for an army to surrender thus, had been compelled to bow their head upon their breast in silence and admit that they had no practicable scheme to offer whereby the conflict might be recommenced on the morrow.

In the Place Turenne and Place du Rivage, Delaherche succeeded with the greatest difficulty in working his way through the press. As he passed the Hotel of the Golden Cross a sorrowful vision greeted his eyes, that of the generals seated in the dining room, gloomily silent, around the empty board; there was nothing left to eat in the house, not even bread. General Bourgain-Desfeuilles, however, who had been storming and vociferating in the kitchen, appeared to have found something, for he suddenly held his peace and ran away swiftly up the stairs, holding in his hands a large paper parcel of a greasy aspect. Such was the crowd assembled there, to stare through the lighted windows upon the guests assembled around that famine-stricken table d'hote, that the manufacturer was obliged to make vigorous play with his elbows, and was frequently driven back by some wild rush of the mob and lost all the distance, and more, that he had just gained. In the Grande Rue, however, the obstacles became actually impassable, and there was a moment when he was inclined to give up in despair; a complete battery seemed to have been driven in there and the guns and *materiel* piled, pell-mell, on top of one another. Deciding finally to take the bull by the horns, he leaped to the axle of a piece and so pursued his way, jumping from wheel to wheel, straddling the guns, at the imminent risk of breaking his legs, if not his neck. Afterward it was some horses that blocked his way, and he made himself lowly and stooped, creeping among the feet and underneath the bellies of the sorry jades, who were ready to die of inanition, like their masters. Then, when after a quarter of an hour's laborious effort he reached the junction of the Rue Saint-Michel, he was terrified at the prospect of the dangers and obstacles that he had still to face, and which, instead of diminishing, seemed to be increasing, and made up his mind to turn down the street above mentioned, which would take him into the Rue des Laboureurs; he hoped that by taking these usually quiet and deserted passages he should escape the crowd and reach his home in safety. As luck would have it he almost directly came upon a house of ill-fame to which a band of drunken soldiers were in process of laying siege, and considering that a stray shot, should one reach him in the fracas, would be equally as unpleasant as one intended for him, he made haste to retrace his steps. Resolving to have done with it he pushed on to the end of the Grande Rue, now gaining a few feet by balancing himself, rope-walker fashion, along the pole of some vehicle, now climbing over an army wagon that barred his way. At the Place du College he was carried along-bodily on the shoulders of the throng for a space of thirty paces; he fell to the ground, narrowly escaped a set of fractured ribs, and saved himself only by the proximity of a friendly iron railing, by the bars of which he pulled himself to his feet. And when at last he reached the Rue Maqua, inundated with perspiration, his clothing almost torn from his back, he found that he had been more than an hour in coming from the Sous-Prefecture, a distance which in ordinary times he was accustomed to accomplish in less than five minutes.

Major Bouroche, with the intention of keeping the ambulance and garden from being overrun with intruders, had caused two sentries to be mounted at the door. This measure was a source of great comfort to Delaherche, who had begun to contemplate the possibilities of his house being subjected to pillage. The sight of the ambulance in the garden, dimly lighted by a few candles and exhaling its fetid, feverish emanations, caused him a fresh constriction of the heart; then, stumbling over the body of a soldier who was stretched in slumber on the stone pavement of the walk, he supposed him to be one of the fugitives who had managed to find his way in there from outside, until, calling to mind the 7th corps treasure that had been deposited there and the sentry who had been set over it, he saw how matters stood: the poor fellow, stationed there since early morning, had been overlooked by his superiors and had succumbed to his fatigue. Besides, the house seemed quite deserted; the ground floor was black as Egypt, and the doors stood wide open. The servants were doubtless all at the ambulance, for there was no one in the kitchen, which was faintly illuminated by the light of a wretched little smoky lamp. He lit a candle and ascended the main staircase very softly, in order not to awaken his wife and mother, whom he had begged to go to bed early after a day where the stress, both mental and physical, had been so intense.

On entering his study, however, he beheld a sight that caused his eyes to dilate with astonishment. Upon the sofa on which Captain Beaudoin had snatched a few hours' repose the day before a soldier lay outstretched; and he could not understand the reason of it until he had looked and recognized young Maurice Levasseur, Henriette's brother. He was still more surprised when, on turning his head, he perceived, stretched on the floor and wrapped in a bed quilt, another soldier, that Jean, whom he had seen for a moment just before the battle. It was plain that the poor fellows, in their distress and fatigue after the conflict, not knowing where else to bestow themselves, had sought refuge there; they were crushed, annihilated, like dead men. He did not linger there, but pushed on to his wife's chamber, which was the next room on the corridor. A lamp was burning on a table in a corner; the profound silence seemed to shudder. Gilberte had thrown herself crosswise on the bed, fully dressed, doubtless in order to be prepared for any catastrophe, and was sleeping peacefully, while, seated on a chair at her side with her head declined and resting lightly on the very edge of the mattress, Henriette was also slumbering, with a fitful, agitated sleep, while big tears welled up beneath her swollen eyelids. He contemplated them silently for a moment, strongly tempted to awake and question the young woman in order to ascertain what she knew. Had she succeeded in reaching Bazeilles? and why was it that she was back there? Perhaps she would be able to give him some tidings of his dyehouse were he to ask her? A feeling of compassion stayed him, however, and he was about to leave the room when his mother, ghost-like, appeared at the threshold of the open door and beckoned him to follow her.

As they were passing through the dining room he expressed his surprise.

"What, have you not been abed to-night?"

She shook her head, then said below her breath:

"I cannot sleep; I have been sitting in an easy-chair beside the colonel. He is very feverish; he awakes at every instant, almost, and then plies me with questions. I don't know how to answer them. Come in and see him, you."

M. de Vineuil had fallen asleep again. His long face, now brightly red, barred by the sweeping mustache that fell across it like a snowy avalanche, was scarce distinguishable on the pillow. Mme. Delaherche had placed a newspaper before the lamp and that corner of the room was lost in semi-darkness, while all the intensity of the bright lamplight was concentrated on her where she sat, uncompromisingly erect, in her fauteuil, her hands crossed before her in her lap, her vague eyes bent on space, in sorrowful reverie.

"I think he must have heard you," she murmured; "he is awaking again."

It was so; the colonel, without moving his head, had reopened his eyes and bent them on Delaherche. He recognized him, and immediately asked in a voice that his exhausted condition made tremulous:

"It is all over, is it not? We have capitulated."

The manufacturer, who encountered the look his mother cast on him at that moment, was on the point of equivocating. But what good would it do? A look of discouragement passed across his face.

"What else remained to do? A single glance at the streets of the city would convince you. General de Wimpffen has just set out for Prussian general headquarters to discuss conditions."

M. de Vineuil's eyes closed again, his long frame was shaken with a protracted shiver of supremely bitter grief, and this deep, long-drawn moan escaped his lips:

"Ah! merciful God, merciful God!" And without opening his eyes he went on in faltering, broken accents: "Ah! the plan I spoke of yesterday -they should have adopted it. Yes, I knew the country; I spoke of my apprehensions to the general, but even him they would not listen to. Occupy all the heights up there to the north, from Saint-Menges to Fleigneux, with your army looking down on and commanding Sedan, able at any time to move on Vrigne-aux-Bois, mistress of Saint-Albert's pass-and there we are; our positions are impregnable, the Mezieres road is under our control-"

His speech became more confused as he proceeded; he stammered a few more unintelligible words, while the vision of the battle that had been born of his fever little by little grew blurred and dim and at last was effaced by slumber. He slept, and in his sleep perhaps the honest officer's dreams were dreams of victory.

"Does the major speak favorably of his case?" Delaherche inquired in a whisper.

Madame Delaherche nodded affirmatively.

"Those wounds in the foot are dreadful things, though," he went on. "I suppose he is likely to be laid up for a long time, isn't he?"

She made him no answer this time, as if all her being, all her faculties were concentrated on contemplating the great calamity of their defeat. She was of another age; she was a survivor of that strong old race of frontier burghers who defended their towns so valiantly in the good days gone by. The clean-cut lines of her stern, set face, with its fleshless, uncompromising nose and thin lips, which the brilliant light of the lamp brought out in high relief against the darkness of the room, told the full extent of her stifled rage and grief and the wound sustained by her antique patriotism, the revolt of which refused even to let her sleep.

About that time Delaherche became conscious of a sensation of isolation, accompanied by a most uncomfortable feeling of physical distress. His hunger was asserting itself again, a griping, intolerable hunger, and he persuaded himself that it was debility alone that was thus robbing him of courage and resolution. He tiptoed softly from the room and, with his candle, again made his way down to the kitchen, but the spectacle he witnessed there was even still more cheerless; the range cold and fireless, the closets empty, the floor strewn with a disorderly litter of towels, napkins, dish-clouts and women's aprons; as if the hurricane of disaster had swept through that place as well, bearing away on its wings all the charm and cheer that appertain naturally to the things we eat and drink. At first he thought he was not going to discover so much as a crust, what was left over of the bread having all found its way to the ambulance in the form of soup. At last, however, in the dark corner of a cupboard he came across the remainder of the beans from yesterday's dinner, where they had been forgotten, and ate them. He accomplished his luxurious repast without the formality of sitting down, without the accompaniment of salt and butter, for which he did not care to trouble himself to ascend to the floor above, desirous only to get away as speedily as possible from that dismal kitchen, where the blinking, smoking little lamp perfumed the air with fumes of petroleum.

It was not much more than ten o'clock, and Delaherche had no other occupation than to speculate on the various probabilities connected with the signing of the capitulation. A persistent apprehension haunted him; a dread lest the conflict might be renewed, and the horrible thought of what the consequences must be in such an event, of which he could not speak, but which rested on his bosom like an incubus. When he had reascended to his study, where he found Maurice and Jean in exactly the same position he had left them in, it was all in vain that he settled himself comfortably in his favorite easy-chair; sleep would not come to him; just as he was on the point of losing himself the crash of a shell would arouse him with a great start. It was the frightful cannonade of the day, the echoes of which were still ringing in his ears; and he would listen breathlessly for a moment, then sit and shudder at the equally appalling silence by which he was now surrounded. As he could not sleep he preferred to move about; he wandered aimlessly among the rooms, taking care to avoid that in which his mother was sitting by the colonel's bedside, for the steady gaze with which she watched him as he tramped nervously up and down had finally had the effect of disconcerting him. Twice he returned to see if Henriette had not awakened, and he paused an instant to glance at his wife's pretty face, so calmly peaceful, on which seemed to be flitting something like the faint shadow of a smile. Then, knowing not what to do, he went downstairs again, came back, moved about from room to room, until it was nearly two in the morning, wearying his ears with trying to decipher some meaning in the sounds that came to him from without.

This condition of affairs could not last. Delaherche resolved to return once more to the Sous-Prefecture, feeling assured that all rest would be quite out of the question for him so long as his ignorance continued. A feeling of despair seized him, however, when he went downstairs and looked out upon the densely crowded street, where the confusion seemed to be worse than ever; never would he have the strength to fight his way to the Place Turenne and back again through obstacles the mere memory of which caused every bone in his body to ache again. And he was mentally discussing matters, when who should come up but Major Bouroche, panting, perspiring, and swearing.

"Tonnerre de Dieu! I wonder if my head's on my shoulders or not!"

He had been obliged to visit the Hotel de Ville to see the mayor about his supply of chloroform, and urge him to issue a requisition for a quantity, for he had many operations to perform, his stock of the drug was exhausted, and he was afraid, he said, that he should be compelled to carve up the poor devils without putting them to sleep.

"Well?" inquired Delaherche.

"Well, they can't even tell whether the apothecaries have any or not!"

But the manufacturer was thinking of other things than chloroform. "No, no," he continued. "Have they brought matters to a conclusion yet? Have they signed the agreement with the Prussians?"

The major made a gesture of impatience. "There is nothing concluded," he cried. "It appears that those scoundrels are making demands out of all reason. Ah, well; let 'em commence afresh, then, and we'll all leave our bones here. That will be best!"

Delaherche's face grew very pale as he listened. "But are you quite sure these things are so?"

"I was told them by those fellows of the municipal council, who are in permanent session at the city hall. An officer had been dispatched from the Sous-Prefecture to lay the whole affair before them."

And he went on to furnish additional details. The interview had taken place at the Chateau de Bellevue, near Donchery, and the participants were General de Wimpffen, General von Moltke, and Bismarck. A stern and inflexible man was that von Moltke, a terrible man to deal with! He began by demonstrating that he was perfectly acquainted with the hopeless situation of the French army; it was destitute of ammunition and subsistence, demoralization and disorder pervaded its ranks, it was utterly powerless to break the iron circle by which it was girt about; while on the other hand the German armies occupied commanding positions from which they could lay the city in ashes in two hours. Coldly, unimpassionedly, he stated his terms: the entire French army to surrender arms and baggage and be treated as prisoners of war. Bismarck took no part in the discussion beyond giving the general his support, occasionally showing his teeth, like a big mastiff, inclined to be pacific on the whole, but quite ready to rend and tear should there be occasion for it. General de Wimpffen in reply protested with all the force he had at his command against these conditions, the most severe that ever were imposed on a vanquished army. He spoke of his personal grief and ill-fortune, the bravery of the troops, the danger there was in driving a proud nation to extremity; for three hours he spoke with all the energy and eloquence of despair, alternately threatening and entreating, demanding that they should content themselves with interning their prisoners in France, or even in Algeria; and in the end the only concession granted was, that the officers might retain their swords, and those among them who should enter into a solemn arrangement, attested by a written parole, to serve no more during the war, might return to their homes. Finally, the armistice to be prolonged until the next morning at ten o'clock; if at that time the terms had not been accepted, the Prussian batteries would reopen fire and the city would be burned.

"That's stupid!" exclaimed Delaherche; "they have no right to burn a city that has done nothing to deserve it!"

The major gave him still further food for anxiety by adding that some officers whom he had met at the Hotel de l'Europe were talking of making a sortie *en masse* just before daylight. An extremely excited state of feeling had prevailed since the tenor of the German demands had become known, and measures the most extravagant were proposed and discussed. No one seemed to be deterred by the consideration that it would be dishonorable to break the truce, taking advantage of the darkness and giving the enemy no notification, and the wildest, most visionary schemes were offered; they would resume the march on Carignan, hewing their way through the Bavarians, which they could do in the black night; they would recapture the plateau of Illy by a surprise; they would raise the blockade of the Mezieres road, or, by a determined, simultaneous rush, would force the German lines and throw themselves into Belgium. Others there were, indeed, who, feeling the hopelessness of their position, said nothing; they would have accepted any terms, signed any paper, with a glad cry of relief, simply to have the affair ended and done with.

"Good-night!" Bouroche said in conclusion. "I am going to try to sleep a couple of hours; I need it badly."

When left by himself Delaherche could hardly breathe. What, could it be true that they were going to fight again, were going to burn and raze Sedan! It was certainly to be, soon as the morrow's sun should be high enough upon the hills to light the horror of the sacrifice. And once again he almost unconsciously climbed the steep ladder that led to the roofs and found himself standing among the chimneys, at the edge of the narrow terrace that overlooked the city; but at that hour of the night the darkness was intense and he could distinguish absolutely nothing amid the swirling waves of the Cimmerian sea that lay beneath him. Then the buildings of the factory below were the first objects which, one by one, disentangled themselves from the shadows and stood out before his vision in indistinct masses, which he had no difficulty in recognizing: the engine-house, the shops, the drying rooms, the storehouses, and when he reflected that within twenty-four hours there would remain of that imposing block of buildings, his fortune and his pride, naught save charred timbers and crumbling walls, he overflowed with pity for himself. He raised his glance thence once more to the horizon, and sent it traveling in a circuit around that profound, mysterious veil of blackness behind which lay slumbering the menace of the morrow. To the south, in the direction of Bazeilles, a few quivering little flames that rose fitfully on the air told where had been the site of the unhappy village, while toward the north the farmhouse in the wood of la Garenne, that had been fired late in the afternoon, was burning still, and the trees about were dyed of a deep red with the ruddy blaze. Beyond the intermittent flashing of those two baleful fires no light to be seen; the brooding silence unbroken by any sound save those half-heard mutterings that pass through the air like harbingers of evil; about them, everywhere, the unfathomable abyss, dead and lifeless. Off there in the distance, very far away, perhaps, perhaps upon the ramparts, was a sound of someone weeping. It was all in vain that he strained his eyes to pierce the veil, to see something of Liry, la Marfee, the batteries of Frenois, and Wadelincourt, that encircling belt of bronze monsters of which he could instinctively feel the presence there, with their outstretched necks and yawning, ravenous muzzles. And as he recalled his glance and let it fall upon the city that lay around and beneath him, he heard its frightened breathing. It was not alone the unquiet slumbers of the soldiers who had fallen in the streets, the blending of inarticulate sounds produced by that gathering of guns, men, and horses; what he fancied he could distinguish was the insomnia, the alarmed watchfulness of his bourgeois neighbors, who, no more than he, could sleep, quivering with feverish terrors, awaiting anxiously the coming of the day. They all must be aware that the capitulation had not been signed, and were all counting the hours, quaking at the thought that should it not be signed the sole resource left them would be to go down into their cellars and wait for their own walls to tumble in on them and crush the life from their bodies. The voice of one in sore straits came up, it seemed to him, from the Rue des Voyards, shouting: "Help! murder!" amid the clash of arms. He bent over the terrace to look, then remained aloft there in the murky thickness of the night where there was not a star to cheer him, wrapped in such an ecstasy of terror that the hairs of his body stood erect.

Below-stairs, at early daybreak, Maurice awoke upon his sofa. He was sore and stiff as if he had been racked; he did not stir, but lay looking listlessly at the windows, which gradually grew white under the light of a cloudy dawn. The hateful memories of the day before all came back to him with that distinctness that characterizes the impressions of our first waking, how they had fought, fled, surrendered. It all rose before his vision, down to the very least detail, and he brooded with horrible anguish on the defeat, whose reproachful echoes seemed to penetrate to the inmost fibers of his being, as if he felt that all the responsibility of it was his. And he went on to reason on the cause of the evil, analyzing himself, reverting to his old habit of bitter and unavailing self-reproach. He would have felt so brave, so glorious had victory remained with them! And now, in defeat, weak and nervous as a woman, he once again gave way to one of those overwhelming fits of despair in which the entire world, seemed to him to be foundering. Nothing was left them; the end of France was come. His frame was shaken by a storm of sobs, he wept hot tears, and joining his hands, the prayers of his childhood rose to his lips in stammering accents.

"O God! take me unto Thee! O God! take unto Thyself all those who are weary and heavy-laden!"

Jean, lying on the floor wrapped in his bed-quilt, began to show some signs of life. Finally, astonished at what he heard, he arose to a sitting posture.

"What is the matter, youngster? Are you ill?" Then, with a glimmering perception of how matters stood, he adopted a more paternal tone. "Come, tell me what the matter is. You must not let yourself be worried by such a little thing as this, you know."

"Ah!" exclaimed Maurice, "it is all up with us, va! we are Prussians now, and we may as well make up our mind to it."

As the peasant, with the hard-headedness of the uneducated, expressed surprise to hear him talk thus, he endeavored to make it clear to him that, the race being degenerate and exhausted, it must disappear and make room for a newer and more vigorous strain. But the other, with an obstinate shake of the head, would not listen to the explanation.

"What! would you try to make me believe that my bit of land is no longer mine? that I would permit the Prussians to take it from me while I am alive and my two arms are left to me? Come, come!"

Then painfully, in such terms as he could command, he went on to tell how affairs looked to him. They had received an all-fired good basting, that was sure as sure could be! but they were not all dead yet, he didn't believe; there were some left, and those would suffice to rebuild the house if they only behaved themselves, working hard and not drinking up what they earned. When a family has trouble, if its members work and put by a little something, they will pull through, in spite of all the bad luck in the world. And further, it is not such a bad thing to get a good cuffing once in a way; it sets one thinking. And, great heavens! if a man has something rotten about him, if he has gangrene in his arms or legs that is spreading all the time, isn't it better to take a hatchet and lop them off rather than die as he would from cholera?

"All up, all up! Ah, no, no! no, no!" he repeated several times. "It is not all up with me, I know very well it is not."

And notwithstanding his seedy condition and demoralized appearance, his hair all matted and pasted to his head by the blood that had flowed from his wound, he drew himself up defiantly, animated by a keen desire to live, to take up the tools of his trade or put his hand to the plow, in order, to use his own expression, to "rebuild the house." He was of the old soil where reason and obstinacy grow side by side, of the land of toil and thrift.

"All the same, though," he continued, "I am sorry for the Emperor. Affairs seemed to be going on well; the farmers were getting a good price for their grain. But surely it was bad judgment on his part to allow himself to become involved in this business!"

Maurice, who was still in "the blues," spoke regretfully: "Ah, the Emperor! I always liked him in my heart, in spite of my republican ideas. Yes, I had it in the blood, on account of my grandfather, I suppose. And now that that limb is rotten and we shall have to lop it off, what is going to become of us?"

His eyes began to wander, and his voice and manner evinced such distress that Jean became alarmed and was about to rise and go to him, when Henriette came into the room. She had just awakened on hearing the sound of voices in the room adjoining hers. The pale light of a cloudy morning now illuminated the apartment.

"You come just in time to give him a scolding," he said, with an affectation of liveliness. "He is not a good boy this morning."

But the sight of his sister's pale, sad face and the recollection of her affliction had had a salutary effect on Maurice by determining a sudden crisis of tenderness. He opened his arms and took her to his bosom, and when she rested her head upon his shoulder, when he held her locked in a close embrace, a feeling of great gentleness pervaded him and they mingled their tears.

"Ah, my poor, poor darling, why have I not more strength and courage to console you! for my sorrows are as nothing compared with yours. That good, faithful Weiss, the husband who loved you so fondly! What will become of you? You have always been the victim; always, and

224

never a murmur from your lips. Think of the sorrow I have already caused you, and who can say that I shall not cause you still more in the future!"

She was silencing him, placing her hand upon his mouth, when Delaherche came into the room, beside himself with indignation. While still on the terrace he had been seized by one of those uncontrollable nervous fits of hunger that are aggravated by fatigue, and had descended to the kitchen in quest of something warm to drink, where he had found, keeping company with his cook, a relative of hers, a carpenter of Bazeilles, whom she was in the act of treating to a bowl of hot wine. This person, who had been one of the last to leave the place while the conflagrations were at their height, had told him that his dyehouse was utterly destroyed, nothing left of it but a heap of ruins.

"The robbers, the thieves! Would you have believed it, hein?" he stammered, addressing Jean and Maurice. "There is no hope left; they mean to burn Sedan this morning as they burned Bazeilles yesterday. I'm ruined, I'm ruined!" The scar that Henriette bore on her forehead attracted his attention, and he remembered that he had not spoken to her yet. "It is true, you went there, after all; you got that wound- Ah! poor Weiss!"

And seeing by the young woman's tears that she was acquainted with her husband's fate, he abruptly blurted out the horrible bit of news that the carpenter had communicated to him among the rest.

"Poor Weiss! it seems they burned him. Yes, after shooting all the civilians who were caught with arms in their hands, they threw their bodies into the flames of a burning house and poured petroleum over them."

Henriette was horror-stricken as she listened. Her tears burst forth, her frame was shaken by her sobs. My God, my God, not even the poor comfort of going to claim her dear dead and give him decent sepulture; his ashes were to be scattered by the winds of heaven! Maurice had again clasped her in his arms and spoke to her endearingly, calling her his poor Cinderella, beseeching her not to take the matter so to heart, a brave woman as she was.

After a time, during which no word was spoken, Delaherche, who had been standing at the window watching the growing day, suddenly turned and addressed the two soldiers:

"By the way, I was near forgetting. What I came up here to tell you is this: down in the courtyard, in the shed where the treasure chests were deposited, there is an officer who is about to distribute the money among the men, so as to keep the Prussians from getting it. You had better go down, for a little money may be useful to you, that is, provided we are all alive a few hours hence."

The advice was good, and Maurice and Jean acted on it, having first prevailed on Henriette to take her brother's place on the sofa. If she could not go to sleep again, she would at least be securing some repose. As for Delaherche, he passed through the adjoining chamber, where Gilberte with her tranquil, pretty face was slumbering still as soundly as a child, neither the sound of conversation nor even Henriette's sobs having availed to make her change her position. From there he went to the apartment where his mother was watching at Colonel de Vineuil's bedside, and thrust his head through the door; the old lady was asleep in her fauteuil, while the colonel, his eyes closed, was like a corpse. He opened them to their full extent and asked:

"Well, it's all over, isn't it?"

Irritated by the question, which detained him at the very moment when he thought he should be able to slip away unobserved, Delaherche gave a wrathful look and murmured, sinking his voice:

"Oh, yes, all over! until it begins again! There is nothing signed."

The colonel went on in a voice scarcely higher than a whisper; delirium was setting in.

"Merciful God, let me die before the end! I do not hear the guns. Why have they ceased firing? Up there at Saint-Menges, at Fleigneux, we have command of all the roads; should the Prussians dare turn Sedan and attack us, we will drive them into the Meuse. The city is there, an insurmountable obstacle between us and them; our positions, too, are the stronger. Forward! the 7th corps will lead, the 12th will protect the retreat-"

And his fingers kept drumming on the counterpane with a measured movement, as if keeping time with the trot of the charger he was riding in his vision. Gradually the motion became slower and slower as his words became more indistinct and he sank off into slumber. It ceased, and he lay motionless and still, as if the breath had left his body.

"Lie still and rest," Delaherche whispered; "when I have news I will return."

Then, having first assured himself that he had not disturbed his mother's slumber, he slipped away and disappeared.

Jean and Maurice, on descending to the shed in the courtyard, had found there an officer of the pay department, seated on a common kitchen chair behind a little unpainted pine table, who, without pen, ink, or paper, without taking receipts or indulging in formalities of any kind, was dispensing fortunes. He simply stuck his hand into the open mouth of the bags filled with bright gold pieces, and as the sergeants of the 7th corps passed in line before him he filled their kepis, never counting what he bestowed with such rapid liberality. The understanding was that the sergeants were subsequently to divide what they received with the surviving men of their half-sections. Each of them received his portion awkwardly, as if it had been a ration of meat or coffee, then stalked off in an embarrassed, self-conscious sort of way, transferring the contents of the *kepi* to his trousers' pockets so as not to display his wealth to the world at large. And not a word was spoken; there was not a sound to be heard but the crystalline chink and rattle of the coin as it was received by those poor devils, dumfounded to see the responsibility of such riches thrust on them when there was not a place in the city where they could purchase a loaf of bread or a quart of wine.

When Jean and Maurice appeared before him the officer, who was holding outstretched his hand filled, as usual, with louis, drew it back.

"Neither of you fellows is a sergeant. No one except sergeants is entitled to receive the money." Then, in haste to be done with his task, he changed his mind: "Never mind, though; here, you corporal, take this. Step lively, now. Next man!"

And he dropped the gold coins into the *kepi* that Jean held out to him. The latter, oppressed by the magnitude of the amount, nearly six hundred francs, insisted that Maurice should take one-half. No one could say what might happen; they might be parted from each other.

They made the division in the garden, before the ambulance, and when they had concluded their financial business they entered, having recognized on the straw near the entrance the drummer-boy of their company, Bastian, a fat, good-natured little fellow, who had had the ill-luck to receive a spent ball in the groin about five o'clock the day before, when the battle was ended. He had been dying by inches for the last twelve hours.

In the dim, white light of morning, at that hour of awakening, the sight of the ambulance sent a chill of horror through them. Three more patients had died during the night, without anyone being aware of it, and the attendants were hurriedly bearing away the corpses in order to make room for others. Those who had been operated on the day before opened wide their eyes in their somnolent, semi-conscious state, and looked with dazed astonishment on that vast dormitory of suffering, where the victims of the knife, only half-slaughtered, rested on their straw. It was in vain that some attempts had been made the night before to clean up

226

the room after the bloody work of the operations; there were great splotches of blood on the ill-swept floor; in a bucket of water a great sponge was floating, stained with red, for all the world like a human brain; a hand, its fingers crushed and broken, had been overlooked and lay on the floor of the shed. It was the parings and trimmings of the human butcher shop, the horrible waste and refuse that ensues upon a day of slaughter, viewed in the cold, raw light of dawn.

Bouroche, who, after a few hours of repose, had already resumed his duties, stopped in front of the wounded drummer-boy, Bastian, then passed on with an imperceptible shrug of his shoulders. A hopeless case; nothing to be done. The lad had opened his eyes, however, and emerging from the comatose state in which he had been lying, was eagerly watching a sergeant who, his *kepi* filled with gold in his hand, had come into the room to see if there were any of his men among those poor wretches. He found two, and to each of them gave twenty francs. Other sergeants came in, and the gold began to fall in showers upon the straw, among the dying men. Bastian, who had managed to raise himself, stretched out his two hands, even then shaking in the final agony.

"Don't forget me! don't forget me!"

The sergeant would have passed on and gone his way, as Bouroche had done. What good could money do there? Then yielding to a kindly impulse, he threw some coins, never stopping to count them, into the poor hands that were already cold.

"Don't forget me! don't forget me!"

Bastian fell backward on his straw. For a long time he groped with stiffening fingers for the elusive gold, which seemed to avoid him. And thus he died.

"The gentleman has blown his candle out; good-night!" said a little, black, wizened zouave, who occupied the next bed. "It's vexatious, when one has the wherewithal to pay for wetting his whistle!"

He had his left foot done up in splints. Nevertheless he managed to raise himself on his knees and elbows and in this posture crawl over to the dead man, whom he relieved of all his money, forcing open his hands, rummaging among his clothing and the folds of his capote. When he got back to his place, noticing that he was observed, he simply said:

"There's no use letting the stuff be wasted, is there?"

Maurice, sick at heart in that atmosphere of human distress and suffering, had long since dragged Jean away. As they passed out through the shed where the operations were performed they saw Bouroche preparing to amputate the leg of a poor little man of twenty, without chloroform, he having been unable to obtain a further supply of the anaesthetic. And they fled, running, so as not to hear the poor boy's shrieks.

Delaherche, who came in from the street just then, beckoned to them and shouted:

"Come upstairs, come, quick! we are going to have breakfast. The cook has succeeded in procuring some milk, and it is well she did, for we are all in great need of something to warm our stomachs." And notwithstanding his efforts to do so, he could not entirely repress his delight and exultation. With a radiant countenance he added, lowering his voice: "It is all right this time. General de Wimpffen has set out again for the German headquarters to sign the capitulation."

Ah, how much those words meant to him, what comfort there was in them, what relief! his horrid nightmare dispelled, his property saved from destruction, his daily life to be resumed, under changed conditions, it is true, but still it was to go on, it was not to cease! It was little Rose who had told him of the occurrences of the morning at the Sous-Prefecture; the girl had come hastening through the streets, now somewhat less choked than they had been, to obtain a supply of bread from an aunt of hers who kept a baker's shop in the quarter; it was striking nine o'clock. As early as eight General de Wimpffen had convened another council of war, consisting of more than thirty generals, to whom he related the results that had been reached so far, the hard conditions imposed by the victorious foe, and his own fruitless efforts to secure a mitigation of them. His emotion was such that his hands shook like a leaf, his eyes were suffused with tears. He was still addressing the assemblage when a colonel of the German staff presented himself, on behalf of General von Moltke, to remind them that, unless a decision were arrived at by ten o'clock, their guns would open fire on the city of Sedan. With this horrible alternative before them the council could do nothing save authorize the general to proceed once more to the Chateau of Bellevue and accept the terms of the victors. He must have accomplished his mission by that time, and the entire French army were prisoners of war.

When she had concluded her narrative Rose launched out into a detailed account of the tremendous excitement the tidings had produced in the city. At the Sous-Prefecture she had seen officers tear the epaulettes from their shoulders, weeping meanwhile like children. Cavalrymen had thrown their sabers from the Pont de Meuse into the river; an entire regiment of cuirassiers had passed, each man tossing his blade over the parapet and sorrowfully watching the water close over it. In the streets many soldiers grasped their muskets by the barrel and smashed them against a wall, while there were artillerymen who removed the mechanism from the mitrailleuses and flung it into the sewer. Some there were who buried or burned the regimental standards. In the Place Turenne an old sergeant climbed upon a gate-post and harangued the throng as if he had suddenly taken leave of his senses, reviling the leaders, stigmatizing them as poltroons and cowards. Others seemed as if dazed, shedding big tears in silence, and others also, it must be confessed (and it is probable that they were in the majority), betrayed by their laughing eyes and pleased expression the satisfaction they felt at the change in affairs. There was an end to their suffering at last; they were prisoners of war, they could not be obliged to fight any more! For so many days they had been distressed by those long, weary marches, with never food enough to satisfy their appetite! And then, too, they were the weaker; what use was there in fighting? If their chiefs had betrayed them, had sold them to the enemy, so much the better; it would be the sooner ended! It was such a delicious thing to think of, that they were to have white bread to eat, were to sleep between sheets!

As Delaherche was about to enter the dining room in company with Maurice and Jean, his mother called to him from above.

"Come up here, please; I am anxious about the colonel."

M. de Vineuil, with wide-open eyes, was talking rapidly and excitedly of the subject that filled his bewildered brain.

"The Prussians have cut us off from Mezieres, but what matters it! See, they have outmarched us and got possession of the plain of Donchery; soon they will be up with the wood of la Falizette and flank us there, while more of them are coming up along the valley of the Givonne. The frontier is behind us; let us kill as many of them as we can and cross it at a bound. Yesterday, yes, that is what I would have advised-"

At that moment his burning eyes lighted on Delaherche. He recognized him; the sight seemed to sober him and dispel the hallucination under which he was laboring, and coming back to the terrible reality, he asked for the third time:

"It is all over, is it not?"

The manufacturer explosively blurted out the expression of his satisfaction; he could not restrain it.

228

"Ah, yes, God be praised! it is all over, completely over. The capitulation must be signed by this time."

The colonel raised himself at a bound to a sitting posture, notwithstanding his bandaged foot; he took his sword from the chair by the bedside where it lay and made an attempt to break it, but his hands trembled too violently, and the blade slipped from his fingers.

"Look out! he will cut himself!" Delaherche cried in alarm. "Take that thing away from him; it is dangerous!"

Mme. Delaherche took possession of the sword. With a feeling of compassionate respect for the poor colonel's grief and despair she did not conceal it, as her son bade her do, but with a single vigorous effort snapped it across her knee, with a strength of which she herself would never have supposed her poor old hands capable. The colonel laid himself down again, casting a look of extreme gentleness upon his old friend, who went back to her chair and seated herself in her usual rigid attitude.

In the dining room the cook had meantime served bowls of hot coffee and milk for the entire party. Henriette and Gilberte had awakened, the latter, completely restored by her long and refreshing slumber, with bright eyes and smiling face; she embraced most tenderly her friend, whom she pitied, she said, from the bottom of her heart. Maurice seated himself beside his sister, while Jean, who was unused to polite society, but could not decline the invitation that was extended to him, was Delaherche's right-hand neighbor. It was Mme. Delaherche's custom not to come to the table with the family; a servant carried her a bowl, which she drank while sitting by the colonel. The party of five, however, who sat down together, although they commenced their meal in silence, soon became cheerful and talkative. Why should they not rejoice and be glad to find themselves there, safe and sound, with food before them to satisfy their hunger, when the country round about was covered with thousands upon thousands of poor starving wretches? In the cool, spacious dining room the snow-white tablecloth was a delight to the eye and the steaming *cafe au lait* seemed delicious.

They conversed, Delaherche, who had recovered his assurance and was again the wealthy manufacturer, the condescending patron courting popularity, severe only toward those who failed to succeed, spoke of Napoleon III., whose face as he saw it last continued to haunt his memory. He addressed himself to Jean, having that simple-minded young man as his neighbor. "Yes, sir, the Emperor has deceived me, and I don't hesitate to say so. His henchmen may put in the plea of mitigating circumstances, but it won't go down, sir; he is evidently the first, the only cause of our misfortunes."

He had quite forgotten that only a few months before he had been an ardent Bonapartist and had labored to ensure the success of the plebiscite, and now he who was henceforth to be known as the Man of Sedan was not even worthy to be pitied; he ascribed to him every known iniquity.

"A man of no capacity, as everyone is now compelled to admit; but let that pass, I say nothing of that. A visionary, a theorist, an unbalanced mind, with whom affairs seemed to succeed as long as he had luck on his side. And there's no use, don't you see, sir, in attempting to work on our sympathies and excite our commiseration by telling us that he was deceived, that the opposition refused him the necessary grants of men and money. It is he who has deceived us, he whose crimes and blunders have landed us in the horrible muddle where we are."

Maurice, who preferred to say nothing on the subject, could not help smiling, while Jean, embarrassed by the political turn the conversation had taken and fearful lest he might make some ill-timed remark, simply replied:

"They say he is a brave man, though."

But those few words, modestly expressed, fairly made Delaherche jump. All his past fear and alarm, all the mental anguish he had suffered, burst from his lips in a cry of concentrated passion, closely allied to hatred.

"A brave man, forsooth; and what does that amount to! Are you aware, sir, that my factory was struck three times by Prussian shells, and that it is no fault of the Emperor's that it was not burned! Are you aware that I, I shall lose a hundred thousand francs by this idiotic business! No, no; France invaded, pillaged, and laid waste, our industries compelled to shut down, our commerce ruined; it is a little too much, I tell you! One brave man like that is quite sufficient; may the Lord preserve us from any more of them! He is down in the blood and mire, and there let him remain!"

And he made a forcible gesture with his closed fist as if thrusting down and holding under the water some poor wretch who was struggling to save himself, then finished his coffee, smacking his lips like a true gourmand. Gilberte waited on Henriette as if she had been a child, laughing a little involuntary laugh when the latter made some exhibition of absent-mindedness. And when at last the coffee had all been drunk they still lingered on in the peaceful quiet of the great cool dining room.

And at that same hour Napoleon III. was in the weaver's lowly cottage on the Donchery road. As early as five o'clock in the morning he had insisted on leaving the Sous-Prefecture; he felt ill at ease in Sedan, which was at once a menace and a reproach to him, and moreover he thought he might, in some measure, alleviate the sufferings of his tender heart by obtaining more favorable terms for his unfortunate army. His object was to have a personal interview with the King of Prussia. He had taken his place in a hired caleche and been driven along the broad highway, with its row of lofty poplars on either side, and this first stage of his journey into exile, accomplished in the chill air of early dawn, must have reminded him forcibly of the grandeur that had been his and that he was putting behind him forever. It was on this road that he had his encounter with Bismarck, who came hurrying to meet him in an old cap and coarse, greased boots, with the sole object of keeping him occupied and preventing him from seeing the King until the capitulation should have been signed. The King was still at Vendresse, some nine miles away. Where was he to go? What roof would afford him shelter while he waited? In his own country, so far away, the Palace of the Tuileries had disappeared from his sight, swallowed up in the bosom of a storm-cloud, and he was never to see it more. Sedan seemed already to have receded into the distance, leagues and leagues, and to be parted from him by a river of blood. In France there were no longer imperial chateaus, nor official residences, nor even a chimney-nook in the house of the humblest functionary, where he would have dared to enter and claim hospitality. And it was in the house of the weaver that he determined to seek shelter, the squalid cottage that stood close to the roadside, with its scanty kitchen-garden inclosed by a hedge and its front of a single story with little forbidding windows. The room above-stairs was simply whitewashed and had a tiled floor; the only furniture was a common pine table and two straw-bottomed chairs. He spent two hours there, at first in company with Bismarck, who smiled to hear him speak of generosity, after that alone in silent misery, flattening his ashy face against the panes, taking his last look at French soil and at the Meuse, winding in and out, so beautiful, among the broad fertile fields.

Then the next day and the days that came after were other wretched stages of that journey; the Chateau of Bellevue, a pretty bourgeois retreat overlooking the river, where he rested that night, where he shed tears after his interview with King William; the sorrowful departure, that most miserable flight in a hired caleche over remote roads to the north of the city, which he avoided, not caring to face the wrath of the vanquished troops and the starving citizens, making a wide circuit over cross-roads by Floing, Fleigneux, and Illy and crossing the stream on a bridge of boats, laid down by the Prussians at Iges; the tragic encounter, the story of which has been so often told, that occurred on the corpse-cumbered plateau of Illy: the miserable Emperor, whose state was such that his horse could not be allowed to trot, had sunk under some more than usually violent attack of his complaint, mechanically smoking, perhaps, his everlasting cigarette, when a band of haggard, dusty, blood-stained prisoners, who were being conducted from Fleigneux to Sedan, were forced to leave the road to let the carriage pass and stood watching it from the ditch; those who were at the head of the line merely eyed him in silence; presently a hoarse, sullen murmur began to make itself heard, and finally, as the caleche proceeded down the line, the men burst out with a storm of yells and cat-calls, shaking their fists and calling down maledictions on the head of him who had been their ruler. After that came the interminable journey across the battlefield, as far as Givonne, amid scenes of havoc and devastation, amid the dead, who lay with staring eyes upturned that seemed to be full of menace; came, too, the bare, dreary fields, the great silent forest, then the frontier, running along the summit of a ridge, marked only by a stone, facing a wooden post that seemed ready to fall, and beyond the soil of Belgium, the end of all, with its road bordered with gloomy hemlocks descending sharply into the narrow valley.

And that first night of exile, that he spent at a common inn, the Hotel de la Poste at Bouillon, what a night it was! When the Emperor showed himself at his window in deference to the throng of French refugees and sight-seers that filled the place, he was greeted with a storm of hisses and hostile murmurs. The apartment assigned him, the three windows of which opened on the public square and on the Semoy, was the typical tawdry bedroom of the provincial inn with its conventional furnishings: the chairs covered with crimson damask, the mahogany armoire a glace, and on the mantel the imitation bronze clock, flanked by a pair of conch shells and vases of artificial flowers under glass covers. On either side of the door was a little single bed, to one of which the wearied aide-de-camp betook himself at nine o'clock and was immediately wrapped in soundest slumber. On the other the Emperor, to whom the god of sleep was less benignant, tossed almost the whole night through, and if he arose to try to quiet his excited nerves by walking, the sole distraction that his eyes encountered was a pair of engravings that were hung to right and left of the chimney, one depicting Rouget de Lisle singing the Marseillaise, the other a crude representation of the Last Judgment, the dead rising from their graves at the sound of the Archangel's trump, the resurrection of the victims of the battlefield, about to appear before their God to bear witness against their rulers.

The imperial baggage train, cause in its day of so much scandal, had been left behind at Sedan, where it rested in ignominious hiding behind the Sous-Prefet's lilac bushes. It puzzled the authorities somewhat to devise means for ridding themselves of what was to them a bete noire, for getting it away from the city unseen by the famishing multitude, upon whom the sight of its flaunting splendor would have produced much the same effect that a red rag does on a maddened bull. They waited until there came an unusually dark night, when horses, carriages, and baggage-wagons, with their silver stew-pans, plate, linen, and baskets of fine wines, all trooped out of Sedan in deepest mystery and shaped their course for Belgium, noiselessly, without beat of drum, over the least frequented roads like a thief stealing away in the night.

PART THIRD

1

All the long, long day of the battle Silvine, up on Remilly hill, where Father Fouchard's little farm was situated, but her heart and soul absent with Honore amid the dangers of the conflict, never once took her eyes from off Sedan, where the guns were roaring. The following day, moreover, her anxiety was even greater still, being increased by her inability to obtain any definite tidings, for the Prussians who were guarding the roads in the vicinity refused to answer questions, as much from reasons of policy as because they knew but very little themselves. The bright sun of the day before was no longer visible, and showers had fallen, making the valley look less cheerful than usual in the wan light.

Toward evening Father Fouchard, who was also haunted by a sensation of uneasiness in the midst of his studied taciturnity, was standing on his doorstep reflecting on the probable outcome of events. His son had no place in his thoughts, but he was speculating how he best might convert the misfortunes of others into fortune for himself, and as he revolved these considerations in his mind he noticed a tall, strapping young fellow, dressed in the peasant's blouse, who had been strolling up and down the road for the last minute or so, looking as if he did not know what to do with himself. His astonishment on recognizing him was so great that he called him aloud by name, notwithstanding that three Prussians happened to be passing at the time.

"Why, Prosper! Is that you?"

The chasseur d'Afrique imposed silence on him with an emphatic gesture; then, coming closer, he said in an undertone:

"Yes, it is I. I have had enough of fighting for nothing, and I cut my lucky. Say, Father Fouchard, you don't happen to be in need of a laborer on your farm, do you?"

All the old man's prudence came back to him in a twinkling. He *was* looking for someone to help him, but it would be better not to say so at once.

"A lad on the farm? faith, no—not just now. Come in, though, all the same, and have a glass. I shan't leave you out on the road when you're in trouble, that's sure."

Silvine, in the kitchen, was setting the pot of soup on the fire, while little Charlot was hanging by her skirts, frolicking and laughing. She did not recognize Prosper at first, although they had formerly served together in the same household, and it was not until she came in, bringing a bottle of wine and two glasses, that she looked him squarely in the face. She uttered a cry of joy and surprise; her sole thought was of Honore.

"Ah, you were there, weren't you? Is Honore all right?"

Prosper's answer was ready to slip from his tongue; he hesitated. For the last two days he had been living in a dream, among a rapid succession of strange, ill-defined events which left behind them no precise memory, as a man starts, half-awakened, from a slumber peopled with fantastic visions. It was true, doubtless, he believed he had seen Honore lying upon a cannon, dead, but he would not have cared to swear to it; what use is there in afflicting people when one is not certain?

"Honore," he murmured, "I don't know, I couldn't say."

232

She continued to press him with her questions, looking at him steadily.

"You did not see him, then?"

He waved his hands before him with a slow, uncertain motion and an expressive shake of the head.

"How can you expect one to remember! There were such lots of things, such lots of things. Look you, of all that d---d battle, if I was to die for it this minute, I could not tell you that much-no, not even the place where I was. I believe men get to be no better than idiots, 'pon my word I do!" And tossing off a glass of wine, he sat gloomily silent, his vacant eyes turned inward on the dark recesses of his memory. "All that I remember is that it was beginning to be dark when I recovered consciousness. I went down while we were charging, and then the sun was very high. I must have been lying there for hours, my right leg caught under poor old Zephyr, who had received a piece of shell in the middle of his chest. There was nothing to laugh at in my position, I can tell you; the dead comrades lying around me in piles, not a living soul in sight, and the certainty that I should have to kick the bucket too unless someone came to put me on my legs again. Gently, gently, I tried to free my leg, but it was no use; Zephyr's weight must have been fully up to that of the five hundred thousand devils. He was warm still. I patted him, I spoke to him, saying all the pretty things I could think of, and here's a thing, do you see, that I shall never forget as long as I live: he opened his eyes and made an effort to raise his poor old head, which was resting on the ground beside my own. Then we had a talk together: 'Poor old fellow,' says I, 'I don't want to say a word to hurt your feelings, but you must want to see me croak with you, you hold me down so hard.' Of course he didn't say he did; he couldn't, but for all that I could read in his great sorrowful eyes how bad he felt to have to part with me. And I can't say how the thing happened, whether he intended it or whether it was part of the death struggle, but all at once he gave himself a great shake that sent him rolling away to one side. I was enabled to get on my feet once more, but ah! in what a pickle; my leg was swollen and heavy as a leg of lead. Never mind, I took Zephyr's head in my arms and kept on talking to him, telling him all the kind thoughts I had in my heart, that he was a good horse, that I loved him dearly, that I should never forget him. He listened to me, he seemed to be so pleased! Then he had another long convulsion, and so he died, with his big vacant eyes fixed on me till the last. It is very strange, though, and I don't suppose anyone will believe me; still, it is the simple truth that great, big tears were standing in his eyes. Poor old Zephyr, he cried just like a man-"

At this point Prosper's emotion got the better of him; tears choked his utterance and he was obliged to break off. He gulped down another glass of wine and went on with his narrative in disjointed, incomplete sentences. It kept growing darker and darker, until there was only a narrow streak of red light on the horizon at the verge of the battlefield; the shadows of the dead horses seemed to be projected across the plain to an infinite distance. The pain and stiffness in his leg kept him from moving; he must have remained for a long time beside Zephyr. Then, with his fears as an incentive, he had managed to get on his feet and hobble away; it was an imperative necessity to him not to be alone, to find comrades who would share his fears with him and make them less. Thus from every nook and corner of the battlefield, from hedges and ditches and clumps of bushes, the wounded who had been left behind dragged themselves painfully in search of companionship, forming when possible little bands of four or five, finding it less hard to agonize and die in the company of their fellow-beings. In the wood of la Garenne Prosper fell in with two men of the 43d regiment; they were not wounded, but had burrowed in the underbrush like rabbits, waiting for the coming of the night. When they learned that he was familiar with the roads they communicated to him their plan, which was to traverse the woods under cover of the darkness and make their escape into Belgium. At first he declined to share their undertaking, for he would have preferred to proceed direct to Remilly, where he was certain to find a refuge, but where was he to obtain the blouse and trousers that he required as a disguise? to say nothing of the impracticability of getting past the numerous Prussian pickets and outposts that filled the valley all the way from la Garenne to Remilly. He therefore ended by consenting to act as guide to the two comrades. His leg was less stiff than it had been, and they were so fortunate as to secure a loaf of bread at a farmhouse. Nine o'clock was striking from the church of a village in the distance as they resumed their way. The only point where they encountered any danger worth mentioning was at la Chapelle, where they fell directly into the midst of a Prussian advanced post before they were aware of it; the enemy flew to arms and blazed away into the darkness, while they, throwing themselves on the ground and alternately crawling and running until the fire slackened, ultimately regained the shelter of the trees. After that they kept to the woods, observing the utmost vigilance. At a bend in the road, they crept up behind an out-lying picket and, leaping on his back, buried a knife in his throat. Then the road was free before them and they no longer had to observe precaution; they went ahead, laughing and whistling. It was about three in the morning when they reached a little Belgian village, where they knocked up a worthy farmer, who at once opened his barn to them; they snuggled among the hay and slept soundly until morning.

The sun was high in the heavens when Prosper awoke. As he opened his eyes and looked about him, while the two comrades were still snoring, he beheld their entertainer engaged in hitching a horse to a great carriole loaded with bread, rice, coffee, sugar, and all sorts of eatables, the whole concealed under sacks of charcoal, and a little questioning elicited from the good man the fact that he had two married daughters living at Raucourt, in France, whom the passage of the Bavarian troops had left entirely destitute, and that the provisions in the carriole were intended for them. He had procured that very morning the safe-conduct that was required for the journey. Prosper was immediately seized by an uncontrollable desire to take a seat in that carriole and return to the country that he loved so and for which his heart was yearning with such a violent nostalgia. It was perfectly simple; the farmer would have to pass through Remilly to reach Raucourt; he would alight there. The matter was arranged in three minutes; he obtained a loan of the longed-for blouse and trousers, and the farmer gave out, wherever they stopped, that he was his servant; so that about six o'clock he got down in front of the church, not having been stopped more than two or three times by the German outposts.

They were all silent for a while, then: "No, I had enough of it!" said Prosper. "If they had but set us at work that amounted to something, as out there in Africa! but this going up the hill only to come down again, the feeling that one is of no earthly use to anyone, that is no kind of a life at all. And then I should be lonely, now that poor Zephyr is dead; all that is left me to do is to go to work on a farm. That will be better than living among the Prussians as a prisoner, don't you think so? You have horses, Father Fouchard; try me, and see whether or not I will love them and take good care of them."

The old fellow's eyes gleamed, but he touched glasses once more with the other and concluded the arrangement without any evidence of eagerness.

"Very well; I wish to be of service to you as far as lies in my power; I will take you. As regards the question of wages, though, you must not speak of it until the war is over, for really I am not in need of anyone and the times are too hard."

Silvine, who had remained seated with Charlot on her lap, had never once taken her eyes from Prosper's face. When she saw him rise with the intention of going to the stable and making immediate acquaintance with its four-footed inhabitants, she again asked:

"Then you say you did not see Honore?"

The question repeated thus abruptly made him start, as if it had suddenly cast a flood of light in upon an obscure corner of his memory. He hesitated for a little, but finally came to a decision and spoke.

"See here, I did not wish to grieve you just now, but I don't believe Honore will ever come back."

"Never come back-what do you mean?"

"Yes, I believe that the Prussians did his business for him. I saw him lying across his gun, his head erect, with a great wound just beneath the heart."

There was silence in the room. Silvine's pallor was frightful to behold, while Father Fouchard displayed his interest in the narrative by replacing upon the table his glass, into which he had just poured what wine remained in the bottle.

"Are you quite certain?" she asked in a choking voice.

"Dame! as certain as one can be of a thing he has seen with his own two eyes. It was on a little hillock, with three trees in a group right beside it; it seems to me I could go to the spot blindfolded."

If it was true she had nothing left to live for. That lad who had been so good to her, who had forgiven her her fault, had plighted his troth and was to marry her when he came home at the end of the campaign! and they had robbed her of him, they had murdered him, and he was lying out there on the battlefield with a wound under the heart! She had never known how strong her love for him had been, and now the thought that she was to see him no more, that he who was hers was hers no longer, aroused her almost to a pitch of madness and made her forget her usual tranquil resignation. She set Charlot roughly down upon the floor, exclaiming:

"Good! I shall not believe that story until I see the evidence of it, until I see it with my own eyes. Since you know the spot you shall conduct me to it. And if it is true, if we find him, we will bring him home with us."

Her tears allowed her to say no more; she bowed her head upon the table, her frame convulsed by long-drawn, tumultuous sobs that shook her from head to foot, while the child, not knowing what to make of such unusual treatment at his mother's hands, also commenced to weep violently. She caught him up and pressed him to her heart, with distracted, stammering words:

"My poor child! my poor child!"

Consternation was depicted on old Fouchard's face. Appearances notwithstanding, he did love his son, after a fashion of his own. Memories of the past came back to him, of days long vanished, when his wife was still living and Honore was a boy at school, and two big tears appeared in his small red eyes and trickled down his old leathery cheeks. He had not wept before in more than ten years. In the end he grew angry at the thought of that son who was his and upon whom he was never to set eyes again; he rapped out an oath or two.

"Nom de Dieu! it is provoking all the same, to have only one boy, and that he should be taken from you!"

When their agitation had in a measure subsided, however, Fouchard was annoyed that Silvine still continued to talk of going to search for Honore's body out there on the battlefield. She made no further noisy demonstration, but harbored her purpose with the dogged silence of despair, and he failed to recognize in her the docile, obedient servant who was wont to perform her daily tasks without a murmur; her great, submissive eyes, in which lay the chief beauty of her face, had assumed an expression of stern determination, while beneath her thick brown hair her cheeks and brow wore a pallor that was like death. She had torn off the red kerchief that was knotted about her neck, and was entirely in black, like a widow in her weeds. It was all in vain that he tried to impress on her the difficulties of the undertaking, the dangers she would be subjected to, the little hope there was of recovering the corpse; she did not even take the trouble to answer him, and he saw clearly that unless he seconded her in her plan she would start out alone and do some unwise thing, and this aspect of the case worried him on account of the complications that might arise between him and the Prussian authorities. He therefore finally decided to go and lay the matter before the mayor of Remilly, who was a kind of distant cousin of his, and they two between them concocted a story: Silvine was to pass as the actual widow of Honore, Prosper became her brother, so that the Bavarian colonel, who had his quarters in the Hotel of the Maltese Cross down in the lower part of the village, made no difficulty about granting a pass which authorized the brother and sister to bring home the body of the husband, provided they could find it. By this time it was night; the only concession that could be obtained from the young woman was that she would delay starting on her expedition until morning.

When morning came old Fouchard could not be prevailed on to allow one of his horses to be taken, fearing he might never set eyes on it again. What assurance had he that the Prussians would not confiscate the entire equipage? At last he consented, though with very bad grace, to loan her the donkey, a little gray animal, and his cart, which, though small, would be large enough to hold a dead man. He gave minute instructions to Prosper, who had had a good night's sleep, but was anxious and thoughtful at the prospect of the expedition now that, being rested and refreshed, he attempted to remember something of the battle. At the last moment Silvine went and took the counterpane from her own bed, folding and spreading it on the floor of the cart. Just as she was about to start she came running back to embrace Charlot.

"I entrust him to your care, Father Fouchard; keep an eye on him and see that he doesn't get hold of the matches."

"Yes, yes; never fear!"

They were late in getting off; it was near seven o'clock when the little procession, the donkey, hanging his head and drawing the narrow cart, leading, descended the steep hill of Remilly. It had rained heavily during the night, and the roads were become rivers of mud; great lowering clouds hung in the heavens, imparting an air of cheerless desolation to the scene.

Prosper, wishing to save all the distance he could, had determined on taking the route that lay through the city of Sedan, but before they reached Pont-Maugis a Prussian outpost halted the cart and held it for over an hour, and finally, after their pass had been referred, one after another, to four or five officials, they were told they might resume their journey, but only on condition of taking the longer, roundabout route by way of Bazeilles, to do which they would have to turn into a cross-road on their left. No reason was assigned; their object was probably to avoid adding to the crowd that encumbered the streets of the city. When Silvine crossed the Meuse by the railroad bridge, that ill-starred bridge that the French had failed to destroy and which, moreover, had been the cause of such slaughter among the Bavarians, she beheld the corpse of an artilleryman floating lazily down with the sluggish current. It caught among some rushes near the bank, hung there a moment, then swung clear and started afresh on its downward way.

Bazeilles, through which they passed from end to end at a slow walk, afforded a spectacle of ruin and desolation, the worst that war can perpetrate when it sweeps with devastating force, like a cyclone, through a land. The dead had been removed; there was not a single corpse to be seen in the village streets, and the rain had washed away the blood; pools of reddish water were to be seen here and there in the roadway, with repulsive, frowzy-looking debris, matted

236

masses that one could not help associating in his mind with human hair. But what shocked and saddened one more than all the rest was the ruin that was visible everywhere; that charming village, only three days before so bright and smiling, with its pretty houses standing in their well-kept gardens, now razed, demolished, annihilated, nothing left of all its beauties save a few smoke-stained walls. The church was burning still, a huge pyre of smoldering beams and girders, whence streamed continually upward a column of dense black smoke that, spreading in the heavens, overshadowed the city like a gigantic funeral pall. Entire streets had been swept away, not a house left on either side, nor any trace that houses had ever been there, save the calcined stone-work lying in the gutter in a pasty mess of soot and ashes, the whole lost in the viscid, ink-black mud of the thoroughfare. Where streets intersected the corner houses were razed down to their foundations, as if they had been carried away bodily by the fiery blast that blew there. Others had suffered less; one in particular, owing to some chance, had escaped almost without injury, while its neighbors on either hand, literally torn to pieces by the iron hail, were like gaunt skeletons. An unbearable stench was everywhere, noticeable, the nauseating odor that follows a great fire, aggravated by the penetrating smell of petroleum, that had been used without stint upon floors and walls. Then, too, there was the pitiful, mute spectacle of the household goods that the people had endeavored to save, the poor furniture that had been thrown from windows and smashed upon the sidewalk, crazy tables with broken legs, presses with cloven sides and split doors, linen, also, torn and soiled, that was trodden under foot; all the sorry crumbs, the unconsidered trifles of the pillage, of which the destruction was being completed by the dissolving rain. Through the breach in a shattered house-front a clock was visible, securely fastened high up on the wall above the mantel-shelf, that had miraculously escaped intact.

"The beasts! the pigs!" growled Prosper, whose blood, though he was no longer a soldier, ran hot at the sight of such atrocities.

He doubled his fists, and Silvine, who was white as a ghost, had to exert the influence of her glance to calm him every time they encountered a sentry on their way. The Bavarians had posted sentinels near all the houses that were still burning, and it seemed as if those men, with loaded muskets and fixed bayonets, were guarding the fires in order that the flames might finish their work. They drove away the mere sightseers who strolled about in the vicinity, and the persons who had an interest there as well, employing first a menacing gesture, and in case that was not sufficient, uttering a single brief, guttural word of command. A young woman, her hair streaming about her shoulders, her gown plastered with mud, persisted in hanging about the smoking ruins of a little house, of which she desired to search the hot ashes, notwithstanding the prohibition of the sentry. The report ran that the woman's little baby had been burned with the house. And all at once, as the Bavarian was roughly thrusting her aside with his heavy hand, she turned on him, vomiting in his face all her despair and rage, lashing him with taunts and insults that were redolent of the gutter, with obscene words which likely afforded her some consolation in her grief and distress. He could not have understood her, for he drew back a pace or two, eying her with apprehension. Three comrades came running up and relieved him of the fury, whom they led away screaming at the top of her voice. Before the ruins of another house a man and two little girls, all three so weary and miserable that they could not stand, lay on the bare ground, sobbing as if their hearts would break; they had seen their little all go up in smoke and flame, and had no place to go, no place to lay their head. But just then a patrol went by, dispersing the knots of idlers, and the street again assumed its deserted aspect, peopled only by the stern, sullen sentries, vigilant to see that their iniquitous instructions were enforced.

"The beasts! the pigs!" Prosper repeated in a stifled voice. "How I should like, oh! how I should like to kill a few of them!"

237

Silvine again made him be silent. She shuddered. A dog, shut up in a carriage-house that the flames had spared and forgotten there for the last two days, kept up an incessant, continuous howling, in a key so inexpressibly mournful that a brooding horror seemed to pervade the low, leaden sky, from which a drizzling rain had now begun to fall. They were then just abreast of the park of Montivilliers, and there they witnessed a most horrible sight. Three great covered carts, those carts that pass along the streets in the early morning before it is light and collect the city's filth and garbage, stood there in a row, loaded with corpses; and now, instead of refuse, they were being filled with dead, stopping wherever there was a body to be loaded, then going on again with the heavy rumbling of their wheels to make another stop further on, threading Bazeilles in its every nook and corner until their hideous cargo overflowed. They were waiting now upon the public road to be driven to the place of their discharge, the neighboring potter's field. Feet were seen projecting from the mass into the air. A head, half-severed from its trunk, hung over the side of the vehicle. When the three lumbering vans started again, swaying and jolting over the inequalities of the road, a long, white hand was hanging outward from one of them; the hand caught upon the wheel, and little by little the iron tire destroyed it, eating through skin and flesh clean down to the bones.

By the time they reached Balan the rain had ceased, and Prosper prevailed on Silvine to eat a bit of the bread he had had the foresight to bring with them. When they were near Sedan, however, they were brought to a halt by another Prussian post, and this time the consequences threatened to be serious; the officer stormed at them, and even refused to restore their pass, which he declared, in excellent French, to be a forgery. Acting on his orders some soldiers had run the donkey and the little cart under a shed. What were they to do? were they to be forced to abandon their undertaking? Silvine was in despair, when all at once she thought of M. Dubreuil, Father Fouchard's relative, with whom she had some slight acquaintance and whose place, the Hermitage, was only a few hundred yards distant, on the summit of the eminence that overlooked the faubourg. Perhaps he might have some influence with the military, seeing that he was a citizen of the place. As they were allowed their freedom, conditionally upon abandoning their equipage, she left the donkey and cart under the shed and bade Prosper accompany her. They ascended the hill on a run, found the gate of the Hermitage standing wide open, and on turning into the avenue of secular elms beheld a spectacle that filled them with amazement.

"The devil!" said Prosper; "there are a lot of fellows who seem to be taking things easy!"

On the fine-crushed gravel of the terrace, at the bottom of the steps that led to the house, was a merry company. Arranged in order around a marble-topped table were a sofa and some easy-chairs in sky-blue satin, forming a sort of fantastic open-air drawing-room, which must have been thoroughly soaked by the rain of the preceding day. Two zouaves, seated in a lounging attitude at either end of the sofa, seemed to be laughing boisterously. A little infantryman, who occupied one of the fauteuils, his head bent forward, was apparently holding his sides to keep them from splitting. Three others were seated in a negligent pose, their elbows resting on the arms of their chairs, while a chasseur had his hand extended as if in the act of taking a glass from the table. They had evidently discovered the location of the cellar, and were enjoying themselves.

"But how in the world do they happen to be here?" murmured Prosper, whose stupefaction increased as he drew nearer to them. "Have the rascals forgotten there are Prussians about?"

But Silvine, whose eyes had dilated far beyond their natural size, suddenly uttered an exclamation of horror. The soldiers never moved hand or foot; they were stone dead. The two zouaves were stiff and cold; they both had had the face shot away, the nose was gone, the eyes were torn from their sockets. If there appeared to be a laugh on the face of him who was holding his sides, it was because a bullet had cut a great furrow through the lower portion of his countenance, smashing all his teeth. The spectacle was an unimaginably horrible one, those poor wretches laughing and conversing in their attitude of manikins, with glassy eyes and open mouths, when Death had laid his icy hand on them and they were never more to know the warmth and motion of life. Had they dragged themselves, still living, to that place, so as to die in one another's company? or was it not rather a ghastly prank of the Prussians, who had collected the bodies and placed them in a circle about the table, out of derision for the traditional gayety of the French nation?

"It's a queer start, though, all the same," muttered Prosper, whose face was very pale. And casting a look at the other dead who lay scattered about the avenue, under the trees and on the turf, some thirty brave fellows, among them Lieutenant Rochas, riddled with wounds and surrounded still by the shreds of the flag, he added seriously and with great respect: "There must have been some very pretty fighting about here! I don't much believe we shall find the bourgeois for whom you are looking."

Silvine entered the house, the doors and windows of which had been battered in and afforded admission to the damp, cold air from without. It was clear enough that there was no one there; the masters must have taken their departure before the battle. She continued to prosecute her search, however, and had entered the kitchen, when she gave utterance to another cry of terror. Beneath the sink were two bodies, fast locked in each other's arms in mortal embrace, one of them a zouave, a handsome, brown-bearded man, the other a huge Prussian with red hair. The teeth of the former were set in the latter's cheek, their arms, stiff in death, had not relaxed their terrible hug, binding the pair with such a bond of everlasting hate and fury that ultimately it was found necessary to bury them in a common grave.

Then Prosper made haste to lead Silvine away, since they could accomplish nothing in that house where Death had taken up his abode, and upon their return, despairing, to the post where the donkey and cart had been detained, it so chanced that they found, in company with the officer who had treated them so harshly, a general on his way to visit the battlefield. This gentleman requested to be allowed to see the pass, which he examined attentively and restored to Silvine; then, with an expression of compassion on his face, he gave directions that the poor woman should have her donkey returned to her and be allowed to go in quest of her husband's body. Stopping only long enough to thank her benefactor, she and her companion, with the cart trundling after them, set out for the Fond de Givonne, obedient to the instructions that were again given them not to pass through Sedan.

After that they bent their course to the left in order to reach the plateau of Illy by the road that crosses the wood of la Garenne, but here again they were delayed; twenty times they nearly abandoned all hope of getting through the wood, so numerous were the obstacles they encountered. At every step their way was barred by huge trees that had been laid low by the artillery fire, stretched on the ground like mighty giants fallen. It was the part of the forest that had suffered so severely from the cannonade, where the projectiles had plowed their way through the secular growths as they might have done through a square of the Old Guard, meeting in either case with the sturdy resistance of veterans. Everywhere the earth was cumbered with gigantic trunks, stripped of their leaves and branches, pierced and mangled, even as mortals might have been, and this wholesale destruction, the sight of the poor limbs, maimed, slaughtered and weeping tears of sap, inspired the beholder with the sickening horror of a human battlefield. There were corpses of men there, too; soldiers, who had stood fraternally by the trees and fallen with them. A lieutenant, from whose mouth exuded a bloody froth, had been tearing up the grass by handfuls in his agony, and his stiffened fingers were still buried in the ground. A little farther on a captain, prone on his stomach, had raised his head to vent his anguish in yells and screams, and death had caught and fixed him in that strange attitude. Others seemed to be slumbering among the herbage, while a zouave; whose blue sash had taken fire, had had his hair and beard burned completely from his head. And several times it happened, as they traversed those woodland glades, that they had to remove a body from the path before the donkey could proceed on his way. Presently they came to a little valley, where the sights of horror abruptly ended. The battle had evidently turned at this point and expended its force in another direction, leaving this peaceful nook of nature untouched. The trees were all uninjured; the carpet of velvety moss was undefiled by blood. A little brook coursed merrily among the duckweed, the path that ran along its bank was shaded by tall beeches. A penetrating charm, a tender peacefulness pervaded the solitude of the lovely spot, where the living waters gave up their coolness to the air and the leaves whispered softly in the silence.

Prosper had stopped to let the donkey drink from the stream.

"Ah, how pleasant it is here!" he involuntarily exclaimed in his delight.

Silvine cast an astonished look about her, as if wondering how it was that she, too, could feel the influence of the peaceful scene. Why should there be repose and happiness in that hidden nook, when surrounding it on every side were sorrow and affliction? She made a gesture of impatience.

"Quick, quick, let us be gone. Where is the spot? Where did you tell me you saw Honore?"

And when, at some fifty paces from there, they at last came out on the plateau of Illy, the level plain unrolled itself in its full extent before their vision. It was the real, the true battlefield that they beheld now, the bare fields stretching away to the horizon under the wan, cheerless sky, whence showers were streaming down continually. There were no piles of dead visible; all the Prussians must have been buried by this time, for there was not a single one to be seen among the corpses of the French that were scattered here and there, along the roads and in the fields, as the conflict had swayed in one direction or another. The first that they encountered was a sergeant, propped against a hedge, a superb man, in the bloom of his youthful vigor; his face was tranquil and a smile seemed to rest on his parted lips. A hundred paces further on, however, they beheld another, lying across the road, who had been mutilated most frightfully, his head almost entirely shot away, his shoulders covered with great splotches of brain matter. Then, as they advanced further into the field, after the single bodies, distributed here and there, they came across little groups; they saw seven men aligned in single rank, kneeling and with their muskets at the shoulder in the position of aim, who had been hit as they were about to fire, while close beside them a subaltern had also fallen as he was in the act of giving the word of command. After that the road led along the brink of a little ravine, and there they beheld a spectacle that aroused their horror to the highest pitch as they looked down into the chasm, into which an entire company seemed to have been blown by the fiery blast; it was choked with corpses, a landslide, an avalanche of maimed and mutilated men, bent and twisted in an inextricable tangle, who with convulsed fingers had caught at the yellow clay of the bank to save themselves in their descent, fruitlessly. And a dusky flock of ravens flew away, croaking noisily, and swarms of flies, thousands upon thousands of them, attracted by the odor of fresh blood, were buzzing over the bodies and returning incessantly.

"Where is the spot?" Silvine asked again.

They were then passing a plowed field that was completely covered with knapsacks. It was manifest that some regiment had been roughly handled there, and the men, in a moment of panic, had relieved themselves of their burdens. The debris of every sort with which the ground was thickly strewn served to explain the episodes of the conflict. There was a stubble field where the scattered kepis, resembling huge poppies, shreds of uniforms, epaulettes, and sword-belts told the story of one of those infrequent hand-to-hand contests in the fierce artillery duel that had lasted twelve hours. But the objects that were encountered most frequently, at every step, in fact, were abandoned weapons, sabers, bayonets, and, more particularly, chassepots; and so numerous were they that they seemed to have sprouted from the earth, a harvest that had matured in a single ill-omened day. Porringers and buckets, also, were scattered along the roads, together with the heterogeneous contents of knapsacks, rice, brushes, clothing, cartridges. The fields everywhere presented an uniform scene of devastation: fences destroyed, trees blighted as if they had been struck by lightning, the very soil itself torn by shells, compacted and hardened by the tramp of countless feet, and so maltreated that it seemed as if seasons must elapse before it could again become productive. Everything had been drenched and soaked by the rain of the preceding day; an odor arose and hung in the air persistently, that odor of the battlefield that smells like fermenting straw and burning cloth, a mixture of rottenness and gunpowder.

Silvine, who was beginning to weary of those fields of death over which she had tramped so many long miles, looked about her with increasing distrust and uneasiness.

"Where is the spot? where is it?"

But Prosper made no answer; he also was becoming uneasy. What distressed him even more than the sights of suffering among his fellow-soldiers was the dead horses, the poor brutes that lay outstretched upon their side, that were met with in great numbers. Many of them presented a most pitiful spectacle, in all sorts of harrowing attitudes, with heads torn from the body, with lacerated flanks from which the entrails protruded. Many were resting on their back, with their four feet elevated in the air like signals of distress. The entire extent of the broad plain was dotted with them. There were some that death had not released after their two days' agony; at the faintest sound they would raise their head, turning it eagerly from right to left, then let it fall again upon the ground, while others lay motionless and momentarily gave utterance to that shrill scream which one who has heard it can never forget, the lament of the dying horse, so piercingly mournful that earth and heaven seemed to shudder in unison with it. And Prosper, with a bleeding heart, thought of poor Zephyr, and told himself that perhaps he might see him once again.

Suddenly he became aware that the ground was trembling under the thundering hoof-beats of a headlong charge. He turned to look, and had barely time to shout to his companion:

"The horses, the horses! Get behind that wall!"

From the summit of a neighboring eminence a hundred riderless horses, some of them still bearing the saddle and master's kit, were plunging down upon them at break-neck speed. They were cavalry mounts that had lost their masters and remained on the battlefield, and instinct had counseled them to associate together in a band. They had had neither hay nor oats for two days, and had cropped the scanty grass from off the plain, shorn the hedge-rows of leaves and twigs, gnawed the bark from the trees, and when they felt the pangs of hunger pricking at their vitals like a keen spur, they started all together at a mad gallop and charged across the deserted, silent fields, crushing the dead out of all human shape, extinguishing the last spark of life in the wounded.

The band came on like a whirlwind; Silvine had only time to pull the donkey and cart to one side where they would be protected by the wall.

"Mon Dieu! we shall be killed!"

But the horses had taken the obstacle in their stride and were already scouring away in the distance on the other side with a rumble like that of a receding thunder-storm; striking into a sunken road they pursued it as far as the corner of a little wood, behind which they were lost to sight.

241

Silvine, when she had brought the cart back into the road, insisted that Prosper should answer her question before they proceeded further.

"Come, where is it? You told me you could find the spot with your eyes bandaged; where is it? We have reached the ground."

He, drawing himself up and anxiously scanning the horizon in every direction, seemed to become more and more perplexed.

"There were three trees, I must find those three trees in the first place. Ah, dame! see here, one's sight is not of the clearest when he is fighting, and it is no such easy matter to remember afterward the roads one has passed over!"

Then perceiving people to his left, two men and a woman, it occurred to him to question them, but the woman ran away at his approach and the men repulsed him with threatening gestures; and he saw others of the same stripe, clad in sordid rags, unspeakably filthy, with the ill-favored faces of thieves and murderers, and they all shunned him, slinking away among the corpses like jackals or other unclean, creeping beasts. Then he noticed that wherever these villainous gentry passed the dead behind them were shoeless, their bare, white feet exposed, devoid of covering, and he saw how it was: they were the tramps and thugs who followed the German armies for the sake of plundering the dead, the detestable crew who followed in the wake of the invasion in order that they might reap their harvest from the field of blood. A tall, lean fellow arose in front of him and scurried away on a run, a sack slung across his shoulder, the watches and small coins, proceeds of his robberies, jingling in his pockets.

A boy about fourteen or fifteen years old, however, allowed Prosper to approach him, and when the latter, seeing him to be French, rated him soundly, the boy spoke up in his defense. What, was it wrong for a poor fellow to earn his living? He was collecting chassepots, and received five sous for every chassepot he brought in. He had run away from his village that morning, having eaten nothing since the day before, and engaged himself to a contractor from Luxembourg, who had an arrangement with the Prussians by virtue of which he was to gather the muskets from the field of battle, the Germans fearing that should the scattered arms be collected by the peasants of the frontier, they might be conveyed into Belgium and thence find their way back to France. And so it was that there was quite a flock of poor devils hunting for muskets and earning their five sous, rummaging among the herbage, like the women who may be seen in the meadows, bent nearly double, gathering dandelions.

"It's a dirty business," Prosper growled.

"What would you have! A chap must eat," the boy replied. "I am not robbing anyone."

Then, as he did not belong to that neighborhood and could not give the information that Prosper wanted, he pointed out a little farmhouse not far away where he had seen some people stirring.

Prosper thanked him and was moving away to rejoin Silvine when he caught sight of a chassepot, partially buried in a furrow. His first thought was to say nothing of his discovery; then he turned about suddenly and shouted, as if he could not help it:

"Hallo! here's one; that will make five sous more for you."

As they approached the farmhouse Silvine noticed other peasants engaged with spades and picks in digging long trenches; but these men were under the direct command of Prussian officers, who, with nothing more formidable than a light walking-stick in their hands, stood by, stiff and silent, and superintended the work. They had requisitioned the inhabitants of all the villages of the vicinity in this manner, fearing that decomposition might be hastened, owing to the rainy weather. Two cart-loads of dead bodies were standing near, and a gang of men was unloading them, laying the corpses side by side in close contiguity to one another, not searching them, not even looking at their faces, while two men followed after, equipped with great shovels, and covered the row with a layer of earth, so thin that the ground had already begun to crack beneath the showers. The work was so badly and hastily done that before two weeks should have elapsed each of those fissures would be breathing forth pestilence. Silvine could not resist the impulse to pause at the brink of the trench and look at those pitiful corpses as they were brought forward, one after another. She was possessed by a horrible fear that in each fresh body the men brought from the cart she might recognize Honore. Was not that he, that poor wretch whose left eye had been destroyed? No! Perhaps that one with the fractured jaw was he? The one thing certain to her mind was that if she did not make haste to find him, wherever he might be on that boundless, indeterminate plateau, they would pick him up and bury him in a common grave with the others. She therefore hurried to rejoin Prosper, who had gone on to the farmhouse with the cart.

"Mon Dieu! how is it that you are not better informed? Where is the place? Ask the people, question them."

There were none but Prussians at the farm, however, together with a woman servant and her child, just come in from the woods, where they had been near perishing of thirst and hunger. The scene was one of patriarchal simplicity and well-earned repose after the fatigues of the last few days. Some of the soldiers had hung their uniforms from a clothes-line and were giving them a thorough brushing, another was putting a patch on his trousers, with great neatness and dexterity, while the cook of the detachment had built a great fire in the middle of the courtyard on which the soup was boiling in a huge pot from which ascended a most appetizing odor of cabbage and bacon. There is no denying that the Prussians generally displayed great moderation toward the inhabitants of the country after the conquest, which was made the easier to them by the spirit of discipline that prevailed among the troops. These men might have been taken for peaceable citizens just come in from their daily avocations, smoking their long pipes. On a bench beside the door sat a stout, red-bearded man, who had taken up the servant's child, a little urchin five or six years old, and was dandling it and talking baby-talk to it in German, delighted to see the little one laugh at the harsh syllables which it could not understand.

Prosper, fearing there might be more trouble in store for them, had turned his back on the soldiers immediately on entering, but those Prussians were really good fellows; they smiled at the little donkey, and did not even trouble themselves to ask for a sight of the pass.

Then ensued a wild, aimless scamper across the bosom of the great, sinister plain. The sun, now sinking rapidly toward the horizon, showed its face for a moment from between two clouds. Was night to descend and surprise them in the midst of that vast charnel-house? Another shower came down; the sun was obscured, the rain and mist formed an impenetrable barrier about them, so that the country around, roads, fields, trees, was shut out from their vision. Prosper knew not where they were; he was lost, and admitted it: his memory was all astray, he could recall nothing precise of the occurrences of that terrible day but one before. Behind them, his head lowered almost to the ground, the little donkey trotted along resignedly, dragging the cart, with his customary docility. First they took a northerly course, then they returned toward Sedan. They had lost their bearings and could not tell in which direction they were going; twice they noticed that they were passing localities that they had passed before and retraced their steps. They had doubtless been traveling in a circle, and there came a moment when in their exhaustion and despair they stopped at a place where three roads met, without courage to pursue their search further, the rain pelting down on them, lost and utterly miserable in the midst of a sea of mud.

But they heard the sound of groans, and hastening to a lonely little house on their left, found there, in one of the bedrooms, two wounded men. All the doors were standing open; the two unfortunates had succeeded in dragging themselves thus far and had thrown themselves on the beds, and for the two days that they had been alternately shivering and burning, their wounds having received no attention, they had seen no one, not a living soul. They were tortured by a consuming thirst, and the beating of the rain against the window-panes added to their torment, but they could not move hand or foot. Hence, when they heard Silvine approaching, the first word that escaped their lips was: "Drink! Give us to drink!" that longing, pathetic cry, with which the wounded always pursue the by-passer whenever the sound of footsteps arouses them from their lethargy. There were many cases similar to this, where men were overlooked in remote corners, whither they had fled for refuge. Some were picked up even five and six days later, when their sores were filled with maggots and their sufferings had rendered them delirious.

When Silvine had given the wretched men a drink Prosper, who, in the more sorely injured of the twain, had recognized a comrade of his regiment, a chasseur d'Afrique, saw that they could not be far from the ground over which Margueritte's division had charged, inasmuch as the poor devil had been able to drag himself to that house. All the information he could get from him, however, was of the vaguest; yes, it was over that way; you turned to the left, after passing a big field of potatoes.

Immediately she was in possession of this slender clue Silvine insisted on starting out again. An inferior officer of the medical department chanced to pass with a cart just then, collecting the dead; she hailed him and notified him of the presence of the wounded men, then, throwing the donkey's bridle across her arm, urged him along over the muddy road, eager to reach the designated spot, beyond the big potato field. When they had gone some distance she stopped, yielding to her despair.

"My God, where is the place! Where can it be?"

Prosper looked about him, taxing his recollection fruitlessly.

"I told you, it is close beside the place where we made our charge. If only I could find my poor Zephyr-"

And he cast a wistful look on the dead horses that lay around them. It had been his secret hope, his dearest wish, during the entire time they had been wandering over the plateau, to see his mount once more, to bid him a last farewell.

"It ought to be somewhere in this vicinity," he suddenly said. "See! over there to the left, there are the three trees. You see the wheel-tracks? And, look, over yonder is a broken-down caisson. We have found the spot; we are here at last!"

Quivering with emotion, Silvine darted forward and eagerly scanned the faces of two corpses, two artillerymen who had fallen by the roadside.

"He is not here! He is not here! You cannot have seen aright. Yes, that is it; some delusion must have cheated your eyes." And little by little an air-drawn hope, a wild delight crept into her mind. "If you were mistaken, if he should be alive! And be sure he is alive, since he is not here!"

Suddenly she gave utterance to a low, smothered cry. She had turned, and was standing on the very position that the battery had occupied. The scene was most frightful, the ground torn and fissured as by an earthquake and covered with wreckage of every description, the dead lying as they had fallen in every imaginable attitude of horror, arms bent and twisted, legs doubled under them, heads thrown back, the lips parted over the white teeth as if their last breath had been expended in shouting defiance to the foe. A corporal had died with his hands pressed convulsively to his eyes, unable longer to endure the dread spectacle. Some gold coins that a lieutenant carried in a belt about his body had been spilled at the same time as his life-blood, and lay scattered among his entrails. There were Adolphe, the driver, and the gunner, Louis, clasped in each other's arms in a fierce embrace, their sightless orbs starting from their sockets, mated even in death. And there, at last, was Honore, recumbent on his disabled gun as on a bed of honor, with the great rent in his side that had let out his young life, his face, unmutilated and beautiful in its stern anger, still turned defiantly toward the Prussian batteries.

"Oh! my friend," sobbed Silvine, "my friend, my friend–"

She had fallen to her knees on the damp, cold ground, her hands joined as if in prayer, in an outburst of frantic grief. The word friend, the only name by which it occurred to her to address him, told the story of the tender affection she had lost in that man, so good, so loving, who had forgiven her, had meant to make her his wife, despite the ugly past. And now all hope was dead within her bosom, there was nothing left to make life desirable. She had never loved another; she would put away her love for him at the bottom of her heart and hold it sacred there. The rain had ceased; a flock of crows that circled above the three trees, croaking dismally, affected her like a menace of evil. Was he to be taken from her again, her cherished dead, whom she had recovered with such difficulty? She dragged herself along upon her knees, and with a trembling hand brushed away the hungry flies that were buzzing about her friend's wide-open eyes.

She caught sight of a bit of blood-stained paper between Honore's stiffened fingers. It troubled her; she tried to gain possession of the paper, pulling at it gently, but the dead man would not surrender it, seemingly tightening his hold on it, guarding it so jealously that it could not have been taken from him without tearing it in bits. It was the letter she had written him, that he had always carried next his heart, and that he had taken from its hiding place in the moment of his supreme agony, as if to bid her a last farewell. It seemed so strange, was such a revelation, that he should have died thinking of her; when she saw what it was a profound delight filled her soul in the midst of her affliction. Yes, surely, she would leave it with him, the letter that was so dear to him! she would not take it from him, since he was so bent on carrying it with him to the grave. Her tears flowed afresh, but they were beneficent tears this time, and brought healing and comfort with them. She arose and kissed his hands, kissed him on the forehead, uttering meanwhile but that one word, which was in itself a prolonged caress:

"My friend! my friend–"

Meantime the sun was declining; Prosper had gone and taken the counterpane from the cart, and between them they raised Honore's body, slowly, reverently, and laid it on the bed-covering, which they had stretched upon the ground; then, first wrapping him in its folds, they bore him to the cart. It was threatening to rain again, and they had started on their return, forming, with the donkey, a sorrowful little cortege on the broad bosom of the accursed plain, when a deep rumbling as of thunder was heard in the distance. Prosper turned his head and had only time to shout:

"The horses! the horses!"

It was the starving, abandoned cavalry mounts making another charge. They came up this time in a deep mass across a wide, smooth field, manes and tails streaming in the wind, froth flying from their nostrils, and the level rays of the fiery setting sun sent the shadow of the infuriated herd clean across the plateau. Silvine rushed forward and planted herself before the cart, raising her arms above her head as if her puny form might have power to check them. Fortunately the ground fell off just at that point, causing them to swerve to the left; otherwise they would have crushed donkey, cart, and all to powder. The earth trembled, and their hoofs sent a volley of clods and small stones flying through the air, one of which struck the donkey on the head and wounded him. The last that was seen of them they were tearing down a ravine.

"It's hunger that starts them off like that," said Prosper. "Poor beasts!"

Silvine, having bandaged the donkey's ear with her handkerchief, took him again by the bridle, and the mournful little procession began to retrace its steps across the plateau, to cover the two leagues that lay between it and Remilly. Prosper had turned and cast a look on the dead horses, his heart heavy within him to leave the field without having seen Zephyr.

A little below the wood of la Garenne, as they were about to turn off to the left to take the road that they had traversed that morning, they encountered another German post and were again obliged to exhibit their pass. And the officer in command, instead of telling them to avoid Sedan, ordered them to keep straight on their course and pass through the city; otherwise they would be arrested. This was the most recent order; it was not for them to question it. Moreover, their journey would be shortened by a mile and a quarter, which they did not regret, weary and foot-sore as they were.

When they were within Sedan, however, they found their progress retarded owing to a singular cause. As soon as they had passed the fortifications their nostrils were saluted by such a stench, they were obliged to wade through such a mass of abominable filth, reaching almost to their knees, as fairly turned their stomachs. The city, where for three days a hundred thousand men had lived without the slightest provision being made for decency or cleanliness, had become a cesspool, a foul sewer, and this devil's broth was thickened by all sorts of solid matter, rotting hay and straw, stable litter, and the excreta of animals. The carcasses of the horses, too, that were knocked on the head, skinned, and cut up in the public squares, in full view of everyone, had their full share in contaminating the atmosphere; the entrails lay decaying in the hot sunshine, the bones and heads were left lying on the pavement, where they attracted swarms of flies. Pestilence would surely break out in the city unless they made haste to rid themselves of all that carrion, of that stratum of impurity, which, in the Rue de Minil, the Rue Maqua, and even on the Place Turenne, reached a depth of twelve inches. The Prussian authorities had taken the matter up, and their placards were to be seen posted about the city, requisitioning the inhabitants, irrespective of rank, laborers, merchants, bourgeois, magistrates, for the morrow; they were ordered to assemble, armed with brooms and shovels, and apply themselves to the task, and were warned that they would be subjected to heavy penalties if the city was not clean by night. The President of the Tribunal had taken time by the forelock, and might even then be seen scraping away at the pavement before his door and loading the results of his labors upon a wheelbarrow with a fire-shovel.

Silvine and Prosper, who had selected the Grande Rue as their route for traversing the city, advanced but slowly through that lake of malodorous slime. In addition to that the place was in a state of ferment and agitation that made it necessary for them to pull up almost at every moment. It was the time that the Prussians had selected for searching the houses in order to unearth those soldiers, who, determined that they would not give themselves up, had hidden themselves away. When, at about two o'clock of the preceding day, General de Wimpffen had returned from the chateau of Bellevue after signing the capitulation, the report immediately began to circulate that the surrendered troops were to be held under guard in the peninsula of Iges until such time as arrangements could be perfected for sending them off to Germany. Some few officers had expressed their intention of taking advantage of that stipulation which accorded them their liberty conditionally on their signing an agreement not to serve again during the campaign. Only one general, so it was said, Bourgain-Desfeuilles, alleging his rheumatism as a reason, had bound himself by that pledge, and when, that very morning, his carriage had driven up to the door of the Hotel of the Golden Cross and he had taken his seat in it to leave the city, the people had hooted and hissed him unmercifully. The operation of disarming had been going on since break of day; the manner of its performance was, the troops defiled by battalions on the Place Turenne, where each man deposited his musket and bayonet on the pile, like a mountain of old iron, which kept rising higher and higher, in a corner of the place. There was a Prussian detachment there under the command of a young officer, a tall, pale youth, wearing a sky-blue tunic and a cap adorned with a cock's feather, who superintended operations with a lofty but soldier-like air, his hands encased in white gloves. A zouave, in a fit of insubordination, having refused to give up his chassepot, the officer ordered that he be taken away, adding, in the same even tone of voice: "And let him be shot forthwith!" The rest of the battalion continued to defile with a sullen and dejected air, throwing down their arms mechanically, as if in haste to have the ceremony ended. But who could estimate the number of those who had disarmed themselves voluntarily, those whose muskets lay scattered over the country, out yonder on the field of battle? And how many, too, within the last twenty-four hours had concealed themselves, flattering themselves with the hope that they might escape in the confusion that reigned everywhere! There was scarcely a house but had its crew of those headstrong idiots who refused to respond when called on, hiding away in corners and shamming death; the German patrols that were sent through the city even discovered them stowed away under beds. And as many, even after they were unearthed, stubbornly persisted in remaining in the cellars whither they had fled for shelter, the patrols were obliged to fire on them through the coal-holes. It was a man-hunt, a brutal and cruel battue, during which the city resounded with rifle-shots and outlandish oaths.

At the Pont du Meuse they found a throng which the donkey was unable to penetrate and were brought to a stand-still. The officer commanding the guard at the bridge, suspecting they were endeavoring to carry on an illicit traffic in bread or meat, insisted on seeing with his own eyes what was contained in the cart; drawing aside the covering, he gazed for an instant on the corpse with a feeling expression, then motioned them to go their way. Still, however, they were unable to get forward, the crowd momentarily grew denser and denser; one of the first detachments of French prisoners was being conducted to the peninsula of Iges under escort of a Prussian guard. The sorry band streamed on in long array, the men in their tattered, dirty uniforms crowding one another, treading on one another's heels, with bowed heads and sidelong, hang-dog looks, the dejected gait and bearing of the vanquished to whom had been left not even so much as a knife with which to cut their throat. The harsh, curt orders of the guard urging them forward resounded like the cracking of a whip in the silence, which was unbroken save for the plashing of their coarse shoes through the semi-liquid mud. Another shower began to fall, and there could be no more sorrowful sight than that band of disheartened soldiers, shuffling along through the rain, like beggars and vagabonds on the public highway.

All at once Prosper, whose heart was beating as if it would burst his bosom with repressed sorrow and indignation, nudged Silvine and called her attention to two soldiers who were passing at the moment. He had recognized Maurice and Jean, trudging along with their companions, like brothers, side by side. They were near the end of the line, and as there was now no impediment in their way, he was enabled to keep them in view as far as the Faubourg of Torcy, as they traversed the level road which leads to Iges between gardens and truck farms.

"Ah!" murmured Silvine, distressed by what she had just seen, fixing her eyes on Honore's body, "it may be that the dead have the better part!"

Night descended while they were at Wadelincourt, and it was pitchy dark long before they reached Remilly. Father Fouchard was greatly surprised to behold the body of his son, for he had felt certain that it would never be recovered. He had been attending to business during the day, and had completed an excellent bargain; the market price for officers' chargers was twenty francs, and he had bought three for forty-five francs.

2

The crush was so great as the column of prisoners was leaving Torcy that Maurice, who had stopped a moment to buy some tobacco, was parted from Jean, and with all his efforts was unable thereafter to catch up with his regiment through the dense masses of men that filled the road. When he at last reached the bridge that spans the canal which intersects the peninsula of Iges at its base, he found himself in a mixed company of chasseurs d'Afrique and troops of the infanterie de marine.

There were two pieces of artillery stationed at the bridge, their muzzles turned upon the interior of the peninsula; it was a place easy of access, but from which exit would seem to be attended with some difficulties. Immediately beyond the canal was a comfortable house, where the Prussians had established a post, commanded by a captain, upon which devolved the duty of receiving and guarding the prisoners. The formalities observed were not excessive; they merely counted the men, as if they had been sheep, as they came streaming in a huddle across the bridge, without troubling themselves overmuch about uniforms or organizations, after which the prisoners were free of the fields and at liberty to select their dwelling-place wherever chance and the road they were on might direct.

The first thing that Maurice did was to address a question to a Bavarian officer, who was seated astride upon a chair, enjoying a tranquil smoke.

"The 106th of the line, sir, can you tell me where I shall find it?"

Either the officer was unlike most German officers and did not understand French, or thought it a good joke to mystify a poor devil of a soldier. He smiled and raised his hand, indicating by his motion that the other was to keep following the road he was pursuing.

Although Maurice had spent a good part of his life in the neighborhood he had never before been on the peninsula; he proceeded to explore his new surroundings, as a mariner might do when cast by a tempest on the shore of a desolate island. He first skirted the Tour a Glaire, a very handsome country-place, whose small park, situated as it was on the bank of the Meuse, possessed a peculiarly attractive charm. After that the road ran parallel with the river, of which the sluggish current flowed on the right hand at the foot of high, steep banks. The way from there was a gradually ascending one, until it wound around the gentle eminence that occupied the central portion of the peninsula, and there were abandoned quarries there and excavations in the ground, in which a network of narrow paths had their termination. A little further on was a mill, seated on the border of the stream. Then the road curved and pursued a descending course until it entered the village of Iges, which was built on the hillside and connected by a ferry with the further shore, just opposite the rope-walk at Saint-Albert. Last of all came meadows and cultivated fields, a broad expanse of level, treeless country, around which the river swept in a wide, circling bend. In vain had Maurice scrutinized every inch of uneven ground on the hillside; all he could distinguish there was cavalry and artillery, preparing their quarters for the night. He made further inquiries, applying among others to a corporal of chasseurs d'Afrique, who could give him no information. The prospect for finding his regiment looked bad; night was coming down, and, leg-weary and disheartened, he seated himself for a moment on a stone by the wayside.

As he sat there, abandoning himself to the sensation of loneliness and despair that crept over him, he beheld before him, across the Meuse, the accursed fields where he had fought the day but one before. Bitter memories rose to his mind, in the fading light of that day of gloom and rain, as he surveyed the saturated, miry expanse of country that rose from the river's bank and was lost on the horizon. The defile of Saint-Albert, the narrow road by which the Prussians had gained their rear, ran along the bend of the stream as far as the white cliffs of the quarries of Montimont. The summits of the trees in the wood of la Falizette rose in rounded, fleecy masses over the rising ground of Seugnon. Directly before his eyes, a little to the left, was Saint-Menges, the road from which descended by a gentle slope and ended at the ferry; there, too, were the mamelon of Hattoy in the center, and Illy, in the far distance, in the background, and Fleigneux, almost hidden in its shallow vale, and Floing, less remote, on the right. He recognized the plateau where he had spent interminable hours among the cabbages, and the eminences that the reserve artillery had struggled so gallantly to hold, where he had seen Honore meet his death on his dismounted gun. And it was as if the baleful scene were again before him with all its abominations, steeping his mind in horror and disgust, until he was sick at heart.

The reflection that soon it would be quite dark and it would not do to loiter there, however, caused him to resume his researches. He said to himself that perhaps the regiment was encamped somewhere beyond the village on the low ground, but the only ones he encountered there were some prowlers, and he decided to make the circuit of the peninsula, following the bend of the stream. As he was passing through a field of potatoes he was sufficiently thoughtful to dig a few of the tubers and put them in his pockets; they were not ripe, but he had nothing better, for Jean, as luck would have it, had insisted on carrying both the two loaves of bread that Delaherche had given them when they left his house. He was somewhat surprised at the number of horses he met with, roaming about the uncultivated lands, that fell off in an easy descent from the central elevation to the Meuse, in the direction of Donchery. Why should they have brought all those animals with them? how were they to be fed? And now it was night in earnest, and quite dark, when he came to a small piece of woods on the water's brink, in which he was surprised to find the cent-gardes of the Emperor's escort, providing for their creature comforts and drying themselves before roaring fires. These gentlemen, who had a separate encampment to themselves, had comfortable tents; their kettles were boiling merrily, there was a milch cow tied to a tree. It did not take Maurice long to see that he was not regarded with favor in that quarter, poor devil of an infantryman that he was, with his ragged, mud-stained uniform. They graciously accorded him permission to roast his potatoes in the ashes of their fires, however, and he withdrew to the shelter of a tree, some hundred yards away, to eat them. It was no longer raining; the sky was clear, the stars were shining brilliantly in the dark blue vault. He saw that he should have to spend the night in the open air and defer his researches until the morrow. He was so utterly used up that he could go no further; the trees would afford him some protection in case it came on to rain again.

The strangeness of his situation, however, and the thought of his vast prison house, open to the winds of heaven, would not let him sleep. It had been an extremely clever move on the part of the Prussians to select that place of confinement for the eighty thousand men who constituted the remnant of the army of Chalons. The peninsula was approximately three miles long by one wide, affording abundant space for the broken fragments of the vanquished host, and Maurice could not fail to observe that it was surrounded on every side by water, the bend of the Meuse encircling it on the north, east and west, while on the south, at the base, connecting the two arms of the loop at the point where they drew together most closely, was the canal. Here alone was an outlet, the bridge, that was defended by two guns; wherefore it may be seen that the guarding of the camp was a comparatively easy task, notwithstanding its great extent. He had already taken note of the chain of sentries on the farther bank, a soldier being stationed by the waterside at every fifty paces, with orders to fire on any man who should attempt to escape by swimming. In the rear the different posts were connected by patrols of uhlans, while further in the distance, scattered over the broad fields, were the dark lines of the Prussian regiments; a threefold living, moving wall, immuring the captive army.

Maurice, in his sleeplessness, lay gazing with wide-open eyes into the blackness of the night, illuminated here and there by the smoldering watch-fires; the motionless forms of the sentinels were dimly visible beyond the pale ribbon of the Meuse. Erect they stood, duskier spots against the dusky shadows, beneath the faint light of the twinkling stars, and at regular intervals their guttural call came to his ears, a menacing watch-cry that was drowned in the hoarse murmur of the river in the distance. At sound of those unmelodious phrases in a foreign tongue, rising on the still air of a starlit night in the sunny land of France, the vision of the past again rose before him: all that he had beheld in memory an hour before, the plateau of Illy cumbered still with dead, the accursed country round about Sedan that had been the scene of such dire disaster; and resting on the ground in that cool, damp corner of a wood, his head pillowed on a root, he again yielded to the feeling of despair that had overwhelmed him the day before while lying on Delaherche's sofa. And that which, intensifying the suffering of his wounded pride, now harassed and tortured him, was the question of the morrow, the feverish longing to know how deep had been their fall, how great the wreck and ruin sustained by their world of yesterday. The Emperor had surrendered his sword to King William; was not, therefore, the abominable war ended? But he recalled the remark he had heard made by two of the Bavarians of the guard who had escorted the prisoners to Iges: "We're all in France, we're all bound for Paris!" In his semi-somnolent, dreamy state the vision of what was to be suddenly rose before his eyes: the empire overturned and swept away amid a howl of universal execration, the republic proclaimed with an outburst of patriotic fervor, while the legend of '92 would incite men to emulate the glorious past, and, flocking to the standards, drive from the country's soil the hated foreigner with armies of brave volunteers. He reflected confusedly upon all the aspects of the case, and speculations followed one another in swift succession through his poor wearied brain: the harsh terms imposed by the victors, the bitterness of defeat, the determination of the vanquished to resist even to the last drop of blood, the fate of those eighty thousand men, his companions, who were to be captives for weeks, months, years, perhaps, first on the peninsula and afterward in German fortresses. The foundations were giving way, and everything was going down, down to the bottomless depths of perdition.

The call of the sentinels, now loud, now low, seemed to sound more faintly in his ears and to be receding in the distance, when suddenly, as he turned on his hard couch, a shot rent the deep silence. A hollow groan rose on the calm air of night, there was a splashing in the water, the brief struggle of one who sinks to rise no more. It was some poor wretch who had attempted to escape by swimming the Meuse and had received a bullet in his brain.

The next morning Maurice was up and stirring with the sun. The sky was cloudless; he was desirous to rejoin Jean and his other comrades of the company with the least possible delay. For a moment he had an idea of going to see what there was in the interior of the peninsula, then resolved he would first complete its circuit. And on reaching the canal his eyes were greeted with the sight of the 106th-or rather what was left of it-a thousand men, encamped along the river bank among some waste lands, with no protection save a row of slender poplars. If he had only turned to the left the night before instead of pursuing a straight course he could have been with his regiment at once. And he noticed that almost all the line regiments were collected along that part of the bank that extends from the Tour a Glaire to the Chateau of Villette-another bourgeois country place, situated more in the direction of Donchery and surrounded by a few hovels-all of them having selected their bivouac near the bridge, sole issue from their prison, as sheep will instinctively huddle together close to the door of their fold, knowing that sooner or later it will be opened for them.

Jean uttered a cry of pleasure. "Ah, so it's you, at last! I had begun to think you were in the river."

He was there with what remained of the squad, Pache and Lapoulle, Loubet and Chouteau. The last named had slept under doorways in Sedan until the attention of the Prussian provost guard had finally restored them to their regiment. The corporal, moreover, was the only surviving officer of the company, death having taken away Sergeant Sapin, Lieutenant Rochas and Captain Beaudoin, and although the victors had abolished distinction of rank among the prisoners, deciding that obedience was due to the German officers alone, the four men had, nevertheless, rallied to him, knowing him to be a leader of prudence and experience, upon whom they could rely in circumstances of difficulty. Thus it was that peace and harmony reigned among them that morning, notwithstanding the stupidity of some and the evil designs of others. In the first place, the night before he had found them a place to sleep in that was comparatively dry, where they had stretched themselves on the ground, the only thing they had left in the way of protection from the weather being the half of a shelter-tent. After that he had managed to secure some wood and a kettle, in which Loubet made coffee for them, the comforting warmth of which had fortified their stomachs. The rain had ceased, the day gave promise of being bright and warm, they had a small supply of biscuit and bacon left, and then, as Chouteau said, it was a comfort to have no orders to obey, to have their fill of loafing. They were prisoners, it was true, but there was plenty of room to move about. Moreover, they would be away from there in two or three days. Under these circumstances the day, which was Sunday, the 4th, passed pleasantly enough.

Maurice, whose courage had returned to him now that he was with the comrades once more, found nothing to annoy him except the Prussian bands, which played all the afternoon beyond the canal. Toward evening there was vocal music, and the men sang in chorus. They could be seen outside the chain of sentries, walking to and fro in little groups and singing solemn melodies in a loud, ringing voice in honor of the Sabbath.

"Confound those bands!" Maurice at last impatiently exclaimed. "They will drive me wild!"

Jean, whose nerves were less susceptible, shrugged his shoulders.

"Dame! they have reason to feel good; and then perhaps they think it affords us pleasure. It hasn't been such a bad day; don't let's find fault."

As night approached, however, the rain began to fall again. Some of the men had taken possession of what few unoccupied houses there were on the peninsula, others were provided with tents that they erected, but by far the greater number, without shelter of any sort, destitute of blankets even, were compelled to pass the night in the open air, exposed to the pouring rain.

About one o'clock Maurice, who had been sleeping soundly as a result of his fatigue, awoke and found himself in the middle of a miniature lake. The trenches, swollen by the heavy downpour, had overflowed and inundated the ground where he lay. Chouteau's and Loubet's wrath vented itself in a volley of maledictions, while Pache shook Lapoulle, who, unmindful of his ducking, slept through it all as if he was never to wake again. Then Jean, remembering the row of poplars on the bank of the canal, collected his little band and ran thither for shelter; and there they passed the remainder of that wretched night, crouching with their backs to the trees, their legs doubled under them, so as to expose as little of their persons as might be to the big drops.

The next day, and the day succeeding it, the weather was truly detestable, what with the continual showers, that came down so copiously and at such frequent intervals that the men's clothing had not time to dry on their backs. They were threatened with famine, too; there was not a biscuit left in camp, and the coffee and bacon were exhausted. During those two days, Monday and Tuesday, they existed on potatoes that they dug in the adjacent fields, and even

252

those vegetables had become so scarce toward the end of the second day that those soldiers who had money paid as high as five sous apiece for them. It was true that the bugles sounded the call for "distribution"; the corporal had nearly run his legs off trying to be the first to reach a great shed near the Tour a Glaire, where it was reported that rations of bread were to be issued, but on the occasion of a first visit he had waited there three hours and gone away empty-handed, and on a second had become involved in a quarrel with a Bavarian. It was well known that the French officers were themselves in deep distress and powerless to assist their men; had the German staff driven the vanquished army out there in the mud and rain with the intention of letting them starve to death? Not the first step seemed to have been taken, not an effort had been made, to provide for the subsistence of those eighty thousand men in that hell on earth that the soldiers subsequently christened Camp Misery, a name that the bravest of them could never hear mentioned in later days without a shudder.

On his return from his wearisome and fruitless expedition to the shed, Jean forgot his usual placidity and gave way to anger.

"What do they mean by calling us up when there's nothing for us? I'll be hanged if I'll put myself out for them another time!"

And yet, whenever there was a call, he hurried off again. It was inhuman to sound the bugles thus, merely because regulations prescribed certain calls at certain hours, and it had another effect that was near breaking Maurice's heart. Every time that the trumpets sounded the French horses, that were running free on the other side of the canal, came rushing up and dashed into the water to rejoin their squadron, as excited at the well-known sound as they would be at the touch of the spur; but in their exhausted condition they were swept away by the current and few attained the shore. It was a cruel sight to see their struggles; they were drowned in great numbers, and their bodies, decomposing and swelling in the hot sunshine, drifted on the bosom of the canal. As for those of them that got to land, they seemed as if stricken with sudden madness, galloping wildly off and hiding among the waste places of the peninsula.

"More bones for the crows to pick!" sorrowfully said Maurice, remembering the great droves of horses that he had encountered on a previous occasion. "If we remain here a few days we shall all be devouring one another. Poor brutes!"

The night between Tuesday and Wednesday was most terrible of all, and Jean, who was beginning to feel seriously alarmed for Maurice's feverish state, made him wrap himself in an old blanket that they had purchased from a zouave for ten francs, while he, with no protection save his water-soaked capote, cheerfully took the drenching of the deluge which that night pelted down without cessation. Their position under the poplars had become untenable; it was a streaming river of mud, the water rested in deep puddles on the surface of the saturated ground. What was worst of all was that they had to suffer on an empty stomach, the evening meal of the six men having consisted of two beets which they had been compelled to eat raw, having no dry wood to make a fire with, and the sweet taste and refreshing coolness of the vegetables had quickly been succeeded by an intolerable burning sensation. Some cases of dysentery had appeared among the men, caused by fatigue, improper food and the persistent humidity of the atmosphere. More than ten times that night did Jean stretch forth his hand to see that Maurice had not uncovered himself in the movements of his slumber, and thus he kept watch and ward over his friend-his back supported by the same tree-trunk, his legs in a pool of water-with tenderness unspeakable. Since the day that on the plateau of Illy his comrade had carried him off in his arms and saved him from the Prussians he had repaid the debt a hundred-fold. He stopped not to reason on it; it was the free gift of all his being, the total forgetfulness of self for love of the other, the finest, most delicate, grandest

253

exhibition of friendship possible, and that, too, in a peasant, whose lot had always been the lowly one of a tiller of the soil and who had never risen far above the earth, who could not find words to express what he felt, acting purely from instinct, in all simplicity of soul. Many a time already he had taken the food from his mouth, as the men of the squad were wont to say; now he would have divested himself of his skin if with it he might have covered the other, to protect his shoulders, to warm his feet. And in the midst of the savage egoism that surrounded them, among that aggregation of suffering humanity whose worst appetites were inflamed and intensified by hunger, he perhaps owed it to his complete abnegation of self that he had preserved thus far his tranquillity of mind and his vigorous health, for he among them all, his great strength unimpaired, alone maintained his composure and something like a level head.

After that distressful night Jean determined to carry into execution a plan that he had been reflecting over since the day previous.

"See here, little one, we can get nothing to eat, and everyone seems to have forgotten us here in this beastly hole; now unless we want to die the death of dogs, it behooves us to stir about a bit. How are your legs?"

The sun had come out again, fortunately, and Maurice was warmed and comforted.

"Oh, my legs are all right!"

"Then we'll start off on an exploring expedition. We've money in our pockets, and the deuce is in it if we can't find something to buy. And we won't bother our heads about the others; they don't deserve it. Let them take care of themselves."

The truth was that Loubet and Chouteau had disgusted him by their trickiness and low selfishness, stealing whatever they could lay hands on and never dividing with their comrades, while no good was to be got out of Lapoulle, the brute, and Pache, the sniveling devotee.

The pair, therefore, Maurice and Jean, started out by the road along the Meuse which the former had traversed once before, on the night of his arrival. At the Tour a Glaire the park and dwelling-house presented a sorrowful spectacle of pillage and devastation, the trim lawns cut up and destroyed, the trees felled, the mansion dismantled. A ragged, dirty crew of soldiers, with hollow cheeks and eyes preternaturally bright from fever, had taken possession of the place and were living like beasts in the filthy chambers, not daring to leave their quarters for a moment lest someone else might come along and occupy them. A little further on they passed the cavalry and artillery, encamped on the hillsides, once so conspicuous by reason of the neatness and jauntiness of their appearance, now run to seed like all the rest, their organization gone, demoralized by that terrible, torturing hunger that drove the horses wild and sent the men straggling through the fields in plundering bands. Below them, to the right, they beheld an apparently interminable line of artillerymen and chasseurs d'Afrique defiling slowly before the mill; the miller was selling them flour, measuring out two handfuls into their handkerchiefs for a franc. The prospect of the long wait that lay before them, should they take their place at the end of the line, determined them to pass on, in the hope that some better opportunity would present itself at the village of Iges; but great was their consternation when they reached it to find the little place as bare and empty as an Algerian village through which has passed a swarm of locusts; not a crumb, not a fragment of anything eatable, neither bread, nor meat, nor vegetables, the wretched inhabitants utterly destitute. General Lebrun was said to be there, closeted with the mayor. He had been endeavoring, ineffectually, to arrange for an issue of bonds, redeemable at the close of the war, in order to facilitate the victualing of the troops. Money had ceased to have any value when there was nothing that it could purchase. The day before two francs had been paid for a biscuit, seven francs for a bottle of wine, a small glass of brandy was twenty sous, a pipeful of tobacco ten sous. And now officers, sword in hand, had to stand guard before the general's house and the neighboring hovels, for bands of marauders were constantly passing, breaking down doors and stealing even the oil from the lamps and drinking it.

Three zouaves invited Maurice and Jean to join them. Five would do the work more effectually than three.

"Come along. There are horses dying in plenty, and if we can but get some dry wood-"

Then they fell to work on the miserable cabin of a poor peasant, smashing the closet doors, tearing the thatch from the roof. Some officers, who came up on a run, threatened them with their revolvers and put them to flight.

Jean, who saw that the few villagers who had remained at Iges were no better off than the soldiers, perceived he had made a mistake in passing the mill without buying some flour.

"There may be some left; we had best go back."

But Maurice was so reduced from inanition and was beginning to suffer so from fatigue that he left him behind in a sheltered nook among the quarries, seated on a fragment of rock, his face turned upon the wide horizon of Sedan. He, after waiting in line for two long hours, finally returned with some flour wrapped in a piece of rag. And they ate it uncooked, dipping it up in their hands, unable to devise any other way. It was not so very bad; It had no particular flavor, only the insipid taste of dough. Their breakfast, such as it was, did them some good, however. They were even so fortunate as to discover a little pool of rain-water, comparatively pure, in a hollow of a rock, at which they quenched their thirst with great satisfaction.

But when Jean proposed that they should spend the remainder of the afternoon there, Maurice negatived the motion with a great display of violence.

"No, no; not here! I should be ill if I were to have that scene before my eyes for any length of time-" With a hand that trembled he pointed to the remote horizon, the hill of Hattoy, the plateaux of Floing and Illy, the wood of la Garenne, those abhorred, detested fields of slaughter and defeat. "While you were away just now I was obliged to turn my back on it, else I should have broken out and howled with rage. Yes, I should have howled like a dog tormented by boys-you can't imagine how it hurts me; it drives me crazy!"

Jean looked at him in surprise; he could not understand that pride, sensitive as a raw sore, that made defeat so bitter to him; he was alarmed to behold in his eyes that wandering, flighty look that he had seen there before. He affected to treat the matter lightly.

"Good! we'll seek another country; that's easy enough to do."

Then they wandered as long as daylight lasted, wherever the paths they took conducted them. They visited the level portion of the peninsula in the hope of finding more potatoes there, but the artillerymen had obtained a plow and turned up the ground, and not a single potato had escaped their sharp eyes. They retraced their steps, and again they passed through throngs of listless, glassy-eyed, starving soldiers, strewing the ground with their debilitated forms, falling by hundreds in the bright sunshine from sheer exhaustion. They were themselves many times overcome by fatigue and forced to sit down and rest; then their deep-seated sensation of suffering would bring them to their feet again and they would recommence their wandering, like animals impelled by instinct to move on perpetually in quest of pasturage. It seemed to them to last for years, and yet the moments sped by rapidly. In the more inland region, over Donchery way, they received a fright from the horses and sought the protection of a wall, where they remained a long time, too exhausted to rise, watching with vague, lack-luster eyes the wild course of the crazed beasts as they raced athwart the red western sky where the sun was sinking.

As Maurice had foreseen, the thousands of horses that shared the captivity of the army, and for which it was impossible to provide forage, constituted a peril that grew greater day by day. At first they had nibbled the vegetation and gnawed the bark off trees, then had attacked the fences and whatever wooden structures they came across, and now they seemed ready to devour one another. It was a frequent occurrence to see one of them throw himself upon another and tear out great tufts from his mane or tail, which he would grind between his teeth, slavering meanwhile at the mouth profusely. But it was at night that they became most terrible, as if they were visited by visions of terror in the darkness. They collected in droves, and, attracted by the straw, made furious rushes upon what few tents there were, overturning and demolishing them. It was to no purpose that the men built great fires to keep them away; the device only served to madden them the more. Their shrill cries were so full of anguish, so dreadful to the ear, that they might have been mistaken for the howls of wild beasts. Were they driven away, they returned, more numerous and fiercer than before. Scarce a moment passed but out in the darkness could be heard the shriek of anguish of some unfortunate soldier whom the crazed beasts had crushed in their wild stampede.

The sun was still above the horizon when Jean and Maurice, on their way back to the camp, were astonished by meeting with the four men of the squad, lurking in a ditch, apparently for no good purpose. Loubet hailed them at once, and Chouteau constituted himself spokesman:

"We are considering ways and means for dining this evening. We shall die if we go on this way; it is thirty-six hours since we have had anything to put in our stomach-so, as there are horses plenty, and horse-meat isn't such bad eating-"

"You'll join us, won't you, corporal?" said Loubet, interrupting, "for, with such a big, strong animal to handle, the more of us there are the better it will be. See, there is one, off yonder, that we've been keeping an eye on for the last hour; that big bay that is in such a bad way. He'll be all the easier to finish."

And he pointed to a horse that was dying of starvation, on the edge of what had once been a field of beets. He had fallen on his flank, and every now and then would raise his head and look about him pleadingly, with a deep inhalation that sounded like a sigh.

"Ah, how long we have to wait!" grumbled Lapoulle, who was suffering torment from his fierce appetite. "I'll go and kill him-shall I?"

But Loubet stopped him. Much obliged! and have the Prussians down on them, who had given notice that death would be the penalty for killing a horse, fearing that the carcass would breed a pestilence. They must wait until it was dark. And that was the reason why the four men were lurking in the ditch, waiting, with glistening, hungry eyes fixed on the dying brute.

"Corporal," asked Pache, in a voice that faltered a little, "you have lots of ideas in your head; couldn't you kill him painlessly?"

Jean refused the cruel task with a gesture of disgust. What, kill that poor beast that was even then in its death agony! oh, no, no! His first impulse had been to fly and take Maurice with him, that neither of them might be concerned in the revolting butchery; but looking at his companion and beholding him so pale and faint, he reproached himself for such an excess of sensibility. What were animals created for after all, mon Dieu, unless to afford sustenance to man! They could not allow themselves to starve when there was food within reach. And it rejoiced him to see Maurice cheer up a little at the prospect of eating; he said in his easy, good-natured way:

"Faith, you're wrong there; I've no ideas in my head, and if he has got to be killed without pain-"

"Oh! that's all one to me," interrupted Lapoulle. "I'll show you."

The two newcomers seated themselves in the ditch and joined the others in their expectancy. Now and again one of the men would rise and make certain that the horse was still there, its neck outstretched to catch the cool exhalations of the Meuse and the last rays of the setting sun, as if bidding farewell to life. And when at last twilight crept slowly o'er the scene the six men were erect upon their feet, impatient that night was so tardy in its coming, casting furtive, frightened looks about them to see they were not observed.

"Ah, zut!" exclaimed Chouteau, "the time is come!"

Objects were still discernible in the fields by the uncertain, mysterious light "between dog and wolf," and Lapoulle went forward first, followed by the five others. He had taken from the ditch a large, rounded boulder, and, with it in his two brawny hands, rushing upon the horse, commenced to batter at his skull as with a club. At the second blow, however, the horse, stung by the pain, attempted to get on his feet. Chouteau and Loubet had thrown themselves across his legs and were endeavoring to hold him down, shouting to the others to help them. The poor brute's cries were almost human in their accent of terror and distress; he struggled desperately to shake off his assailants, and would have broken them like a reed had he not been half dead with inanition. The movements of his head prevented the blows from taking effect; Lapoulle was unable to despatch him.

"Nom de Dieu! how hard his bones are! Hold him, somebody, until I finish him."

Jean and Maurice stood looking at the scene in silent horror; they heard not Chouteau's appeals for assistance; were powerless to raise a hand. And Pache, in a sudden outburst of piety and pity, dropped on his knees, joined his hands, and began to mumble the prayers that are repeated at the bedside of the dying.

"Merciful God, have pity on him. Let him, good Lord, depart in peace~"

Again Lapoulle struck ineffectually, with no other effect than to destroy an ear of the wretched creature, that threw back its head and gave utterance to a loud, shrill scream.

"Hold on!" growled Chouteau; "this won't do; he'll get us all in the lockup. We must end the matter. Hold him fast, Loubet."

He took from his pocket a penknife, a small affair of which the blade was scarcely longer than a man's finger, and casting himself prone on the animal's body and passing an arm about its neck, began to hack away at the live flesh, cutting away great morsels, until he found and severed the artery. He leaped quickly to one side; the blood spurted forth in a torrent, as when the plug is removed from a fountain, while the feet stirred feebly and convulsive movements ran along the skin, succeeding one another like waves of the sea. It was near five minutes before the horse was dead. His great eyes, dilated wide and filled with melancholy and affright, were fixed upon the wan-visaged men who stood waiting for him to die; then they grew dim and the light died from out them.

"Merciful God," muttered Pache, still on his knees, "keep him in thy holy protection-succor him, Lord, and grant him eternal rest."

Afterward, when the creature's movements had ceased, they were at a loss to know where the best cut lay and how they were to get at it. Loubet, who was something of a Jack-of-all-trades, showed them what was to be done in order to secure the loin, but as he was a tyro at the butchering business and, moreover, had only his small penknife to work with, he quickly lost his way amid the warm, quivering flesh. And Lapoulle, in his impatience, having attempted to be of assistance by making an incision in the belly, for which there was no necessity whatever, the scene of bloodshed became truly sickening. They wallowed in the gore and entrails that covered the ground about them, like a pack of ravening wolves collected around the carcass of

257

their prey, fleshing their keen fangs in it.

"I don't know what cut that may be," Loubet said at last, rising to his feet with a huge lump of meat in his hands, "but by the time we've eaten it, I don't believe any of us will be hungry."

Jean and Maurice had averted their eyes in horror from the disgusting spectacle; still, however, the pangs of hunger were gnawing at their vitals, and when the band slunk rapidly away, so as not to be caught in the vicinity of the incriminating carcass, they followed it. Chouteau had discovered three large beets, that had somehow been overlooked by previous visitors to the field, and carried them off with him. Loubet had loaded the meat on Lapoulle's shoulders so as to have his own arms free, while Pache carried the kettle that belonged to the squad, which they had brought with them on the chance of finding something to cook in it. And the six men ran as if their lives were at stake, never stopping to take breath, as if they heard the pursuers at their heels.

Suddenly Loubet brought the others to a halt.

"It's idiotic to run like this; let's decide where we shall go to cook the stuff."

Jean, who was beginning to recover his self-possession, proposed the quarries. They were only three hundred yards distant, and in them were secret recesses in abundance where they could kindle a fire without being seen. When they reached the spot, however, difficulties of every description presented themselves. First, there was the question of wood; fortunately a laborer, who had been repairing the road, had gone home and left his wheelbarrow behind him; Lapoulle quickly reduced it to fragments with the heel of his boot. Then there was no water to be had that was fit to drink; the hot sunshine had dried up all the pools of rain-water. True there was a pump at the Tour a Glaire, but that was too far away, and besides it was never accessible before midnight; the men forming in long lines with their bowls and porringers, only too happy when, after waiting for hours, they could escape from the jam with their supply of the precious fluid unspilled. As for the few wells in the neighborhood, they had been dry for the last two days, and the bucket brought up nothing save mud and slime. Their sole resource appeared to be the water of the Meuse, which was parted from them by the road.

"I'll take the kettle and go and fill it," said Jean.

The others objected.

"No, no! We don't want to be poisoned; it is full of dead bodies!"

They spoke the truth. The Meuse was constantly bringing down corpses of men and horses; they could be seen floating with the current at any moment of the day, swollen and of a greenish hue, in the early stages of decomposition. Often they were caught in the weeds and bushes on the bank, where they remained to poison the atmosphere, swinging to the tide with a gentle, tremulous motion that imparted to them a semblance of life. Nearly every soldier who had drunk that abominable water had suffered from nausea and colic, often succeeded afterward by dysentery. It seemed as if they must make up their mind to use it, however, as there was no other; Maurice explained that there would be no danger in drinking it after it was boiled.

"Very well, then; I'll go," said Jean. And he started, taking Lapoulle with him to carry the kettle.

By the time they got the kettle filled and on the fire it was quite dark. Loubet had peeled the beets and thrown them into the water to cook-a feast fit for the gods, he declared it would be-and fed the fire with fragments of the wheelbarrow, for they were all suffering so from hunger that they could have eaten the meat before the pot began to boil. Their huge shadows danced fantastically in the firelight on the rocky walls of the quarry. Then they found it impossible longer to restrain their appetite, and threw themselves upon the unclean mess, tearing the flesh with eager, trembling fingers and dividing it among them, too impatient even to make use of the knife. But, famishing as they were, their stomachs revolted; they felt the want of salt, they could not swallow that tasteless, sickening broth, those chunks of half-cooked, viscid meat that had a taste like clay. Some among them had a fit of vomiting. Pache was very ill. Chouteau and Loubet heaped maledictions on that infernal old nag, that had caused them such trouble to get him to the pot and then given them the colic. Lapoulle was the only one among them who ate abundantly, but he was in a very bad way that night when, with his three comrades, he returned to their resting-place under the poplars by the canal.

On their way back to camp Maurice, without uttering a word, took advantage of the darkness to seize Jean by the arm and drag him into a by-path. Their comrades inspired him with unconquerable disgust; he thought he should like to go and sleep in the little wood where he had spent his first night on the peninsula. It was a good idea, and Jean commended it highly when he had laid himself down on the warm, dry ground, under the shelter of the dense foliage. They remained there until the sun was high in the heavens, and enjoyed a sound, refreshing slumber, which restored to them something of their strength.

The following day was Thursday, but they had ceased to note the days; they were simply glad to observe that the weather seemed to be coming off fine again. Jean overcame Maurice's repugnance and prevailed on him to return to the canal, to see if their regiment was not to move that day. Not a day passed now but detachments of prisoners, a thousand to twelve hundred strong, were sent off to the fortresses in Germany. The day but one before they had seen, drawn up in front of the Prussian headquarters, a column of officers of various grades, who were going to Pont-a-Mousson, there to take the railway. Everyone was possessed with a wild, feverish longing to get away from that camp where they had seen such suffering. Ah! if it but might be their turn! And when they found the 106th still encamped on the bank of the canal, in the inevitable disorder consequent upon such distress, their courage failed them and they despaired.

Jean and Maurice that day thought they saw a prospect of obtaining something to eat. All the morning a lively traffic had been going on between the prisoners and the Bavarians on the other side of the canal; the former would wrap their money in a handkerchief and toss it across to the opposite shore, the latter would return the handkerchief with a loaf of coarse brown bread, or a plug of their common, damp tobacco. Even soldiers who had no money were not debarred from participating in this commerce, employing, instead of currency, their white uniform gloves, for which the Germans appeared to have a weakness. For two hours packages were flying across the canal in its entire length under this primitive system of exchanges. But when Maurice dispatched his cravat with a five-franc piece tied in it to the other bank, the Bavarian who was to return him a loaf of bread gave it, whether from awkwardness or malice, such an ineffectual toss that it fell in the water. The incident elicited shouts of laughter from the Germans. Twice again Maurice repeated the experiment, and twice his loaf went to feed the fishes. At last the Prussian officers, attracted by the uproar, came running up and prohibited their men from selling anything to the prisoners, threatening them with dire penalties and punishments in case of disobedience. The traffic came to a sudden end, and Jean had hard work to pacify Maurice, who shook his fists at the scamps, shouting to them to give him back his five-franc pieces.

This was another terrible day, notwithstanding the warm, bright sunshine. Twice the bugle sounded and sent Jean hurrying off to the shed whence rations were supposed to be issued, but on each occasion he only got his toes trod on and his ribs racked in the crush. The Prussians, whose organization was so wonderfully complete, continued to manifest the same brutal inattention to the necessities of the vanquished army. On the representations of Generals Douay and Lebrun, they had indeed sent in a few sheep as well as some wagon-loads of bread, but so little care was taken to guard them that the sheep were carried off bodily and the wagons pillaged as soon as they reached the bridge, the consequence of which was that the troops who were encamped a hundred yards further on were no better off than before; it was only the worst element, the plunderers and bummers, who benefited by the provision trains. And thereon Jean, who, as he said, saw how the trick was done, brought Maurice with him to the bridge to keep an eye on the victuals.

It was four o'clock, and they had not had a morsel to eat all that beautiful bright Thursday, when suddenly their eyes were gladdened by the sight of Delaherche. A few among the citizens of Sedan had with infinite difficulty obtained permission to visit the prisoners, to whom they carried provisions, and Maurice had on several occasions expressed his surprise at his failure to receive any tidings of his sister. As soon as they recognized Delaherche in the distance, carrying a large basket and with a loaf of bread under either arm, they darted forward fast as their legs could carry them, but even thus they were too late; a crowding, jostling mob closed in, and in the confusion the dazed manufacturer was relieved of his basket and one of his loaves, which vanished from his sight so expeditiously that he was never able to tell the manner of their disappearance.

"Ah, my poor friends!" he stammered, utterly crestfallen in his bewilderment and stupefaction, he who but a moment before had come through the gate with a smile on his lips and an air of good-fellowship, magnanimously forgetting his superior advantages in his desire for popularity.

Jean had taken possession of the remaining loaf and saved it from the hungry crew, and while he and Maurice, seated by the roadside, were making great inroads in it, Delaherche opened his budget of news for their benefit. His wife, the Lord be praised! was very well, but he was greatly alarmed for the colonel, who had sunk into a condition of deep prostration, although his mother continued to bear him company from morning until night.

"And my sister?" Maurice inquired.

"Ah, yes! your sister; true. She insisted on coming with me; it was she who brought the two loaves of bread. She had to remain over yonder, though, on the other side of the canal; the sentries wouldn't let her pass the gate. You know the Prussians have strictly prohibited the presence of women in the peninsula."

Then he spoke of Henriette, and of her fruitless attempts to see her brother and come to his assistance. Once in Sedan chance had brought her face to face with Cousin Gunther, the man who was captain in the Prussian Guards. He had passed her with his haughty, supercilious air, pretending not to recognize her. She, also, with a sensation of loathing, as if she were in the presence of one of her husband's murderers, had hurried on with quickened steps; then, with a sudden change of purpose for which she could not account, had turned back and told him all the manner of Weiss's death, in harsh accents of reproach. And he, thus learning how horribly a relative had met his fate, had taken the matter coolly; it was the fortune of war; the same thing might have happened to himself. His face, rendered stoically impassive by the discipline of the soldier, had barely betrayed the faintest evidence of interest. After that, when she informed him that her brother was a prisoner and besought him to use his influence to obtain for her an opportunity of seeing him, he had excused himself on the ground that he was powerless in the matter; the instructions were explicit and might not be disobeyed. He appeared to place the regimental orderly book on a par with the Bible. She left him with the clearly defined impression that he believed he was in the country for the sole purpose of sitting in judgment on the French people, with all the intolerance and arrogance of the hereditary enemy, swollen by his personal hatred for the nation whom it had devolved on him to chastise.

"And now," said Delaherche in conclusion, "you won't have to go to bed supperless to-night;

you have had a little something to eat. The worst is that I am afraid I shall not be able to secure another pass."

He asked them if there was anything he could do for them outside, and obligingly consented to take charge of some pencil-written letters confided to him by other soldiers, for the Bavarians had more than once been seen to laugh as they lighted their pipes with missives which they had promised to forward. Then, when Jean and Maurice had accompanied him to the gate, he exclaimed:

"Look! over yonder, there's Henriette! Don't you see her waving her handkerchief?"

True enough, among the crowd beyond the line of sentinels they distinguished a little, thin, pale face, a white dot that trembled in the sunshine. Both were deeply affected, and, with moist eyes, raising their hands above their head, answered her salutation by waving them frantically in the air.

The following day was Friday, and it was then that Maurice felt that his cup of horror was full to overflowing. After another night of tranquil slumber in the little wood he was so fortunate as to secure another meal, Jean having come across an old woman at the Chateau of Villette who was selling bread at ten francs the pound. But that day they witnessed a spectacle of which the horror remained imprinted on their minds for many weeks and months.

The day before Chouteau had noticed that Pache had ceased complaining and was going about with a careless, satisfied air, as a man might do who had dined well. He immediately jumped at the conclusion that the sly fox must have a concealed treasure somewhere, the more so that he had seen him absent himself for near an hour that morning and come back with a smile lurking on his face and his mouth filled with unswallowed food. It must be that he had had a windfall, had probably joined some marauding party and laid in a stock of provisions. And Chouteau labored with Loubet and Lapoulle to stir up bad feeling against the comrade, with the latter more particularly. Hein! wasn't he a dirty dog, if he had something to eat, not to go snacks with the comrades! He ought to have a lesson that he would remember, for his selfishness.

"To-night we'll keep a watch on him, don't you see. We'll learn whether he dares to stuff himself on the sly, when so many poor devils are starving all around him."

"Yes, yes, that's the talk! we'll follow him," Lapoulle angrily declared. "We'll see about it!"

He doubled his fists; he was like a crazy man whenever the subject of eating was mentioned in his presence. His enormous appetite caused him to suffer more than the others; his torment at times was such that he had been known to stuff his mouth with grass. For more than thirty-six hours, since the night when they had supped on horseflesh and he had contracted a terrible dysentery in consequence, he had been without food, for he was so little able to look out for himself that, notwithstanding his bovine strength, whenever he joined the others in a marauding raid he never got his share of the booty. He would have been willing to give his blood for a pound of bread.

As it was beginning to be dark Pache stealthily made his way to the Tour a Glaire and slipped into the park, while the three others cautiously followed him at a distance.

"It won't do to let him suspect anything," said Chouteau. "Be on your guard in case he should look around."

But when he had advanced another hundred paces Pache evidently had no idea there was anyone near, for he began to hurry forward at a swift gait, not so much as casting a look behind. They had no difficulty in tracking him to the adjacent quarries, where they fell on him as he was in the act of removing two great flat stones, to take from the cavity beneath part of a loaf of bread. It was the last of his store; he had enough left for one more meal.

261

"You dirty, sniveling priest's whelp!" roared Lapoulle, "so that is why you sneak away from us! Give me that; it's my share!"

Why should he give his bread? Weak and puny as he was, his slight form dilated with anger, while he clutched the loaf against his bosom with all the strength he could master. For he also was hungry.

"Let me alone. It's mine."

Then, at sight of Lapoulle's raised fist, he broke away and ran, sliding down the steep banks of the quarries, making his way across the bare fields in the direction of Donchery, the three others after him in hot pursuit. He gained on them, however, being lighter than they, and possessed by such overmastering fear, so determined to hold on to what was his property, that his speed seemed to rival the wind. He had already covered more than half a mile and was approaching the little wood on the margin of the stream when he encountered Jean and Maurice, who were on their way back to their resting-place for the night. He addressed them an appealing, distressful cry as he passed; while they, astounded by the wild hunt that went fleeting by, stood motionless at the edge of a field, and thus it was that they beheld the ensuing tragedy.

As luck would have it, Pache tripped over a stone and fell. In an instant the others were on top of him-shouting, swearing, their passion roused to such a pitch of frenzy that they were like wolves that had run down their prey.

"Give me that," yelled Lapoulle, "or by G-d I'll kill you!"

And he had raised his fist again when Chouteau, taking from his pocket the penknife with which he had slaughtered the horse and opening it, placed it in his hand.

"Here, take it! the knife!"

But Jean meantime had come hurrying up, desirous to prevent the mischief he saw brewing, losing his wits like the rest of them, indiscreetly speaking of putting them all in the guardhouse; whereon Loubet, with an ugly laugh, told him he must be a Prussian, since they had no longer any commanders, and the Prussians were the only ones who issued orders.

"Nom de Dieu!" Lapoulle repeated, "will you give me that?"

Despite the terror that blanched his cheeks Pache hugged the bread more closely to his bosom, with the obstinacy of the peasant who never cedes a jot or tittle of that which is his.

"No!"

Then in a second all was over; the brute drove the knife into the other's throat with such violence that the wretched man did not even utter a cry. His arms relaxed, the bread fell to the ground, into the pool of blood that had spurted from the wound.

At sight of the imbecile, uncalled-for murder, Maurice, who had until then been a silent spectator of the scene, appeared as if stricken by a sudden fit of madness. He raved and gesticulated, shaking his fist in the face of the three men and calling them murderers, assassins, with a violence that shook his frame from head to foot. But Lapoulle seemed not even to hear him. Squatted on the ground beside the corpse, he was devouring the bloodstained bread, an expression of stupid ferocity on his face, with a loud grinding of his great jaws, while Chouteau and Loubet, seeing him thus terrible in the gratification of his wild-beast appetite, did not even dare claim their portion.

By this time night had fallen, a pleasant night with a clear sky thick-set with stars, and Maurice and Jean, who had regained the shelter of their little wood, presently perceived Lapoulle wandering up and down the river bank. The two others had vanished, had doubtless returned to the encampment by the canal, their mind troubled by reason of the corpse they left behind them. He, on the other hand, seemed to dread going to rejoin the comrades. When he was more himself and his brutish, sluggish intellect showed him the full extent of his crime, he had evidently experienced a twinge of anguish that made motion a necessity, and not daring to return to the interior of the peninsula, where he would have to face the body of his victim, had sought the bank of the stream, where he was now tramping to and fro with uneven, faltering steps. What was going on within the recesses of that darkened mind that guided the actions of that creature, so degraded as to be scarce higher than the animal? Was it the awakening of remorse? or only the fear lest his crime might be discovered? He could not remain there; he paced his beat as a wild beast shambles up and down its cage, with a sudden and ever-increasing longing to fly, a longing that ached and pained like a physical hurt, from which he felt he should die, could he do nothing to satisfy it. Quick, quick, he must fly, must fly at once, from that prison where he had slain a fellow-being. And yet, the coward in him, it may be, gaining the supremacy, he threw himself on the ground, and for a long time lay crouched among the herbage.

And Maurice said to Jean in his horror and disgust:

"See here, I cannot remain longer in this place; I tell you plainly I should go mad. I am surprised that the physical part of me holds out as it does; my bodily health is not so bad, but the mind is going; yes! it is going, I am certain of it. If you leave me another day in this hell I am lost. I beg you, let us go away, let us start at once!"

And he went on to propound the wildest schemes for getting away. They would swim the Meuse, would cast themselves on the sentries and strangle them with a cord he had in his pocket, or would beat out their brains with rocks, or would buy them over with the money they had left and don their uniform to pass through the Prussian lines.

"My dear boy, be silent!" Jean sadly answered; "it frightens me to hear you talk so wildly. Is there any reason in what you say, are any of your plans feasible? Wait; to-morrow we'll see about it. Be silent!"

He, although his heart, no less than his friend's, was wrung by the horrors that surrounded them on every side, had preserved his mental balance amid the debilitating effects of famine, among the grisly visions of that existence than which none could approach more nearly the depth of human misery. And as his companion's frenzy continued to increase and he talked of casting himself into the Meuse, he was obliged to restrain him, even to the point of using violence, scolding and supplicating, tears standing in his eyes. Then suddenly he said:

"See! look there!"

A splash was heard coming from the river, and they saw it was Lapoulle, who had finally decided to attempt to escape by the stream, first removing his capote in order that it might not hinder his movements; and his white shirt made a spot of brightness that was distinctly visible upon the dusky bosom of the moving water. He was swimming up-stream with a leisurely movement, doubtless on the lookout for a place where he might land with safety, while on the opposite shore there was no difficulty in discerning the shadowy forms of the sentries, erect and motionless in the semi-obscurity. There came a sudden flash that tore the black veil of night, a report that went with bellowing echoes and spent itself among the rocks of Montimont. The water boiled and bubbled for an instant, as it does under the wild efforts of an unpracticed oarsman. And that was all; Lapoulle's body, the white spot on the dusky stream, floated away, lifeless, upon the tide.

The next day, which was Saturday, Jean aroused Maurice as soon as it was day and they returned to the camp of the 106th, with the hope that they might move that day, but there were no orders; it seemed as though the regiment's existence were forgotten. Many of the troops had been sent away, the peninsula was being depopulated, and sickness was terribly prevalent among those who were left behind. For eight long days disease had been germinating in that hell on earth; the rains had ceased, but the blazing, scorching sunlight had only wrought a change of evils. The excessive heat completed the exhaustion of the men and gave to the numerous cases of dysentery an alarmingly epidemic character. The excreta of that army of sick poisoned the air with their noxious emanations. No one could approach the Meuse or the canal, owing to the overpowering stench that rose from the bodies of drowned soldiers and horses that lay festering among the weeds. And the horses, that dropped in the fields from inanition, were decomposing so rapidly and forming such a fruitful source of pestilence that the Prussians, commencing to be alarmed on their own account, had provided picks and shovels and forced the prisoners to bury them.

That day, however, was the last on which they suffered from famine. As their numbers were so greatly reduced and provisions kept pouring in from every quarter, they passed at a single bound from the extreme of destitution to the most abundant plenty. Bread, meat, and wine, even, were to be had without stint; eating went on from morning till night, until they were ready to drop. Darkness descended, and they were eating still; in some quarters the gorging was continued until the next morning. To many it proved fatal.

That whole day Jean made it his sole business to keep watch over Maurice, who he saw was ripe for some rash action. He had been drinking; he spoke of his intention of cuffing a Prussian officer in order that he might be sent away. And at night Jean, having discovered an unoccupied corner in the cellar of one of the outbuildings at the Tour a Glaire, thought it advisable to go and sleep there with his companion, thinking that a good night's rest would do him good, but it turned out to be the worst night in all their experience, a night of terror during which neither of them closed an eye. The cellar was inhabited by other soldiers; lying in the same corner were two who were dying of dysentery, and as soon as it was fairly dark they commenced to relieve their sufferings by moans and inarticulate cries, a hideous death-rattle that went on uninterruptedly until morning. These sounds finally became so horrific there in the intense darkness, that the others who were resting there, wishing to sleep, allowed their anger to get the better of them and shouted to the dying men to be silent. They did not hear; the rattle went on, drowning all other sounds, while from without came the drunken clamor of those who were eating and drinking still, with insatiable appetite.

Then commenced for Maurice a period of agony unspeakable. He would have fled from the awful sounds that brought the cold sweat of anguish in great drops to his brow, but when he arose and attempted to grope his way out he trod on the limbs of those extended there, and finally fell to the ground, a living man immured there in the darkness with the dying. He made no further effort to escape from this last trial. The entire frightful disaster arose before his mind, from the time of their departure from Rheims to the crushing defeat of Sedan. It seemed to him that in that night, in the inky blackness of that cellar, where the groans of two dying soldiers drove sleep from the eyelids of their comrades, the ordeal of the army of Chalons had reached its climax. At each of the stations of its passion the army of despair, the expiatory band, driven forward to the sacrifice, had spent its life-blood in atonement for the faults of others; and now, unhonored amid disaster, covered with contumely, it was enduring martyrdom in that cruel scourging, the severity of which it had done nothing to deserve. He felt it was too much; he was heartsick with rage and grief, hungering for justice, burning with a fierce desire to be avenged on destiny.

When daylight appeared one of the soldiers was dead, the other was lingering on in protracted agony.

"Come along, little one," Jean gently said; "we'll go and get a breath of fresh air; it will do us good."

But when the pair emerged into the pure, warm morning air and, pursuing the river bank, were near the village of Iges, Maurice grew flightier still, and extending his hand toward the vast expanse of sunlit battlefield, the plateau of Illy in front of them, Saint-Menges to the left, the wood of la Garenne to the right, he cried:

"No, I cannot, I cannot bear to look on it! The sight pierces my heart and drives me mad. Take me away, oh! take me away, at once, at once!"

It was Sunday once more; the bells were pealing from the steeples of Sedan, while the music of a German military band floated on the air in the distance. There were still no orders for their regiment to move, and Jean, alarmed to see Maurice's deliriousness increasing, determined to attempt the execution of a plan that he had been maturing in his mind for the last twenty-four hours. On the road before the tents of the Prussians another regiment, the 5th of the line, was drawn up in readiness for departure. Great confusion prevailed in the column, and an officer, whose knowledge of the French language was imperfect, had been unable to complete the roster of the prisoners. Then the two friends, having first torn from their uniform coat the collar and buttons in order that the number might not betray their identity, quietly took their place in the ranks and soon had the satisfaction of crossing the bridge and leaving the chain of sentries behind them. The same idea must have presented itself to Loubet and Chouteau, for they caught sight of them somewhat further to the rear, peering anxiously about them with the guilty eyes of murderers.

Ah, what comfort there was for them in that first blissful moment! Outside their prison the sunlight was brighter, the air more bracing; it was like a resurrection, a bright renewal of all their hopes. Whatever evil fortune might have in store for them, they dreaded it not; they snapped their fingers at it in their delight at having seen the last of the horrors of Camp Misery.

3

That morning Maurice and Jean listened for the last time to the gay, ringing notes of the French bugles, and now they were on their way to Pont-a-Mousson, marching in the ranks of the convoy of prisoners, which was guarded front and rear by platoons of Prussian infantry, while a file of men with fixed bayonets flanked the column on either side. Whenever they came to a German post they heard only the lugubrious, ear-piercing strains of the Prussian trumpets.

Maurice was glad to observe that the column took the left-hand road and would pass through Sedan; perhaps he would have an opportunity of seeing his sister Henriette. All the pleasure, however, that he had experienced at his release from that foul cesspool where he had spent nine days of agony was dashed to the ground and destroyed during the three-mile march from the peninsula of Iges to the city. It was but another form of his old distress to behold that array of prisoners, shuffling timorously through the dust of the road, like a flock of sheep with the dog at their heels. There is no spectacle in all the world more pitiful than that of a column of vanquished troops being marched off into captivity under guard of their conquerors, without arms, their empty hands hanging idly at their sides; and these men, clad in rags and tatters, besmeared with the filth in which they had lain for more than a week, gaunt and wasted after their long fast, were more like vagabonds than soldiers; they resembled loathsome, horribly dirty tramps, whom the gendarmes would have picked up along the highways and consigned to the lockup. As they passed through the Faubourg of Torcy, where men paused on the sidewalks and women came to their doors to regard them with mournful, compassionate interest, the blush of shame rose to Maurice's cheek, he hung his head and a bitter taste came to his mouth.

Jean, whose epidermis was thicker and mind more practical, thought only of their stupidity in not having brought off with them a loaf of bread apiece. In the hurry of their abrupt departure they had even gone off without breakfasting, and hunger soon made its presence felt by the nerveless sensation in their legs. Others among the prisoners appeared to be in the same boat, for they held out money, begging the people of the place to sell them something to eat. There was one, an extremely tall man, apparently very ill, who displayed a gold piece, extending it above the heads of the soldiers of the escort; and he was almost frantic that he could purchase nothing. Just at that time Jean, who had been keeping his eyes open, perceived a bakery a short distance ahead, before which were piled a dozen loaves of bread; he immediately got his money ready and, as the column passed, tossed the baker a five-franc piece and endeavored to secure two of the loaves; then, when the Prussian who was marching at his side pushed him back roughly into the ranks, he protested, demanding that he be allowed to recover his money from the baker. But at that juncture the captain commanding the detachment, a short, bald-headed man with a brutal expression of face, came hastening up; he raised his revolver over Jean's head as if about to strike him with the butt, declaring with an oath that he would brain the first man that dared to lift a finger. And the rest of the captives continued to shamble on, stirring up the dust of the road with their shuffling feet, with eyes averted and shoulders bowed, cowed and abjectly submissive as a drove of cattle.

"Oh! how good it would seem to slap the fellow's face just once!" murmured Maurice, as if he meant it. "How I should like to let him have just one from the shoulder, and drive his teeth down his dirty throat!"

And during the remainder of their march he could not endure to look on that captain, with his ugly, supercilious face.

They had entered Sedan and were crossing the Pont de Meuse, and the scenes of violence and brutality became more numerous than ever. A woman darted forward and would have embraced a boyish young sergeant -likely she was his mother-and was repulsed with a blow from a musket-butt that felled her to the ground. On the Place Turenne the guards hustled and maltreated some citizens because they cast provisions to the prisoners. In the Grande Rue one of the convoy fell in endeavoring to secure a bottle that a lady extended to him, and was assisted to his feet with kicks. For a week now Sedan had witnessed the saddening spectacle of the defeated driven like cattle through its streets, and seemed no more accustomed to it than at the beginning; each time a fresh detachment passed the city was stirred to its very depths by a movement of pity and indignation.

Jean had recovered his equanimity; his thoughts, like Maurice's, reverted to Henriette, and the idea occurred to him that they might see Delaherche somewhere among the throng. He gave his friend a nudge of the elbow.

"Keep your eyes open if we pass through their street presently, will you?"

They had scarce more than struck into the Rue Maqua, indeed, when they became aware of several pairs of eyes turned on the column from one of the tall windows of the factory, and as they drew nearer recognized Delaherche and his wife Gilberte, their elbows resting on the railing of the balcony, and behind them the tall, rigid form of old Madame Delaherche. They had a supply of bread with them, and the manufacturer was tossing the loaves down into the hands that were upstretched with tremulous eagerness to receive them. Maurice saw at once that his sister was not there, while Jean anxiously watched the flying loaves, fearing there might none be left for them. They both had raised their arms and were waving them frantically above their head, shouting meanwhile with all the force of their lungs:

"Here we are! This way, this way!"

The Delaherches seemed delighted to see them in the midst of their surprise. Their faces, pallid with emotion, suddenly brightened, and they displayed by the warmth of their gestures the pleasure they experienced in the encounter. There was one solitary loaf left, which Gilberte insisted on throwing with her own hands, and pitched it into Jean's extended arms in such a charmingly awkward way that she gave a winsome laugh at her own expense. Maurice, unable to stop on account of the pressure from the rear, turned his head and shouted, in a tone of anxious inquiry:

"And Henriette? Henriette?"

Delaherche replied with a long farrago, but his voice was inaudible in the shuffling tramp of so many feet. He seemed to understand that the young man had failed to catch his meaning, for he gesticulated like a semaphore; there was one gesture in particular that he repeated several times, extending his arm with a sweeping motion toward the south, apparently intending to convey the idea of some point in the remote distance: Off there, away off there. Already the head of the column was wheeling into the Rue du Minil, the facade of the factory was lost to sight, together with the kindly faces of the three Delaherches; the last the two friends saw of them was the fluttering of the white handkerchief with which Gilberte waved them a farewell.

"What did he say?" asked Jean.

Maurice, in a fever of anxiety, was still looking to the rear where there was nothing to be seen. "I don't know; I could not understand him; I shall have no peace of mind until I hear from her."

And the trailing, shambling line crept slowly onward, the Prussians urging on the weary men with the brutality of conquerors; the column left the city by the Minil gate in straggling, long-drawn array, hastening their steps, like sheep at whose heels the dogs are snapping.

When they passed through Bazeilles Jean and Maurice thought of Weiss, and cast their eyes about in an effort to distinguish the site of the little house that had been defended with such bravery. While they were at Camp Misery they had heard the woeful tale of slaughter and conflagration that had blotted the pretty village from existence, and the abominations that they now beheld exceeded all they had dreamed of or imagined. At the expiration of twelve days the ruins were smoking still; the tottering walls had fallen in, there were not ten houses standing. It afforded them some small comfort, however, to meet a procession of carts and wheelbarrows loaded with Bavarian helmets and muskets that had been collected after the conflict. That evidence of the chastisement that had been inflicted on those murderers and incendiaries went far toward mitigating the affliction of defeat.

The column was to halt at Douzy to give the men an opportunity to eat breakfast. It was not without much suffering that they reached that place; already the prisoners' strength was giving out, exhausted as they were by their ten days of fasting. Those who the day before had availed of the abundant supplies to gorge themselves were seized with vertigo, their enfeebled legs refused to support their weight, and their gluttony, far from restoring their lost strength, was a further source of weakness to them. The consequence was that, when the train was halted in a meadow to the left of the village, these poor creatures flung themselves upon the ground with no desire to eat. Wine was wanting; some charitable women who came, bringing a few bottles, were driven off by the sentries. One of them in her affright fell and sprained her ankle, and there ensued a painful scene of tears and hysterics, during which the Prussians confiscated the bottles and drank their contents amid jeers and insulting laughter. This tender compassion of the peasants for the poor soldiers who were being led away into captivity was manifested constantly along the route, while it was said the harshness they displayed toward the generals amounted almost to cruelty. At that same Douzy, only a few days previously, the villagers had hooted and reviled a number of paroled officers who were on their way to Pont-a-Mousson. The roads were not safe for general officers; men wearing the blouse-escaped soldiers, or deserters, it may be-fell on them with pitch-forks and endeavored to take their life as traitors, credulously pinning their faith to that legend of bargain and sale which, even twenty years later, was to continue to shed its opprobrium upon those leaders who had commanded armies in that campaign.

Maurice and Jean ate half their bread, and were so fortunate as to have a mouthful of brandy with which to wash it down, thanks to the kindness of a worthy old farmer. When the order was given to resume their advance, however, the distress throughout the convoy was extreme. They were to halt for the night at Mouzon, and although the march was a short one, it seemed as if it would tax the men's strength more severely than they could bear; they could not get on their feet without giving utterance to cries of pain, so stiff did their tired legs become the moment they stopped to rest. Many removed their shoes to relieve their galled and bleeding feet. Dysentery continued to rage; a man fell before they had gone half a mile, and they had to prop him against a wall and leave him. A little further on two others sank at the foot of a hedge, and it was night before an old woman came along and picked them up. All were stumbling, tottering, and dragging themselves along, supporting their forms with canes, which the Prussians, perhaps in derision, had suffered them to cut at the margin of a wood. They were a straggling array of tramps and beggars, covered with sores, haggard, emaciated, and footsore; a sight to bring tears to the eyes of the most stony-hearted. And the guards continued to be as brutally strict as ever; those who for any purpose attempted to leave the ranks were driven back with blows, and the platoon that brought up the rear had orders to prod with their bayonets those who hung back. A sergeant having refused to go further, the captain summoned two of his men and instructed them to seize him, one by either arm, and in

268

this manner the wretched man was dragged over the ground until he agreed to walk. And what made the whole thing more bitter and harder to endure was the utter insignificance of that little pimply-faced, bald-headed officer, so insufferably consequential in his brutality, who took advantage of his knowledge of French to vituperate the prisoners in it in curt, incisive words that cut and stung like the lash of a whip.

"Oh!" Maurice furiously exclaimed, "to get the puppy in my hands and drain him of his blood, drop by drop!"

His powers of endurance were almost exhausted, but it was his rage that he had to choke down, even more than his fatigue, that was cause of his suffering. Everything exasperated him and set on edge his tingling nerves; the harsh notes of the Prussian trumpets particularly, which inspired him with a desire to scream each time he heard them. He felt he should never reach the end of their cruel journey without some outbreak that would bring down on him the utmost severity of the guard. Even now, when traversing the smallest hamlets, he suffered horribly and felt as if he should die with shame to behold the eyes of the women fixed pityingly on him; what would it be when they should enter Germany, and the populace of the great cities should crowd the streets to laugh and jeer at them as they passed? And he pictured to himself the cattle cars into which they would be crowded for transportation, the discomforts and humiliations they would have to suffer on the journey, the dismal life in German fortresses under the leaden, wintry sky. No, no; he would have none of it; better to take the risk of leaving his bones by the roadside on French soil than go and rot off yonder, for months and months, perhaps, in the dark depths of a casemate.

"Listen," he said below his breath to Jean, who was walking at his side; "we will wait until we come to a wood; then we'll break through the guards and run for it among the trees. The Belgian frontier is not far away; we shall have no trouble in finding someone to guide us to it."

Jean, accustomed as he was to look at things coolly and calculate chances, put his veto on the mad scheme, although he, too, in his revolt, was beginning to meditate the possibilities of an escape.

"Have you taken leave of your senses! the guard will fire on us, and we shall both be killed."

But Maurice replied there was a chance the soldiers might not hit them, and then, after all, if their aim should prove true, it would not matter so very much.

"Very well!" rejoined Jean, "but what is going to become of us afterward, dressed in uniform as we are? You know perfectly well that the country is swarming in every direction with Prussian troops; we could not go far unless we had other clothes to put on. No, no, my lad, it's too risky; I'll not let you attempt such an insane project."

And he took the young man's arm and held it pressed against his side, as if they were mutually sustaining each other, continuing meanwhile to chide and soothe him in a tone that was at once rough and affectionate.

Just then the sound of a whispered conversation close behind them caused them to turn and look around. It was Chouteau and Loubet, who had left the peninsula of Iges that morning at the same time as they, and whom they had managed to steer clear of until the present moment. Now the two worthies were close at their heels, and Chouteau must have overheard Maurice's words, his plan for escaping through the mazes of a forest, for he had adopted it on his own behalf. His breath was hot upon their neck as he murmured:

"Say, comrades, count us in on that. That's a capital idea of yours, to skip the ranch. Some of the boys have gone already, and sure we're not going to be such fools as to let those bloody pigs drag us away like dogs into their infernal country. What do you say, eh? Shall we four make a break for liberty?"

269

Maurice's excitement was rising to fever-heat again; Jean turned and said to the tempter:

"If you are so anxious to get away, why don't you go? there's nothing to prevent you. What are you up to, any way?"

He flinched a little before the corporal's direct glance, and allowed the true motive of his proposal to escape him.

"Dame! it would be better that four should share the undertaking. One or two of us might have a chance of getting off."

Then Jean, with an emphatic shake of the head, refused to have anything whatever to do with the matter; he distrusted the gentleman, he said, as he was afraid he would play them some of his dirty tricks. He had to exert all his authority with Maurice to retain him on his side, for at that very moment an opportunity presented itself for attempting the enterprise; they were passing the border of a small but very dense wood, separated from the road only by the width of a field that was covered by a thick growth of underbrush. Why should they not dash across that field and vanish in the thicket? was there not safety for them in that direction?

Loubet had so far said nothing. His mind was made up, however, that he was not going to Germany to run to seed in one of their dungeons, and his nose, mobile as a hound's, was sniffing the atmosphere, his shifty eyes were watching for the favorable moment. He would trust to his legs and his mother wit, which had always helped him out of his scrapes thus far. His decision was quickly made.

"Ah, zut! I've had enough of it; I'm off!"

He broke through the line of the escort, and with a single bound was in the field, Chouteau following his example and running at his side. Two of the Prussian soldiers immediately started in pursuit, but the others seemed dazed, and it did not occur to them to send a ball after the fugitives. The entire episode was so soon over that it was not easy to note its different phases. Loubet dodged and doubled among the bushes and it appeared as if he would certainly succeed in getting off, while Chouteau, less nimble, was on the point of being captured, but the latter, summoning up all his energies in a supreme burst of speed, caught up with his comrade and dexterously tripped him; and while the two Prussians were lumbering up to secure the fallen man, the other darted into the wood and vanished. The guard, finally remembering that they had muskets, fired a few ineffectual shots, and there was some attempt made to search the thicket, which resulted in nothing.

Meantime the two soldiers were pummeling poor Loubet, who had not regained his feet. The captain came running up, beside himself with anger, and talked of making an example, and with this encouragement kicks and cuffs and blows from musket-butts continued to rain down upon the wretched man with such fury that when at last they stood him on his feet he was found to have an arm broken and his skull fractured. A peasant came along, driving a cart, in which he was placed, but he died before reaching Mouzon.

"You see," was all that Jean said to Maurice.

The two friends cast a look in the direction of the wood that sufficiently expressed their sentiments toward the scoundrel who had gained his freedom by such base means, while their hearts were stirred with feelings of deepest compassion for the poor devil whom he had made his victim, a guzzler and a toper, who certainly did not amount to much, but a merry, good-natured fellow all the same, and nobody's fool. And that was always the way with those who kept bad company, Jean moralizingly observed: they might be very fly, but sooner or later a bigger rascal was sure to come along and make a meal of them.

Notwithstanding this terrible lesson Maurice, upon reaching Mouzon, was still possessed by his unalterable determination to attempt an escape. The prisoners were in such an exhausted condition when they reached the place that the Prussians had to assist them to set up the few tents that were placed at their disposal. The camp was formed near the town, on low and marshy ground, and the worst of the business was that another convoy having occupied the spot the day before, the field was absolutely invisible under the superincumbent filth; it was no better than a common cesspool, of unimaginable foulness. The sole means the men had of self-protection was to scatter over the ground some large flat stones, of which they were so fortunate as to find a number in the vicinity. By way of compensation they had a somewhat less hard time of it that evening; the strictness of their guardians was relaxed a little once the captain had disappeared, doubtless to seek the comforts of an inn. The sentries began by winking at the irregularity of the proceeding when some children came along and commenced to toss fruit, apples and pears, over their heads to the prisoners; the next thing was they allowed the people of the neighborhood to enter the lines, so that in a short time the camp was swarming with impromptu merchants, men and women, offering for sale bread, wine, cigars, even. Those who had money had no trouble in supplying their needs so far as eating, drinking, and smoking were concerned. A bustling animation prevailed in the dim twilight; it was like a corner of the market place in a town where a fair is being held.

But Maurice drew Jean behind their tent and again said to him in his nervous, flighty way:

"I can't stand it; I shall make an effort to get away as soon as it is dark. To-morrow our course will take us away from the frontier; it will be too late."

"Very well, we'll try it," Jean replied, his powers of resistance exhausted, his imagination, too, seduced by the pleasing idea of freedom. "They can't do more than kill us."

After that he began to scrutinize more narrowly the venders who surrounded him on every side. There were some among the comrades who had succeeded in supplying themselves with blouse and trousers, and it was reported that some of the charitable people of the place had regular stocks of garments on hand, designed to assist prisoners in escaping. And almost immediately his attention was attracted to a pretty girl, a tall blonde of sixteen with a pair of magnificent eyes, who had on her arm a basket containing three loaves of bread. She was not crying her wares like the rest; an anxious, engaging smile played on her red lips, her manner was hesitating. He looked her steadily in the face; their glances met and for an instant remained confounded. Then she came up, with the embarrassed smile of a girl unaccustomed to such business.

"Do you wish to buy some bread?"

He made no reply, but questioned her by an imperceptible movement of the eyelids. On her answering yes, by an affirmative nod of the head, he asked in a very low tone of voice:

"There is clothing?"

"Yes, under the loaves."

Then she began to cry her merchandise aloud: "Bread! bread! who'll buy my bread?" But when Maurice would have slipped a twenty-franc piece into her fingers she drew back her hand abruptly and ran away, leaving the basket with them. The last they saw of her was the happy, tender look in her pretty eyes, as in the distance she turned and smiled on them.

When they were in possession of the basket Jean and Maurice found difficulties staring them in the face. They had strayed away from their tent, and in their agitated condition felt they should never succeed in finding it again. Where were they to bestow themselves? and how effect their change of garments? It seemed to them that the eyes of the entire assemblage were focused on the basket, which Jean carried with an awkward air, as if it contained dynamite, and that its contents must be plainly visible to everyone. It would not do to waste time, however; they must be up and doing. They stepped into the first vacant tent they came to, where each of them hurriedly slipped on a pair of trousers and donned a blouse, having first deposited their discarded uniforms in the basket, which they placed on the ground in a dark corner of the tent and abandoned to its fate. There was a circumstance that gave them no small uneasiness, however; they found only one head-covering, a knitted woolen cap, which Jean insisted Maurice should wear. The former, fearing his bare-headedness might excite suspicion, was hanging about the precincts of the camp on the lookout for a covering of some description, when it occurred to him to purchase his hat from an extremely dirty old man who was selling cigars.

"Brussels cigars, three sous apiece, two for five!"

Customs regulations were in abeyance since the battle of Sedan, and the imports of Belgian merchandise had been greatly stimulated. The old man had been making a handsome profit from his traffic, but that did not prevent him from driving a sharp bargain when he understood the reason why the two men wanted to buy his hat, a greasy old affair of felt with a great hole in its crown. He finally consented to part with it for two five-franc pieces, grumbling that he should certainly have a cold in his head.

Then Jean had another idea, which was neither more nor less than to buy out the old fellow's stock in trade, the two dozen cigars that remained unsold. The bargain effected, he pulled his hat down over his eyes and began to cry in the itinerant hawker's drawling tone:

"Here you are, Brussels cigars, two for three sous, two for three sous!"

Their safety was now assured. He signaled Maurice to go on before. It happened to the latter to discover an umbrella lying on the grass; he picked it up and, as a few drops of rain began to fall just then, opened it tranquilly as they were about to pass the line of sentries.

"Two for three sous, two for three sous, Brussels cigars!"

It took Jean less than two minutes to dispose of his stock of merchandise. The men came crowding about him with chaff and laughter: a reasonable fellow, that; he didn't rob poor chaps of their money! The Prussians themselves were attracted by such unheard-of bargains, and he was compelled to trade with them. He had all the time been working his way toward the edge of the enceinte, and his last two cigars went to a big sergeant with an immense beard, who could not speak a word of French.

"Don't walk so fast, confound it!" Jean breathed in a whisper behind Maurice's back. "You'll have them after us."

Their legs seemed inclined to run away with them, although they did their best to strike a sober gait. It caused them a great effort to pause a moment at a cross-roads, where a number of people were collected before an inn. Some villagers were chatting peaceably with German soldiers, and the two runaways made a pretense of listening, and even hazarded a few observations on the weather and the probability of the rain continuing during the night. They trembled when they beheld a man, a fleshy gentleman, eying them attentively, but as he smiled with an air of great good-nature they thought they might venture to address him, asking in a whisper:

"Can you tell us if the road to Belgium is guarded, sir?"

"Yes, it is; but you will be safe if you cross this wood and afterward cut across the fields, to the left."

Once they were in the wood, in the deep, dark silence of the slumbering trees, where no sound reached their ears, where nothing stirred and they believed their safety was assured them, they sank into each other's arms in an uncontrollable impulse of emotion. Maurice was sobbing violently, while big tears trickled slowly down Jean's cheeks. It was the natural revulsion of their overtaxed feelings after the long-protracted ordeal they had passed through, the joy and delight of their mutual assurance that their troubles were at an end, and that thenceforth suffering and they were to be strangers. And united by the memory of what they had endured together in ties closer than those of brotherhood, they clasped each other in a wild embrace, and the kiss that they exchanged at that moment seemed to them to possess a savor and a poignancy such as they had never experienced before in all their life; a kiss such as they never could receive from lips of woman, sealing their undying friendship, giving additional confirmation to the certainty that thereafter their two hearts would be but one, for all eternity.

When they had separated at last: "Little one," said Jean, in a trembling voice, "it is well for us to be here, but we are not at the end. We must look about a bit and try to find our bearings."

Maurice, although he had no acquaintance with that part of the frontier, declared that all they had to do was to pursue a straight course, whereon they resumed their way, moving among the trees in Indian file with the greatest circumspection, until they reached the edge of the thicket. There, mindful of the injunction of the kind-hearted villager, they were about to turn to the left and take a short cut across the fields, but on coming to a road, bordered with a row of poplars on either side they beheld directly in their path the watch-fire of a Prussian detachment. The bayonet of the sentry, pacing his beat, gleamed in the ruddy light, the men were finishing their soup and conversing; the fugitives stood not upon the order of their going, but plunged into the recesses of the wood again, in mortal terror lest they might be pursued. They thought they heard the sound of voices, of footsteps on their trail, and thus for over an hour they wandered at random among the copses, until all idea of locality was obliterated from their brain; now racing like affrighted animals through the underbrush, again brought up all standing, the cold sweat trickling down their face, before a tree in which they beheld a Prussian. And the end of it was that they again came out on the poplar-bordered road not more than ten paces from the sentry, and quite near the soldiers, who were toasting their toes in tranquil comfort.

"Hang the luck!" grumbled Jean. "This must be an enchanted wood."

This time, however, they had been heard. The sound of snapping twigs and rolling stones betrayed them. And as they did not answer the challenge of the sentry, but made off at the double-quick, the men seized their muskets and sent a shower of bullets crashing through the thicket, into which the fugitives had plunged incontinently.

"Nom de Dieu!" ejaculated Jean, with a stifled cry of pain.

He had received something that felt like the cut of a whip in the calf of his left leg, but the impact was so violent that it drove him up against a tree.

"Are you hurt?" Maurice anxiously inquired.

"Yes, and in the leg, worse luck!"

They both stood holding their breath and listening, in dread expectancy of hearing their pursuers clamoring at their heels; but the firing had ceased and nothing stirred amid the intense stillness that had again settled down upon the wood and the surrounding country. It was evident that the Prussians had no inclination to beat up the thicket.

Jean, who was doing his best to keep on his feet; forced back a groan. Maurice sustained him with his arm.

"Can't you walk?"

"I should say not!" He gave way to a fit of rage, he, always so self-contained. He clenched his fists, could have thumped himself. "God in Heaven, if this is not hard luck! to have one's legs knocked from under him at the very time he is most in need of them! It's too bad, too bad, by my soul it is! Go on, you, and put yourself in safety!"

But Maurice laughed quietly as he answered:

"That is silly talk!"

He took his friend's arm and helped him along, for neither of them had any desire to linger there. When, laboriously and by dint of heroic effort, they had advanced some half-dozen paces further, they halted again with renewed alarm at beholding before them a house, standing at the margin of the wood, apparently a sort of farmhouse. Not a light was visible at any of the windows, the open courtyard gate yawned upon the dark and deserted dwelling. And when they plucked up their courage a little and ventured to enter the courtyard, great was their surprise to find a horse standing there with a saddle on his back, with nothing to indicate the why or wherefore of his being there. Perhaps it was the owner's intention to return, perhaps he was lying behind a bush with a bullet in his brain. They never learned how it was.

But Maurice had conceived a new scheme, which appeared to afford him great satisfaction.

"See here, the frontier is too far away; we should never succeed in reaching it without a guide. What do you say to changing our plan and going to Uncle Fouchard's, at Remilly? I am so well acquainted with every inch of the road that I'm sure I could take you there with my eyes bandaged. Don't you think it's a good idea, eh? I'll put you on this horse, and I suppose Uncle Fouchard will grumble, but he'll take us in."

Before starting he wished to take a look at the injured leg. There were two orifices; the ball appeared to have entered the limb and passed out, fracturing the tibia in its course. The flow of blood had not been great; he did nothing more than bandage the upper part of the calf tightly with his handkerchief.

"Do you fly, and leave me here," Jean said again.

"Hold your tongue; you are silly!"

When Jean was seated firmly in the saddle Maurice took the bridle and they made a start. It was somewhere about eleven o'clock, and he hoped to make the journey in three hours, even if they should be unable to proceed faster than a walk. A difficulty that he had not thought of until then, however, presented itself to his mind and for a moment filled him with consternation: how were they to cross the Meuse in order to get to the left bank? The bridge at Mouzon would certainly be guarded. At last he remembered that there was a ferry lower down the stream, at Villers, and trusting to luck to befriend him, he shaped his course for that village, striking across the meadows and tilled fields of the right bank. All went well enough at first; they had only to dodge a cavalry patrol which forced them to hide in the shadow of a wall and remain there half an hour. Then the rain began to come down in earnest and his progress became more laborious, compelled as he was to tramp through the sodden fields beside the horse, which fortunately showed itself to be a fine specimen of the equine race, and perfectly gentle. On reaching Villers he found that his trust in the blind goddess, Fortune, had not been misplaced; the ferryman, who, at that late hour, had just returned from setting a Bavarian officer across the river, took them at once and landed them on the other shore without delay or accident.

And it was not until they reached the village, where they narrowly escaped falling into the clutches of the pickets who were stationed along the entire length of the Remilly road, that their dangers and hardships really commenced; again they were obliged to take to the fields, feeling their way along blind paths and cart-tracks that could scarcely be discerned in the darkness. The most trivial obstacle sufficed to drive them a long way out of their course. They squeezed through hedges, scrambled down and up the steep banks of ditches, forced a passage for themselves through the densest thickets. Jean, in whom a low fever had developed under the drizzling rain, had sunk down crosswise on his saddle in a condition of semi-consciousness, holding on with both hands by the horse's mane, while Maurice, who had slipped the bridle over his right arm, had to steady him by the legs to keep him from tumbling to the ground. For more than a league, for two long, weary hours that seemed like an eternity, did they toil onward in this fatiguing way; floundering, stumbling, slipping in such a manner that it seemed at every moment as if men and beast must land together in a heap at the bottom of some descent. The spectacle they presented was one of utter, abject misery, besplashed with mud, the horse trembling in every limb, the man upon his back a helpless mass, as if at his last gasp, the other, wild-eyed and pale as death, keeping his feet only by an effort of fraternal love. Day was breaking; it was not far from five o'clock when at last they came to Remilly.

In the courtyard of his little farmhouse, which was situated at the extremity of the pass of Harancourt, overlooking the village, Father Fouchard was stowing away in his carriole the carcasses of two sheep that he had slaughtered the day before. The sight of his nephew, coming to him at that hour and in that sorry plight, caused him such perturbation of spirit that, after the first explanatory words, he roughly cried:

"You want me to take you in, you and your friend? and then settle matters with the Prussians afterward, I suppose. I'm much obliged to you, but no! I might as well die right straight off and have done with it."

He did not go so far, however, as to prohibit Maurice and Prosper from taking Jean from the horse and laying him on the great table in the kitchen. Silvine ran and got the bolster from her bed and slipped it beneath the head of the wounded man, who was still unconscious. But it irritated the old fellow to see the man lying on his table; he grumbled and fretted, saying that the kitchen was no place for him; why did they not take him away to the hospital at once? since there fortunately was a hospital at Remilly, near the church, in the old schoolhouse; and there was a big room in it, with everything nice and comfortable.

"To the hospital!" Maurice hotly replied, "and have the Prussians pack him off to Germany as soon as he is well, for you know they treat all the wounded as prisoners of war. Do you take me for a fool, uncle? I did not bring him here to give him up."

Things were beginning to look dubious, the uncle was threatening to pitch them out upon the road, when someone mentioned Henriette's name.

"What about Henriette?" inquired the young man.

And he learned that his sister had been an inmate of the house at Remilly for the last two days; her affliction had weighed so heavily on her that life at Sedan, where her existence had hitherto been a happy one, was become a burden greater than she could bear. Chancing to meet with Doctor Dalichamp of Raucourt, with whom she was acquainted, her conversation with him had been the means of bringing her to take up her abode with Father Fouchard, in whose house she had a little bedroom, in order to devote herself entirely to the care of the sufferers in the neighboring hospital. That alone, she said, would serve to quiet her bitter memories. She paid her board and was the means of introducing many small comforts into the life of the farmhouse, which caused Father Fouchard to regard her with an eye of favor. The weather was always fine with him, provided he was making money.

"Ah! so my sister is here," said Maurice. "That must have been what M. Delaherche wished to tell me, with his gestures that I could not understand. Very well; if she is here, that settles it; we shall remain."

Notwithstanding his fatigue he started off at once in quest of her at the ambulance, where she had been on duty during the preceding night, while the uncle cursed his luck that kept him from being off with the carriole to sell his mutton among the neighboring villages, so long as the confounded business that he had got mixed up in remained unfinished.

When Maurice returned with Henriette they caught the old man making a critical examination of the horse, that Prosper had led away to the stable. The animal seemed to please him; he was knocked up, but showed signs of strength and endurance. The young man laughed and told his uncle he might have him as a gift if he fancied him, while Henriette, taking her relative aside, assured him Jean should be no expense to him; that she would take charge of him and nurse him, and he might have the little room behind the cow-stables, where no Prussian would ever think to look for him. And Father Fouchard, still wearing a very sulky face and but half convinced that there was anything to be made out of the affair, finally closed the discussion by jumping into his carriole and driving off, leaving her at liberty to act as she pleased.

It took Henriette but a few minutes, with the assistance of Silvine and Prosper, to put the room in order; then she had Jean brought in and they laid him on a cool, clean bed, he giving no sign of life during the operation save to mutter some unintelligible words. He opened his eyes and looked about him, but seemed not to be conscious of anyone's presence in the room. Maurice, who was just beginning to be aware how utterly prostrated he was by his fatigue, was drinking a glass of wine and eating a bit of cold meat, left over from the yesterday's dinner, when Doctor Dalichamp came in, as was his daily custom previous to visiting the hospital, and the young man, in his anxiety for his friend, mustered up his strength to follow him, together with his sister, to the bedside of the patient.

The doctor was a short, thick-set man, with a big round head, on which the hair, as well as the fringe of beard about his face, had long since begun to be tinged with gray. The skin of his ruddy, mottled face was tough and indurated as a peasant's, spending as he did most of his time in the open air, always on the go to relieve the sufferings of his fellow-creatures; while the large, bright eyes, the massive nose, indicative of obstinacy, and the benignant if somewhat sensual mouth bore witness to the lifelong charities and good works of the honest country doctor; a little brusque at times, not a man of genius, but whom many years of practice in his profession had made an excellent healer.

When he had examined Jean, still in a comatose state, he murmured:

"I am very much afraid that amputation will be necessary."

The words produced a painful impression on Maurice and Henriette. Presently, however, he added:

"Perhaps we may be able to save the leg, but it will require the utmost care and attention, and will take a very long time. For the moment his physical and mental depression is such that the only thing to do is to let him sleep. To-morrow we shall know more."

Then, having applied a dressing to the wound, he turned to Maurice, whom he had known in bygone days, when he was a boy.

"And you, my good fellow, would be better off in bed than sitting there."

The young man continued to gaze before him into vacancy, as if he had not heard. In the confused hallucination that was due to his fatigue he developed a kind of delirium, a supersensitive nervous excitation that embraced all he had suffered in mind and body since the beginning of the campaign. The spectacle of his friend's wretched state, his own condition, scarce less pitiful, defeated, his hands tied, good for nothing, the reflection that all those heroic efforts had culminated in such disaster, all combined to incite him to frantic rebellion against destiny. At last he spoke.

"It is not ended; no, no! we have not seen the end, and I must go away. Since *he* must lie there on his back for weeks, for months, perhaps, I cannot stay; I must go, I must go at once. You

276

will assist me, won't you, doctor? you will supply me with the means to escape and get back to Paris?"

Pale and trembling, Henriette threw her arms about him and caught him to her bosom.

"What words are those you speak? enfeebled as you are, after all the suffering you have endured! but think not I shall let you go; you shall stay here with me! Have you not paid the debt you owe your country? and should you not think of me, too, whom you would leave to loneliness? of me, who have nothing now in all the wide world save you?"

Their tears flowed and were mingled. They held each other in a wild tumultuous embrace, with that fond affection which, in twins, often seems as if it antedated existence. But for all that his exaltation did not subside, but assumed a higher pitch.

"I tell you I must go. Should I not go I feel I should die of grief and shame. You can have no idea how my blood boils and seethes in my veins at the thought of remaining here in idleness. I tell you that this business is not going to end thus, that we must be avenged. On whom, on what? Ah! that I cannot tell; but avenged we must and shall be for such misfortune, in order that we may yet have courage to live on!"

Doctor Dalichamp, who had been watching the scene with intense interest, cautioned Henriette by signal to make no reply. Maurice would doubtless be more rational after he should have slept; and sleep he did, all that day and all the succeeding night, for more than twenty hours, and never stirred hand or foot. When he awoke next morning, however, he was as inflexible as ever in his determination to go away. The fever had subsided; he was gloomy and restless, in haste to withdraw himself from influences that he feared might weaken his patriotic fervor. His sister, with many tears, made up her mind that he must be allowed to have his way, and Doctor Dalichamp, when he came to make his morning visit, promised to do what he could to facilitate the young man's escape by turning over to him the papers of a hospital attendant who had died recently at Raucourt. It was arranged that Maurice should don the gray blouse with the red cross of Geneva on its sleeve and pass through Belgium, thence to make his way as best he might to Paris, access to which was as yet uninterrupted.

He did not leave the house that day, keeping himself out of sight and waiting for night to come. He scarcely opened his mouth, although he did make an attempt to enlist the new farm-hand in his enterprise.

"Say, Prosper, don't you feel as if you would like to go back and have one more look at the Prussians?"

The ex-chasseur d'Afrique, who was eating a cheese sandwich, stopped and held his knife suspended in the air.

"It don't strike me that it is worth while, from what we were allowed to see of them before. Why should you wish me to go back there, when the only use our generals can find for the cavalry is to send it in after the battle is ended and let it be cut to pieces? No, faith, I'm sick of the business, giving us such dirty work as that to do!" There was silence between them for a moment; then he went on, doubtless to quiet the reproaches of his conscience as a soldier: "And then the work is too heavy here just now; the plowing is just commencing, and then there'll be the fall sowing to be looked after. We must think of the farm work, mustn't we? for fighting is well enough in its way, but what would become of us if we should cease to till the ground? You see how it is; I can't leave my work. Not that I am particularly in love with Father Fouchard, for I doubt very strongly if I shall ever see the color of his money, but the beasties are beginning to take to me, and faith! when I was up there in the Old Field this morning, and gave a look at that d--d Sedan lying yonder in the distance, you can't tell how good it made me feel to be guiding my oxen and driving the plow through the furrow, all alone in the bright

277

sunshine."

As soon as it was fairly dark, Doctor Dalichamp came driving up in his old gig. It was his intention to see Maurice to the frontier. Father Fouchard, well pleased to be rid of one of his guests at least, stepped out upon the road to watch and make sure there were none of the enemy's patrols prowling in the neighborhood, while Silvine put a few stitches in the blouse of the defunct ambulance man, on the sleeve of which the red cross of the corps was prominently displayed. The doctor, before taking his place in the vehicle, examined Jean's leg anew, but could not as yet promise that he would be able to save it. The patient was still in a profound lethargy, recognizing no one, never opening his mouth to speak, and Maurice was about to leave him without the comfort of a farewell, when, bending over to give him a last embrace, he saw him open his eyes to their full extent; the lips parted, and in a faint voice he said:

"You are going away?" And in reply to their astonished looks: "Yes, I heard what you said, though I could not stir. Take the remainder of the money, then. Put your hand in my trousers' pocket and take it."

Each of them had remaining nearly two hundred francs of the sum they had received from the corps paymaster.

But Maurice protested. "The money!" he exclaimed. "Why, you have more need of it than I, who have the use of both my legs. Two hundred francs will be abundantly sufficient to see me to Paris, and to get knocked in the head afterward won't cost me a penny. I thank you, though, old fellow, all the same, and good-by and good-luck to you; thanks, too, for having always been so good and thoughtful, for, had it not been for you, I should certainly be lying now at the bottom of some ditch, like a dead dog."

Jean made a deprecating gesture. "Hush. You owe me nothing; we are quits. Would not the Prussians have gathered me in out there the other day had you not picked me up and carried me off on your back? and yesterday again you saved me from their clutches. Twice have I been beholden to you for my life, and now I am in your debt. Ah, how unhappy I shall be when I am no longer with you!" His voice trembled and tears rose to his eyes. "Kiss me, dear boy!"

They embraced, and, as it had been in the wood the day before, that kiss set the seal to the brotherhood of dangers braved in each other's company, those few weeks of soldier's life in common that had served to bind their hearts together with closer ties than years of ordinary friendship could have done. Days of famine, sleepless nights, the fatigue of the weary march, death ever present to their eyes, these things made the foundation on which their affection rested. When two hearts have thus by mutual gift bestowed themselves the one upon the other and become fused and molten into one, is it possible ever to sever the connection? But the kiss they had exchanged the day before, among the darkling shadows of the forest, was replete with the joy of their new-found safety and the hope that their escape awakened in their bosom, while this was the kiss of parting, full of anguish and doubt unutterable. Would they meet again some day? and how, under what circumstances of sorrow or of gladness?

Doctor Dalichamp had clambered into his gig and was calling to Maurice. The young man threw all his heart and soul into the embrace he gave his sister Henriette, who, pale as death in her black mourning garments, looked on his face in silence through her tears.

"He whom I leave to your care is my brother. Watch over him, love him as I love him!"

4

Jean's chamber was a large room, with floor of brick and whitewashed walls, that had once done duty as a store-room for the fruit grown on the farm. A faint, pleasant odor of pears and apples lingered there still, and for furniture there was an iron bedstead, a pine table and two chairs, to say nothing of a huge old walnut clothes-press, tremendously deep and wide, that looked as if it might hold an army. A lazy, restful quiet reigned there all day long, broken only by the deadened sounds that came from the adjacent stables, the faint lowing of the cattle, the occasional thud of a hoof upon the earthen floor. The window, which had a southern aspect, let in a flood of cheerful sunlight; all the view it afforded was a bit of hillside and a wheat field, edged by a little wood. And this mysterious chamber was so well hidden from prying eyes that never a one in all the world would have suspected its existence.

As it was to be her kingdom, Henriette constituted herself lawmaker from the beginning. The regulation was that no one save she and the doctor should have access to Jean; this in order to avert suspicion. Silvine, even, was never to set foot in the room unless by direction. Early each morning the two women came in and put things to rights, and after that, all the long day, the door was as impenetrable as if it had been a wall of stone. And thus it was that Jean found himself suddenly secluded from the world, after many weeks of tumultuous activity, seeing no face save that of the gentle woman whose footfall on the floor gave back no sound. She appeared to him, as he had beheld her for the first time down yonder in Sedan, like an apparition, with her somewhat large mouth, her delicate, small features, her hair the hue of ripened grain, hovering about his bedside and ministering to his wants with an air of infinite goodness.

The patient's fever was so violent during the first few days that Henriette scarce ever left him. Doctor Dalichamp dropped in every morning on his way to the hospital and examined and dressed the wound. As the ball had passed out, after breaking the tibia, he was surprised that the case presented no better aspect; he feared there was a splinter of the bone remaining there that he had not succeeded in finding with the probe, and that might make resection necessary. He mentioned the matter to Jean, but the young man could not endure the thought of an operation that would leave him with one leg shorter than the other and lame him permanently. No, no! he would rather die than be a cripple for life. So the good doctor, leaving the wound to develop further symptoms, confined himself for the present to applying a dressing of lint saturated with sweet oil and phenic acid having first inserted a drain-an India rubber tube-to carry off the pus. He frankly told his patient, however, that unless he submitted to an operation he must not hope to have the use of his limb for a very long time. Still, after the second week, the fever subsided and the young man's general condition was improved, so long as he could be content to rest quiet in his bed.

Then Jean's and Henriette's relations began to be established on a more systematic basis. Fixed habits commenced to prevail; it seemed to them that they had never lived otherwise-that they were to go on living forever in that way. All the hours and moments that she did not devote to the ambulance were spent with him; she saw to it that he had his food and drink at proper intervals. She assisted him to turn in bed with a strength of wrist that no one, seeing her slender arms, would have supposed was in her. At times they would converse; but as a general thing, especially in the earlier days, they had not much to say. They never seemed to tire of each other's company, though. On the whole it was a very pleasant life they led in that calm, restful atmosphere, he with the horrible scenes of the battlefield still fresh in his memory, she in her widow's weeds, her heart bruised and bleeding with the great loss she had sustained. At first he had experienced a sensation of embarrassment, for he felt she was his superior, almost a lady, indeed, while he had never been aught more than a common soldier and a peasant. He could barely read and write. When finally he came to see that she affected no airs of superiority, but treated him on the footing of an equal, his confidence returned to him in a measure and he showed himself in his true colors, as a man of intelligence by reason of his sound, unpretentious common sense. Besides, he was surprised at times to think he could note a change was gradually coming over him; it seemed to him that his mind was less torpid than it had been, that it was clearer and more active, that he had novel ideas in his head, and more of them; could it be that the abominable life he had been leading for the last two months, his horrible sufferings, physical and moral, had exerted a refining influence on him? But that which assisted him most to overcome his shyness was to find that she was really not so very much wiser than he. She was but a little child when, at her mother's death, she became the household drudge, with her three men to care for, as she herself expressed it-her grandfather, her father, and her brother-and she had not had the time to lay in a large stock of learning. She could read and write, could spell words that were not too long, and "do sums," if they were not too intricate; and that was the extent of her acquirement. And if she continued to intimidate him still, if he considered her far and away the superior of all other women upon earth, it was because he knew the ineffable tenderness, the goodness of heart, the unflinching courage, that animated that frail little body, who went about her duties silently and met them as if they had been pleasures.

They had in Maurice a subject of conversation that was of common interest to them both and of which they never wearied. It was to Maurice's friend, his brother, to whom she was devoting herself thus tenderly, the brave, kind man, so ready with his aid in time of trouble, who she felt had made her so many times his debtor. She was full to overflowing with a sentiment of deepest gratitude and affection, that went on widening and deepening as she came to know him better and recognize his sterling qualities of head and heart, and he, whom she was tending like a little child, was actuated by such grateful sentiments that he would have liked to kiss her hands each time she gave him a cup of bouillon. Day by day did this bond of tender sympathy draw them nearer to each other in that profound solitude amid which they lived, harassed by an anxiety that they shared in common. When he had utterly exhausted his recollections of the dismal march from Rheims to Sedan, to the particulars of which she never seemed to tire of listening, the same question always rose to their lips: what was Maurice doing then? why did he not write? Could it be that the blockade of Paris was already complete, and was that the reason why they received no news? They had as yet had but one letter from him, written at Rouen, three days after his leaving them, in which he briefly stated that he had reached that city on his way to Paris, after a long and devious journey. And then for a week there had been no further word; the silence had remained unbroken.

In the morning, after Doctor Dalichamp had attended to his patient, he liked to sit a while and chat, putting his cares aside for the moment. Sometimes he also returned at evening and made a longer visit, and it was in this way that they learned what was going on in the great world outside their peaceful solitude and the terrible calamities that were desolating their country. He was their only source of intelligence; his heart, which beat with patriotic ardor, overflowed with rage and grief at every fresh defeat, and thus it was that his sole topic of conversation was the victorious progress of the Prussians, who, since Sedan, had spread themselves over France like the waves of some black ocean. Each day brought its own tidings of disaster, and resting disconsolately on one of the two chairs that stood by the bedside, he would tell in mournful tones and with trembling gestures of the increasing gravity of the situation. Oftentimes he came with his pockets stuffed with Belgian newspapers, which he would leave behind him when he went away. And thus the echoes of defeat, days, weeks, after the event, reverberated in that quiet room, serving to unite yet more closely in community of sorrow the two poor sufferers who were shut within its walls.

It was from some of those old newspapers that Henriette read to Jean the occurrences at Metz, the Titanic struggle that was three times renewed, separated on each occasion by a day's interval. The story was already five weeks old, but it was new to him, and he listened with a bleeding heart to the repetition of the miserable narrative of defeat to which he was not a stranger. In the deathly stillness of the room the incidents of the woeful tale unfolded themselves as Henriette, with the sing-song enunciation of a schoolgirl, picked out her words and sentences. When, after Froeschwiller and Spickeren, the 1st corps, routed and broken into fragments, had swept away with it the 5th, the other corps stationed along the frontier *en echelon* from Metz to Bitche, first wavering, then retreating in their consternation at those reverses, had ultimately concentrated before the intrenched camp on the right bank of the Moselle. But what waste of precious time was there, when they should not have lost a moment in retreating on Paris, a movement that was presently to be attended with such difficulty! The Emperor had been compelled to turn over the supreme command to Marshal Bazaine, to whom everyone looked with confidence for a victory. Then, on the 14th[*] came the affair of Borny, when the army was attacked at the moment when it was at last about to cross the stream, having to sustain the onset of two German armies: Steinmetz's, which was encamped in observation in front of the intrenched camp, and Prince Frederick Charles's, which had passed the river higher up and come down along the left bank in order to bar the French from access to their country; Borny, where the firing did not begin until it was three o'clock; Borny, that barren victory, at the end of which the French remained masters of their positions, but which left them astride the Moselle, tied hand and foot, while the turning movement of the second German army was being successfully accomplished. After that, on the 16th, was the battle of Rezonville; all our corps were at last across the stream, although, owing to the confusion that prevailed at the junction of the Mars-la-Tour and Etain roads, which the Prussians had gained possession of early in the morning by a brilliant movement of their cavalry and artillery, the 3d and 4th corps were hindered in their march and unable to get up; a slow, dragging, confused battle, which, up to two o'clock, Bazaine, with only a handful of men opposed to him, should have won, but which he wound up by losing, thanks to his inexplicable fear of being cut off from Metz; a battle of immense extent, spreading over leagues of hill and plain, where the French, attacked in front and flank, seemed willing to do almost anything except advance, affording the enemy time to concentrate and to all appearances co-operating with them to ensure the success of the Prussian plan, which was to force their withdrawal to the other side of the river. And on the 18th, after their retirement to the intrenched camp, Saint-Privat was fought, the culmination of the gigantic struggle, where the line of battle extended more than eight miles in length, two hundred thousand Germans with seven hundred guns arrayed against a hundred and twenty thousand French with but five hundred guns, the Germans facing toward Germany, the French toward France, as if invaders and invaded had inverted their roles in the singular tactical movements that had been going on; after two o'clock the conflict was most sanguinary, the Prussian Guard being repulsed with tremendous slaughter and Bazaine, with a left wing that withstood the onsets of the enemy like a wall of adamant, for a long time victorious, up to the moment, at the approach of evening, when the weaker right wing was compelled by the terrific losses it had sustained to abandon Saint-Privat, involving in its rout the remainder of the army, which, defeated and driven back under the walls of Metz, was thenceforth to be imprisoned in a circle of flame and iron.

[*] August.-TR.

As Henriette pursued her reading Jean momentarily interrupted her to say:

"Ah, well! and to think that we fellows, after leaving Rheims, were looking for Bazaine! They were always telling us he was coming; now I can see why he never came!"

The marshal's despatch, dated the 19th, after the battle of Saint-Privat, in which he spoke of resuming his retrograde movement by way of Montmedy, that despatch which had for its effect the advance of the army of Chalons, would seem to have been nothing more than the report of a defeated general, desirous to present matters under their most favorable aspect, and it was not until a considerably later period, the 29th, when the tidings of the approach of this relieving army had reached him through the Prussian lines, that he attempted a final effort, on the right bank this time, at Noiseville, but in such a feeble, half-hearted way that on the 1st of September, the day when the army of Chalons was annihilated at Sedan, the army of Metz fell back to advance no more, and became as if dead to France. The marshal, whose conduct up to that time may fairly be characterized as that of a leader of only moderate ability, neglecting his opportunities and failing to move when the roads were open to him, after that blockaded by forces greatly superior to his own, was now about to be seduced by alluring visions of political greatness and become a conspirator and a traitor.

But in the papers that Doctor Dalichamp brought them Bazaine was still the great man and the gallant soldier, to whom France looked for her salvation.

And Jean wanted certain passages read to him again, in order that he might more clearly understand how it was that while the third German army, under the Crown Prince of Prussia, had been leading them such a dance, and the first and second were besieging Metz, the latter were so strong in men and guns that it had been possible to form from them a fourth army, which, under the Crown Prince of Saxony, had done so much to decide the fortune of the day at Sedan. Then, having obtained the information he desired, resting on that bed of suffering to which his wound condemned him, he forced himself to hope in spite of all.

"That's how it is, you see; we were not so strong as they! No one can ever get at the rights of such matters while the fighting is going on. Never mind, though; you have read the figures as the newspapers give them: Bazaine has a hundred and fifty thousand men with him, he has three hundred thousand small arms and more than five hundred pieces of artillery; take my word for it, he is not going to let himself be caught in such a scrape as we were. The fellows all say he is a tough man to deal with; depend on it he's fixing up a nasty dose for the enemy, and he'll make 'em swallow it."

Henriette nodded her head and appeared to agree with him, in order to keep him in a cheerful frame of mind. She could not follow those complicated operations of the armies, but had a presentiment of coming, inevitable evil. Her voice was fresh and clear; she could have gone on reading thus for hours; only too glad to have it in her power to relieve the tedium of his long day, though at times, when she came to some narrative of slaughter, her eyes would fill with tears that made the words upon the printed page a blur. She was doubtless thinking of her husband's fate, how he had been shot down at the foot of the wall and his body desecrated by the touch of the Bavarian officer's boot.

"If it gives you such pain," Jean said in surprise, "you need not read the battles; skip them."

But, gentle and self-sacrificing as ever, she recovered herself immediately.

"No, no; don't mind my weakness; I assure you it is a pleasure to me."

One evening early in October, when the wind was blowing a small hurricane outside, she came in from the ambulance and entered the room with an excited air, saying:

"A letter from Maurice! the doctor just gave it me."

With each succeeding morning the twain had been becoming more and more alarmed that the young man sent them no word, and now that for a whole week it had been rumored everywhere that the investment of Paris was complete, they were more disturbed in mind than ever, despairing of receiving tidings, asking themselves what could have happened him after he left Rouen. And now the reason of the long silence was made clear to them: the letter that he had addressed from Paris to Doctor Dalichamp on the 18th, the very day that ended railway communication with Havre, had gone astray and had only reached them at last by a miracle, after a long and circuitous journey.

"Ah, the dear boy!" said Jean, radiant with delight. "Read it to me, quick!"

The wind was howling and shrieking more dismally than ever, the window of the apartment strained and rattled as if someone were trying to force an entrance. Henriette went and got the little lamp, and placing it on the table beside the bed applied herself to the reading of the missive, so close to Jean that their faces almost touched. There was a sensation of warmth and comfort in the peaceful room amid the roaring of the storm that raged without.

It was a long letter of eight closely filled pages, in which Maurice first told how, soon after his arrival on the 16th, he had had the good fortune to get into a line regiment that was being recruited up to its full strength. Then, reverting to facts of history, he described in brief but vigorous terms the principal events of that month of terror: how Paris, recovering her sanity in a measure after the madness into which the disasters of Wissembourg and Froeschwiller had driven her, had comforted herself with hopes of future victories, had cheered herself with fresh illusions, such as lying stories of the army's successes, the appointment of Bazaine to the chief command, the levee en masse, bogus dispatches, which the ministers themselves read from the tribune, telling of hecatombs of slaughtered Prussians. And then he went on to tell how, on the 3d of September, the thunderbolt had a second time burst over the unhappy capital: all hope gone, the misinformed, abused, confiding city dazed by that crushing blow of destiny, the cries: "Down with the Empire!" that resounded at night upon the boulevards, the brief and gloomy session of the Chamber at which Jules Favre read the draft of the bill that conceded the popular demand. Then on the next day, the ever-memorable 4th of September, was the upheaval of all things, the second Empire swept from existence in atonement for its mistakes and crimes, the entire population of the capital in the streets, a torrent of humanity a half a million strong filling the Place de la Concorde and streaming onward in the bright sunshine of that beautiful Sabbath day to the great gates of the Corps Legislatif, feebly guarded by a handful of troops, who up-ended their muskets in the air in token of sympathy with the populace-smashing in the doors, swarming into the assembly chambers, whence Jules Favre, Gambetta and other deputies of the Left were even then on the point of departing to proclaim the Republic at the Hotel de Ville; while on the Place Saint-Germain-l'Auxerrois a little wicket of the Louvre opened timidly and gave exit to the Empress-regent, attired in black garments and accompanied by a single female friend, both the women trembling with affright and striving to conceal themselves in the depths of the public cab, which went jolting with its scared inmates from the Tuileries, through whose apartments the mob was at that moment streaming. On the same day Napoleon III. left the inn at Bouillon, where he had passed his first night of exile, bending his way toward Wilhelmshohe.

Here Jean, a thoughtful expression on his face, interrupted Henriette.

"Then we have a republic now? So much the better, if it is going to help us whip the Prussians!"

But he shook his head; he had always been taught to look distrustfully on republics when he was a peasant. And then, too, it did not seem to him a good thing that they should be of differing minds when the enemy was fronting them. After all, though, it was manifest there had to be a change of some kind, since everyone knew the Empire was rotten to the core and

283

the people would have no more of it.

Henriette finished the letter, which concluded with a mention of the approach of the German armies. On the 13th, the day when a committee of the Government of National Defense had established its quarters at Tours, their advanced guards had been seen at Lagny, to the east of Paris. On the 14th and 15th they were at the very gates of the city, at Creteil and Joinville-le-Pont. On the 18th, however, the day when Maurice wrote, he seemed to have ceased to believe in the possibility of maintaining a strict blockade of Paris; he appeared to be under the influence of one of his hot fits of blind confidence, characterising the siege as a senseless and impudent enterprise that would come to an ignominious end before they were three weeks older, relying on the armies that the provinces would surely send to their relief, to say nothing of the army of Metz, that was already advancing by way of Verdun and Rheims. And the links of the iron chain that their enemies had forged for them had been riveted together; it encompassed Paris, and now Paris was a city shut off from all the world, whence no letter, no word of tidings longer came, the huge prison-house of two millions of living beings, who were to their neighbors as if they were not.

Henriette was oppressed by a sense of melancholy. "Ah, merciful heaven!" she murmured, "how long will all this last, and shall we ever see him more!"

A more furious blast bent the sturdy trees out-doors and made the timbers of the old farmhouse creak and groan. Think of the sufferings the poor fellows would have to endure should the winter be severe, fighting in the snow, without bread, without fire!

"Bah!" rejoined Jean, "that's a very nice letter of his, and it's a comfort to have heard from him. We must not despair."

Thus, day by day, the month of October ran its course, with gray melancholy skies, and if ever the wind went down for a short space it was only to bring the clouds back in darker, heavier masses. Jean's wound was healing very slowly; the outflow from the drain was not the "laudable pus" which would have permitted the doctor to remove the appliance, and the patient was in a very enfeebled state, refusing, however, to be operated on in his dread of being left a cripple. An atmosphere of expectant resignation, disturbed at times by transient misgivings for which there was no apparent cause, pervaded the slumberous little chamber, to which the tidings from abroad came in vague, indeterminate shape, like the distorted visions of an evil dream. The hateful war, with its butcheries and disasters, was still raging out there in the world, in some quarter unknown to them, without their ever being able to learn the real course of events, without their being conscious of aught save the wails and groans that seemed to fill the air from their mangled, bleeding country. And the dead leaves rustled in the paths as the wind swept them before it beneath the gloomy sky, and over the naked fields brooded a funereal silence, broken only by the cawing of the crows, presage of a bitter winter.

A principal subject of conversation between them at this time was the hospital, which Henriette never left except to come and cheer Jean with her company. When she came in at evening he would question her, making the acquaintance of each of her charges, desirous to know who would die and who recover; while she, whose heart and soul were in her occupation, never wearied, but related the occurrences of the day in their minutest details.

"Ah," she would always say, "the poor boys, the poor boys!"

It was not the ambulance of the battlefield, where the blood from the wounded came in a fresh, bright stream, where the flesh the surgeon's knife cut into was firm and healthy; it was the decay and rottenness of the hospital, where the odor of fever and gangrene hung in the air, damp with the exhalations of the lingering convalescents and those who were dying by inches. Doctor Dalichamp had had the greatest difficulty in procuring the necessary beds, sheets and pillows, and every day he had to accomplish miracles to keep his patients alive, to obtain for them bread, meat and desiccated vegetables, to say nothing of bandages, compresses and other appliances. As the Prussian officers in charge of the military hospital in Sedan had refused him everything, even chloroform, he was accustomed to send to Belgium for what he required. And yet he had made no discrimination between French and Germans; he was even then caring for a dozen Bavarian soldiers who had been brought in there from Bazeilles. Those bitter adversaries who but a short time before had been trying to cut each other's throat now lay side by side, their passions calmed by suffering. And what abodes of distress and misery they were, those two long rooms in the old schoolhouse of Remilly, where, in the crude light that streamed through the tall windows, some thirty beds in each were arranged on either side of a narrow passage.

As late even as ten days after the battle wounded men had been discovered in obscure corners, where they had been overlooked, and brought in for treatment. There were four who had crawled into a vacant house at Balan and remained there, without attendance, kept from starving in some way, no one could tell how, probably by the charity of some kind-hearted neighbor, and their wounds were alive with maggots; they were as dead men, their system poisoned by the corruption that exuded from their wounds. There was a purulency, that nothing could check or overcome, that hovered over the rows of beds and emptied them. As soon as the door was passed one's nostrils were assailed by the odor of mortifying flesh. From drains inserted in festering sores fetid matter trickled, drop by drop. Oftentimes it became necessary to reopen old wounds in order to extract a fragment of bone that had been overlooked. Then abscesses would form, to break out after an interval in some remote portion of the body. Their strength all gone, reduced to skeletons, with ashen, clayey faces, the miserable wretches suffered the torments of the damned. Some, so weakened they could scarcely draw their breath, lay all day long upon their back, with tight shut, darkened eyes, like corpses in which decomposition had already set in; while others, denied the boon of sleep, tossing in restless wakefulness, drenched with the cold sweat that streamed from every pore, raved like lunatics, as if their suffering had made them mad. And whether they were calm or violent, it mattered not; when the contagion of the fever reached them, then was the end at hand, the poison doing its work, flying from bed to bed, sweeping them all away in one mass of corruption.

But worst of all was the condemned cell, the room to which were assigned those who were attacked by dysentery, typhus or small-pox. There were many cases of black small-pox. The patients writhed and shrieked in unceasing delirium, or sat erect in bed with the look of specters. Others had pneumonia and were wasting beneath the stress of their frightful cough. There were others again who maintained a continuous howling and were comforted only when their burning, throbbing wound was sprayed with cold water. The great hour of the day, the one that was looked forward to with eager expectancy, was that of the doctor's morning visit, when the beds were opened and aired and an opportunity was afforded their occupants to stretch their limbs, cramped by remaining long in one position. And it was the hour of dread and terror as well, for not a day passed that, as the doctor went his rounds, he was not pained to see on some poor devil's skin the bluish spots that denoted the presence of gangrene. The operation would be appointed for the following day, when a few more inches of the leg or arm would be sliced away. Often the gangrene kept mounting higher and higher, and amputation had to be repeated until the entire limb was gone.

Every evening on her return Henriette answered Jean's questions in the same tone of compassion:

"Ah, the poor boys, the poor boys!"

And her particulars never varied; they were the story of the daily recurring torments of that earthly hell. There had been an amputation at the shoulder-joint, a foot had been taken off, a humerus resected; but would gangrene or purulent contagion be clement and spare the patient? Or else they had been burying some one of their inmates, most frequently a Frenchman, now and then a German. Scarcely a day passed but a coarse coffin, hastily knocked together from four pine boards, left the hospital at the twilight hour, accompanied by a single one of the attendants, often by the young woman herself, that a fellow-creature might not be laid away in his grave like a dog. In the little cemetery at Remilly two trenches had been dug, and there they slumbered, side by side, French to the right, Germans to the left, their enmity forgotten in their narrow bed.

Jean, without ever having seen them, had come to feel an interest in certain among the patients. He would ask for tidings of them.

"And 'Poor boy,' how is he getting on today?"

This was a little soldier, a private in the 5th of the line, not yet twenty years old, who had doubtless enlisted as a volunteer. The by-name: "Poor boy" had been given him and had stuck because he always used the words in speaking of himself, and when one day he was asked the reason he replied that that was the name by which his mother had always called him. Poor boy he was, in truth, for he was dying of pleurisy brought on by a wound in his left side.

"Ah, poor fellow," replied Henriette, who had conceived a special fondness for this one of her charges, "he is no better; he coughed all the afternoon. It pained my heart to hear him."

"And your bear, Gutman, how about him?" pursued Jean, with a faint smile. "Is the doctor's report more favorable?"

"Yes, he thinks he may be able to save his life. But the poor man suffers dreadfully."

Although they both felt the deepest compassion for him, they never spoke of Gutman but a smile of gentle amusement came to their lips. Almost immediately upon entering on her duties at the hospital the young woman had been shocked to recognize in that Bavarian soldier the features: big blue eyes, red hair and beard and massive nose, of the man who had carried her away in his arms the day they shot her husband at Bazeilles. He recognized her as well, but could not speak; a musket ball, entering at the back of the neck, had carried away half his tongue. For two days she recoiled with horror, an involuntary shudder passed through her frame, each time she had to approach his bed, but presently her heart began to melt under the imploring, very gentle looks with which he followed her movements in the room. Was he not the blood-splashed monster, with eyes ablaze with furious rage, whose memory was ever present to her mind? It cost her an effort to recognize him now in that submissive, uncomplaining creature, who bore his terrible suffering with such cheerful resignation. The nature of his affliction, which is not of frequent occurrence, enlisted for him the sympathies of the entire hospital. It was not even certain that his name was Gutman; he was called so because the only sound he succeeded in articulating was a word of two syllables that resembled that more than it did anything else. As regarded all other particulars concerning him everyone was in the dark; it was generally believed, however, that he was married and had children. He seemed to understand a few words of French, for he would answer questions that were put to him with an emphatic motion of the head: "Married?" yes, yes! "Children?" yes, yes! The interest and excitement he displayed one day that he saw some flour induced them to believe he might have been a miller. And that was all. Where was the mill, whose wheel had ceased to turn? In what distant Bavarian village were the wife and children now weeping their lost husband and father? Was he to die, nameless, unknown, in that foreign country, and leave his dear ones forever ignorant of his fate?

"Today," Henriette told Jean one evening, "Gutman kissed his hand to me. I cannot give him a drink of water, or render him any other trifling service, but he manifests his gratitude by the most extravagant demonstrations. Don't smile; it is too terrible to be buried thus alive before one's time has come."

Toward the end of October Jean's condition began to improve. The doctor thought he might venture to remove the drain, although he still looked apprehensive whenever he examined the wound, which, nevertheless appeared to be healing as rapidly as could be expected. The convalescent was able to leave his bed, and spent hours at a time pacing his room or seated at the window, looking out on the cheerless, leaden sky. Then time began to hang heavy on his hands; he spoke of finding something to do, asked if he could not be of service on the farm. Among the secret cares that disturbed his mind was the question of money, for he did not suppose he could have lain there for six long weeks and not exhaust his little fortune of two hundred francs, and if Father Fouchard continued to afford him hospitality it must be that Henriette had been paying his board. The thought distressed him greatly; he did not know how to bring about an explanation with her, and it was with a feeling of deep satisfaction that he accepted the position of assistant at the farm, with the understanding that he was to help Silvine with the housework, while Prosper was to be continued in charge of the out-door labors.

Notwithstanding the hardness of the times Father Fouchard could well afford to take on another hand, for his affairs were prospering. While the whole country was in the throes of dissolution and bleeding at every limb, he had succeeded in so extending his butchering business that he was now slaughtering three and even four times as many animals as he had ever done before. It was said that since the 31st of August he had been carrying on a most lucrative business with the Prussians. He who on the 30th had stood at his door with his cocked gun in his hand and refused to sell a crust of bread to the starving soldiers of the 7th corps had on the following day, upon the first appearance of the enemy, opened up as dealer in all kinds of supplies, had disinterred from his cellar immense stocks of provisions, had brought back his flocks and herds from the fastnesses where he had concealed them; and since that day he had been one of the heaviest purveyors of meat to the German armies, exhibiting consummate address in bargaining with them and in getting his money promptly for his merchandise. Other dealers at times suffered great inconvenience from the insolent arbitrariness of the victors, whereas he never sold them a sack of flour, a cask of wine or a quarter of beef that he did not get his pay for it as soon as delivered in good hard cash. It made a good deal of talk in Remilly; people said it was scandalous on the part of a man whom the war had deprived of his only son, whose grave he never visited, but left to be cared for by Silvine; but nevertheless they all looked up to him with respect as a man who was making his fortune while others, even the shrewdest, were having a hard time of it to keep body and soul together. And he, with a sly leer out of his small red eyes, would shrug his shoulders and growl in his bull-headed way:

"Who talks of patriotism! I am more a patriot than any of them. Would you call it patriotism to fill those bloody Prussians' mouths gratis? What they get from me they have to pay for. Folks will see how it is some of these days!"

On the second day of his employment Jean remained too long on foot, and the doctor's secret fears proved not to be unfounded; the wound opened, the leg became greatly inflamed and swollen, he was compelled to take to his bed again. Dalichamp suspected that the mischief was due to a spicule of bone that the two consecutive days of violent exercise had served to liberate. He explored the wound and was so fortunate as to find the fragment, but there was a shock attending the operation, succeeded by a high fever, which exhausted all Jean's strength. He had never in his life been reduced to a condition of such debility: his recovery promised to be a work of time, and faithful Henriette resumed her position as nurse and companion in the little chamber, where winter with icy breath now began to make its presence felt. It was early

November, already the east wind had brought on its wings a smart flurry of snow, and between those four bare walls, on the uncarpeted floor where even the tall, gaunt old clothes-press seemed to shiver with discomfort, the cold was extreme. As there was no fireplace in the room they determined to set up a stove, of which the purring, droning murmur assisted to brighten their solitude a bit.

The days wore on, monotonously, and that first week of the relapse was to Jean and Henriette the dreariest and saddest in all their long, unsought intimacy. Would their suffering never end? were they to hope for no surcease of misery, the danger always springing up afresh? At every moment their thoughts sped away to Maurice, from whom they had received no further word. They were told that others were getting letters, brief notes written on tissue paper and brought in by carrier-pigeons. Doubtless the bullet of some hated German had slain the messenger that, winging its way through the free air of heaven, was bringing them their missive of joy and love. Everything seemed to retire into dim obscurity, to die and be swallowed up in the depths of the premature winter. Intelligence of the war only reached them a long time after the occurrence of events, the few newspapers that Doctor Dalichamp still continued to supply them with were often a week old by the time they reached their hands. And their dejection was largely owing to their want of information, to what they did not know and yet instinctively felt to be the truth, to the prolonged death-wail that, spite of all, came to their ears across the frozen fields in the deep silence that lay upon the country.

One morning the doctor came to them in a condition of deepest discouragement. With a trembling hand he drew from his pocket a Belgian newspaper and threw it on the bed, exclaiming:

"Alas, my friends, poor France is murdered; Bazaine has played the traitor!"

Jean, who had been dozing, his back supported by a couple of pillows, suddenly became wide-awake.

"What, a traitor?"

"Yes, he has surrendered Metz and the army. It is the experience of Sedan over again, only this time they drain us of our last drop of life-blood." Then taking up the paper and reading from it: "One hundred and fifty thousand prisoners, one hundred and fifty-three eagles and standards, one hundred and forty-one field guns, seventy-six machine guns, eight hundred casemate and barbette guns, three hundred thousand muskets, two thousand military train wagons, material for eighty-five batteries-"

And he went on giving further particulars: how Marshal Bazaine had been blockaded in Metz with the army, bound hand and foot, making no effort to break the wall of adamant that surrounded him; the doubtful relations that existed between him and Prince Frederick Charles, his indecision and fluctuating political combinations, his ambition to play a great role in history, but a role that he seemed not to have fixed upon himself; then all the dirty business of parleys and conferences, and the communications by means of lying, unsavory emissaries with Bismarck, King William and the Empress-regent, who in the end put her foot down and refused to negotiate with the enemy on the basis of a cession of territory; and, finally, the inevitable catastrophe, the completion of the web that destiny had been weaving, famine in Metz, a compulsory capitulation, officers and men, hope and courage gone, reduced to accept the bitter terms of the victor. France no longer had an army.

"In God's name!" Jean ejaculated in a deep, low voice. He had not fully understood it all, but until then Bazaine had always been for him the great captain, the one man to whom they were to look for salvation. "What is left us to do now? What will become of them at Paris?"

The doctor was just coming to the news from Paris, which was of a disastrous character. He called their attention to the fact that the paper from which he was reading was dated November 5. The surrender of Metz had been consummated on the 27th of October, and the tidings were not known in Paris until the 30th. Coming, as it did, upon the heels of the reverses recently sustained at Chevilly, Bagneux and la Malmaison, after the conflict at Bourget and the loss of that position, the intelligence had burst like a thunderbolt over the desperate populace, angered and disgusted by the feebleness and impotency of the government of National Defense. And thus it was that on the following day, the 31st, the city was threatened with a general insurrection, an immense throng of angry men, a mob ripe for mischief, collecting on the Place de l'Hotel de Ville, whence they swarmed into the halls and public offices, making prisoners the members of the Government, whom the National Guard rescued later in the day only because they feared the triumph of those incendiaries who were clamoring for the commune. And the Belgian journal wound up with a few stinging comments on the great City of Paris, thus torn by civil war when the enemy was at its gates. Was it not the presage of approaching decomposition, the puddle of blood and mire that was to engulf a world?

"That's true enough!" said Jean, whose face was very white. "They've no business to be squabbling when the Prussians are at hand!"

But Henriette, who had said nothing as yet, always making it her rule to hold her tongue when politics were under discussion, could not restrain a cry that rose from her heart. Her thoughts were ever with her brother.

"Mon Dieu, I hope that Maurice, with all the foolish ideas he has in his head, won't let himself get mixed up in this business!"

They were all silent in their distress; and it was the doctor, who was ardently patriotic, who resumed the conversation.

"Never mind; if there are no more soldiers, others will grow. Metz has surrendered, Paris may surrender, even; but it don't follow from that that France is wiped out. Yes, the strong-box is all right, as our peasants say, and we will live on in spite of all."

It was clear, however, that he was hoping against hope. He spoke of the army that was collecting on the Loire, whose initial performances, in the neighborhood of Arthenay, had not been of the most promising; it would become seasoned and would march to the relief of Paris. His enthusiasm was aroused to boiling pitch by the proclamations of Gambetta, who had left Paris by balloon on the 7th of October and two days later established his headquarters at Tours, calling on every citizen to fly to arms, and instinct with a spirit at once so virile and so sagacious that the entire country gave its adhesion to the dictatorial powers assumed for the public safety. And was there not talk of forming another army in the North, and yet another in the East, of causing soldiers to spring from the ground by sheer force of faith? It was to be the awakening of the provinces, the creation of all that was wanting by exercise of indomitable will, the determination to continue the struggle until the last sou was spent, the last drop of blood shed.

"Bah!" said the doctor in conclusion as he arose to go, "I have many a time given up a patient, and a week later found him as lively as a cricket."

Jean smiled. "Doctor, hurry up and make a well man of me, so I can go back to my post down yonder."

But those evil tidings left Henriette and him in a terribly disheartened state. There came another cold wave, with snow, and when the next day Henriette came in shivering from the hospital she told her friend that Gutman was dead. The intense cold had proved fatal to many among the wounded; it was emptying the rows of beds. The miserable man whom the loss of his tongue had condemned to silence had lain two days in the throes of death. During his last hour she had remained seated at his bedside, unable to resist the supplication of his pleading gaze. He seemed to be speaking to her with his tearful eyes, trying to tell, it may be, his real name and the name of the village, so far away, where a wife and little ones were watching for his return. And he had gone from them a stranger, known of none, sending her a last kiss with his uncertain, stiffening fingers, as if to thank her once again for all her gentle care. She was the only one who accompanied the remains to the cemetery, where the frozen earth, the unfriendly soil of the stranger's country, rattled with a dull, hollow sound on the pine coffin, mingled with flakes of snow.

The next day, again, Henriette said upon her return at evening:

"'Poor boy' is dead." She could not keep back her tears at mention of his name. "If you could but have seen and heard him in his pitiful delirium! He kept calling me: 'Mamma! mamma!' and stretched his poor thin arms out to me so entreatingly that I had to take him on my lap. His suffering had so wasted him that he was no heavier than a boy of ten, poor fellow. And I held and soothed him, so that he might die in peace; yes, I held him in my arms, I whom he called his mother and who was but a few years older than himself. He wept, and I myself could not restrain my tears; you can see I am weeping still-" Her utterance was choked with sobs; she had to pause. "Before his death he murmured several times the name which he had given himself: 'Poor boy, poor boy!' Ah, how just the designation! poor boys they are indeed, some of them so young and all so brave, whom your hateful war maims and mangles and causes to suffer so before they are laid away at last in their narrow bed!"

Never a day passed now but Henriette came in at night in this anguished state, caused by some new death, and the suffering of others had the effect of bringing them together even more closely still during the sorrowful hours that they spent, secluded from all the world, in the silent, tranquil chamber. And yet those hours were full of sweetness, too, for affection, a feeling which they believed to be a brother's and sister's love, had sprung up in those two hearts which little by little had come to know each other's worth. To him, with his observant, thoughtful nature, their long intimacy had proved an elevating influence, while she, noting his unfailing kindness of heart and evenness of temper, had ceased to remember that he was one of the lowly of the earth and had been a tiller of the soil before he became a soldier. Their understanding was perfect; they made a very good couple, as Silvine said with her grave smile. There was never the least embarrassment between them; when she dressed his leg the calm serenity that dwelt in the eyes of both was undisturbed. Always attired in black, in her widow's garments, it seemed almost as if she had ceased to be a woman.

But during those long afternoons when Jean was left to himself he could not help giving way to speculation. The sentiment he experienced for his friend was one of boundless gratitude, a sort of religious reverence, which would have made him repel the idea of love as if it were a sort of sacrilege. And yet he told himself that had he had a wife like her, so gentle, so loving, so helpful, his life would have been an earthly paradise. His great misfortune, his unhappy marriage, the evil years he had spent at Rognes, his wife's tragic end, all the sad past, arose before him with a softened feeling of regret, with an undefined hope for the future, but without distinct purpose to try another effort to master happiness. He closed his eyes and dropped off into a doze, and then he had a confused vision of being at Remilly, married again and owner of a bit of land, sufficient to support a family of honest folks whose wants were not extravagant. But it was all a dream, lighter than thistle-down; he knew it could never, never be. He believed his heart to be capable of no emotion stronger than friendship, he loved Henriette as he did solely because he was Maurice's brother. And then that vague dream of marriage had come to be in some measure a comfort to him, one of those fancies of the imagination that we know is never to be realized and with which we fondle ourselves in our hours of melancholy.

For her part, such thoughts had never for a moment presented themselves to Henriette's mind. Since the day of the horrible tragedy at Bazeilles her bruised heart had lain numb and lifeless in her bosom, and if consolation in the shape of a new affection had found its way thither, it could not be otherwise than without her knowledge; the latent movement of the seed deep-buried in the earth, which bursts its sheath and germinates, unseen of human eye. She failed even to perceive the pleasure it afforded her to remain for hours at a time by Jean's bedside, reading to him those newspapers that never brought them tidings save of evil. Never had her pulses beat more rapidly at the touch of his hand, never had she dwelt in dreamy rapture on the vision of the future with a longing to be loved once more. And yet it was in that chamber alone that she found comfort and oblivion. When she was there, busying herself with noiseless diligence for her patient's well-being, she was at peace; it seemed to her that soon her brother would return and all would be well, they would all lead a life of happiness together and never more be parted. And it appeared to her so natural that things should end thus that she talked of their relations without the slightest feeling of embarrassment, without once thinking to question her heart more closely, unaware that she had already made the chaste surrender of it.

But as she was on the point of leaving for the hospital one afternoon she looked into the kitchen as she passed and saw there a Prussian captain and two other officers, and the icy terror that filled her at the sight, then, for the first time, opened her eyes to the deep affection she had conceived for Jean. It was plain that the men had heard of the wounded man's presence at the farm and were come to claim him; he was to be torn from them and led away captive to the dungeon of some dark fortress deep in Germany. She listened tremblingly, her heart beating tumultuously.

The captain, a big, stout man, who spoke French with scarce a trace of foreign accent, was rating old Fouchard soundly.

"Things can't go on in this way; you are not dealing squarely by us. I came myself to give you warning, once for all, that if the thing happens again I shall take other steps to remedy it; and I promise you the consequences will not be agreeable."

Though entirely master of all his faculties the old scamp assumed an air of consternation, pretending not to understand, his mouth agape, his arms describing frantic circles on the air.

"How is that, sir, how is that?"

"Oh, come, there's no use attempting to pull the wool over my eyes; you know perfectly well that the three beeves you sold me on Sunday last were rotten-yes, diseased, and rotten through and through; they must have been where there was infection, for they poisoned my men; there are two of them in such a bad way that they may be dead by this time for all I know."

Fouchard's manner was expressive of virtuous indignation. "What, my cattle diseased! why, there's no better meat in all the country; a sick woman might feed on it to build her up!" And he whined and sniveled, thumping himself on the chest and calling God to witness he was an honest man; he would cut off his right hand rather than sell bad meat. For more than thirty years he had been known throughout the neighborhood, and not a living soul could say he had ever been wronged in weight or quality. "They were as sound as a dollar, sir, and if your men had the belly-ache it was because they ate too much-unless some villain hocussed the pot-"

And so he ran on, with such a flux of words and absurd theories that finally the captain, his patience exhausted, cut him short.

"Enough! You have had your warning; see you profit by it! And there is another matter: we have our suspicions that all you people of this village give aid and comfort to the francs-tireurs of the wood of Dieulet, who killed another of our sentries day before yesterday. Mind what I say; be careful!"

When the Prussians were gone Father Fouchard shrugged his shoulders with a contemptuous sneer. Why, yes, of course he sold them carcasses that had never been near the slaughter house; that was all they would ever get to eat from him. If a peasant had a cow die on his hands of the rinderpest, or if he found a dead ox lying in the ditch, was not the carrion good enough for those dirty Prussians? To say nothing of the pleasure there was in getting a big price out of them for tainted meat at which a dog would turn up his nose. He turned and winked slyly at Henriette, who was glad to have her fears dispelled, muttering triumphantly:

"Say, little girl, what do you think now of the wicked people who go about circulating the story that I am not a patriot? Why don't they do as I do, eh? sell the blackguards carrion and put their money in their pocket. Not a patriot! why, good Heavens! I shall have killed more of them with my diseased cattle than many a soldier with his chassepot!"

When the story reached Jean's ears, however, he was greatly disturbed. If the German authorities suspected that the people of Remilly were harboring the francs-tireurs from Dieulet wood they might at any time come and beat up his quarters and unearth him from his retreat. The idea that he should be the means of compromising his hosts or bringing trouble to Henriette was unendurable to him. Yielding to the young woman's entreaties, however, he consented to delay his departure yet for a few days, for his wound was very slow in healing and he was not strong enough to go away and join one of the regiments in the field, either in the North or on the Loire.

From that time forward, up to the middle of December, the stress of their anxiety and mental suffering exceeded even what had gone before. The cold was grown to be so intense that the stove no longer sufficed to heat the great, barn-like room. When they looked from their window on the crust of snow that covered the frozen earth they thought of Maurice, entombed down yonder in distant Paris, that was now become a city of death and desolation, from which they scarcely ever received reliable intelligence. Ever the same questions were on their lips: what was he doing, why did he not let them hear from him? They dared not voice their dreadful doubts and fears; perhaps he was ill, or wounded; perhaps even he was dead. The scanty and vague tidings that continued to reach them occasionally through the newspapers were not calculated to reassure them. After numerous lying reports of successful sorties, circulated one day only to be contradicted the next, there was a rumor of a great victory gained by General Ducrot at Champigny on the 2d of December; but they speedily learned that on the following day the general, abandoning the positions he had won, had been forced to recross the Marne and send his troops into cantonments in the wood of Vincennes. With each new day the Parisians saw themselves subjected to fresh suffering and privation: famine was beginning to make itself felt; the authorities, having first requisitioned horned cattle, were now doing the same with potatoes, gas was no longer furnished to private houses, and soon the fiery flight of the projectiles could be traced as they tore through the darkness of the unlighted streets. And so it was that neither of them could draw a breath or eat a mouthful without being haunted by the image of Maurice and those two million living beings, imprisoned in their gigantic sepulcher.

From every quarter, moreover, from the northern as well as from the central districts, most discouraging advices continued to arrive. In the north the 22d army corps, composed of gardes mobiles, depot companies from various regiments and such officers and men as had not been involved in the disasters of Sedan and Metz, had been forced to abandon Amiens and retreat on Arras, and on the 5th of December Rouen had also fallen into the hands of the enemy, after a mere pretense of resistance on the part of its demoralized, scanty garrison. In the center the victory of Coulmiers, achieved on the 3d of November by the army of the Loire, had resuscitated for a moment the hopes of the country: Orleans was to be reoccupied, the Bavarians were to be put to flight, the movement by way of Etampes was to culminate in the relief of Paris; but on December 5 Prince Frederick Charles had retaken Orleans and cut in two the army of the Loire, of which three corps fell back on Bourges and Vierzon, while the remaining two, commanded by General Chanzy, retired to Mans, fighting and falling back alternately for a whole week, most gallantly. The Prussians were everywhere, at Dijon and at Dieppe, at Vierzon as well as at Mans. And almost every morning came the intelligence of some fortified place that had capitulated, unable longer to hold out under the bombardment. Strasbourg had succumbed as early as the 28th of September, after standing forty-six days of siege and thirty-seven of shelling, her walls razed and her buildings riddled by more than two hundred thousand projectiles. The citadel of Laon had been blown into the air; Toul had surrendered; and following them, a melancholy catalogue, came Soissons with its hundred and twenty-eight pieces of artillery, Verdun, which numbered a hundred and thirty-six, Neufbrisach with a hundred, La Fere with seventy, Montmedy, sixty-five. Thionville was in flames, Phalsbourg had only opened her gates after a desperate resistance that lasted eighty days. It seemed as if all France were doomed to burn and be reduced to ruins by the never-ceasing cannonade.

One morning that Jean manifested a fixed determination to be gone, Henriette seized both his hands and held them tight clasped in hers.

"Ah, no! I beg you, do not go and leave me here alone. You are not strong enough; wait a few days yet, only a few days. I will let you go, I promise you I will, whenever the doctor says you are well enough to go and fight."

5

The cold was intense on that December evening. Silvine and Prosper, together with little Charlot, were alone in the great kitchen of the farmhouse, she busy with her sewing, he whittling away at a whip that he proposed should be more than usually ornate. It was seven o'clock; they had dined at six, not waiting for Father Fouchard, who they supposed had been detained at Raucourt, where there was a scarcity of meat, and Henriette, whose turn it was to watch that night at the hospital, had just left the house, after cautioning Silvine to be sure to replenish Jean's stove with coal before she went to bed.

Outside a sky of inky blackness overhung the white expanse of snow. No sound came from the village, buried among the drifts; all that was to be heard in the kitchen was the scraping of Prosper's knife as he fashioned elaborate rosettes and lozenges on the dogwood stock. Now and then he stopped and cast a glance at Charlot, whose flaxen head was nodding drowsily. When the child fell asleep at last the silence seemed more profound than ever. The mother noiselessly changed the position of the candle that the light might not strike the eyes of her little one; then sitting down to her sewing again, she sank into a deep reverie. And Prosper, after a further period of hesitation, finally mustered up courage to disburden himself of what he wished to say.

"Listen, Silvine; I have something to tell you. I have been watching for an opportunity to speak to you in private-"

Alarmed by his preface, she raised her eyes and looked him in the face.

"This is what it is. You'll forgive me for frightening you, but it is best you should be forewarned. In Remilly this morning, at the corner by the church, I saw Goliah; I saw him as plain as I see you sitting there. Oh, no! there can be no mistake; I was not dreaming!"

Her face suddenly became white as death; all she was capable of uttering was a stifled moan:

"My God! my God!"

Prosper went on, in words calculated to give her least alarm, and related what he had learned during the day by questioning one person and another. No one doubted now that Goliah was a spy, that he had formerly come and settled in the country with the purpose of acquainting himself with its roads, its resources, the most insignificant details pertaining to the life of its inhabitants. Men reminded one another of the time when he had worked for Father Fouchard on his farm and of his sudden disappearance; they spoke of the places he had had subsequently to that over toward Beaumont and Raucourt. And now he was back again, holding a position of some sort at the military post of Sedan, its duties apparently not very well defined, going about from one village to another, denouncing this man, fining that, keeping an eye to the filling of the requisitions that made the peasants' lives a burden to them. That very morning he had frightened the people of Remilly almost out of their wits in relation to a delivery of flour, alleging it was short in weight and had not been furnished within the specified time.

"You are forewarned," said Prosper in conclusion, "and now you'll know what to do when he shows his face here-"

She interrupted him with a terrified cry.

"Do you think he will come here?"

"Dame! it appears to me extremely probable he will. It would show great lack of curiosity if he didn't, since he knows he has a young one here that he has never seen. And then there's you, besides, and you're not so very homely but he might like to have another look at you."

She gave him an entreating glance that silenced his rude attempt at gallantry. Charlot, awakened by the sound of their voices, had raised his head. With the blinking eyes of one suddenly aroused from slumber he looked about the room, and recalled the words that some idle fellow of the village had taught him; and with the solemn gravity of a little man of three he announced:

"Dey're loafers, de Prussians!"

His mother went and caught him frantically in her arms and seated him on her lap. Ah! the poor little waif, at once her delight and her despair, whom she loved with all her soul and who brought the tears to her eyes every time she looked on him, flesh of her flesh, whom it wrung her heart to hear the urchins with whom he consorted in the street tauntingly call "the little Prussian!" She kissed him, as if she would have forced the words back into his mouth.

"Who taught my darling such naughty words? It's not nice; you must not say them again, my loved one."

Whereon Charlot, with the persistency of childhood, laughing and squirming, made haste to reiterate:

"Dey're dirty loafers, de Prussians!"

And when his mother burst into tears he clung about her neck and also began to howl dismally. Mon Dieu, what new evil was in store for her! Was it not enough that she had lost in Honore the one single hope of her life, the assured promise of oblivion and future happiness? and was that man to appear upon the scene again to make her misery complete?

"Come," she murmured, "come along, darling, and go to bed. Mamma will kiss her little boy all the same, for he does not know the sorrow he causes her."

And she went from the room, leaving Prosper alone. The good fellow, not to add to her embarrassment, had averted his eyes from her face and was apparently devoting his entire attention to his carving.

Before putting Charlot to bed it was Silvine's nightly custom to take him in to say good-night to Jean, with whom the youngster was on terms of great friendship. As she entered the room that evening, holding her candle before her, she beheld the convalescent seated upright in bed, his open eyes peering into the obscurity. What, was he not asleep? Faith, no; he had been ruminating on all sorts of subjects in the silence of the winter night; and while she was cramming the stove with coal he frolicked for a moment with Charlot, who rolled and tumbled on the bed like a young kitten. He knew Silvine's story, and had a very kindly feeling for the meek, courageous girl whom misfortune had tried so sorely, mourning the only man she had ever loved, her sole comfort that child of shame whose existence was a daily reproach to her. When she had replaced the lid on the stove, therefore, and came to the bedside to take the boy from his arms, he perceived by her red eyes that she had been weeping. What, had she been having more trouble? But she would not answer his question: some other day she would tell him what it was if it seemed worth the while. Mon Dieu! was not her life one of continual suffering now?

Silvine was at last lugging Charlot away in her arms when there arose from the courtyard of the farm a confused sound of steps and voices. Jean listened in astonishment.

"What is it? It can't be Father Fouchard returning, for I did not hear his wagon wheels." Lying on his back in his silent chamber, with nothing to occupy his mind, he had become acquainted with every detail of the routine of home life on the farm, of which the sounds were all familiar to his ears. Presently he added: "Ah, I see; it is those men again, the francs-tireurs from Dieulet, after something to eat."

295

"Quick, I must be gone!" said Silvine, hurrying from the room and leaving him again in darkness. "I must make haste and see they get their loaves."

A loud knocking was heard at the kitchen door and Prosper, who was beginning to tire of his solitude, was holding a hesitating parley with the visitors. He did not like to admit strangers when the master was away, fearing he might be held responsible for any damage that might ensue. His good luck befriended him in this instance, however, for just then Father Fouchard's carriole came lumbering up the acclivity, the tramp of the horse's feet resounding faintly on the snow that covered the road. It was the old man who welcomed the newcomers.

"Ah, good! it's you fellows. What have you on that wheelbarrow?"

Sambuc, lean and hungry as a robber and wrapped in the folds of a blue woolen blouse many times too large for him, did not even hear the farmer; he was storming angrily at Prosper, his honest brother, as he called him, who had only then made up his mind to unbar the door.

"Say, you! do you take us for beggars that you leave us standing in the cold in weather such as this?"

But Prosper did not trouble himself to make any other reply than was expressed in a contemptuous shrug of the shoulders, and while he was leading the horse off to the stable old Fouchard, bending over the wheelbarrow, again spoke up.

"So, it's two dead sheep you've brought me. It's lucky it's freezing weather, otherwise we should know what they are by the smell."

Cabasse and Ducat, Sambuc's two trusty henchmen, who accompanied him in all his expeditions, raised their voices in protest.

"Oh!" cried the first, with his loud-mouthed Provencal volubility, "they've only been dead three days. They're some of the animals that died on the Raffins farm, where the disease has been putting in its fine work of late."

"Procumbit humi bos," spouted the other, the ex-court officer whose excessive predilection for the ladies had got him into difficulties, and who was fond of airing his Latin on occasion.

Father Fouchard shook his head and continued to disparage their merchandise, declaring it was too "high." Finally he took the three men into the kitchen, where he concluded the business by saying:

"After all, they'll have to take it and make the best of it. It comes just in season, for there's not a cutlet left in Raucourt. When a man's hungry he'll eat anything, won't he?" And very well pleased at heart, he called to Silvine, who just then came in from putting Charlot to bed: "Let's have some glasses; we are going to drink to the downfall of old Bismarck."

Fouchard maintained amicable relations with these francs-tireurs from Dieulet wood, who for some three months past had been emerging at nightfall from the fastnesses where they made their lurking place, killing and robbing a Prussian whenever they could steal upon him unawares, descending on the farms and plundering the peasants when there was a scarcity of the other kind of game. They were the terror of all the villages in the vicinity, and the more so that every time a provision train was attacked or a sentry murdered the German authorities avenged themselves on the adjacent hamlets, the inhabitants of which they accused of abetting the outrages, inflicting heavy penalties on them, carrying off their mayors as prisoners, burning their poor hovels. Nothing would have pleased the peasants more than to deliver Sambuc and his band to the enemy, and they were only deterred from doing so by their fear of being shot in the back at a turn in the road some night should their attempt fail of success.

It had occurred to Fouchard to inaugurate a traffic with them. Roaming about the country in every direction, peering with their sharp eyes into ditches and cattle sheds, they had become his purveyors of dead animals. Never an ox or a sheep within a radius of three leagues was stricken down by disease but they came by night with their barrow and wheeled it away to him, and he paid them in provisions, most generally in bread, that Silvine baked in great batches expressly for the purpose. Besides, if he had no great love for them, he experienced a secret feeling of admiration for the francs-tireurs, a set of handy rascals who went their way and snapped their fingers at the world, and although he was making a fortune from his dealings with the Prussians, he could never refrain from chuckling to himself with grim, savage laughter as often as he heard that one of them had been found lying at the roadside with his throat cut.

"Your good health!" said he, touching glasses with the three men. Then, wiping his mouth with the back of his hand: "Say, have you heard of the fuss they're making over the two headless uhlans that they picked up over there near Villecourt? Villecourt was burned yesterday, you know; they say it was the penalty the village had to pay for harboring you. You'll have to be prudent, don't you see, and not show yourselves about here for a time. I'll see the bread is sent you somewhere."

Sambuc shrugged his shoulders and laughed contemptuously. What did he care for the Prussians, the dirty cowards! And all at once he exploded in a fit of anger, pounding the table with his fist.

"Tonnerre de Dieu! I don't mind the uhlans so much; they're not so bad, but it's the other one I'd like to get a chance at once-you know whom I mean, the other fellow, the spy, the man who used to work for you."

"Goliah?" said Father Fouchard.

Silvine, who had resumed her sewing, dropped it in her lap and listened with intense interest.

"That's his name, Goliah! Ah, the brigand! he is as familiar with every inch of the wood of Dieulet as I am with my pocket, and he's like enough to get us pinched some fine morning. I heard of him today at the Maltese Cross making his boast that he would settle our business for us before we're a week older. A dirty hound, he is, and he served as guide to the Prussians the day before the battle of Beaumont; I leave it to these fellows if he didn't."

"It's as true as there's a candle standing on that table!" attested Cabasse.

"Per silentia amica lunoe," added Ducat, whose quotations were not always conspicuous for their appositeness.

But Sambuc again brought his heavy fist down upon the table. "He has been tried and adjudged guilty, the scoundrel! If ever you hear of his being in the neighborhood just send me word, and his head shall go and keep company with the heads of the two uhlans in the Meuse; yes, by G-d! I pledge you my word it shall."

There was silence. Silvine was very white, and gazed at the men with unwinking, staring eyes.

"Those are things best not be talked too much about," old Fouchard prudently declared. "Your health, and good-night to you."

They emptied the second bottle, and Prosper, who had returned from the stable, lent a hand to load upon the wheelbarrow, whence the dead sheep had been removed, the loaves that Silvine had placed in an old grain-sack. But he turned his back and made no reply when his brother and the other two men, wheeling the barrow before them through the snow, stalked away and were lost to sight in the darkness, repeating:

"Good-night, good-night! an plaisir!"

They had breakfasted the following morning, and Father Fouchard was alone in the kitchen when the door was thrown open and Goliah in the flesh entered the room, big and burly, with the ruddy hue of health on his face and his tranquil smile. If the old man experienced anything in the nature of a shock at the suddenness of the apparition he let no evidence of it escape him. He peered at the other through his half-closed lids while he came forward and shook his former employer warmly by the hand.

"How are you, Father Fouchard?"

Then only the old peasant seemed to recognize him.

"Hallo, my boy, is it you? You've been filling out; how fat you are!"

And he eyed him from head to foot as he stood there, clad in a sort of soldier's greatcoat of coarse blue cloth, with a cap of the same material, wearing a comfortable, prosperous air of self-content. His speech betrayed no foreign accent, moreover; he spoke with the slow, thick utterance of the peasants of the district.

"Yes, Father Fouchard, it's I in person. I didn't like to be in the neighborhood without dropping in just to say how-do-you-do to you."

The old man could not rid himself of a feeling of distrust. What was the fellow after, anyway? Could he have heard of the francs-tireurs' visit to the farmhouse the night before? That was something he must try to ascertain. First of all, however, it would be best to treat him politely, as he seemed to have come there in a friendly spirit.

"Well, my lad, since you are so pleasant we'll have a glass together for old times' sake."

He went himself and got a bottle and two glasses. Such expenditure of wine went to his heart, but one must know how to be liberal when he has business on hand. The scene of the preceding night was repeated, they touched glasses with the same words, the same gestures.

"Here's to your good health, Father Fouchard."

"And here's to yours, my lad."

Then Goliath unbent and his face assumed an expression of satisfaction; he looked about him like a man pleased with the sight of objects that recalled bygone times. He did not speak of the past, however, nor, for the matter of that, did he speak of the present. The conversation ran on the extremely cold weather, which would interfere with farming operations; there was one good thing to be said for the snow, however: it would kill off the insects. He barely alluded, with a slightly pained expression, to the partially concealed hatred, the affright and scorn, with which he had been received in the other houses of Remilly. Every man owes allegiance to his country, doesn't he? It is quite clear he should serve his country as well as he knows how. In France, however, no one looked at the matter in that light; there were things about which people had very queer notions. And as the old man listened and looked at that broad, innocent, good-natured face, beaming with frankness and good-will, he said to himself that surely that excellent fellow had had no evil designs in coming there.

"So you are all alone today, Father Fouchard?"

"Oh, no; Silvine is out at the barn, feeding the cows. Would you like to see her?"

Goliath laughed. "Well, yes. To be quite frank with you, it was on Silvine's account that I came."

Old Fouchard felt as if a great load had been taken off his mind; he went to the door and shouted at the top of his voice:

"Silvine! Silvine! There's someone here to see you."

And he went away about his business without further apprehension, since the lass was there to look out for the property. A man must be in a bad way, he reflected, to let a fancy for a girl keep such a hold on him after such a length of time, years and years.

When Silvine entered the room she was not surprised to find herself in presence of Goliah, who remained seated and contemplated her with his broad smile, in which, however, there was a trace of embarrassment. She had been expecting him, and stood stock-still immediately she stepped across the doorsill, nerving herself and bracing all her faculties. Little Charlot came running up and hid among her petticoats, astonished and frightened to see a strange man there. Then succeeded a few seconds of awkward silence.

"And this is the little one, then?" Goliah asked at last in his most dulcet tone.

"Yes," was Silvine's curt, stern answer.

Silence again settled down upon the room. He had known there was a child, although he had gone away before the birth of his offspring, but this was the first time he had laid eyes on it. He therefore wished to explain matters, like a young man of sense who is confident he can give good reasons for his conduct.

"Come, Silvine, I know you cherish bitter feelings against me-and yet there is no reason why you should. If I went away, if I have been cause to you of so much suffering, you might have told yourself that perhaps it was because I was not my own master. When a man has masters over him he must obey them, mustn't he? If they had sent me off on foot to make a journey of a hundred leagues I should have been obliged to go. And, of course, I couldn't say a word to you about it; you have no idea how bad it made me feel to go away as I did without bidding you good-by. I won't say to you now that I felt certain I should return to you some day; still, I always fully expected that I should, and, as you see, here I am again-"

She had turned away her head and was looking through the window at the snow that carpeted the courtyard, as if resolved to hear no word he said. Her persistent silence troubled him; he interrupted his explanations to say:

"Do you know you are prettier than ever!"

True enough, she was very beautiful in her pallor, with her magnificent great eyes that illuminated all her face. The heavy coils of raven hair that crowned her head seemed the outward symbol of the inward sorrow that was gnawing at her heart.

"Come, don't be angry! you know that I mean you no harm. If I did not love you still I should not have come back, that's very certain. Now that I am here and everything is all right once more we shall see each other now and then, shan't we?"

She suddenly stepped a pace backward, and looking him squarely in the face:

"Never!"

"Never!-and why? Are you not my wife, is not that child ours?"

She never once took her eyes from off his face, speaking with impressive slowness:

"Listen to me; it will be better to end that matter once for all. You knew Honore; I loved him, he was the only man who ever had my love. And now he is dead; you robbed me of him, you murdered him over there on the battlefield, and never again will I be yours. Never!"

She raised her hand aloft as if invoking heaven to record her vow, while in her voice was such depth of hatred that for a moment he stood as if cowed, then murmured:

"Yes, I heard that Honore was dead; he was a very nice young fellow. But what could you expect? Many another has died as well; it is the fortune of war. And then it seemed to me that once he was dead there would no longer be a barrier between us, and let me remind you, Silvine, that after all I was never brutal toward you-"

But he stopped short at sight of her agitation; she seemed as if about to tear her own flesh in her horror and distress.

"Oh! that is just it; yes, it is that which seems as if it would drive me wild. Why, oh! why did I yield when I never loved you? Honore's departure left me so broken down, I was so sick in mind and body that never have I been able to recall any portion of the circumstances; perhaps it was because you talked to me of him and appeared to love him. My God! the long nights I have spent thinking of that time and weeping until the fountain of my tears was dry! It is dreadful to have done a thing that one had no wish to do and afterward be unable to explain the reason of it. And he had forgiven me, he had told me that he would marry me in spite of all when his time was out, if those hateful Prussians only let him live. And you think I will return to you. No, never, never! not if I were to die for it!"

Goliah's face grew dark. She had always been so submissive, and now he saw she was not to be shaken in her fixed resolve. Notwithstanding his easy-going nature he was determined he would have her, even if he should be compelled to use force, now that he was in a position to enforce his authority, and it was only his inherent prudence, the instinct that counseled him to patience and diplomacy, that kept him from resorting to violent measures now. The hard-fisted colossus was averse to bringing his physical powers into play; he therefore had recourse to another method for making her listen to reason.

"Very well; since you will have nothing more to do with me I will take away the child."

"What do you mean?"

Charlot, whose presence had thus far been forgotten by them both, had remained hanging to his mother's skirts, struggling bravely to keep down his rising sobs as the altercation waxed more warm. Goliah, leaving his chair, approached the group.

"You're my boy, aren't you? You're a good little Prussian. Come along with me."

But before he could lay hands on the child Silvine, all a-quiver with excitement, had thrown her arms about it and clasped it to her bosom.

"He, a Prussian, never! He's French, was born in France!"

"You say he's French! Look at him, and look at me; he's my very image. Can you say he resembles you in any one of his features?"

She turned her eyes on the big, strapping lothario, with his curling hair and beard and his broad, pink face, in which the great blue eyes gleamed like globes of polished porcelain; and it was only too true, the little one had the same yellow thatch, the same rounded cheeks, the same light eyes; every feature of the hated race was reproduced faithfully in him. A tress of her jet black hair that had escaped from its confinement and wandered down upon her shoulder in the agitation of the moment showed her how little there was in common between the child and her.

"I bore him; he is mine!" she screamed in fury. "He's French, and will grow up to be a Frenchman, knowing no word of your dirty German language; and some day he shall go and help to kill the whole pack of you, to avenge those whom you have murdered!"

Charlot, tightening his clasp about her neck, began to cry, shrieking:

"Mammy, mammy, I'm 'fraid! take me away!"

Then Goliah, doubtless because he did not wish to create a scandal, stepped back, and in a harsh, stern voice, unlike anything she had ever heard from his lips before, made this declaration:

"Bear in mind what I am about to tell you, Silvine. I know all that happens at this farm. You harbor the francs-tireurs from the wood of Dieulet, among them that Sambuc who is brother to your hired man; you supply the bandits with provisions. And I know that that hired man, Prosper, is a chasseur d'Afrique and a deserter, and belongs to us by rights. Further, I know that you are concealing on your premises a wounded man, another soldier, whom a word from me would suffice to consign to a German fortress. What do you think: am I not well informed?"

She was listening to him now, tongue-tied and terror-stricken, while little Charlot kept piping in her ear with lisping voice:

"Oh! mammy, mammy, take me away, I'm 'fraid!"

"Come," resumed Goliah, "I'm not a bad fellow, and I don't like quarrels and bickering, as you are well aware, but I swear by all that's holy I will have them all arrested, Father Fouchard and the rest, unless you consent to admit me to your chamber on Monday next. I will take the child, too, and send him away to Germany to my mother, who will be very glad to have him; for you have no further right to him, you know, if you are going to leave me. You understand me, don't you? The folks will all be gone, and all I shall have to do will be to come and carry him away. I am the master; I can do what pleases me-come, what have you to say?"

But she made no answer, straining the little one more closely to her breast as if fearing he might be torn from her then and there, and in her great eyes was a look of mingled terror and execration.

"It is well; I give you three days to think the matter over. See to it that your bedroom window that opens on the orchard is left open. If I do not find the window open next Monday evening at seven o'clock I will come with a detail the following day and arrest the inmates of the house and then will return and bear away the little one. Think of it well; au revoir, Silvine."

He sauntered quietly away, and she remained standing, rooted to her place, her head filled with such a swarming, buzzing crowd of terrible thoughts that it seemed to her she must go mad. And during the whole of that long day the tempest raged in her. At first the thought occurred to her instinctively to take her child in her arms and fly with him, wherever chance might direct, no matter where; but what would become of them when night should fall and envelop them in darkness? how earn a livelihood for him and for herself? Then she determined she would speak to Jean, would notify Prosper, and Father Fouchard himself, and again she hesitated and changed her mind: was she sufficiently certain of the friendship of those people that she could be sure they would not sacrifice her to the general safety, she who was cause that they were menaced all with such misfortune? No, she would say nothing to anyone; she would rely on her own efforts to extricate herself from the peril she had incurred by braving that bad man. But what scheme could she devise; mon Dieu! how could she avert the threatened evil, for her upright nature revolted; she could never have forgiven herself had she been the instrument of bringing disaster to so many people, to Jean in particular, who had always been so good to Charlot.

The hours passed, one by one; the next day's sun went down, and still she had decided upon nothing. She went about her household duties as usual, sweeping the kitchen, attending to the cows, making the soup. No word fell from her lips, and rising ever amid the ominous silence she preserved, her hatred of Goliah grew with every hour and impregnated her nature with its poison. He had been her curse; had it not been for him she would have waited for Honore, and Honore would be living now, and she would be happy. Think of his tone and manner when he made her understand he was the master! He had told her the truth, moreover; there were no longer gendarmes or judges to whom she could apply for protection; might made right. Oh, to be the stronger! to seize and overpower him when he came, he who talked of seizing others! All she considered was the child, flesh of her flesh; the chance-met father was naught, never had been aught, to her. She had no particle of wifely feeling toward him, only a sentiment of concentrated rage, the deep-seated hatred of the vanquished for the victor, when she thought of him. Rather than surrender the child to him she would have killed it, and killed herself afterward. And as she had told him, the child he had left her as a gift of hate she would have wished were already grown and capable of defending her; she looked into the future and beheld him with a musket, slaughtering hecatombs of Prussians. Ah, yes! one Frenchman more to assist in wreaking vengeance on the hereditary foe!

There was but one day remaining, however; she could not afford to waste more time in arriving at a decision. At the very outset, indeed, a hideous project had presented itself among the whirling thoughts that filled her poor, disordered mind: to notify the francs-tireurs, to give Sambuc the information he desired so eagerly; but the idea had not then assumed definite form and shape, and she had put it from her as too atrocious, not suffering herself even to consider it: was not that man the father of her child? she could not be accessory to his murder. Then the thought returned, and kept returning at more frequently recurring intervals, little by little forcing itself upon her and enfolding her in its unholy influence; and now it had entire possession of her, holding her captive by the strength of its simple and unanswerable logic. The peril and calamity that overhung them all would vanish with that man; he in his grave, Jean, Prosper, Father Fouchard would have nothing more to fear, while she herself would retain possession of Charlot and there would be never a one in all the world to challenge her right to him. All that day she turned and re-turned the project in her mind, devoid of further strength to bid it down, considering despite herself the murder in its different aspects, planning and arranging its most minute details. And now it was become the one fixed, dominant idea, making a portion of her being, that she no longer stopped to reason on, and when finally she came to act, in obedience to that dictate of the inevitable, she went forward as in a dream, subject to the volition of another, a someone within her whose presence she had never known till then.

Father Fouchard had taken alarm, and on Sunday he dispatched a messenger to the francs-tireurs to inform them that their supply of bread would be forwarded to the quarries of Boisville, a lonely spot a mile and a quarter from the house, and as Prosper had other work to do the old man sent Silvine with the wheelbarrow. It was manifest to the young woman that Destiny had taken the matter in its hands; she spoke, she made an appointment with Sambuc for the following evening, and there was no tremor in her voice, as if she were pursuing a course marked out for her from which she could not depart. The next day there were still other signs which proved that not only sentient beings, but inanimate objects as well, favored the crime. In the first place Father Fouchard was called suddenly away to Raucourt, and knowing he could not get back until after eight o'clock, instructed them not to wait dinner for him. Then Henriette, whose night off it was, received word from the hospital late in the afternoon that the nurse whose turn it was to watch was ill and she would have to take her place; and as Jean never left his chamber under any circumstances, the only remaining person from whom interference was to be feared was Prosper. It revolted the chasseur d'Afrique, the idea of killing a man that way, three against one, but when his brother arrived, accompanied by his faithful myrmidons, the disgust he felt for the villainous crew was lost in his detestation of the Prussians; sure he wasn't going to put himself out to save one of the dirty hounds, even if they did do him up in a way that was not according to rule; and he settled matters with his conscience by going to bed and burying his head under the blankets, that he might hear nothing that would tempt him to act in accordance with his soldierly instincts.

It lacked a quarter of seven, and Charlot seemed determined not to go to sleep. As a general thing his head declined upon the table the moment he had swallowed his last mouthful of soup.

"Come, my darling, go to sleep," said Silvine, who had taken him to Henriette's room; "mamma has put you in the nice lady's big bed."

302

But the child was excited by the novelty of the situation; he kicked and sprawled upon the bed, bubbling with laughter and animal spirits.

"No, no-stay, little mother-play, little mother."

She was very gentle and patient, caressing him tenderly and repeating:

"Go to sleep, my darling; shut your eyes and go to sleep, to please mamma."

And finally slumber overtook him, with a happy laugh upon his lips. She had not taken the trouble to undress him; she covered him warmly and left the room, and so soundly was he in the habit of sleeping that she did not even think it necessary to turn the key in the door.

Silvine had never known herself to be so calm, so clear and alert of mind. Her decision was prompt, her movements were light, as if she had parted company with her material frame and were acting under the domination of that other self, that inner being which she had never known till then. She had already let in Sambuc, with Cabasse and Ducat, enjoining upon them the exercise of the strictest caution, and now she conducted them to her bedroom and posted them on either side the window, which she threw open wide, notwithstanding the intense cold. The darkness was profound; barely a faint glimmer of light penetrated the room, reflected from the bosom of the snow without. A deathlike stillness lay on the deserted fields, the minutes lagged interminably. Then, when at last the deadened sound was heard of footsteps drawing near, Silvine withdrew and returned to the kitchen, where she seated herself and waited, motionless as a corpse, her great eyes fixed on the flickering flame of the solitary candle.

And the suspense was long protracted, Goliah prowling warily about the house before he would risk entering. He thought he could depend on the young woman, and had therefore come unarmed save for a single revolver in his belt, but he was haunted by a dim presentiment of evil; he pushed open the window to its entire extent and thrust his head into the apartment, calling below his breath:

"Silvine! Silvine!"

Since he found the window open to him it must be that she had thought better of the matter and changed her mind. It gave him great pleasure to have it so, although he would rather she had been there to welcome him and reassure his fears. Doubtless Father Fouchard had summoned her away; some odds and ends of work to finish up. He raised his voice a little:

"Silvine! Silvine!"

No answer, not a sound. And he threw his leg over the window-sill and entered the room, intending to get into bed and snuggle away among the blankets while waiting, it was so bitter cold.

All at once there was a furious rush, with the noise of trampling, shuffling feet, and smothered oaths and the sound of labored breathing. Sambuc and his two companions had thrown themselves on Goliah, and notwithstanding their superiority in numbers they found it no easy task to overpower the giant, to whom his peril lent tenfold strength. The panting of the combatants, the straining of sinews and cracking of joints, resounded for a moment in the obscurity. The revolver, fortunately, had fallen to the floor in the struggle. Cabasse's choking, inarticulate voice was heard exclaiming: "The cords, the cords!" and Ducat handed to Sambuc the coil of thin rope with which they had had the foresight to provide themselves. Scant ceremony was displayed in binding their hapless victim; the operation was conducted to the accompaniment of kicks and cuffs. The legs were secured first, then the arms were firmly pinioned to the sides, and finally they wound the cord at random many times around the Prussian's body, wherever his contortions would allow them to place it, with such an affluence of loops and knots that he had the appearance of being enmeshed in a gigantic net. To his unintermitting outcries Ducat's voice responded: "Shut your jaw!" and Cabasse silenced him more effectually by gagging him with an old blue handkerchief. Then, first waiting a moment to get their breath, they carried him, an inert mass, to the kitchen and deposited him upon the big table, beside the candle.

"Ah, the Prussian scum!" exclaimed Sambuc, wiping the sweat from his forehead, "he gave us trouble enough! Say, Silvine, light another candle, will you, so we can get a good view of the d--d pig and see what he looks like."

Silvine arose, her wide-dilated eyes shining bright from out her colorless face. She spoke no word, but lit another candle and came and placed it by Goliah's head on the side opposite the other; he produced the effect, thus brilliantly illuminated, of a corpse between two mortuary tapers. And in that brief moment their glances met; his was the wild, agonized look of the supplicant whom his fears have overmastered, but she affected not to understand, and withdrew to the sideboard, where she remained standing with her icy, unyielding air.

"The beast has nearly chewed my finger off," growled Cabasse, from whose hand blood was trickling. "I'm going to spoil his ugly mug for him."

He had taken the revolver from the floor and was holding it poised by the barrel in readiness to strike, when Sambuc disarmed him.

"No, no! none of that. We are not murderers, we francs-tireurs; we are judges. Do you hear, you dirty Prussian? we're going to try you; and you need have no fear, your rights shall be respected. We can't let you speak in your own defense, for if we should unmuzzle you you would split our ears with your bellowing, but I'll see that you have a lawyer presently, and a famous good one, too!"

He went and got three chairs and placed them in a row, forming what it pleased him to call the court, he sitting in the middle with one of his followers on either hand. When all three were seated he arose and commenced to speak, at first ironically aping the gravity of the magistrate, but soon launching into a tirade of blood-thirsty invective.

"I have the honor to be at the same time President of the Court and Public Prosecutor. That, I am aware, is not strictly in order, but there are not enough of us to fill all the roles. I accuse you, therefore, of entering France to play the spy on us, recompensing us for our hospitality with the most abominable treason. It is to you to whom we are principally indebted for our recent disasters, for after the battle of Nouart you guided the Bavarians across the wood of Dieulet by night to Beaumont. No one but a man who had lived a long time in the country and was acquainted with every path and cross-road could have done it, and on this point the conviction of the court is unalterable; you were seen conducting the enemy's artillery over roads that had become lakes of liquid mud, where eight horses had to be hitched to a single gun to drag it out of the slough. A person looking at those roads would hesitate to believe that an army corps could ever have passed over them. Had it not been for you and your criminal action in settling among us and betraying us the surprise of Beaumont would have never been, we should not have been compelled to retreat on Sedan, and perhaps in the end we might have come off victorious. I will say nothing of the disgusting career you have been pursuing since then, coming here in disguise, terrorizing and denouncing the poor country people, so that they tremble at the mention of your name. You have descended to a depth of depravity beyond which it is impossible to go, and I demand from the court sentence of death."

Silence prevailed in the room. He had resumed his seat, and finally, rising again, said:

"I assign Ducat to you as counsel for the defense. He has been sheriff's officer, and might have made his mark had it not been for his little weakness. You see that I deny you nothing; we are disposed to treat you well."

Goliah, who could not stir a finger, bent his eyes on his improvised defender. It was in his eyes alone that evidence of life remained, eyes that burned intensely with ardent supplication under the ashy brow, where the sweat of anguish stood in big drops, notwithstanding the cold.

Ducat arose and commenced his plea. "Gentlemen, my client, to tell the truth, is the most noisome blackguard that I ever came across in my life, and I should not have been willing to appear in his defense had I not a mitigating circumstance to plead, to wit: they are all that way in the country he came from. Look at him closely; you will read his astonishment in his eyes; he does not understand the gravity of his offense. Here in France we may employ spies, but no one would touch one of them unless with a pair of pincers, while in that country espionage is considered a highly honorable career and an extremely meritorious manner of serving the state. I will even go so far as to say, gentlemen, that possibly they are not wrong; our noble sentiments do us honor, but they have also the disadvantage of bringing us defeat. If I may venture to speak in the language of Cicero and Virgil, *quos vult perdere Jupiter dementat.* You will understand the allusion, gentlemen."

And he took his seat again, while Sambuc resumed:

"And you, Cabasse, have you nothing to say either for or against the defendant?"

"All I have to say," shouted the Provencal, "is that we are wasting a deal of breath in settling that scoundrel's hash. I've had my little troubles in my lifetime, and plenty of 'em, but I don't like to see people trifle with the affairs of the law; it's unlucky. Let him die, I say!"

Sambuc rose to his feet with an air of profound gravity.

"This you both declare to be your verdict, then-death?"

"Yes, yes! death!"

The chairs were pushed back, he advanced to the table where Goliah lay, saying:

"You have been tried and sentenced; you are to die."

The flame of the two candles rose about their unsnuffed wicks and flickered in the draught, casting a fitful, ghastly light on Goliah's distorted features. The fierce efforts he made to scream for mercy, to vociferate the words that were strangling him, were such that the handkerchief knotted across his mouth was drenched with spume, and it was a sight most horrible to see, that strong man reduced to silence, voiceless already as a corpse, about to die with that torrent of excuse and entreaty pent in his bosom.

Cabasse cocked the revolver. "Shall I let him have it?" he asked.

"No, no!" Sambuc shouted in reply; "he would be only too glad." And turning to Goliah: "You are not a soldier; you are not worthy of the honor of quitting the world with a bullet in your head. No, you shall die the death of a spy and the dirty pig that you are."

He looked over his shoulder and politely said:

"Silvine, if it's not troubling you too much, I would like to have a tub."

During the whole of the trial scene Silvine had not moved a muscle. She had stood in an attitude of waiting, with drawn, rigid features, as if mind and body had parted company, conscious of nothing but the one fixed idea that had possessed her for the last two days. And when she was asked for a tub she received the request as a matter of course and proceeded at once to comply with it, disappearing into the adjoining shed, whence she returned with the big tub in which she washed Charlot's linen.

"Hold on a minute! place it under the table, close to the edge."

305

She placed the vessel as directed, and as she rose to her feet her eyes again encountered Goliah's. In the look of the poor wretch was a supreme prayer for mercy, the revolt of the man who cannot bear the thought of being stricken down in the pride of his strength. But in that moment there was nothing of the woman left in her; nothing but the fierce desire for that death for which she had been waiting as a deliverance. She retreated again to the buffet, where she remained standing in silent expectation.

Sambuc opened the drawer of the table and took from it a large kitchen knife, the one that the household employed to slice their bacon.

"So, then, as you are a pig, I am going to stick you like a pig."

He proceeded in a very leisurely manner, discussing with Cabasse, and Ducat the proper method of conducting the operation. They even came near quarreling, because Cabasse alleged that in Provence, the country he came from, they hung pigs up by the heels to stick them, at which Ducat expressed great indignation, declaring that the method was a barbarous and inconvenient one.

"Bring him well forward to the edge of the table, his head over the tub, so as to avoid soiling the floor."

They drew him forward, and Sambuc went about his task in a tranquil, decent manner. With a single stroke of the keen knife he slit the throat crosswise from ear to ear, and immediately the blood from the severed carotid artery commenced to drip, drip into the tub with the gentle plashing of a fountain. He had taken care not to make the incision too deep; only a few drops spurted from the wound, impelled by the action of the heart. Death was the slower in coming for that, but no convulsion was to be seen, for the cords were strong and the body was utterly incapable of motion. There was no death-rattle, not a quiver of the frame. On the face alone was evidence of the supreme agony, on that terror-distorted mask whence the blood retreated drop by drop, leaving the skin colorless, with a whiteness like that of linen. The expression faded from the eyes; they became dim, the light died from out them.

"Say, Silvine, we shall want a sponge, too."

She made no reply, standing riveted to the floor in an attitude of unconsciousness, her arms folded tightly across her bosom, her throat constricted as by the clutch of a mailed hand, gazing on the horrible spectacle. Then all at once she perceived that Charlot was there, grasping her skirts with his little hands; he must have awaked and managed to open the intervening doors, and no one had seen him come stealing in, childlike, curious to know what was going on. How long had he been there, half-concealed behind his mother? From beneath his shock of yellow hair his big blue eyes were fixed on the trickling blood, the thin red stream that little by little was filling the tub. Perhaps he had not understood at first and had found something diverting in the sight, but suddenly he seemed to become instinctively aware of all the abomination of the thing; he gave utterance to a sharp, startled cry:

"Oh, mammy! oh, mammy! I'm 'fraid, take me away!"

It gave Silvine a shock, so violent that it convulsed her in every fiber of her being. It was the last straw; something seemed to give way in her, the excitement that had sustained her for the last two days while under the domination of her one fixed idea gave way to horror. It was the resurrection of the dormant woman in her; she burst into tears, and with a frenzied movement caught Charlot up and pressed him wildly to her heart. And she fled with him, running with distracted terror, unable to see or hear more, conscious of but one overmastering need, to find some secret spot, it mattered not where, in which she might cast herself upon the ground and seek oblivion.

It was at this crisis that Jean rose from his bed and, softly opening his door, looked out into the passage. Although he generally gave but small attention to the various noises that reached him from the farmhouse, the unusual activity that prevailed this evening, the trampling of feet, the shouts and cries, in the end excited his curiosity. And it was to the retirement of his sequestered chamber that Silvine, sobbing and disheveled, came for shelter, her form convulsed by such a storm of anguish that at first he could not grasp the meaning of the rambling, inarticulate words that fell from her blanched lips. She kept constantly repeating the same terrified gesture, as if to thrust from before her eyes some hideous, haunting vision. At last he understood, the entire abominable scene was pictured clearly to his mind: the traitorous ambush, the slaughter, the mother, her little one clinging to her skirts, watching unmoved the murdered father, whose life-blood was slowly ebbing; and it froze his marrow—the peasant and the soldier was sick at heart with anguished horror. Ah, hateful, cruel war! that changed all those poor folks to ravening wolves, bespattering the child with the father's blood! An accursed sowing, to end in a harvest of blood and tears!

Resting on the chair where she had fallen, covering with frantic kisses little Charlot, who clung, sobbing, to her bosom, Silvine repeated again and again the one unvarying phrase, the cry of her bleeding heart.

"Ah, my poor child, they will no more say you are a Prussian! Ah, my poor child, they will no more say you are a Prussian!"

Meantime Father Fouchard had returned and was in the kitchen. He had come hammering at the door with the authority of the master, and there was nothing left to do but open to him. The surprise he experienced was not exactly an agreeable one on beholding the dead man outstretched on his table and the blood-filled tub beneath. It followed naturally, his disposition not being of the mildest, that he was very angry.

"You pack of rascally slovens! say, couldn't you have gone outdoors to do your dirty work? Do you take my place for a shambles, eh? coming here and ruining the furniture with such goings-on?" Then, as Sambuc endeavored to mollify him and explain matters, the old fellow went on with a violence that was enhanced by his fears: "And what do you suppose I am to do with the carcass, pray? Do you consider it a gentlemanly thing to do, to come to a man's house like this and foist a stiff off on him without so much as saying by your leave? Suppose a patrol should come along, what a nice fix I should be in! but precious little you fellows care whether I get my neck stretched or not. Now listen: do you take that body at once and carry it away from here; if you don't, by G-d, you and I will have a settlement! You hear me; take it by the head, take it by the heels, take it any way you please, but get it out of here and don't let there be a hair of it remaining in this room at the end of three minutes from now!"

In the end Sambuc prevailed on Father Fouchard to let him have a sack, although it wrung the old miser's heartstrings to part with it. He selected one that was full of holes, remarking that anything was good enough for a Prussian. Cabasse and Ducat had all the trouble in the world to get Goliath into it; it was too short and too narrow for the long, broad body, and the feet protruded at its mouth. Then they carried their burden outside and placed it on the wheelbarrow that had served to convey to them their bread.

"You'll not be troubled with him any more, I give you my word of honor!" declared Sambuc. "We'll go and toss him into the Meuse."

"Be sure and fasten a couple of big stones to his feet," recommended Fouchard, "so the lubber shan't come up again."

And the little procession, dimly outlined against the white waste of snow, started and soon was buried in the blackness of the night, giving no sound save the faint, plaintive creaking of the barrow.

In after days Sambuc swore by all that was good and holy he had obeyed the old man's directions, but none the less the corpse came to the surface and was discovered two days afterward by the Prussians among the weeds at Pont-Maugis, and when they saw the manner of their countryman's murder, his throat slit like a pig, their wrath and fury knew no bounds. Their threats were terrible, and were accompanied by domiciliary visits and annoyances of every kind. Some of the villagers must have blabbed, for there came a party one night and arrested Father Fouchard and the Mayor of Remilly on the charge of giving aid and comfort to the francs-tireurs, who were manifestly the perpetrators of the crime. And Father Fouchard really came out very strong under those untoward circumstances, exhibiting all the impassability of a shrewd old peasant, who knew the value of silence and a tranquil demeanor. He went with his captors without the least sign of perturbation, without even asking them for an explanation. The truth would come out. In the country roundabout it was whispered that he had already made an enormous fortune from the Prussians, sacks and sacks of gold pieces, that he buried away somewhere, one by one, as he received them.

All these stories were a terrible source of alarm to Henriette when she came to hear of them. Jean, fearing he might endanger the safety of his hosts, was again eager to get away, although the doctor declared he was still too weak, and she, saddened by the prospect of their approaching separation, insisted on his delaying his departure for two weeks. At the time of Father Fouchard's arrest Jean had escaped a like fate by hiding in the barn, but he was liable to be taken and led away captive at any moment should there be further searches made. She was also anxious as to her uncle's fate, and so she resolved one morning to go to Sedan and see the Delaherches, who had, it was said, a Prussian officer of great influence quartered in their house.

"Silvine," she said, as she was about to start, "take good care of our patient; see he has his bouillon at noon and his medicine at four o'clock."

The maid of all work, ever busy with her daily recurring tasks, was again the submissive and courageous woman she had been of old; she had the care of the farm now, moreover, in the absence of the master, while little Charlot was constantly at her heels, frisking and gamboling around her.

"Have no fear, madame, he shall want for nothing. I am here and will look out for him."

6

Life had fallen back into something like its accustomed routine with the Delaherches at their house in the Rue Maqua after the terrible shock of the capitulation, and for nearly four months the long days had been slowly slipping by under the depressing influence of the Prussian occupation.

There was one corner, however, of the immense structure that was always closed, as if it had no occupant: it was the chamber that Colonel de Vineuil still continued to inhabit, at the extreme end of the suite where the master and his family spent their daily life. While the other windows were thrown open, affording evidence by sight and sound of the activity that prevailed within, those of that room were dark and lifeless, their blinds invariably drawn. The colonel had complained that the daylight hurt his eyes; no one knew whether or not this was strictly true, but a lamp was kept burning at his bedside day and night to humor him in his fancy. For two long months he had kept his bed, although Major Bouroche asserted there was nothing more serious than a contusion of the ankle and a fragment of bone chipped away; the wound refused to heal and complications of various kinds had ensued. He was able to get up now, but was in such a state of utter mental prostration, his mysterious ailment had taken such firm hold upon his system, that he was content to spend his days in idleness, stretched on a lounge before a great wood fire. He had wasted away until he was little more than a shadow, and still the physician who was attending him could find no lesion to account for that lingering death. He was slowly fading away, like the flame of a lamp in which the supply of oil is giving out.

Mme. Delaherche, the mother, had immured herself there with him on the day succeeding the occupation. No doubt they understood each other, and had expressed in two words, once for all, their common purpose to seclude themselves in that apartment so long as there should be Prussians quartered in the house. They had afforded compulsory hospitality to many of the enemy for various lengths of time; one, a Captain, M. Gartlauben, was there still, had taken up his abode with them permanently. But never since that first day had mention of those things passed the colonel's and the old lady's lips. Notwithstanding her seventy-eight years she was up every morning soon as it was day and came and took her position in the fauteuil that was awaiting her in the chimney nook opposite her old friend. There, by the steady, tranquil lamplight, she applied herself industriously to knitting socks for the children of the poor, while he, his eyes fixed on the crumbling brands, with no occupation for body or mind, was as one already dead, in a state of constantly increasing stupor. They certainly did not exchange twenty words in the course of a day; whenever she, who still continued to go about the house at intervals, involuntarily allowed some bit of news from the outer world to escape her lips, he silenced her with a gesture, so that no tidings of the siege of Paris, the disasters on the Loire and all the daily renewed horrors of the invasion had gained admission there. But the colonel might stop his ears and shut out the light of day as he would in his self-appointed tomb; the air he breathed must have brought him through key-hole and crevices intelligence of the calamity that was everywhere throughout the land, for every new day beheld him sinking, slowly dying, despite his determination not to know the evil news.

While matters were in this condition at one end of the house Delaherche, who was never contented unless occupied, was bustling about and making attempts to start up his business once more, but what with the disordered condition of the labor market and the pecuniary embarrassment of many among his customers, he had so far only put a few looms in motion. Then it occurred to him, as a means of killing the time that hung heavy on his hands, to make a complete inventory of his business and perfect certain changes and improvements that he had long had in mind. To assist him in his labors he had just then at his disposal a young man, the son of an old business acquaintance, who had drifted in on him after the battle. Edmond Lagarde, who, although he was twenty-three years old, would not have been taken for more than eighteen, had grown to man's estate in his father's little dry-goods shop at Passy; he was a sergeant in the 5th line regiment and had fought with great bravery throughout the campaign, so much so that he had been knocked over near the Minil gate about five o'clock, when the battle was virtually ended, his left arm shattered by one of the last shots fired that day, and Delaherche, when the other wounded were removed from the improvised ambulance in the drying room, had good-naturedly received him as an inmate of his house. It was under these circumstances that Edmond was now one of the family, having an apartment in the house and taking his meals at the common table, and, now that his wound was healed, acting as a sort of secretary to the manufacturer while waiting for a chance to get back to Paris. He had signed a parole binding himself not to attempt to leave the city, and owing to this and to his protector's influence the Prussian authorities did not interfere with him. He was fair, with blue eyes, and pretty as a woman; so timid withal that his face assumed a beautiful hue of rosy red whenever anyone spoke to him. He had been his mother's darling; she had impoverished herself, expending all the profits of their little business to send him to college. And he adored Paris and bewailed his compulsory absence from it when talking to Gilberte, did this wounded cherub, whom the young woman had displayed great good-fellowship in nursing.

Finally, their household had received another addition in the person of M. de Gartlauben, a captain in the German landwehr, whose regiment had been sent to Sedan to supply the place of troops dispatched to service in the field. He was a personage of importance, notwithstanding his comparatively modest rank, for he was nephew to the governor-general, who, from his headquarters at Rheims, exercised unlimited power over all the district. He, too, prided himself on having lived at Paris, and seized every occasion ostentatiously to show he was not ignorant of its pleasures and refinements; concealing beneath this film of varnish his inborn rusticity, he assumed as well as he was able the polish of one accustomed to good society. His tall, portly form was always tightly buttoned in a close-fitting uniform, and he lied outrageously about his age, never being able to bring himself to own up to his forty-five years. Had he had more intelligence he might have made himself an object of greater dread, but as it was his over-weening vanity, kept him in a continual state of satisfaction with himself, for never could such a thing have entered his mind as that anyone could dare to ridicule him.

At a subsequent period he rendered Delaherche services that were of inestimable value. But what days of terror and distress were those that followed upon the heels of the capitulation! the city, overrun with German soldiery, trembled in momentary dread of pillage and conflagration. Then the armies of the victors streamed away toward the valley of the Seine, leaving behind them only sufficient men to form a garrison, and the quiet that settled upon the place was that of a necropolis: the houses all closed, the shops shut, the streets deserted as soon as night closed in, the silence unbroken save for the hoarse cries and heavy tramp of the patrols. No letters or newspapers reached them from the outside world; Sedan was become a dungeon, where the immured citizens waited in agonized suspense for the tidings of disaster with which the air was instinct. To render their misery complete they were threatened with famine; the city awoke one morning from its slumbers to find itself destitute of bread and meat and the country roundabout stripped naked, as if a devouring swarm of locusts had passed that way, by the hundreds of thousands of men who for a week past had been pouring along its roads and across its fields in a devastating torrent. There were provisions only for two days, and the authorities were compelled to apply to Belgium for relief; all supplies now came from their neighbors across the frontier, whence the customs guards had disappeared, swept away like all else in the general cataclysm. Finally there were never-ending vexations and annoyances, a conflict that commenced to rage afresh each morning between the Prussian governor and his underlings, quartered at the Sous-Prefecture, and the Municipal Council, which was in permanent session at the Hotel de Ville. It was all in vain that the city fathers fought like heroes, discussing, objecting, protesting, contesting the ground inch by inch; the inhabitants had to succumb to the exactions that constantly became more burdensome, to the whims and unreasonableness of the stronger.

In the beginning Delaherche suffered great tribulation from the officers and soldiers who were billeted on him. It seemed as if representatives from every nationality on the face of the globe presented themselves at his door, pipe in mouth. Not a day passed but there came tumbling in upon the city two or three thousand men, horse, foot and dragoons, and although they were by rights entitled to nothing more than shelter and firing, it was often found expedient to send out in haste and get them provisions. The rooms they occupied were left in a shockingly filthy condition. It was not an infrequent occurrence that the officers came in drunk and made themselves even more obnoxious than their men. Such strict discipline was maintained, however, that instances of violence and marauding were rare; in all Sedan there were but two cases reported of outrages committed on women. It was not until a later period, when Paris displayed such stubbornness in her resistance, that, exasperated by the length to which the struggle was protracted, alarmed by the attitude of the provinces and fearing a general rising of the populace, the savage war which the francs-tireurs had inaugurated, they laid the full weight of their heavy hand upon the suffering people.

Delaherche had just had an experience with a lodger who had been quartered on him, a captain of cuirassiers, who made a practice of going to bed with his boots on and when he went away left his apartment in an unmentionably filthy condition, when in the last half of September Captain de Gartlauben came to his door one evening when it was raining in torrents. The first hour he was there did not promise well for the pleasantness of their future relations; he carried matters with a high hand, insisting that he should be given the best bedroom, trailing the scabbard of his sword noisily up the marble staircase; but encountering Gilberte in the corridor he drew in his horns, bowed politely, and passed stiffly on. He was courted with great obsequiousness, for everyone was well aware that a word from him to the colonel commanding the post of Sedan would suffice to mitigate a requisition or secure the release of a friend or relative. It was not very long since his uncle, the governor-general at Rheims, had promulgated a particularly detestable and cold-blooded order, proclaiming martial law and decreeing the penalty of death to whomsoever should give aid and comfort to the enemy, whether by acting for them as a spy, by leading astray German troops that had been entrusted to their guidance, by destroying bridges and artillery, or by damaging the railroads and telegraph lines. The enemy meant the French, of course, and the citizens scowled and involuntarily doubled their fists as they read the great white placard nailed against the door of post headquarters which attributed to them as a crime their best and most sacred aspirations. It was so hard, too, to have to receive their intelligence of German victories through the cheering of the garrison! Hardly a day passed over their heads that they were spared this bitter humiliation; the soldiers would light great fires and sit around them, feasting and drinking all night long, while the townspeople, who were not allowed to be in the streets after nine o'clock, listened to the tumult from the depths of their darkened houses, crazed with suspense, wondering what new catastrophe had befallen. It was on one of these occasions, somewhere about the middle of October, that M. de Gartlauben for the first time proved himself to be possessed of some delicacy of feeling. Sedan had been jubilant all that day with renewed hopes, for there was a rumor that the army of the Loire, then marching to the relief of Paris, had gained a great victory; but how many times before had the best of news been converted into tidings of disaster! and sure enough, early in the evening it became known for certain that the Bavarians had taken Orleans. Some soldiers had collected in a house across the way from the factory in the Rue Maqua, and were so boisterous in their rejoicings that the Captain, noticing Gilberte's annoyance, went and silenced them, remarking that he himself thought their uproar ill-timed.

Toward the close of the month M. de Gartlauben was in position to render some further trifling services. The Prussian authorities, in the course of sundry administrative reforms inaugurated by them, had appointed a German Sous-Prefect, and although this step did not put an end to the exactions to which the city was subjected, the new official showed himself to be comparatively reasonable. One of the most frequent among the causes of difference that were constantly springing up between the officers of the post and the municipal council was that which arose from the custom of requisitioning carriages for the use of the staff, and there was a great hullaballoo raised one morning that Delaherche failed to send his caleche and pair to the Sous-Prefecture: the mayor was arrested and the manufacturer would have gone to keep him company up in the citadel had it not been for M. de Gartlauben, who promptly quelled the rising storm. Another day he secured a stay of proceedings for the city, which had been mulcted in the sum of thirty thousand francs to punish it for its alleged dilatoriness in rebuilding the bridge of Villette, a bridge that the Prussians themselves had destroyed: a disastrous piece of business that was near being the ruin of Sedan. It was after the surrender at Metz, however, that Delaherche contracted his main debt of gratitude to his guest. The terrible news burst on the citizens like a thunderclap, dashing to the ground all their remaining hopes, and early in the ensuing week the streets again began to be encumbered with the countless hosts of the German forces, streaming down from the conquered fortress: the army of Prince Frederick Charles moving on the Loire, that of General Manteuffel, whose destination was Amiens and Rouen, and other corps on the march to reinforce the besiegers before Paris. For several days the houses were full to overflowing with soldiers, the butchers' and bakers' shops were swept clean, to the last bone, to the last crumb; the streets were pervaded by a greasy, tallowy odor, as after the passage of the great migratory bands of olden times. The buildings in the Rue Maqua, protected by a friendly influence, escaped the devastating irruption, and were only called on to give shelter to a few of the leaders, men of education and refinement.

Owing to these circumstances, Delaherche at last began to lay aside his frostiness of manner. As a general thing the bourgeois families shut themselves in their apartments and avoided all communication with the officers who were billeted on them; but to him, who was of a sociable nature and liked to extract from life what enjoyment it had to offer, this enforced sulkiness in the end became unbearable. His great, silent house, where the inmates lived apart from one another in a chill atmosphere of distrust and mutual dislike, damped his spirits terribly. He began by stopping M. de Gartlauben on the stairs one day to thank him for his favors, and thus by degrees it became a regular habit with the two men to exchange a few words when they met. The result was that one evening the Prussian captain found himself seated in his host's study before the fireplace where some great oak logs were blazing, smoking a cigar and amicably discussing the news of the day. For the first two weeks of their new intimacy Gilberte did not make her appearance in the room; he affected to ignore her existence, although, at every faintest sound, his glance would be directed expectantly upon the door of the connecting apartment. It seemed to be his object to keep his position as an enemy as much as possible in the background, trying to show he was not narrow-minded or a bigoted patriot, laughing and joking pleasantly over certain rather ridiculous requisitions. For example, a demand was made one day for a coffin and a shroud; that shroud and coffin afforded him no end of amusement. As regarded other things, such as coal, oil, milk, sugar, butter, bread, meat, to say nothing of clothing, stoves and lamps-all the necessaries of daily life, in a word-he shrugged his shoulders: mon Dieu! what would you have? No doubt it was vexatious; he was even willing to admit that their demands were excessive, but that was how it was in war times; they had to keep themselves alive in the enemy's country. Delaherche, who was very sore over these incessant requisitions, expressed his opinion of them with frankness, pulling them to pieces mercilessly at their nightly confabs, in much the same way as he might have criticised the cook's kitchen accounts. On only one occasion did their discussion become at all acrimonious, and that was in relation to the impost of a million francs that the Prussian prefet at Rethel had levied on the department of the Ardennes, the alleged pretense of which was to indemnify Germany for damages caused by French ships of war and by the expulsion of Germans domiciled in French territory. Sedan's proportionate share of the assessment was forty-two thousand francs. And he labored strenuously with his visitor to convince him of the iniquity of the imposition; the city was differently circumstanced from the other towns, it had had more than its share of affliction, and should not be burdened with that new exaction. The pair always came out of their discussions better friends than when they went in; one delighted to have had an opportunity of hearing himself talk, the other pleased with himself for having displayed a truly Parisian urbanity.

One evening Gilberte came into the room, with her air of thoughtless gayety. She paused at the threshold, affecting embarrassment. M. de Gartlauben rose, and with much tact presently withdrew, but on repeating his visit the following evening and finding Gilberte there again, he settled himself in his usual seat in the chimney-corner. It was the commencement of a succession of delightful evenings that they passed together in the study of the master of the house, not in the drawing-room-wherein lay a nice distinction. And at a later period when, yielding to their guest's entreaties, the young woman consented to play for him, she did not invite him to the salon, but entered the room alone, leaving the communicating door open. In those bitter winter evenings the old oaks of the Ardennes gave out a grateful warmth from the depths of the great cavernous fireplace; there was a cup of fragrant tea for them about ten o'clock; they laughed and chatted in the comfortable, bright room. And it did not require extra powers of vision to see that M. de Gartlauben was rapidly falling head over ears in love with that sprightly young woman, who flirted with him as audaciously as she had flirted in former days at Charleville with Captain Beaudoin's friends. He began to pay increased attention to his person, displayed a gallantry that verged on the fantastic, was raised to the pinnacle of bliss by the most trifling favor, tormented by the one ever-present anxiety not to appear a barbarian in her eyes, a rude soldier who did not know the ways of women.

And thus it was that in the big, gloomy house in the Rue Maqua a twofold life went on. While at meal-times Edmond, the wounded cherub with the pretty face, lent a listening ear to Delaherche's unceasing chatter, blushing if ever Gilberte asked him to pass her the salt, while at evening M. de Gartlauben, seated in the study, with eyes upturned in silent ecstasy, listened to a sonata by Mozart performed for his benefit by the young woman in the adjoining drawing-room, a stillness as of death continued to pervade the apartment where Colonel de Vineuil and Madame Delaherche spent their days, the blinds tight drawn, the lamp continually burning, like a votive candle illuminating a tomb. December had come and wrapped the city in a winding-sheet of snow; the cruel news seemed all the bitterer for the piercing cold. After General Ducrot's repulse at Champigny, after the loss of Orleans, there was left but one dark, sullen hope: that the soil of France might avenge their defeat, exterminate and swallow up the victors. Let the snow fall thicker and thicker still, let the earth's crust crack and open under the biting frost, that in it the entire German nation might find a grave! And there came another sorrow to wring poor Madame Delaherche's heart. One night when her son was from home, having been suddenly called away to Belgium on business, chancing to pass Gilberte's door she heard within a low murmur of voices and smothered laughter. Disgusted and sick at heart she returned to her own room, where her horror of the abominable thing she suspected the existence of would not let her sleep: it could have been none other but the Prussian whose voice she heard; she had thought she had noticed glances of intelligence passing; she was prostrated by this supreme disgrace. Ah, that woman, that abandoned woman, whom her son had insisted on bringing to the house despite her commands and prayers, whom she had forgiven, by her silence, after Captain Beaudoin's death! And now the thing was repeated, and this time the infamy was even worse. What was she to do? Such an enormity must not go unpunished beneath her roof. Her mind was torn by the conflict that raged there, in her uncertainty as to the course she should pursue. The colonel, desiring to know nothing of what occurred outside his room, always checked her with a gesture when he thought she was about to give him any piece of news, and she had said nothing to him of the matter that had caused her such suffering; but on those days when she came to him with tears standing in her eyes and sat for hours in mournful silence, he would look at her and say to himself that France had sustained yet another defeat.

This was the condition of affairs in the house in the Rue Maqua when Henriette dropped in there one morning to endeavor to secure Delaherche's influence in favor of Father Fouchard. She had heard people speak, smiling significantly as they did so, of the servitude to which Gilberte had reduced Captain de Gartlauben; she was, therefore, somewhat embarrassed when she encountered old Madame Delaherche, to whom she thought it her duty to explain the object of her visit, ascending the great staircase on her way to the colonel's apartment.

"Dear madame, it would be so kind of you to assist us! My uncle is in great danger; they talk of sending him away to Germany."

The old lady, although she had a sincere affection for Henriette, could scarce conceal her anger as she replied:

"I am powerless to help you, my child; you should not apply to me." And she continued, notwithstanding the agitation on the other's face: "You have selected an unfortunate moment for your visit; my son has to go to Belgium to-night. Besides, he could not have helped you; he has no more influence than I have. Go to my daughter-in-law; she is all powerful."

And she passed on toward the colonel's room, leaving Henriette distressed to have unwittingly involved herself in a family drama. Within the last twenty-four hours Madame Delaherche had made up her mind to lay the whole matter before her son before his departure for Belgium, whither he was going to negotiate a large purchase of coal to enable him to put some of his idle looms in motion. She could not endure the thought that the abominable thing should be repeated beneath her eyes while he was absent, and was only waiting to make sure he would not defer his departure until some other day, as he had been doing all the past week. It was a terrible thing to contemplate: the wreck of her son's happiness, the Prussian disgraced and driven from their doors, the wife, too, thrust forth upon the street and her name ignominiously placarded on the walls, as had been threatened would be done with any woman who should dishonor herself with a German.

Gilberte gave a little scream of delight on beholding Henriette.

"Ah, how glad I am to see you! It seems an age since we met, and one grows old so fast in the midst of all these horrors!" Thus running on she dragged her friend to her bedroom, where she seated her on the lounge and snuggled down close beside her. "Come, take off your things; you must stay and breakfast with us. But first we'll talk a bit; you must have such lots and lots of things to tell me! I know that you are without news of your brother. Ah, that poor Maurice, how I pity him, shut up in Paris, with no gas, no wood, no bread, perhaps! And that young man whom you have been nursing, that friend of your brother's-oh! a little bird has told me all about it-isn't it for his sake you are here today?"

Henriette's conscience smote her, and she did not answer. Was it not really for Jean's sake that she had come, in order that, the old uncle being released, the invalid, who had grown so dear to her, might have no further cause for alarm? It distressed her to hear his name mentioned by Gilberte; she could not endure the thought of enlisting in his favor an influence that was of so ambiguous a character. Her inbred scruples of a pure, honest woman made themselves felt, now it seemed to her that the rumors of a liaison with the Prussian captain had some foundation.

"Then I'm to understand that it's in behalf of this young man that you come to us for assistance?" Gilberte insistently went on, as if enjoying her friend's discomfiture. And as the latter, cornered and unable to maintain silence longer, finally spoke of Father Fouchard's arrest: "Why, to be sure! What a silly thing I am-and I was talking of it only this morning! You did well in coming to us, my dear; we must go about your uncle's affair at once and see what we can do for him, for the last news I had was not reassuring. They are on the lookout for someone of whom to make an example."

"Yes, I have had you in mind all along," Henriette hesitatingly replied. "I thought you might be willing to assist me with your advice, perhaps with something more substantial~"

The young woman laughed merrily. "You little goose, I'll have your uncle released inside three days. Don't you know that I have a Prussian captain here in the house who stands ready to obey my every order? Understand, he can refuse me nothing!" And she laughed more heartily than ever, in the giddy, thoughtless triumph of her coquettish nature, holding in her own and patting the hands of her friend, who was so uncomfortable that she could not find words in which to express her thanks, horrified by the avowal that was implied in what she had just heard. But how to account for such serenity, such childlike gayety? "Leave it to me; I'll send you home to-night with a mind at rest."

When they passed into the dining room Henriette was struck by Edmond's delicate beauty, never having seen him before. She eyed him with the pleasure she would have felt in looking at a pretty toy. Could it be possible that that boy had served in the army? and how could they have been so cruel as to break his arm? The story of his gallantry in the field made him even more interesting still, and Delaherche, who had received Henriette with the cordiality of a man to whom the sight of a new face is a godsend, while the servants were handing round the cutlets and the potatoes cooked in their jackets, never seemed to tire of eulogizing his secretary, who was as industrious and well behaved as he was handsome. They made a very pleasant and homelike picture, the four, thus seated around the bright table in the snug, warm dining room.

"So you want us to interest ourselves in Father Fouchard's case, and it's to that we owe the pleasure of your visit, eh?" said the manufacturer. "I'm extremely sorry that I have to go away to-night, but my wife will set things straight for you in a jiffy; there's no resisting her, she has only to ask for a thing to get it." He laughed as he concluded his speech, which was uttered in perfect simplicity of soul, evidently pleased and flattered that his wife possessed such influence, in which he shone with a kind of reflected glory. Then turning suddenly to her: "By the way, my dear, has Edmond told you of his great discovery?"

"No; what discovery?" asked Gilberte, turning her pretty caressing eyes full on the young sergeant.

The cherub blushed whenever a woman looked at him in that way, as if the exquisiteness of his sensations was too much for him. "It's nothing, madame; only a bit of old lace; I heard you saying the other day you wanted some to put on your mauve peignoir. I happened yesterday to

315

come across five yards of old Bruges point, something really handsome and very cheap. The woman will be here presently to show it to you."

She could have kissed him, so delighted was she. "Oh, how nice of you! You shall have your reward."

Then, while a terrine of foie-gras, purchased in Belgium, was being served, the conversation took another turn; dwelling for an instant on the quantities of fish that were dying of poison in the Meuse, and finally coming around to the subject of the pestilence that menaced Sedan when there should be a thaw. Even as early as November, there had been several cases of disease of an epidemic character. Six thousand francs had been expended after the battle in cleansing the city and collecting and burning clothing, knapsacks, haversacks, all the debris that was capable of harboring infection; but, for all that, the surrounding fields continued to exhale sickening odors whenever there came a day or two of warmer weather, so replete were they with half-buried corpses, covered only with a few inches of loose earth. In every direction the ground was dotted with graves; the soil cracked and split in obedience to the forces acting beneath its surface, and from the fissures thus formed the gases of putrefaction issued to poison the living. In those more recent days, moreover, another center of contamination had been discovered, the Meuse, although there had already been removed from it the bodies of more than twelve hundred dead horses. It was generally believed that there were no more human remains left in the stream, until, one day, a garde champetre, looking attentively down into the water where it was some six feet deep, discovered some objects glimmering at the bottom, that at first he took for stones; but they proved to be corpses of men, that had been mutilated in such a manner as to prevent the gas from accumulating in the cavities of the body and hence had been kept from rising to the surface. For near four months they had been lying there in the water among the eel-grass. When grappled for the irons brought them up in fragments, a head, an arm, or a leg at a time; at times the force of the current would suffice to detach a hand or foot and send it rolling down the stream. Great bubbles of gas rose to the surface and burst, still further empoisoning the air.

"We shall get along well enough as long as the cold weather lasts," remarked Delaherche, "but as soon as the snow is off the ground we shall have to go to work in earnest to abate the nuisance; if we don't we shall be wanting graves for ourselves." And when his wife laughingly asked him if he could not find some more agreeable subject to talk about at the table, he concluded by saying: "Well, it will be a long time before any of us will care to eat any fish out of the Meuse."

They had finished their repast, and the coffee was being poured, when the maid came to the door and announced that M. de Gartlauben presented his compliments and wanted to know if he might be allowed to see them for a moment. There was a slight flutter of excitement, for it was the first time he had ever presented himself at that hour of the day. Delaherche, seeing in the circumstance a favorable opportunity for presenting Henriette to him, gave orders that he should be introduced at once. The doughty captain, when he beheld another young woman in the room, surpassed himself in politeness, even accepting a cup of coffee, which he took without sugar, as he had seen many people do at Paris. He had only asked to be received at that unusual hour, he said, that he might tell Madame he had succeeded in obtaining the pardon of one of her proteges, a poor operative in the factory who had been arrested on account of a squabble with a Prussian. And Gilberte thereon seized the opportunity to mention Father Fouchard's case.

"Captain, I wish to make you acquainted with one of my dearest friends, who desires to place herself under your protection. She is the niece of the farmer who was arrested lately at Remilly, as you are aware, for being mixed up with that business of the francs-tireurs."

316

"Yes, yes, I know; the affair of the spy, the poor fellow who was found in a sack with his throat cut. It's a bad business, a very bad business. I am afraid I shall not be able to do anything."

"Oh, Captain, don't say that! I should consider it such a favor!"

There was a caress in the look she cast on him, while he beamed with satisfaction, bowing his head in gallant obedience. Her wish was his law!

"You would have all my gratitude, sir," faintly murmured Henriette, to whose memory suddenly rose the image of her husband, her dear Weiss, slaughtered down yonder at Bazeilles, filling her with invincible repugnance.

Edmond, who had discreetly taken himself off on the arrival of the captain, now reappeared and whispered something in Gilberte's ear. She rose quickly from the table, and, announcing to the company that she was going to inspect her lace, excused herself and followed the young man from the room. Henriette, thus left alone with the two men, went and took a seat by herself in the embrasure of a window, while they remained seated at the table and went on talking in a loud tone.

"Captain, you'll have a *petit verre* with me. You see I don't stand on ceremony with you; I say whatever comes into my head, because I know you to be a fair-minded man. Now I tell you your prefet is all wrong in trying to extort those forty-two thousand francs from the city. Just think once of all our losses since the beginning of the war. In the first place, before the battle, we had the entire French army on our hands, a set of ragged, hungry, exhausted men; and then along came your rascals, and their appetites were not so very poor, either. The passage of those troops through the place, what with requisitions, repairing damages and expenses of all sorts, stood us in a million and a half. Add as much more for the destruction caused by your artillery and by conflagration during the battle; there you have three millions. Finally, I am well within bounds in estimating the loss sustained by our trade and manufactures at two millions. What do you say to that, eh? A grand total of five million francs for a city of thirteen thousand inhabitants! And now you come and ask us for forty-two thousand more as a contribution to the expense of carrying on the war against us! Is it fair, is it reasonable? I leave it to your own sense of justice!"

M. de Gartlauben nodded his head with an air of profundity, and made answer:

"What can you expect? It is the fortune of war, the fortune of war."

To Henriette, seated in her window seat, her ears ringing, and vague, sad images of every sort fleeting through her brain, the time seemed to pass with mortal slowness, while Delaherche asserted on his word of honor that Sedan could never have weathered the crisis produced by the exportation of all their specie had it not been for the wisdom of the local magnates in emitting an issue of paper money, a step that had saved the city from financial ruin.

"Captain, will you have just a drop of cognac more?" and he skipped to another topic. "It was not France that started the war; it was the Emperor. Ah, I was greatly deceived in the Emperor. He need never expect to sit on the throne again; we would see the country dismembered first. Look here! there was just one man in this country last July who saw things as they were, and that was M. Thiers; and his action at the present time in visiting the different capitals of Europe is most wise and patriotic. He has the best wishes of every good citizen; may he be successful!"

He expressed the conclusion of his idea by a gesture, for he would have considered it improper to speak of his desire for peace before a Prussian, no matter how friendly he might be, although the desire burned fiercely in his bosom, as it did in that of every member of the old conservative bourgeoisie who had favored the plebiscite. Their men and money were exhausted, it was time for them to throw up the sponge; and a deep-seated feeling of hatred toward Paris, for the obstinacy with which it held out, prevailed in all the provinces that were in possession of the enemy. He concluded in a lower tone, his allusion being to Gambetta's inflammatory proclamations:

"No, no, we cannot give our suffrages to fools and madmen. The course they advocate would end in general massacre. I, for my part, am for M. Thiers, who would submit the questions at issue to the popular vote, and as for their Republic, great heavens! let them have it if they want it, while waiting for something better; it don't trouble me in the slightest."

Captain de Gartlauben continued to nod his head very politely with an approving air, murmuring:

"To be sure, to be sure-"

Henriette, whose feeling of distress had been increasing, could stand their talk no longer. She could assign no definite reason for the sensation of inquietude that possessed her; it was only a longing to get away, and she rose and left the room quietly in quest of Gilberte, whose absence had been so long protracted. On entering the bedroom, however, she was greatly surprised to find her friend stretched on the lounge, weeping bitterly and manifestly suffering from some extremely painful emotion.

"Why, what is the matter? What has happened you?"

The young woman's tears flowed faster still and she would not speak, manifesting a confusion that sent every drop of blood coursing from her heart up to her face. At last, throwing herself into the arms that were opened to receive her and concealing her face in the other's bosom, she stammered:

"Oh, darling if you but knew. I shall never dare to tell you-and yet I have no one but you, you alone perhaps can tell me what is best to do." A shiver passed through her frame, her voice was scarcely audible. "I was with Edmond-and then-and then Madame Delaherche came into the room and caught me-"

"Caught you! What do you mean?"

"Yes, we were here in the room; he was holding me in his arms and kissing me-" And clasping Henriette convulsively in her trembling arms she told her all. "Oh, my darling, don't judge me severely; I could not bear it! I know I promised you it should never happen again, but you have seen Edmond, you know how brave he is, how handsome! And think once of the poor young man, wounded, ill, with no one to give him a mother's care! And then he has never had the enjoyments that wealth affords; his family have pinched themselves to give him an education. I could not be harsh with him."

Henriette listened, the picture of surprise; she could not recover from her amazement. "What! you don't mean to say it was the little sergeant! Why, my dear, everyone believes the Prussian to be your lover!"

Gilberte straightened herself up with an indignant air, and dried her eyes. "The Prussian my lover? No, thank you! He's detestable; I can't endure him. I wonder what they take me for? What have I ever done that they should suppose I could be guilty of such baseness? No, never! I would rather die than do such a thing!" In the earnestness of her protestations her beauty had assumed an angry and more lofty cast that made her look other than she was. And all at once, sudden as a flash, her coquettish gayety, her thoughtless levity, came back to her face, accompanied by a peal of silvery laughter. "I won't deny that I amuse myself at his expense. He adores me, and I have only to give him a look to make him obey. You have no idea what fun it is to bamboozle that great big man, who seems to think he will have his reward some day."

"But that is a very dangerous game you're playing," Henriette gravely said.

"Oh, do you think so? What risk do I incur? When he comes to see he has nothing to expect he can't do more than be angry with me and go away. But he will never see it! You don't know the man; I read him like a book from the very start: he is one of those men with whom a woman can do what she pleases and incur no danger. I have an instinct that guides me in these matters and which has never deceived me. He is too consumed by vanity; no human consideration will ever drive it into his head that by any possibility a woman could get the better of him. And all he will get from me will be permission to carry away my remembrance, with the consoling thought that he has done the proper thing and behaved himself like a gallant man who has long been an inhabitant of Paris." And with her air of triumphant gayety she added: "But before he leaves he shall cause Uncle Fouchard to be set at liberty, and all his recompense for his trouble shall be a cup of tea sweetened by these fingers."

But suddenly her fears returned to her: she remembered what must be the terrible consequences of her indiscretion, and her eyes were again bedewed with tears.

"Mon Dieu! and Madame Delaherche-how will it all end? She bears me no love; she is capable of telling the whole story to my husband."

Henriette had recovered her composure. She dried her friend's eyes, and made her rise from the lounge and arrange her disordered clothing.

"Listen, my dear; I cannot bring myself to scold you, and yet you know what my sentiments must be. But I was so alarmed by the stories I heard about the Prussian, the business wore such an extremely ugly aspect, that this affair really comes to me as a sort of relief by comparison. Cease weeping; things may come out all right."

Her action was taken none too soon, for almost immediately Delaherche and his mother entered the room. He said that he had made up his mind to take the train for Brussels that afternoon and had been giving orders to have a carriage ready to carry him across the frontier into Belgium; so he had come to say good-by to his wife. Then turning and addressing Henriette:

"You need have no further fears. M. de Gartlauben, just is he was going away, promised me he would attend to your uncle's case, and although I shall not be here, my wife will keep an eye to it."

Since Madame Delaherche had made her appearance in the apartment Gilberte had not once taken her anxious eyes from off her face. Would she speak, would she tell what she had seen, and keep her son from starting on his projected journey? The elder lady, also, soon as she crossed the threshold, had bent her fixed gaze in silence on her daughter-in-law. Doubtless her stern patriotism induced her to view the matter in somewhat the same light that Henriette had viewed it. Mon Dieu! since it was that young man, that Frenchman who had fought so bravely, was it not her duty to forgive, even as she had forgiven once before, in Captain Beaudoin's case? A look of greater softness rose to her eyes; she averted her head. Her son might go; Edmond would be there to protect Gilberte against the Prussian. She even smiled faintly, she whose grim face had never once relaxed since the news of the victory at Coulmiers.

"Au revoir," she said, folding her son in her arms. "Finish up your business quickly as you can and come back to us."

And she took herself slowly away, returning to the prison-like chamber across the corridor, where the colonel, with his dull gaze, was peering into the shadows that lay outside the disk of bright light which fell from the lamp.

Henriette returned to Remilly that same evening, and one morning, three days afterward, had the pleasure to see Father Fouchard come walking into the house, as calmly as if he had merely stepped out to transact some business in the neighborhood. He took a seat by the table and refreshed himself with some bread and cheese, and to all the questions that were put to him replied with cool deliberation, like a man who had never seen anything to alarm him in his situation. What reason had he to be afraid? He had done nothing wrong; it was not he who had killed the Prussian, was it? So he had just said to the authorities: "Investigate the matter; I know nothing about it." And they could do nothing but release him, and the mayor as well, seeing they had no proofs against them. But the eyes of the crafty, sly old peasant gleamed with delight at the thought of how nicely he had pulled the wool over the eyes of those dirty blackguards, who were beginning to higgle with him over the quality of the meat he furnished to them.

December was drawing near its end, and Jean insisted on going away. His leg was quite strong again, and the doctor announced that he was fit to go and join the army. This was to Henriette a subject of profoundest sorrow, which she kept locked in her bosom as well as she was able. No tidings from Paris had reached them since the disastrous battle of Champigny; all they knew was that Maurice's regiment had been exposed to a murderous fire and had suffered severely. Ever that deep, unbroken silence; no letter, never the briefest line for them, when they knew that families in Raucourt and Sedan were receiving intelligence of their loved ones by circuitous ways. Perhaps the pigeon that was bringing them the so eagerly wished-for news had fallen a victim to some hungry bird of prey, perhaps the bullet of a Prussian had brought it to the ground at the margin of a wood. But the fear that haunted them most of all was that Maurice was dead; the silence of the great city off yonder in the distance, uttering no cry in the mortal hug of the investment, was become to them in their agonized suspense the silence of death. They had abandoned all hope of tidings, and when Jean declared his settled purpose to be gone, Henriette only gave utterance to this stifled cry of despair:

"My God! then all is ended, and I am to be left alone!"

It was Jean's desire to go and serve with the Army of the North, which had recently been re-formed under General Faidherbe. Now that General Manteuffel's corps had moved forward to Dieppe there were three departments, cut off from the rest of France, that this army had to defend, le Nord, le Pas-de-Calais, and la Somme, and Jean's plan, not a difficult one to carry into execution, was simply to make for Bouillon and thence complete his journey across Belgian territory. He knew that the 23d corps was being recruited, mainly from such old soldiers of Sedan and Metz as could be gathered to the standards. He had heard it reported that General Faidherbe was about to take the field, and had definitely appointed the next ensuing Sunday as the day of his departure, when news reached him of the battle of Pont-Noyelle, that drawn battle which came so near being a victory for the French.

It was Dr. Dalichamp again in this instance who offered the services of his gig and himself as driver to Bouillon. The good man's courage and kindness were boundless. At Raucourt, where typhus was raging, communicated by the Bavarians, there was not a house where he had not one or more patients, and this labor was additional to his regular attendance at the two hospitals at Raucourt and Remilly. His ardent patriotism, the impulse that prompted him to protest against unnecessary barbarity, had twice led to his being arrested by the Prussians, only to be released on each occasion. He gave a little laugh of satisfaction, therefore, the morning he came with his vehicle to take up Jean, pleased to be the instrument of assisting the escape of another of the victims of Sedan, those poor, brave fellows, as he called them, to whom he gave his professional services and whom he aided with his purse. Jean, who knew of Henriette's straitened circumstances and had been suffering from lack of funds since his relapse, accepted gratefully the fifty francs that the doctor offered him for traveling expenses.

Father Fouchard did things handsomely at the leave-taking, sending Silvine to the cellar for two bottles of wine and insisting that everyone should drink a glass to the extermination of the Germans. He was a man of importance in the country nowadays and had his "plum" hidden away somewhere or other; he could sleep in peace now that the francs-tireurs had disappeared, driven like wild beasts from their lair, and his sole wish was for a speedy conclusion of the war. He had even gone so far in one of his generous fits as to pay Prosper his wages in order to retain his services on the farm, which the young man had no thought of leaving. He touched glasses with Prosper, and also with Silvine, whom he at times was half inclined to marry, knowing what a treasure he had in his faithful, hard-working little servant; but what was the use? he knew she would never leave him, that she would still be there when Charlot should be grown and go in turn to serve his country as a soldier. And touching his glass to Henriette's, Jean's, and the doctor's, he exclaimed:

"Here's to the health of you all! May you all prosper and be no worse off than I am!"

Henriette would not let Jean go away without accompanying him as far as Sedan. He was in citizen's dress, wearing a frock coat and derby hat that the doctor had loaned him. The day was piercingly cold; the sun's rays were reflected from a crust of glittering snow. Their intention had been to pass through the city without stopping, but when Jean learned that his old colonel was still at the Delaherches' he felt an irresistible desire to go and pay his respects to him, and at the same time thank the manufacturer for his many kindnesses. His visit was destined to bring him an additional, a final sorrow, in that city of mournful memories. On reaching the structure in the Rue Maqua they found the household in a condition of the greatest distress and disorder, Gilberte wringing her hands, Madame Delaherche weeping great silent tears, while her son, who had come in from the factory, where work was gradually being resumed, uttered exclamations of surprise. The colonel had just been discovered, stone dead, lying exactly as he had fallen, in a heap on the floor of his chamber. The physician, who was summoned with all haste, could assign no cause for the sudden death; there was no indication of paralysis or heart trouble. The colonel had been stricken down, and no one could tell from what quarter the blow came; but the following morning, when the room was thrown open, a piece of an old newspaper was found, lying on the carpet, that had been wrapped around a book and contained the account of the surrender of Metz.

"My, dear," said Gilberte to Henriette, "as Captain de Gartlauben was coming downstairs just now he removed his hat as he passed the door of the room where my uncle's body is lying. Edmond saw it; he's an extremely well-bred man, don't you think so?"

In all their intimacy Jean had never yet kissed Henriette. Before resuming his seat in the gig with the doctor he endeavored to thank her for all her devoted kindness, for having nursed and loved him as a brother, but somehow the words would not come at his command; he opened his arms and, with a great sob, clasped her in a long embrace, and she, beside herself with the grief of parting, returned his kiss. Then the horse started, he turned about in his seat, there was a waving of hands, while again and again two sorrowful voices repeated in choking accents:

"Farewell! Farewell!"

On her return to Remilly that evening Henriette reported for duty at the hospital. During the silent watches of the night she was visited by another convulsive attack of sobbing, and wept, wept as if her tears would never cease to flow, clasping her hands before her as if between them to strangle her bitter sorrow.

7

On the day succeeding the battle of Sedan the mighty hosts of the two German armies, without the delay of a moment, commenced their march on Paris, the army of the Meuse coming in by the north through the valley of the Marne, while the third army, passing the Seine at Villeneuve-Saint-Georges, turned the city to the south and moved on Versailles; and when, on that bright, warm September morning, General Ducrot, to whom had been assigned the command of the as yet incomplete 14th corps, determined to attack the latter force while it was marching by the flank, Maurice's new regiment, the 115th, encamped in the woods to the left of Meudon, did not receive its orders to advance until the day was lost. A few shells from the enemy sufficed to do the work; the panic started with a regiment of zouaves made up of raw recruits, and quickly spreading to the other troops, all were swept away in a headlong rout that never ceased until they were safe behind the walls of Paris, where the utmost consternation prevailed. Every position in advance of the southern line of fortifications was lost, and that evening the wires of the Western Railway telegraph, the city's sole remaining means of communicating with the rest of France, were cut. Paris was cut off from the world.

The condition of their affairs caused Maurice a terrible dejection. Had the Germans been more enterprising they might have pitched their tents that night in the Place du Carrousel, but with the prudence of their race they had determined that the siege should be conducted according to rule and precept, and had already fixed upon the exact lines of investment, the position of the army of the Meuse being at the north, stretching from Croissy to the Marne, through Epinay, the cordon of the third army at the south, from Chennevieres to Chatillon and Bougival, while general headquarters, with King William, Bismarck, and General von Moltke, were established at Versailles. The gigantic blockade, that no one believed could be successfully completed, was an accomplished fact; the city, with its girdle of fortifications eight leagues and a half in length, embracing fifteen forts and six detached redoubts, was henceforth to be transformed into a huge prison-pen. And the army of the defenders comprised only the 13th corps, commanded by General Vinoy, and the 14th, then in process of reconstruction under General Ducrot, the two aggregating an effective strength of eighty thousand men; to which were to be added fourteen thousand sailors, fifteen thousand of the francs corps, and a hundred and fifteen thousand mobiles, not to mention the three hundred thousand National Guards distributed among the sectional divisions of the ramparts. If this seems like a large force it must be remembered that there were few seasoned and trained soldiers among its numbers. Men were constantly being drilled and equipped; Paris was a great intrenched camp. The preparations for the defense went on from hour to hour with feverish haste; roads were built, houses demolished within the military zone; the two hundred siege guns and the twenty-five hundred pieces of lesser caliber were mounted in position, other guns were cast; an arsenal, complete in every detail, seemed to spring from the earth under the tireless efforts of Dorian, the patriotic war minister. When, after the rupture of the negotiations at Ferrieres, Jules Favre acquainted the country with M. von Bismarck's demands-the cession of Alsace, the garrison of Strasbourg to be surrendered, three milliards of indemnity-a cry of rage went up and the continuation of the war was demanded by acclaim as a condition indispensable to the country's existence. Even with no hope of victory Paris must defend herself in order that France might live.

On a Sunday toward the end of September Maurice was detailed to carry a message to the further end of the city, and what he witnessed along the streets he passed through filled him with new hope. Ever since the defeat of Chatillon it had seemed to him that the courage of the people was rising to a level with the great task that lay before them. Ah! that Paris that he had known so thoughtless, so wayward, so keen in the pursuit of pleasure; he found it now quite changed, simple, earnest, cheerfully brave, ready for every sacrifice. Everyone was in uniform; there was scarce a head that was not decorated with the *kepi* of the National Guard. Business of every sort had come to a sudden standstill, as the hands of a watch cease to move when the mainspring snaps, and at the public meetings, among the soldiers in the guard-room, or where the crowds collected in the streets, there was but one subject of conversation, inflaming the hearts and minds of all-the determination to conquer. The contagious influence of illusion, scattered broadcast, unbalanced weaker minds; the people were tempted to acts of generous folly by the tension to which they were subjected. Already there was a taint of morbid, nervous excitability in the air, a feverish condition in which men's hopes and fears alike became distorted and exaggerated, arousing the worst passions of humanity at the slightest breath of suspicion. And Maurice was witness to a scene in the Rue des Martyrs that produced a profound impression on him, the assault made by a band of infuriated men on a house from which, at one of the upper windows, a bright light had been displayed all through the night, a signal, evidently, intended to reach the Prussians at Bellevue over the roofs of Paris. There were jealous citizens who spent all their nights on their house-tops, watching what was going on around them. The day before a poor wretch had had a narrow escape from drowning at the hands of the mob, merely because he had opened a map of the city on a bench in the Tuileries gardens and consulted it.

And that epidemic of suspicion Maurice, who had always hitherto been so liberal and fair-minded, now began to feel the influence of in the altered views he was commencing to entertain concerning men and things. He had ceased to give way to despair, as he had done after the rout at Chatillon, when he doubted whether the French army would ever muster up sufficient manhood to fight again: the sortie of the 30th of September on l'Hay and Chevilly, that of the 13th of October, in which the mobiles gained possession of Bagneux, and finally that of October 21, when his regiment captured and held for some time the park of la Malmaison, had restored to him all his confidence, that flame of hope that a spark sufficed to light and was extinguished as quickly. It was true the Prussians had repulsed them in every direction, but for all that the troops had fought bravely; they might yet be victorious in the end. It was Paris now that was responsible for the young man's gloomy forebodings, that great fickle city that at one moment was cheered by bright illusions and the next was sunk in deepest despair, ever haunted by the fear of treason in its thirst for victory. Did it not seem as if Trochu and Ducrot were treading in the footsteps of the Emperor and Marshal MacMahon and about to prove themselves incompetent leaders, the unconscious instruments of their country's ruin? The same movement that had swept away the Empire was now threatening the Government of National Defense, a fierce longing of the extremists to place themselves in control in order that they might save France by the methods of '92; even now Jules Favre and his co-members were more unpopular than the old ministers of Napoleon III. had ever been. Since they would not fight the Prussians, they would do well to make way for others, for those revolutionists who saw an assurance of victory in decreeing the levee en masse, in lending an ear to those visionaries who proposed to mine the earth beneath the Prussians' feet, or annihilate them all by means of a new fashioned Greek fire.

Just previous to the 31st of October Maurice was more than usually a victim to this malady of distrust and barren speculation. He listened now approvingly to crude fancies that would formerly have brought a smile of contempt to his lips. Why should he not? Were not imbecility and crime abroad in the land? Was it unreasonable to look for the miraculous when his world was falling in ruins about him? Ever since the time he first heard the tidings of Froeschwiller, down there in front of Mulhausen, he had harbored a deep-seated feeling of rancor in his breast; he suffered from Sedan as from a raw sore, that bled afresh with every new reverse; the memory of their defeats, with all the anguish they entailed, was ever present to his mind; body and mind enfeebled by long marches, sleepless nights, and lack of food, inducing a mental torpor that left them doubtful even if they were alive; and the thought that so much suffering was to end in another and an irremediable disaster maddened him, made of that cultured man an unreflecting being, scarce higher in the scale than a very little child, swayed by each passing impulse of the moment. Anything, everything, destruction, extermination, rather than pay a penny of French money or yield an inch of French soil! The revolution that since the first reverse had been at work within him, sweeping away the legend of Napoleonic glory, the sentimental Bonapartism that he owed to the epic narratives of his grandfather, was now complete. He had ceased to be a believer in Republicanism, pure and simple, considering the remedy not drastic enough; he had begun to dabble in the theories of the extremists, he was a believer in the necessity of the Terror as the only means of ridding them of the traitors and imbeciles who were about to slay the country. And so it was that he was heart and soul with the insurgents when, on the 31st of October, tidings of disaster came pouring in on them in quick succession: the loss of Bourget, that had been captured from the enemy only a few days before by a dashing surprise; M. Thiers' return to Versailles from his visit to the European capitals, prepared to treat for peace, so it was said, in the name of Napoleon III.; and finally the capitulation of Metz, rumors of which had previously been current and which was now confirmed, the last blow of the bludgeon, another Sedan, only attended by circumstances of blacker infamy. And when he learned next day the occurrences at the Hotel de Ville-how the insurgents had been for a brief time successful, how the members of the Government of National Defense had been made prisoners and held until four o'clock in the morning, how finally the fickle populace, swayed at one moment by detestation for the ministers and at the next terrified by the prospect of a successful revolution, had released them-he was filled with regret at the miscarriage of the attempt, at the non-success of the Commune, which might have been their salvation, calling the people to arms, warning them of the country's danger, arousing the cherished memories of a nation that wills it will not perish. Thiers did not dare even to set his foot in Paris, where there was some attempt at illumination to celebrate the failure of the negotiations.

The month of November was to Maurice a period of feverish expectancy. There were some conflicts of no great importance, in which he had no share. His regiment was in cantonments at the time in the vicinity of Saint-Ouen, whence he made his escape as often as he could to satisfy his craving for news. Paris, like him, was awaiting the issue of events in eager suspense. The election of municipal officers seemed to have appeased political passion for the time being, but a circumstance that boded no good for the future was that those elected were rabid adherents of one or another party. And what Paris was watching and praying for in that interval of repose was the grand sortie that was to bring them victory and deliverance. As it had always been, so it was now; confidence reigned everywhere: they would drive the Prussians from their position, would pulverize them, annihilate them. Great preparations were being made in the peninsula of Gennevilliers, the point where there was most likelihood of the operation being attended with success. Then one morning came the joyful tidings of the victory at Coulmiers; Orleans was recaptured, the army of the Loire was marching to the relief of Paris, was even then, so it was reported, in camp at Etampes. The aspect of affairs was entirely changed: all they had to do now was to go and effect a junction with it beyond the Marne. There had been a general reorganization of the forces; three armies had been created, one composed of the battalions of National Guards and commanded by General Clement Thomas, another, comprising the 13th and 14th corps, to which were added a few reliable regiments, selected indiscriminately wherever they could be found, was to form the main column of attack under the lead of General Ducrot, while the third, intended to act as a reserve, was made up entirely of mobiles and turned over to General Vinoy. And when Maurice laid him down to sleep in the wood of Vincennes on the night of the 28th of November, with his comrades of the 115th, he was without a doubt of their success. The three corps of the second army were all there, and it was common talk that their junction with the army of the Loire had been fixed for the following day at Fontainebleau. Then ensued a series of mischances, the usual blunders arising from want of foresight; a sudden rising of the river, which prevented the engineers from laying the pontoon bridge; conflicting orders, which delayed the movement of the troops. The 115th was among the first regiments to pass the river on the following night, and in the neighborhood of ten o'clock, with Maurice in its ranks, it entered Champigny under a destructive fire. The young man was wild with excitement; he fired so rapidly that his chassepot burned his fingers, notwithstanding the intense cold. His sole thought was to push onward, ever onward, surmounting every obstacle until they should join their brothers from the provinces over there across the river. But in front of Champigny and Bry the army fell up against the park walls of Coeuilly and Villiers, that the Prussians had converted into impregnable fortresses, more than a quarter of a mile in length. The men's courage faltered, and after that the action went on in a half-hearted way; the 3d corps was slow in getting up, the 1st and 2d, unable to advance, continued for two days longer to hold Champigny, which they finally abandoned on the night of December 2, after their barren victory. The whole army retired to the wood of Vincennes, where the men's only shelter was the snow-laden branches of the trees, and Maurice, whose feet were frost-bitten, laid his head upon the cold ground and cried.

The gloom and dejection that reigned in the city, after the failure of that supreme effort, beggars the powers of description. The great sortie that had been so long in preparation, the irresistible eruption that was to be the deliverance of Paris, had ended in disappointment, and three days later came a communication from General von Moltke under a flag of truce, announcing that the army of the Loire had been defeated and that the German flag again waved over Orleans. The girdle was being drawn tighter and tighter about the doomed city all whose struggles were henceforth powerless to burst its iron fetters. But Paris seemed to accumulate fresh powers of resistance in the delirium of its despair. It was certain that ere long they would have to count famine among the number of their foes. As early as October the people had been restricted in their consumption of butcher's meat, and in December, of all the immense herds of beeves and flocks of sheep that had been turned loose in the Bois de Boulogne, there was not a single creature left alive, and horses were being slaughtered for food. The stock of flour and wheat, with what was subsequently taken for the public use by forced sale, it was estimated would keep the city supplied with bread for four months. When the flour was all consumed mills were erected in the railway stations to grind the grain. The supply of coal, too, was giving out; it was reserved to bake the bread and for use in the mills and arms factories. And Paris, her streets without gas and lighted by petroleum lamps at infrequent intervals; Paris, shivering under her icy mantle; Paris, to whom the authorities doled out her scanty daily ration of black bread and horse flesh, continued to hope-in spite of all, talking of Faidherbe in the north, of Chanzy on the Loire, of Bourbaki in the east, as if their victorious armies were already beneath the walls. The men and women who stood waiting, their feet in snow and slush, in interminable lines before the bakers' and butchers' shops, brightened up a bit at times at the news of some imaginary success of the army. After the discouragement of each defeat the unquenchable flame of their illusion would burst out and blaze more brightly than ever among those wretched people, whom starvation and every kind of suffering had rendered almost delirious. A soldier on the Place du Chateau d'Eau having spoken of surrender, the by-standers mobbed and were near killing him. While the army, its endurance exhausted, feeling the end was near, called for peace, the populace clamored still for the sortie en masse, the torrential sortie, in which the entire population of the capital, men, women, and children, even, should take part, rushing upon the Prussians like water from a broken dyke and overwhelming them by sheer force of numbers.

And Maurice kept himself apart from his comrades, with an ever-increasing disgust for the life and duties of a soldier, that condemned him to inactivity and uselessness behind the ramparts of Mont-Valerien. He grasped every occasion to get away and hasten to Paris, where his heart was. It was in the midst of the great city's thronging masses alone that he found rest and peace of mind; he tried to force himself to hope as they hoped. He often went to witness the departure of the balloons, which were sent up every other day from the station of the Northern Railway with a freight of despatches and carrier pigeons. They rose when the ropes were cast loose and soon were lost to sight in the cheerless wintry sky, and all hearts were filled with anguish when the wind wafted them in the direction of the German frontier. Many of them were never heard of more. He had himself twice written to his sister Henriette, without ever learning if she had received his letters. The memory of his sister and of Jean, living as they did in that outer, shadowy world from which no tidings ever reached him now, was become so blurred and faint that he thought of them but seldom, as of affections that he had left behind him in some previous existence. The incessant conflict of despair and hope in which he lived occupied all the faculties of his being too fully to leave room for mere human feelings. Then, too, in the early days of January he was goaded to the verge of frenzy by the action of the enemy in shelling the district on the left bank of the river. He had come to credit the Prussians with reasons of humanity for their abstention, which was in fact due simply to the difficulties they experienced in bringing up their guns and getting them in position. Now that a shell had killed two little girls at the Val-de-Grace, his scorn and hatred knew no bounds for those barbarous ruffians who murdered little children and threatened to burn the libraries and museums. After the first days of terror, however, Paris had resumed its life of dogged, unfaltering heroism.

Since the reverse of Champigny there had been but one other attempt, ending in disaster like the rest, in the direction of Bourget; and the evening when the plateau of Avron was evacuated, under the fire of the heavy siege artillery battering away at the forts, Maurice was a sharer in the rage and exasperation that possessed the entire city. The growing unpopularity that threatened to hurl from power General Trochu and the Government of National Defense was so augmented by this additional repulse that they were compelled to attempt a supreme and hopeless effort. What, did they refuse the services of the three hundred thousand National Guards, who from the beginning had been demanding their share in the peril and in the victory! This time it was to be the torrential sortie that had all along been the object of the popular clamor; Paris was to throw open its dikes and drown the Prussians beneath the on-pouring waves of its children. Notwithstanding the certainty of a fresh defeat, there was no way of avoiding a demand that had its origin in such patriotic motives; but in order to limit the slaughter as far as possible, the chiefs determined to employ, in connection with the regular army, only the fifty-nine mobilized battalions of the National Guard. The day preceding the 19th of January resembled some great public holiday; an immense crowd gathered on the boulevards and in the Champs-Elysees to witness the departing regiments, which marched proudly by, preceded by their bands, the men thundering out patriotic airs. Women and children followed them along the sidewalk, men climbed on the benches to wish them Godspeed. The next morning the entire population of the city hurried out to the Arc de Triomphe, and it was almost frantic with delight when at an early hour news came of the capture of Montretout; the tales that were told of the gallant behavior of the National Guard sounded like epics; the Prussians had been beaten all along the line, the French would occupy Versailles before night. As a natural result the consternation was proportionately great when, at nightfall, the inevitable defeat became known. While the left wing was seizing Montretout the center, which had succeeded in carrying the outer wall of Buzanval Park, had encountered a second inner wall, before which it broke. A thaw had set in, the roads were heavy from the effects of a fine, drizzling rain, and the guns, those guns that had been cast by popular subscription and were to the Parisians as the apple of their eye, could not get up. On the right General Ducrot's column was tardy in getting into action and saw nothing of the fight. Further effort was useless, and General Trochu was compelled to order a retreat. Montretout was abandoned, and Saint-Cloud as well, which the Prussians burned, and when it became fully dark the horizon of Paris was illuminated by the conflagration.

Maurice himself this time felt that the end was come. For four hours he had remained in the park of Buzanval with the National Guards under the galling fire from the Prussian intrenchments, and later, when he got back to the city, he spoke of their courage in the highest terms. It was undisputed that the Guards fought bravely on that occasion; after that was it not self-evident that all the disasters of the army were to be attributed solely to the imbecility and treason of its leaders? In the Rue de Rivoli he encountered bands of men shouting: "Hurrah for the Commune! down with Trochu!" It was the leaven of revolution beginning to work again in the popular mind, a fresh outbreak of public opinion, and so formidable this time that the Government of National Defense, in order to preserve its own existence, thought it necessary to compel General Trochu's resignation and put General Vinoy in his place. On that same day Maurice, chancing to enter a hall in Belleville where a public meeting was going on, again heard the *levee en masse* demanded with clamorous shouts. He knew the thing to be chimerical, and yet it set his heart a-beating more rapidly to see such a determined will to conquer. When all is ended, is it not left us to attempt the impossible? All that night he dreamed of miracles.

Then a long week went by, during which Paris lay agonizing without a murmur. The shops had ceased to open their doors; in the lonely streets the infrequent wayfarer never met a carriage. Forty thousand horses had been eaten; dogs, cats and rats were now luxuries, commanding a high price. Ever since the supply of wheat had given out the bread was made from rice and oats, and was black, damp, and slimy, and hard to digest; to obtain the ten ounces that constituted a day's ration involved a wait, often of many hours, in line before the bake-house. Ah, the sorrowful spectacle it was, to see those poor women shivering in the pouring rain, their feet in the ice-cold mud and water! the misery and heroism of the great city that would not surrender! The death rate had increased threefold; the theaters were converted into hospitals. As soon as it became dark the quarters where luxury and vice had formerly held carnival were shrouded in funereal blackness, like the faubourgs of some accursed city, smitten by pestilence. And in that silence, in that obscurity, naught was to be heard save the unceasing roar of the cannonade and the crash of bursting shells, naught to be seen save the red flash of the guns illuminating the wintry sky.

On the 28th of January the news burst on Paris like a thunderclap that for the past two days negotiations had been going on, between Jules Favre and M. von Bismarck, looking to an armistice, and at the same time it learned that there was bread for only ten days longer, a space of time that would hardly suffice to revictual the city. Capitulation was become a matter of material necessity. Paris, stupefied by the hard truths that were imparted to it at that late day, remained sullenly silent and made no sign. Midnight of that day heard the last shot from the German guns, and on the 29th, when the Prussians had taken possession of the forts, Maurice went with his regiment into the camp that was assigned them over by Montrouge, within the fortifications. The life that he led there was an aimless one, made up of idleness and feverish unrest. Discipline was relaxed; the soldiers did pretty much as they pleased, waiting in inactivity to be dismissed to their homes. He, however, continued to hang around the camp in a semi-dazed condition, moody, nervous, irritable, prompt to take offense on the most trivial provocation. He read with avidity all the revolutionary newspapers he could lay hands on; that three weeks' armistice, concluded solely for the purpose of allowing France to elect an assembly that should ratify the conditions of peace, appeared to him a delusion and a snare, another and a final instance of treason. Even if Paris were forced to capitulate, he was with Gambetta for the prosecution of the war in the north and on the line of the Loire. He overflowed with indignation at the disaster of Bourbaki's army in the east, which had been compelled to throw itself into Switzerland, and the result of the elections made him furious: it would be just as he had always predicted; the base, cowardly provinces, irritated by Paris' protracted resistance, would insist on peace at any price and restore the monarchy while the Prussian guns were still directed on the city. After the first sessions, at Bordeaux, Thiers, elected in twenty-six departments and constituted by unanimous acclaim the chief executive, appeared to his eyes a monster of iniquity, the father of lies, a man capable of every crime. The terms of the peace concluded by that assemblage of monarchists seemed to him to put the finishing touch to their infamy, his blood boiled merely at the thought of those hard conditions: an indemnity of five milliards, Metz to be given up, Alsace to be ceded, France's blood and treasure pouring from the gaping wound, thenceforth incurable, that was thus opened in her flank.

Late in February Maurice, unable to endure his situation longer, made up his mind he would desert. A stipulation of the treaty provided that the troops encamped about Paris should be disarmed and returned to their abodes, but he did not wait to see it enforced; it seemed to him that it would break his heart to leave brave, glorious Paris, which only famine had been able to subdue, and so he bade farewell to army life and hired for himself a small furnished room next the roof of a tall apartment house in the Rue des Orties, at the top of the butte des Moulins, whence he had an outlook over the immense sea of roofs from the Tuileries to the Bastille. An old friend, whom he had known while pursuing his law studies, had loaned him a hundred francs. In addition to that he had caused his name to be inscribed on the roster of a battalion of National Guards as soon as he was settled in his new quarters, and his pay, thirty sous a day, would be enough to keep him alive. The idea of going to the country and there leading a tranquil life, unmindful of what was happening to the country, filled him with horror; the letters even that he received from his sister Henriette, to whom he had written immediately after the armistice, annoyed him by their tone of entreaty, their ardent solicitations that he would come home to Remilly and rest. He refused point-blank; he would go later on when the Prussians should be no longer there.

And so Maurice went on leading an idle, vagabondish sort of life, in a state of constant feverish agitation. He had ceased to be tormented by hunger; he devoured the first white bread he got with infinite gusto; but the city was a prison still: German guards were posted at the gates, and no one was allowed to pass them until he had been made to give an account of himself. There had been no resumption of social life as yet; industry and trade were at a standstill; the people lived from day to day, watching to see what would happen next, doing nothing, simply vegetating in the bright sunshine of the spring that was now coming on apace. During the siege there had been the military service to occupy men's minds and tire their limbs, while now the entire population, isolated from all the world, had suddenly been reduced to a state of utter stagnation, mental as well as physical. He did as others did, loitering his time away from morning till night, living in an atmosphere that for months had been vitiated by the germs arising from the half-crazed mob. He read the newspapers and was an assiduous frequenter of public meetings, where he would often smile and shrug his shoulders at the rant and fustian of the speakers, but nevertheless would go away with the most ultra notions teeming in his brain, ready to engage in any desperate undertaking in the defense of what he considered truth and justice. And sitting by the window in his little bedroom, and looking out over the city, he would still beguile himself with dreams of victory; would tell himself that France and the Republic might yet be saved, so long as the treaty of peace remained unsigned.

The 1st of March was the day fixed for the entrance of the Prussians into Paris, and a long-drawn howl of wrath and execration went up from every heart. Maurice never attended a meeting now that he did not hear Thiers, the Assembly, even the men of September 4th themselves, cursed and reviled because they had not spared the great heroic city that crowning degradation. He was himself one night aroused to such a pitch of frenzy that he took the floor and shouted that it was the duty of all Paris to go and die on the ramparts rather than suffer the entrance of a single Prussian. It was quite natural that the spirit of insurrection should show itself thus, should bud and blossom in the full light of day, among that populace that had first been maddened by months of distress and famine and then had found itself reduced to a condition of idleness that afforded it abundant leisure to brood on the suspicions and fancied wrongs that were largely the product of its own disordered imagination. It was one of those moral crises that have been noticed as occurring after every great siege, in which excessive patriotism, thwarted in its aims and aspirations, after having fired men's minds, degenerates into a blind rage for vengeance and destruction. The Central Committee, elected by delegates from the National Guard battalions, had protested against any attempt to disarm their constituents. Then came an immense popular demonstration on the Place de la Bastille, where there were red flags, incendiary speeches and a crowd that overflowed the square, the affair ending with the murder of a poor inoffensive agent of police, who was bound to a plank, thrown into the canal, and then stoned to death. And forty-eight hours later, during the night of the 26th of February, Maurice, awakened by the beating of the long roll and the sound of the tocsin, beheld bands of men and women streaming along the Boulevard des Batignolles and dragging cannon after them. He descended to the street, and laying hold of the rope of a gun along with some twenty others, was told how the people had gone to the Place Wagram and taken the pieces in order that the Assembly might not deliver them to the Prussians. There were seventy of them; teams were wanting, but the strong arms of the mob, tugging at the ropes and pushing at the limbers and axles, finally brought them to the summit of Montmartre with the mad impetuosity of a barbarian horde assuring the safety of its idols. When on March 1 the Prussians took possession of the quarter of the Champs Elysees, which they were to occupy only for one day, keeping themselves strictly within the limits of the barriers, Paris looked on in sullen silence, its streets deserted, its houses closed, the entire city lifeless and shrouded in its dense veil of mourning.

Two weeks more went by, during which Maurice could hardly have told how he spent his time while awaiting the approach of the momentous events of which he had a distinct presentiment. Peace was concluded definitely at last, the Assembly was to commence its regular sessions at Versailles on the 20th of the month; and yet for him nothing was concluded: he felt that they were ere long to witness the beginning of a dreadful drama of atonement. On the 18th of March, as he was about to leave his room, he received a letter from Henriette urging him to come and join her at Remilly, coupled with a playful threat that she would come and carry him off with her if he delayed too long to afford her that great pleasure. Then she went on to speak of Jean, concerning whose affairs she was extremely anxious; she told how, after leaving her late in December to join the Army of the North, he had been seized with a low fever that had kept him long a prisoner in a Belgian hospital, and only the preceding week he had written her that he was about to start for Paris, notwithstanding his enfeebled condition, where he was determined to seek active service once again. Henriette closed her letter by begging her brother to give her a faithful account of how

329

matters were with Jean as soon as he should have seen him. Maurice laid the open letter before him on the table and sank into a confused revery. Henriette, Jean; his sister whom he loved so fondly, his brother in suffering and privation; how absent from his daily thoughts had those dear ones been since the tempest had been raging in his bosom! He aroused himself, however, and as his sister advised him that she had been unable to give Jean the number of the house in the Rue des Orties, promised himself to go that very day to the office where the regimental records were kept and hunt up his friend. But he had barely got beyond his door and was crossing the Rue Saint-Honore when he encountered two fellow-soldiers of his battalion, who gave him an account of what had happened that morning and during the night before at Montmartre, and the three men started off on a run toward the scene of the disturbance.

Ah, that day of the 18th of March, the elation and enthusiasm that it aroused in Maurice! In after days he could never remember clearly what he said and did. First he beheld himself dimly, as through a veil of mist, convulsed with rage at the recital of how the troops had attempted, in the darkness and quiet that precedes the dawn, to disarm Paris by seizing the guns on Montmartre heights. It was evident that Thiers, who had arrived from Bordeaux, had been meditating the blow for the last two days, in order that the Assembly at Versailles might proceed without fear to proclaim the monarchy. Then the scene shifted, and he was on the ground at Montmartre itself-about nine o'clock it was-fired by the narrative of the people's victory: how the soldiery had come sneaking up in the darkness, how the delay in bringing up the teams had given the National Guards an opportunity to fly to arms, the troops, having no heart to fire on women and children, reversing their muskets and fraternizing with the people. Then he had wandered desultorily about the city, wherever chance directed his footsteps, and by midday had satisfied himself that the Commune was master of Paris, without even the necessity of striking a blow, for Thiers and the ministers had decamped from their quarters in the Ministry of Foreign Affairs, the entire government was flying in disorder to Versailles, the thirty thousand troops had been hastily conducted from the city, leaving more than five thousand deserters from their numbers along the line of their retreat. And later, about half-past five in the afternoon, he could recall being at a corner of the exterior boulevard in the midst of a mob of howling lunatics, listening without the slightest evidence of disapproval to the abominable story of the murder of Generals Lecomte and Clement Thomas. Generals, they called themselves; fine generals, they! The leaders they had had at Sedan rose before his memory, voluptuaries and imbeciles; one more, one less, what odds did it make! And the remainder of the day passed in the same state of half-crazed excitement, which served to distort everything to his vision; it was an insurrection that the very stones of the streets seemed to have favored, spreading, swelling, finally becoming master of all at a stroke in the unforeseen fatality of its triumph, and at ten o'clock in the evening delivering the Hotel de Ville over to the members of the Central Committee, who were greatly surprised to find themselves there.

There was one memory, however, that remained very distinct to Maurice's mind: his unexpected meeting with Jean. It was three days now since the latter had reached Paris, without a sou in his pocket, emaciated and enfeebled by the illness that had consigned him to a hospital in Brussels and kept him there two months, and having had the luck to fall in with Captain Ravaud, who had commanded a company in the 106th, he had enlisted at once in his former acquaintance's new company in the 124th. His old rank as corporal had been restored to him, and that evening he had just left the Prince Eugene barracks with his squad on his way to the left bank, where the entire army was to concentrate, when a mob collected about his men and stopped them as they were passing along the boulevard Saint-Martin. The insurgents yelled and shouted, and evidently were preparing to disarm his little band. With perfect coolness he told them to let him alone, that he had no business with them or their affairs; all he wanted was to obey his orders without harming anybody. Then a cry of glad surprise was heard, and Maurice, who had chanced to pass that way, threw himself on the other's neck and gave him a brotherly hug.

"What, is it you! My sister wrote me about you. And just think, no later than this very morning I was going to look you up at the war office!"

Jean's eyes were dim with big tears of pleasure.

"Ah, my dear lad how glad I am to see you once more! I have been looking for you, too, but where could a fellow expect to find you in this confounded great big place?"

To the crowd, continuing their angry muttering, Maurice turned and said:

"Let me talk to them, citizens! They're good fellows; I'll answer for them." He took his friend's hands in his, and lowering his voice: "You'll join us, won't you?"

Jean's face was the picture of surprise. "How, join you? I don't understand." Then for a moment he listened while Maurice railed against the government, against the army, raking up old sores and recalling all their sufferings, telling how at last they were going to be masters, punish dolts and cowards and preserve the Republic. And as he struggled to get the problems the other laid before him through his brain, the tranquil face of the unlettered peasant was clouded with an increasing sorrow. "Ah, no! ah, no! my boy. I can't join you if it's for that fine work you want me. My captain told me to go with my men to Vaugirard, and there I'm going. In spite of the devil and his angels I will go there. That's natural enough; you ought to know how it is yourself." He laughed with frank simplicity and added:

"It's you who'll come along with us."

But Maurice released his hands with an angry gesture of dissent, and thus they stood for some seconds, face to face, one under the influence of that madness that was sweeping all Paris off its feet, the malady that had been bequeathed to them by the crimes and follies of the late reign, the other strong in his ignorance and practical common sense, untainted as yet because he had grown up apart from the contaminating principle, in the land where industry and thrift were honored. They were brothers, however, none the less; the tie that united them was strong, and it was a pang to them both when the crowd suddenly surged forward and parted them.

"Au revoir, Maurice!"

"Au revoir, Jean!"

It was a regiment, the 79th, debouching from a side street, that had caused the movement among the crowd, forcing the rioters back to the sidewalks by the weight of its compact column, closed in mass. There was some hooting, but no one ventured to bar the way against the soldier boys, who went by at double time, well under control of their officers. An opportunity was afforded the little squad of the 124th to make their escape, and they followed in the wake of the larger body.

"Au revoir, Jean!"

"Au revoir, Maurice!"

They waved their hands once more in a parting salute, yielding to the fatality that decreed their separation in that manner, but each none the less securely seated in the other's heart.

The extraordinary occurrences of the next and the succeeding days crowded on the heels of one another in such swift sequence that Maurice had scarcely time to think. On the morning of the 19th Paris awoke without a government, more surprised than frightened to learn that a panic during the night had sent army, ministers, and all the public service scurrying away to

331

Versailles, and as the weather happened to be fine on that magnificent March Sunday, Paris stepped unconcernedly down into the streets to have a look at the barricades. A great white poster, bearing the signature of the Central Committee and convoking the people for the communal elections, attracted attention by the moderation of its language, although much surprise was expressed at seeing it signed by names so utterly unknown. There can be no doubt that at this incipient stage of the Commune Paris, in the bitter memory of what it had endured, in the suspicions by which it was haunted, and in its unslaked thirst for further fighting, was against Versailles. It was a condition of absolute anarchy, moreover, the conflict for the moment being between the mayors and the Central Committee, the former fruitlessly attempting to introduce measures of conciliation, while the latter, uncertain as yet to what extent it could rely on the federated National Guard, continued modestly to lay claim to no higher title than that of defender of the municipal liberties. The shots fired against the pacific demonstration in the Place Vendome, the few corpses whose blood reddened the pavements, first sent a thrill of terror circulating through the city. And while these things were going on, while the insurgents were taking definite possession of the ministries and all the public buildings, the agitation, rage and alarm prevailing at Versailles were extreme, the government there hastening to get together sufficient troops to repel the attack which they felt sure they should not have to wait for long. The steadiest and most reliable divisions of the armies of the North and of the Loire were hurried forward. Ten days sufficed to collect a force of nearly eighty thousand men, and the tide of returning confidence set in so strongly that on the 2d of April two divisions opened hostilities by taking from the federates Puteaux and Courbevoie.

It was not until the day following the events just mentioned that Maurice, starting out with his battalion to effect the conquest of Versailles, beheld, amid the throng of misty, feverish memories that rose to his poor wearied brain, Jean's melancholy face as he had seen it last, and seemed to hear the tones of his last mournful *au revoir*. The military operations of the Versaillese had filled the National Guard with alarm and indignation; three columns, embracing a total strength of fifty thousand men, had gone storming that morning through Bougival and Meudon on their way to seize the monarchical Assembly and Thiers, the murderer. It was the torrential sortie that had been demanded with such insistence during the siege, and Maurice asked himself where he should ever see Jean again unless among the dead lying on the field of battle down yonder. But it was not long before he knew the result; his battalion had barely reached the Plateau des Bergeres, on the road to Reuil, when the shells from Mont-Valerien came tumbling among the ranks. Universal consternation reigned; some had supposed that the fort was held by their comrades of the Guard, while others averred that the commander had promised solemnly to withhold his fire. A wild panic seized upon the men; the battalions broke and rushed back to Paris fast as their legs would let them, while the head of the column, diverted by a flanking movement of General Vinoy, was driven back on Reuil and cut to pieces there.

Then Maurice, who had escaped unharmed from the slaughter, his nerves still quivering with the fury that had inspired him on the battlefield, was filled with fresh detestation for that so-called government of law and order which always allowed itself to be beaten by the Prussians, and could only muster up a little courage when it came to oppressing Paris. And the German armies were still there, from Saint-Denis to Charenton, watching the shameful spectacle of internecine conflict! Thus, in the fierce longing for vengeance and destruction that animated him, he could not do otherwise than sanction the first measures of communistic violence, the building of barricades in the streets and public squares, the arrest of the archbishop, some priests, and former officeholders, who were to be held as hostages. The atrocities that distinguished either side in that horrible conflict were already beginning to manifest themselves, Versailles shooting the prisoners it made, Paris retaliating with a decree that for each one of its soldiers murdered three hostages should forfeit their life. The horror of it, that fratricidal conflict, that wretched nation completing the work of destruction by devouring its own children! And the little reason that remained to Maurice, in the ruin of all the things he had hitherto held sacred, was quickly dissipated in the whirlwind of blind fury that swept all before it. In his eyes the Commune was to be the avenger of all the wrongs they had suffered, the liberator, coming with fire and sword to purify and punish. He was not quite clear in mind about it all, but remembered having read how great and flourishing the old free cities had become, how wealthy provinces had federated and imposed their law upon the world. If Paris should be victorious he beheld her, crowned with an aureole of glory, building up a new France, where liberty and justice should be the watchwords, organizing a new society, having first swept away the rotten debris of the old. It was true that when the result of the elections became known he was somewhat surprised by the strange mixture of moderates, revolutionists, and socialists of every sect and shade to whom the accomplishment of the great work was intrusted; he was acquainted with several of the men and knew them to be of extremely mediocre abilities. Would not the strongest among them come in collision and neutralize one another amid the clashing ideas which they represented? But on the day when the ceremony of the inauguration of the Commune took place before the Hotel de Ville, amid the thunder of artillery and trophies and red banners floating in the air, his boundless hopes again got the better of his fears and he ceased to doubt. Among the lies of some and the unquestioning faith of others, the illusion started into life again with renewed vigor, in the acute crisis of the malady raised to paroxysmal pitch.

During the entire month of April Maurice was on duty in the neighborhood of Neuilly. The gentle warmth of the early spring had brought out the blossoms on the lilacs, and the fighting was conducted among the bright verdure of the gardens; the National Guards came into the city at night with bouquets of flowers stuck in their muskets. The troops collected at Versailles were now so numerous as to warrant their formation in two armies, a first line under the orders of Marshal MacMahon and a reserve commanded by General Vinoy. The Commune had nearly a hundred thousand National Guards mobilized and as many more on the rosters who could be called out at short notice, but fifty thousand were as many as they ever brought into the field at one time. Day by day the plan of attack adopted by the Versaillese became more manifest: after occupying Neuilly they had taken possession of the Chateau of Becon and soon after of Asnieres, but these movements were simply to make the investment more complete, for their intention was to enter the city by the Point-du-Jour soon as the converging fire from Mont-Valerien and Fort d'Issy should enable them to carry the rampart there. Mont-Valerien was theirs already, and they were straining every nerve to capture Issy, utilizing the works abandoned by the Germans for the purpose. Since the middle of April the fire of musketry and artillery had been incessant; at Levallois and Neuilly the fighting never ceased, the skirmishers blazing away uninterruptedly, by night as well as by day. Heavy guns, mounted on armored cars, moved to and fro on the Belt Railway, shelling Asnieres over the roofs of Levallois. It was at Vanves and Issy, however, that the cannonade was fiercest; it shook the windows of Paris as the siege had done when it was at its height. And when finally, on the 9th of May, Fort d'Issy was obliged to succumb and fell into the hands of the Versailles army the defeat of the Commune was assured, and in their frenzy of panic the leaders resorted to most detestable measures.

Maurice favored the creation of a Committee of Public Safety. The warnings of history came to his mind; had not the hour struck for adopting energetic methods if they wished to save the country? There was but one of their barbarities that really pained him, and that was the destruction of the Vendome column; he reproached himself for the feeling as being a childish weakness, but his grandfather's voice still sounded in his ears repeating the old familiar tales of Marengo, Austerlitz, Jena, Eylau, Friedland, Wagram, the Moskowa-those epic narratives that thrilled his pulses yet as often as he thought of them. But that they should demolish the house of the murderer Thiers, that they should retain the hostages as a guarantee and a menace, was not that right and just when the Versaillese were unchaining their fury on Paris, bombarding it, destroying its edifices, slaughtering women and children with their shells? As he saw the end of his dream approaching dark thoughts of ruin and destruction filled his mind. If their ideas of justice and retribution were not to prevail, if they were to be crushed out of them with their life-blood, then perish the world, swept away in one of those cosmic upheavals that are the beginning of a new life. Let Paris sink beneath the waves, let it go up in smoke and flame, like a gigantic funeral pyre, sooner than let it be again delivered over to its former state of vice and misery, to that old vicious social system of abominable injustice. And he dreamed another dark, terrible dream, the great city reduced to ashes, naught to be seen on either side the Seine but piles of smoldering ruins, the festering wound purified and healed with fire, a catastrophe without a name, such as had never been before, whence should arise a new race. Wild stories were everywhere circulated, which interested him intensely, of the mines that were driven under all the quarters of the city, the barrels of powder with which the catacombs were stuffed, the monuments and public buildings ready to be blown into the air at a moment's notice; and all were connected by electric wires in such a way that a single spark would suffice to set them off; there were great stores of inflammable substances, too, especially petroleum, with which the streets and avenues were to be converted into seething lakes of flame. The Commune had sworn that should the Versaillese enter the city not one of them would ever get beyond the barricades that closed the ends of the streets; the pavements would yawn, the houses would sink in ruins, Paris would go up in flames, and bury assailants and assailed under its ashes.

And if Maurice solaced himself with these crazy dreams, it was because of his secret discontent with the Commune itself. He had lost all confidence in its members, he felt it was inefficient, drawn this way and that by so many conflicting elements, losing its head and becoming purposeless and driveling as it saw the near approach of the peril with which it was menaced. Of the social reforms it had pledged itself to it had not been able to accomplish a single one, and it was now quite certain that it would leave behind it no great work to perpetuate its name. But what more than all beside was gnawing at its vitals was the rivalries by which it was distracted, the corroding suspicion and distrust in which each of its members lived. For some time past many of them, the more moderate and the timid, had ceased to attend its sessions. The others shaped their course day by day in accordance with events, trembling at the idea of a possible dictatorship; they had reached that point where the factions of revolutionary assemblages exterminate one another by way of saving the country. Cluzeret had become suspected, then Dombrowski, and Rossel was about to share their fate. Delescluze, appointed Civil Delegate at War, could do nothing of his own volition, notwithstanding his great authority. And thus the grand social effort that they had had in view wasted itself in the ever-widening isolation about those men, whose power had become a nullity, whose actions were the result of their despair.

In Paris there was an increasing feeling of terror. Paris, irritated at first against Versailles, shivering at the recollection of what it had suffered during the siege, was now breaking away from the Commune. The compulsory enrollment, the decree incorporating every man under forty in the National Guard, had angered the more sedate citizens and been the means of bringing about a general exodus: men in disguise and provided with forged papers of Alsatian citizenship made their escape by way of Saint-Denis; others let themselves down into the moat in the darkness of the night with ropes and ladders. The wealthy had long since taken their departure. None of the factories and workshops had opened their doors; trade and commerce there was none; there was no employment for labor; the life of enforced idleness went on amid the alarmed expectancy of the frightful denouement that everyone felt could not be far away. And the people depended for their daily bread on the pay of the National Guards, that dole of thirty sous that was paid from the millions extorted from the Bank of France, the thirty sous for the sake of which alone many men were wearing the uniform, which had been one of the primary causes and the raison d'etre of the insurrection. Whole districts were deserted, the shops closed, the house-fronts lifeless. In the bright May sunshine that flooded the empty streets the few pedestrians beheld nothing moving save the barbaric display of the burial of some federates killed in action, the funeral train where no priest walked, the hearse draped with red flags, followed by a crowd of men and women bearing bouquets of immortelles. The churches were closed and did duty each evening as political club-rooms. The revolutionary journals alone were hawked about the streets; the others had been suppressed. Great Paris was indeed an unhappy city in those days, what with its republican sympathies that made it detest the monarchical Assembly at Versailles and its ever-increasing terror of the Commune, from which it prayed most fervently to be delivered among all the grisly stories that were current, the daily arrests of citizens as hostages, the casks of gunpowder that filled the sewers, where men patrolled by day and night awaiting the signal to apply the torch.

Maurice, who had never been a drinking man, allowed himself to be seduced by the too prevalent habit of over-indulgence. It had become a thing of frequent occurrence with him now, when he was out on picket duty or had to spend the night in barracks, to take a "pony" of brandy, and if he took a second it was apt to go to his head in the alcohol-laden atmosphere that he was forced to breathe. It had become epidemic, that chronic drunkenness, among those men with whom bread was scarce and who could have all the brandy they wanted by asking for it. Toward evening on Sunday, the 21st of May, Maurice came home drunk, for the first time in his life, to his room in the Rue des Orties, where he was in the habit of sleeping occasionally. He had been at Neuilly again that day, blazing away at the enemy and taking a nip now and then with the comrades, to see if it would not relieve the terrible fatigue from which he was suffering. Then, with a light head and heavy legs, he came and threw himself on the bed in his little chamber; it must have been through force of instinct, for he could never remember how he got there. And it was not until the following morning, when the sun was high in the heavens, that he awoke, aroused by the ringing of the alarm bells, the blare of trumpets and beating of drums. During the night the Versaillese, finding a gate undefended, had effected an unresisted entrance at the Point-du-Jour.

When he had thrown on his clothes and hastened down into the street, his musket slung across his shoulder by the strap, a band of frightened soldiers whom he fell in with at the *mairie* of the arrondissement related to him the occurrences of the night, in the midst of a confusion such that at first he had hard work to understand. Fort d'Issy and the great battery at Montretout, seconded by Mont Valerien, for the last ten days had been battering the rampart at the Point-du-Jour, as a consequence of which the Saint-Cloud gate was no longer tenable and an assault had been ordered for the following morning, the 22d; but someone who chanced to pass that way at about five o'clock perceived that the gate was unprotected and immediately notified the guards in the trenches, who were not more than fifty yards away. Two companies of the 37th regiment of regulars were the first to enter the city, and were quickly followed by the entire 4th corps under General Douay. All night long the troops were pouring in in an uninterrupted stream. At seven o'clock Verge's division marched down to the bridge at Grenelle, crossed, and pushed on to the Trocadero. At nine General Clinchamp was master of Passy and la Muette. At three o'clock in the morning the 1st corps had pitched its tents in the Bois de Boulogne, while at about the same hour Bruat's division was passing the Seine to seize the Sevres gate and facilitate the movement of the 2d Corps, General de Cissey's, which occupied the district of Grenelle an hour later. The Versailles army, therefore, on the morning of the 22d, was master of the Trocadero and the Chateau of la Muette on the right bank, and of Grenelle on the left; and great was the rage and consternation that prevailed among the Communists, who were already accusing one another of treason, frantic at the thought of their inevitable defeat.

When Maurice at last understood the condition of affairs his first thought was that the end had come, that all left him was to go forth and meet his death. But the tocsin was pealing, drums were beating, women and children, even, were working on the barricades, the streets were alive with the stir and bustle of the battalions hurrying to assume the positions assigned them in the coming conflict. By midday it was seen that the Versaillese were remaining quiet in their new positions, and then fresh courage returned to the hearts of the soldiers of the Commune, who were resolved to conquer or die. The enemy's army, which they had feared to see in possession of the Tuileries by that time, profiting by the stern lessons of experience and imitating the prudent tactics of the Prussians, conducted its operations with the utmost caution. The Committee of Public Safety and Delescluze, Delegate at War, directed the defense from their quarters in the Hotel de Ville. It was reported that a last proposal for a peaceable arrangement had been rejected by them with disdain. That served to inspire the men with still more courage, the triumph of Paris was assured, the resistance would be as unyielding as the attack was vindictive, in the implacable hate, swollen by lies and cruelties, that inflamed the heart of either army. And that day was spent by Maurice in the quarters of the Champ de Mars and the Invalides, firing and falling back slowly from street to street. He had not been able to find his battalion; he fought in the ranks with comrades who were strangers to him, accompanying them in their march to the left bank without taking heed whither they were going. About four o'clock they had a furious conflict behind a barricade that had been thrown across the Rue de l'Universite, where it comes out on the Esplanade, and it was not until twilight that they abandoned it on learning that Bruat's division, stealing up along the quai, had seized the Corps Legislatif. They had a narrow escape from capture, and it was with great difficulty that they managed to reach the Rue de Lille after a long circuit through the Rue Saint-Dominique and the Rue Bellechasse. At the close of that day the army of Versailles occupied a line which, beginning at the Vanves gate, led past the Corps Legislatif, the Palace of the Elysee, St. Augustine's Church, the Lazare station, and ended at the Asnieres gate.

The next day, Tuesday, the 23d, was warm and bright, and a terrible day it was for Maurice. The few hundred federates with whom he was, and in whose ranks were men of many different battalions, were charged with the defense of the entire quartier, from the *quai* to the Rue Saint-Dominique. Most of them had bivouacked in the gardens of the great mansions that line the Rue de Lille; he had had an unbroken night's rest on a grass-plot at one side of the Palace of the Legion of Honor. It was his belief that soon as it was light enough the troops would move out from their shelter behind the Corps Legislatif and force them back upon the strong barricades in the Rue du Bac, but hour after hour passed and there was no sign of an attack. There was only some desultory firing at long range between parties posted at either end of the streets. The Versaillese, who were not desirous of attempting a direct attack on the front of the formidable fortress into which the insurgents had converted the terrace of the Tuileries, developed their plan of action with great circumspection; two strong columns were sent out to right and left that, skirting the ramparts, should first seize Montmartre and the Observatory and then, wheeling inward, swoop down on the central quarters, surrounding them and capturing all they contained, as a shoal of fish is captured in the meshes of a gigantic net. About two o'clock Maurice heard that the tricolor was floating over Montmartre: the great battery of the Moulin de la Galette had succumbed to the combined attack of three army corps, which hurled their battalions simultaneously on the northern and western faces of the butte through the Rues Lepic, des Saules and du Mont-Cenis; then the waves of the victorious troops had poured back on Paris, carrying the Place Saint-Georges, Notre-Dame de Lorette, the *mairie* in the Rue Drouot and the new Opera House, while on the left bank the turning movement, starting from the cemetery of Mont-Parnasse, had reached the Place d'Enfer and the Horse Market. These tidings of the rapid progress of the hostile army were received by the communards with mingled feelings of rage and terror amounting almost to stupefaction. What, Montmartre carried in two hours; Montmartre, the glorious, the impregnable citadel of the insurrection! Maurice saw that the ranks were thinning about him; trembling soldiers, fearing the fate that was in store for them should they be caught, were slinking furtively away to look for a place where they might wash the powder grime from hands and face and exchange their uniform for a blouse. There was a rumor that the enemy were making ready to attack the Croix-Rouge and take their position in flank. By this time the barricades in the Rues Martignac and Bellechasse had been carried, the red-legs were beginning to make their appearance at the end of the Rue de Lille, and soon all that remained was a little band of fanatics and men with the courage of their opinions, Maurice and some fifty more, who were resolved to sell their lives dearly, killing as many as they could of those Versaillese, who treated the federates like thieves and murderers, dragging away the prisoners they made and shooting them in the rear of the line of battle. Their bitter animosity had broadened and deepened since the days before; it was war to the knife between those rebels dying for an idea and that army, inflamed with reactionary passions and irritated that it was kept so long in the field.

About five o'clock, as Maurice and his companions were finally falling back to seek the shelter of the barricades in the Rue du Bac, descending the Rue de Lille and pausing at every moment to fire another shot, he suddenly beheld volumes of dense black smoke pouring from an open window in the Palace of the Legion of Honor. It was the first fire kindled in Paris, and in the furious insanity that possessed him it gave him a fierce delight. The hour had struck; let the whole city go up in flame, let its people be cleansed by the fiery purification! But a sight that he saw presently filled him with surprise: a band of five or six men came hurrying out of the building, headed by a tall varlet in whom he recognized Chouteau, his former comrade in the squad of the 106th. He had seen him once before, after the 18th of March, wearing a gold-laced kepi; he seemed by his bedizened uniform to have risen in rank, was probably on the staff of some one of the many generals who were never seen where there was fighting going on. He remembered the account somebody had given him of that fellow Chouteau, of his quartering himself in the Palace of the Legion of Honor and living there, guzzling and swilling, in company with a mistress, wallowing with his boots on in the great luxurious beds, smashing the plate-glass mirrors with shots from his revolver, merely for the amusement there was in it. It was even asserted that the woman left the building every morning in one of the state carriages, under pretense of going to the Halles for her day's marketing, carrying off with her great bundles of linen, clocks, and even articles of furniture, the fruit of their thieveries. And Maurice, as he watched him running away with his men, carrying a bucket of petroleum on his arm, experienced a sickening sensation of doubt and felt his faith beginning to waver. How could the terrible work they were engaged in be good, when men like that were the workmen?

Hours passed, and still he fought on, but with a bitter feeling of distress, with no other wish than that he might die. If he had erred, let him at least atone for his error with his blood! The barricade across the Rue de Lille, near its intersection with the Rue du Bac, was a formidable one, composed of bags and casks filled with earth and faced by a deep ditch. He and a scant dozen of other federates were its only defenders, resting in a semi-recumbent position on the ground, infallibly causing every soldier who exposed himself to bite the dust. He lay there, without even changing his position, until nightfall, using up his cartridges in silence, in the dogged sullenness of his despair. The dense clouds of smoke from the Palace of the Legion of Honor were billowing upward in denser masses, the flames undistinguishable as yet in the dying daylight, and he watched the fantastic, changing forms they took as the wind whirled them downward to the street. Another fire had broken out in an hotel not far away. And all at once a comrade came running up to tell him that the enemy, not daring to advance along the street, were making a way for themselves through the houses and gardens, breaking down the walls with picks. The end was close at hand; they might come out in the rear of the barricade at any moment. A shot having been fired from an upper window of a house on the corner, he saw Chouteau and his gang, with their petroleum and their lighted torch, rush with frantic speed to the buildings on either side and climb the stairs, and half an hour later, in the increasing darkness, the entire square was in flames, while he, still prone on the ground behind his shelter, availed himself of the vivid light to pick off any venturesome soldier who stepped from his protecting doorway into the narrow street.

How long did Maurice keep on firing? He could not tell; he had lost all consciousness of time and place. It might be nine o'clock, or ten, perhaps. He continued to load and fire; his condition of hopelessness and gloom was pitiable; death seemed to him long in coming. The detestable work he was engaged in gave him now a sensation of nausea, as the fumes of the wine he has drunk rise and nauseate the drunkard. An intense heat began to beat on him from the houses that were burning on every side-an air that scorched and asphyxiated. The carrefour, with the barricades that closed it in, was become an intrenched camp, guarded by the roaring flames that rose on every side and sent down showers of sparks. Those were the orders, were they not? to fire the adjacent houses before they abandoned the barricades, arrest the progress of the troops by an impassable sea of flame, burn Paris in the face of the enemy advancing to take possession of it. And presently he became aware that the houses in the Rue du Bac were not the only ones that were devoted to destruction; looking behind him he beheld the whole sky suffused with a bright, ruddy glow; he heard an ominous roar in the distance, as if all Paris were bursting into conflagration. Chouteau was no longer to be seen; he had long since fled to save his skin from the bullets. His comrades, too, even those most zealous in the cause, had one by one stolen away, affrighted at the approaching prospect of being outflanked. At last he was left alone, stretched at length between two sand bags, his every faculty bent on defending the front of the barricade, when the soldiers, who had made their way through the gardens in the middle of the block, emerged from a house in the Rue du Bac and pounced on him from the rear.

For two whole days, in the fevered excitement of the supreme conflict, Maurice had not once thought of Jean, nor had Jean, since he entered Paris with his regiment, which had been assigned to Bruat's division, for a single moment remembered Maurice. The day before his duties had kept him in the neighborhood of the Champ de Mars and the Esplanade of the Invalides, and on this day he had remained in the Place du Palais-Bourbon until nearly noon, when the troops were sent forward to clean out the barricades of the quartier, as far as the Rue des Saints-Peres. A feeling of deep exasperation against the rioters had gradually taken possession of him, usually so calm and self-contained, as it had of all his comrades, whose ardent wish it was to be allowed to go home and rest after so many months of fatigue. But of all the atrocities of the Commune that stirred his placid nature and made him forgetful even of his tenderest affections, there were none that angered him as did those conflagrations. What, burn houses, set fire to palaces, and simply because they had lost the battle! Only robbers and murderers were capable of such work as that. And he who but the day before had sorrowed over the summary executions of the insurgents was now like a madman, ready to rend and tear, yelling, shouting, his eyes starting from their sockets.

338

Jean burst like a hurricane into the Rue du Bac with the few men of his squad. At first he could distinguish no one; he thought the barricade had been abandoned. Then, looking more closely, he perceived a communard extended on the ground between two sand bags; he stirred, he brought his piece to the shoulder, was about to discharge it down the Rue du Bac. And impelled by blind fate, Jean rushed upon the man and thrust his bayonet through him, nailing him to the barricade.

Maurice had not had time to turn. He gave a cry and raised his head. The blinding light of the burning buildings fell full on their faces.

"O Jean, dear old boy, is it you?"

To die, that was what he wished, what he had been longing for. But to die by his brother's hand, ah! the cup was too bitter; the thought of death no longer smiled on him.

"Is it you, Jean, old friend?"

Jean, sobered by the terrible shock, looked at him with wild eyes. They were alone; the other soldiers had gone in pursuit of the fugitives. About them the conflagrations roared and crackled and blazed up higher than before; great sheets of white flame poured from the windows, while from within came the crash of falling ceilings. And Jean cast himself on the ground at Maurice's side, sobbing, feeling him, trying to raise him to see if he might not yet be saved.

"My boy, oh! my poor, poor boy!"

VIII.

When at about nine o'clock the train from Sedan, after innumerable delays along the way, rolled into the Saint-Denis station, the sky to the south was lit up by a fiery glow as if all Paris was burning. The light had increased with the growing darkness, and now it filled the horizon, climbing constantly higher up the heavens and tingeing with blood-red hues some clouds, that lay off to the eastward in the gloom which the contrast rendered more opaque than ever.

The travelers alighted, Henriette among the first, alarmed by the glare they had beheld from the windows of the cars as they rushed onward across the darkling fields. The soldiers of a Prussian detachment, moreover, that had been sent to occupy the station, went through the train and compelled the passengers to leave it, while two of their number, stationed on the platform, shouted in guttural French:

"Paris is burning. All out here! this train goes no further. Paris is burning, Paris is burning!"

Henriette experienced a terrible shock. Mon Dieu! was she too late, then? Receiving no reply from Maurice to her two last letters, the alarming news from Paris had filled her with such mortal terror that she determined to leave Remilly and come and try to find her brother in the great city. For months past her life at Uncle Fouchard's had been a melancholy one; the troops occupying the village and the surrounding country had become harsher and more exacting as the resistance of Paris was protracted, and now that peace was declared and the regiments were stringing along the roads, one by one, on their way home to Germany, the country and the cities through which they passed were taxed to their utmost to feed the hungry soldiers. The morning when she arose at daybreak to go and take the train at Sedan, looking out into the courtyard of the farmhouse she had seen a body of cavalry who had slept there all night, scattered promiscuously on the bare ground, wrapped in their long cloaks. They were so numerous that the earth was hidden by them. Then, at the shrill summons of a trumpet call, all had risen to their feet, silent, draped in the folds of those long mantles, and in such serried, close array that she involuntarily thought of the graves of a battlefield opening and giving up their dead at the call of the last trump. And here again at Saint-Denis she encountered the Prussians, and it was from Prussian lips that came that cry which caused her such distress:

"All out here! this train goes no further. Paris is burning!"

Henriette, her little satchel in her hand, rushed distractedly up to the men in quest of information. There had been heavy fighting in Paris for the last two days, they told her, the railway had been destroyed, the Germans were watching the course of events. But she insisted on pursuing her journey at every risk, and catching sight upon the platform of the officer in command of the detachment detailed to guard the station, she hurried up to him.

"Sir, I am terribly distressed about my brother, and am trying to get to him. I entreat you, furnish me with the means to reach Paris." The light from a gas jet fell full on the captain's face she stopped in surprise. "What, Otto, is it you! Oh, mon Dieu, be good to me, since chance has once more brought us together!"

It was Otto Gunther, the cousin, as stiff and ceremonious as ever, tight-buttoned in his Guard's uniform, the picture of a narrow-minded martinet. At first he failed to recognize the little, thin, insignificant-looking woman, with the handsome light hair and the pale, gentle face; it was only by the brave, honest look that filled her eyes that he finally remembered her. His only answer was a slight shrug of the shoulders.

"You know I have a brother in the army," Henriette eagerly went on. "He is in Paris; I fear he has allowed himself to become mixed up with this horrible conflict. O Otto, I beseech you, assist me to continue my journey."

At last he condescended to speak. "But I can do nothing to help you; really I cannot. There have been no trains running since yesterday; I believe the rails have been torn up over by the ramparts somewhere. And I have neither a horse and carriage nor a man to guide you at my disposal."

She looked him in the face with a low, stifled murmur of pain and sorrow to behold him thus obdurate. "Oh, you will do nothing to aid me. My God, to whom then can I turn!"

It was an unlikely story for one of those Prussians to tell, whose hosts were everywhere all-powerful, who had the city at their beck and call, could have requisitioned a hundred carriages and brought a thousand horses from their stables. And he denied her prayer with the haughty air of a victor who has made it a law to himself not to interfere with the concerns of the vanquished, lest thereby he might defile himself and tarnish the luster of his new-won laurels.

"At all events," continued Henriette, "you know what is going on in the city; you won't refuse to tell me that much."

He gave a smile, so faint as scarce to be perceptible. "Paris is burning. Look! come this way, you can see more clearly."

Leaving the station, he preceded her along the track for a hundred steps or so until they came to an iron foot-bridge that spanned the road. When they had climbed the narrow stairs and reached the floor of the structure, resting their elbows on the railing, they beheld the broad level plain outstretched before them, at the foot of the slope of the embankment.

"You see, Paris is burning."

It was in the neighborhood of ten o'clock. The fierce red glare that lit the southern sky was ever mounting higher. The blood-red clouds had disappeared from where they had floated in the east; the zenith was like a great inverted bowl of inky blackness, across which ran the reflections of the distant flames. The horizon was one unbroken line of fire, but to the right they could distinguish spots where the conflagration was raging with greater fury, sending up great spires and pinnacles of flame, of the most vivid scarlet, to pierce the dense opacity above, amid billowing clouds of smoke. It was like the burning of some great forest, where the fire bridges intervening space, and leaps from tree to tree; one would have said the very earth must be calcined and reduced to ashes beneath the heat of Paris' gigantic funeral pyre.

"Look," said Otto, "that eminence that you see profiled in black against the red background is Montmartre. There on the left, at Belleville and la Villette, there has not been a house burned yet; it must be they are selecting the districts of the wealthy for their work; and it spreads, it spreads. Look! there is another conflagration breaking out; watch the flames there to the right, how they seethe and rise and fall; observe the shifting tints of the vapors that rise from the blazing furnace. And others, and others still; the heavens are on fire!"

He did not raise his voice or manifest any sign of feeling, and it froze Henriette's blood that a human being could stand by and witness such a spectacle unmoved. Ah, that those Prussians should be there to see that sight! She saw an insult in his studied calmness, in the faint smile that played upon his lips, as if he had long foreseen and been watching for that unparalleled disaster. So, Paris was burning then at last, Paris, upon whose monuments the German shells had scarce been able to inflict more than a scratch! and he was there to see it burn, and in the spectacle found compensation for all his grievances, the inordinate length to which the siege had been protracted, the bitter, freezing weather, the difficulties they had surmounted only to see them present themselves anew under some other shape, the toil and trouble they had had in mounting their heavy guns, while all the time Germany from behind was reproaching them with their dilatoriness. Nothing in all the glory of their victory, neither the ceded provinces nor the indemnity of five milliards, appealed to him so strongly as did that sight of Paris, in a fit of furious madness, immolating herself and going up in smoke and flame on that beautiful spring night.

"Ah, it was sure to come," he added in a lower voice. "Fine work, my masters!"

It seemed to Henriette as if her heart would break in presence of that dire catastrophe. Her personal grief was lost to sight for some minutes, swallowed up in the great drama of a people's atonement that was being enacted before her eyes. The thought of the lives that would be sacrificed to the devouring flames, the sight of the great capital blazing on the horizon, emitting the infernal light of the cities that were accursed and smitten for their iniquity, elicited from her an involuntary cry of anguish. She clasped her hands, asking:

"Oh, merciful Father, of what have we been guilty that we should be punished thus?"

Otto raised his arm in an oratorical attitude. He was on the point of speaking, with the stern, cold-blooded vehemence of the military bigot who has ever a quotation from Holy Writ at his tongue's end, but glancing at the young woman, the look he encountered from her candid, gentle eyes checked him. Besides, his gesture had spoken for him; it told his hatred for the nation, his conviction that he was in France to mete out justice, delegated by the God of Armies, to chastise a perverse and stiff-necked generation. Paris was burning off there on the horizon in expiation of its centuries of dissolute life, of its heaped-up measure of crime and lust. Once again the German race were to be the saviors of the world, were to purge Europe of the remnant of Latin corruption. He let his arm fall to his side and simply said:

"It is the end of all. There is another quartier doomed, for see, a fresh fire has broken out there to the right. In that direction, that line of flame that creeps onward like a stream of lava-"

Neither spoke for a long time; an awed silence rested on them. The great waves of flame continued to ascend, sending up streamers and ribbons of vivid light high into the heavens. Beneath the sea of fire was every moment extending its boundaries, a tossing, stormy, burning ocean, whence now arose dense clouds of smoke that collected over the city in a huge pall of a somber coppery hue, which was wafted slowly athwart the blackness of the night, streaking the vault of heaven with its accursed rain of ashes and of soot.

Henriette started as if awaking from an evil dream, and, the thought of her brother flowing in again upon her mind, once more became a supplicant.

"Can you do nothing for me? won't you assist me to get to Paris?"

With his former air of unconcern Otto again raised his eyes to the horizon, smiling vaguely.

"What would be the use? since to-morrow morning the city will be a pile of ruins!"

And that was all; she left the bridge, without even bidding him good-by, flying, she knew not whither, with her little satchel, while he remained yet a long time at his post of observation, a motionless figure, rigid and erect, lost in the darkness of the night, feasting his eyes on the spectacle of that Babylon in flames.

Almost the first person that Henriette encountered on emerging from the station was a stout lady who was chaffering with a hackman over his charge for driving her to the Rue Richelieu in Paris, and the young woman pleaded so touchingly, with tears in her eyes, that finally the lady consented to let her occupy a seat in the carriage. The driver, a little swarthy man, whipped up his horse and did not open his lips once during the ride, but the stout lady was extremely loquacious, telling how she had left the city the day but one before after tightly locking and bolting her shop, but had been so imprudent as to leave some valuable papers behind, hidden in a hole in the wall; hence her mind had been occupied by one engrossing thought for the two hours that the city had been burning, how she might return and snatch her property from the flames. The sleepy guards at the barrier allowed the carriage to pass without much difficulty, the worthy lady allaying their scruples with a fib, telling them she was bringing back her niece with her to Paris to assist in nursing her husband, who had been wounded by the Versaillese. It was not until they commenced to make their way along the paved streets that they encountered serious obstacles; they were obliged at every moment to turn out in order to avoid the barricades that were erected across the roadway, and when at last they reached the boulevard Poissoniere the driver declared he would go no further. The two women were therefore forced to continue their way on foot, through the Rue du Sentier, the Rue des Jeuneurs, and all the circumscribing region of the Bourse. As they approached the fortifications the blazing sky had made their way as bright before them as if it had been broad day; now they were surprised by the deserted and tranquil condition of the streets, where the only sound that disturbed the stillness was a dull, distant roar. In the vicinity of the Bourse, however, they were alarmed by the sound of musketry; they slipped along with great caution, hugging the walls. On reaching the Rue Richelieu and finding her shop had not been disturbed, the stout lady was so overjoyed that she insisted on seeing her traveling companion safely housed; they struck through the Rue du Hazard, the Rue Saint-Anne, and finally reached the Rue des Orties. Some federates, whose battalion was still holding the Rue Saint-Anne, attempted to prevent them from passing. It was four o'clock and already quite light when Henriette, exhausted by the fatigue of her long day and the stress of her emotions, reached the old house in the Rue des Orties and found the door standing open. Climbing the dark, narrow staircase, she turned to the left and discovered behind a door a ladder that led upward toward the roof.

Maurice, meantime, behind the barricade in the Rue du Bac, had succeeded in raising himself to his knees, and Jean's heart throbbed with a wild, tumultuous hope, for he believed he had pinned his friend to the earth.

"Oh, my little one, are you alive still? is that great happiness in store for me, brute that I am? Wait a moment, let me see."

He examined the wound with great tenderness by the light of the burning buildings. The bayonet had gone through the right arm near the shoulder, but a more serious part of the business was that it had afterward entered the body between two of the ribs and probably touched the lung. Still, the wounded man breathed without much apparent difficulty, but the right arm hung useless at his side.

"Poor old boy, don't grieve! We shall have time to say good-by to each other, and it is better thus, you see; I am glad to have done with it all. You have done enough for me to make up for this, for I should have died long ago in some ditch, even as I am dying now, had it not been for you."

But Jean, hearing him speak thus, again gave way to an outburst of violent grief.

"Hush, hush! Twice you saved me from the clutches of the Prussians. We were quits; it was my turn to devote my life, and instead of that I have slain you. Ah, tonnerre de Dieu! I must have been drunk not to recognize you; yes, drunk as a hog from glutting myself with blood."

Tears streamed from his eyes at the recollection of their last parting, down there, at Remilly, when they embraced, asking themselves if they should ever meet again, and how, under what circumstances of sorrow or of gladness. It was nothing, then, that they had passed toilsome days and sleepless nights together, with death staring them in the face? It was to bring them to this abominable thing, to this senseless, atrocious fratricide, that their hearts had been fused in the crucible of those weeks of suffering endured in common? No, no, it could not be; he turned in horror from the thought.

"Let's see what I can do, little one; I must save you."

The first thing to be done was to remove him to a place of safety, for the troops dispatched the wounded Communists wherever they found them. They were alone, fortunately; there was not a minute to lose. He first ripped the sleeve from wrist to shoulder with his knife, then took off the uniform coat. Some blood flowed; he made haste to bandage the arm securely with strips that he tore from the lining of the garment for the purpose. After that he staunched as well as he could the wound in the side and fastened the injured arm over it, He luckily had a bit of cord in his pocket, which he knotted tightly around the primitive dressing, thus assuring the immobility of the injured parts and preventing hemorrhage.

"Can you walk?"

"Yes, I think so."

But he did not dare to take him through the streets thus, in his shirt sleeves. Remembering to have seen a dead soldier lying in an adjacent street, he hurried off and presently came back with a capote and a *kepi*. He threw the greatcoat over his friend's shoulders and assisted him to slip his uninjured arm into the left sleeve. Then, when he had put the *kepi* on his head:

"There, now you are one of us-where are we to go?"

That was the question. His reviving hope and courage were suddenly damped by a horrible uncertainty. Where were they to look for a shelter that gave promise of security? the troops were searching the houses, were shooting every Communist they took with arms in his hands. And in addition to that, neither of them knew a soul in that portion of the city to whom they might apply for succor and refuge; not a place where they might hide their heads.

"The best thing to do would be to go home where I live," said Maurice. "The house is out of the way; no one will ever think of visiting it. But it is in the Rue des Orties, on the other side of the river."

Jean gave vent to a muttered oath in his irresolution and despair.

"Nom de Dieu! What are we to do?"

It was useless to think of attempting to pass the Pont Royal, which could not have been more brilliantly illuminated if the noonday sun had been shining on it. At every moment shots were heard coming from either bank of the river. Besides that, the blazing Tuileries lay directly in their path, and the Louvre, guarded and barricaded, would be an insurmountable obstacle.

"That ends it, then; there's no way open," said Jean, who had spent six months in Paris on his return from the Italian campaign.

An idea suddenly flashed across his brain. There had formerly been a place a little below the Pont Royal where small boats were kept for hire; if the boats were there still they would make the venture. The route was a long and dangerous one, but they had no choice, and, further, they must act with decision.

"See here, little one, we're going to clear out from here; the locality isn't healthy. I'll manufacture an excuse for my lieutenant; I'll tell him the communards took me prisoner and I got away."

Taking his unhurt arm he sustained him for the short distance they had to traverse along the Rue du Bac, where the tall houses on either hand were now ablaze from cellar to garret, like huge torches. The burning cinders fell on them in showers, the heat was so intense that the hair on their head and face was singed, and when they came out on the *quai* they stood for a moment dazed and blinded by the terrific light of the conflagrations, rearing their tall crests heavenward, on either side the Seine.

"One wouldn't need a candle to go to bed by here," grumbled Jean, with whose plans the illumination promised to interfere. And it was only when he had helped Maurice down the steps to the left and a little way down stream from the bridge that he felt somewhat easy in mind. There was a clump of tall trees standing on the bank of the stream, whose shadow gave them a measure of security. For near a quarter of an hour the dark forms moving to and fro on the opposite *quai* kept them in a fever of apprehension. There was firing, a scream was heard, succeeded by a loud splash, and the bosom of the river was disturbed. The bridge was evidently guarded.

"Suppose we pass the night in that shed?" suggested Maurice, pointing to the wooden structure that served the boatman as an office.

"Yes, and get pinched to-morrow morning!"

Jean was still harboring his idea. He had found quite a flotilla of small boats there, but they were all securely fastened with chains; how was he to get one loose and secure a pair of oars? At last he discovered two oars that had been thrown aside as useless; he succeeded in forcing a padlock, and when he had stowed Maurice away in the bow, shoved off and allowed the boat to drift with the current, cautiously hugging the shore and keeping in the shadow of the bathing-houses. Neither of them spoke a word, horror-stricken as they were by the baleful spectacle that presented itself to their vision. As they floated down the stream and their horizon widened the enormity of the terrible sight increased, and when they reached the bridge of Solferino a single glance sufficed to embrace both the blazing *quais*.

On their left the palace of the Tuileries was burning. It was not yet dark when the Communists had fired the two extremities of the structure, the Pavilion de Flore and the Pavilion de Marsan, and with rapid strides the flames had gained the Pavilion de l'Horloge in the central portion, beneath which, in the Salle des Marechaux, a mine had been prepared by stacking up casks of powder. At that moment the intervening buildings were belching from their shattered windows dense volumes of reddish smoke, streaked with long ribbons of blue flame. The roofs, yawning as does the earth in regions where volcanic agencies prevail, were seamed with great cracks through which the raging sea of fire beneath was visible. But the grandest, saddest spectacle of all was that afforded by the Pavilion de Flore, to which the torch had been earliest applied and which was ablaze from its foundation to its lofty summit, burning with a deep, fierce roar that could be heard far away. The petroleum with which the floors and hangings had been soaked gave the flames an intensity such that the ironwork of the balconies was seen to twist and writhe in the convolutions of a serpent, and the tall monumental chimneys, with their elaborate carvings, glowed with the fervor of live coals.

Then, still on their left, were, first, the Chancellerie of the Legion of Honor, which was fired at five o'clock in the afternoon and had been burning nearly seven hours, and next, the Palace of the Council of State, a huge rectangular structure of stone, which was spouting torrents of fire from every orifice in each of its two colonnaded stories. The four structures surrounding the great central court had all caught at the same moment, and the petroleum, which here also had been distributed by the barrelful, had poured down the four grand staircases at the four corners of the building in rivers of hellfire. On the facade that faced the river the black line of the mansard was profiled distinctly against the ruddy sky, amid the red tongues that rose to lick its base, while colonnades, entablatures, friezes, carvings, all stood out with startling vividness in the blinding, shimmering glow. So great was the energy of the fire, so terrible its propulsive force, that the colossal structure was in some sort raised bodily from the earth, trembling and rumbling on its foundations, preserving intact only its four massive walls, in the fierce eruption that hurled its heavy zinc roof high in air. Then, close at one side were the d'Orsay barracks, which burned with a flame that seemed to pierce the heavens, so purely white and so unwavering that it was like a tower of light. And finally, back from the river, were still other fires, the seven houses in the Rue du Bac, the twenty-two houses in the Rue de Lille, helping to tinge the sky a deeper crimson, profiling their flames on other flames, in a blood-red ocean that seemed to have no end.

Jean murmured in awed tone:

"Did ever mortal man look on the like of this! the very river is on fire."

Their boat seemed to be sailing on the bosom of an incandescent stream. As the dancing lights of the mighty conflagrations were caught by the ripples of the current the Seine seemed to be pouring down torrents of living coals; flashes of intensest crimson played fitfully across its surface, the blazing brands fell in showers into the water and were extinguished with a hiss. And ever they floated downward with the tide on the bosom of that blood-red stream, between the blazing palaces on either hand, like wayfarers in some accursed city, doomed to destruction and burning on the banks of a river of molten lava.

"Ah!" exclaimed Maurice, with a fresh access of madness at the sight of the havoc he had longed for, "let it burn, let it all go up in smoke!"

But Jean silenced him with a terrified gesture, as if he feared such blasphemy might bring them evil. Where could a young man whom he loved so fondly, so delicately nurtured, so well informed, have picked up such ideas? And he applied himself more vigorously to the oars, for they had now passed the bridge of Solferino and were come out into a wide open space of water. The light was so intense that the river was illuminated as by the noonday sun when it stands vertically above men's heads and casts no shadow. The most minute objects, such as the eddies in the stream, the stones piled on the banks, the small trees along the quais, stood out before their vision with wonderful distinctness. The bridges, too, were particularly noticeable in their dazzling whiteness, and so clearly defined that they could have counted every stone; they had the appearance of narrow gangways thrown across the fiery stream to connect one conflagration with the other. Amid the roar of the flames and the general clamor a loud crash occasionally announced the fall of some stately edifice. Dense clouds of soot hung in the air and settled everywhere, the wind brought odors of pestilence on its wings. And another horror was that Paris, those more distant quarters of the city that lay back from the banks of

345

the Seine, had ceased to exist for them. To right and left of the conflagration that raged with such fierce resplendency was an unfathomable gulf of blackness; all that presented itself to their strained gaze was a vast waste of shadow, an empty void, as if the devouring element had reached the utmost limits of the city and all Paris were swallowed up in everlasting night. And the heavens, too, were dead and lifeless; the flames rose so high that they extinguished the stars.

Maurice, who was becoming delirious, laughed wildly.

"High carnival at the Consoil d'Etat and at the Tuileries to-night! They have illuminated the facades, women are dancing beneath the sparkling chandeliers. Ah, dance, dance and be merry, in your smoking petticoats, with your chignons ablaze-"

And he drew a picture of the feasts of Sodom and Gomorrah, the music, the lights, the flowers, the unmentionable orgies of lust and drunkenness, until the candles on the walls blushed at the shamelessness of the display and fired the palaces that sheltered such depravity. Suddenly there was a terrific explosion. The fire, approaching from either extremity of the Tuileries, had reached the Salle des Marechaux, the casks of powder caught, the Pavilion de l'Horloge was blown into the air with the violence of a powder mill. A column of flame mounted high in the heavens, and spreading, expanded in a great fiery plume on the inky blackness of the sky, the crowning display of the horrid *fete*.

"Bravo!" exclaimed Maurice, as at the end of the play, when the lights are extinguished and darkness settles on the stage.

Again Jean, in stammering, disconnected sentences, besought him to be quiet. No, no, it was not right to wish evils to anyone! And if they invoked destruction, would not they themselves perish in the general ruin? His sole desire was to find a landing place so that he might no longer have that horrid spectacle before his eyes. He considered it best not to attempt to land at the Pont de la Concorde, but, rounding the elbow of the Seine, pulled on until they reached the Quai de la Conference, and even at that critical moment, instead of shoving the skiff out into the stream to take its chances, he wasted some precious moments in securing it, in his instinctive respect for the property of others. While doing this he had seated Maurice comfortably on the bank; his plan was to reach the Rue des Orties through the Place de la Concorde and the Rue Saint-Honore. Before proceeding further he climbed alone to the top of the steps that ascended from the *quai* to explore the ground, and on witnessing the obstacles they would have to surmount his courage was almost daunted. There lay the impregnable fortress of the Commune, the terrace of the Tuileries bristling with cannon, the Rues Royale, Florentin, and Rivoli obstructed by lofty and massive barricades; and this state of affairs explained the tactics of the army of Versailles, whose line that night described an immense arc, the center and apex resting on the Place de la Concorde, one of the two extremities being at the freight depot of the Northern Railway on the right bank, the other on the left bank, at one of the bastions of the ramparts, near the gate of Arcueil. But as the night advanced the Communards had evacuated the Tuileries and the barricades and the regular troops had taken possession of the quartier in the midst of further conflagrations; twelve houses at the junction of the Rue Saint-Honore and the Rue Royale had been burning since nine o'clock in the evening.

When Jean descended the steps and reached the river-bank again he found Maurice in a semi-comatose condition, the effects of the reaction after his hysterical outbreak.

"It will be no easy job. I hope you are going to be able to walk, youngster?"

"Yes, yes; don't be alarmed. I'll get there somehow, alive or dead."

It was not without great difficulty that he climbed the stone steps, and when he reached the level ground of the *quai* at the summit he walked very slowly, supported by his companion's arm, with the shuffling gait of a somnambulist. The day had not dawned yet, but the reflected light from the burning buildings cast a lurid illumination on the wide Place. They made their way in silence across its deep solitude, sick at heart to behold the mournful scene of devastation it presented. At either extremity, beyond the bridge and at the further end of the Rue Royale, they could faintly discern the shadowy outlines of the Palais Bourbon and the Church of the Madeleine, torn by shot and shell. The terrace of the Tuileries had been breached by the fire of the siege guns and was partially in ruins. On the Place itself the bronze railings and ornaments of the fountains had been chipped and defaced by the balls; the colossal statue of Lille lay on the ground shattered by a projectile, while near at hand the statue of Strasbourg, shrouded in heavy veils of crape, seemed to be mourning the ruin that surrounded it on every side. And near the Obelisk, which had escaped unscathed, a gas-pipe in its trench had been broken by the pick of a careless workman, and the escaping gas, fired by some accident, was flaring up in a great undulating jet, with a roaring, hissing sound.

Jean gave a wide berth to the barricade erected across the Rue Royale between the Ministry of Marine and the Garde-Meuble, both of which the fire had spared; he could hear the voices of the soldiers behind the sand bags and casks of earth with which it was constructed. Its front was protected by a ditch, filled with stagnant, greenish water, in which was floating the dead body of a federate, and through one of its embrasures they caught a glimpse of the houses in the carrefour Saint-Honore, which were burning still in spite of the engines that had come in from the suburbs, of which they heard the roar and clatter. To right and left the trees and the kiosks of the newspaper venders were riddled by the storm of bullets to which they had been subjected. Loud cries of horror arose; the firemen, in exploring the cellar of one of the burning houses, had come across the charred bodies of seven of its inmates.

Although the barricade that closed the entrance to the Rue Saint-Florentin and the Rue de Rivoli by its skilled construction and great height appeared even more formidable than the other, Jean's instinct told him they would have less difficulty in getting by it. It was completely evacuated, indeed, and the Versailles troops had not yet entered it. The abandoned guns were resting in the embrasures in peaceful slumber, the only living thing behind that invincible rampart was a stray dog, that scuttled away in haste. But as Jean was making what speed he could along the Rue Saint-Florentin, sustaining Maurice, whose strength was giving out, that which he had been in fear of came to pass; they fell directly into the arms of an entire company of the 88th of the line, which had turned the barricade.

"Captain," he explained, "this is a comrade of mine, who has just been wounded by those bandits. I am taking him to the hospital."

It was then that the capote which he had thrown over Maurice's shoulders stood them in good stead, and Jean's heart was beating like a trip-hammer as at last they turned into the Rue Saint-Honore. Day was just breaking, and the sound of shots reached their ears from the cross-streets, for fighting was going on still throughout the quartier. It was little short of a miracle that they finally reached the Rue des Frondeurs without sustaining any more disagreeable adventure. Their progress was extremely slow; the last four or five hundred yards appeared interminable. In the Rue des Frondeurs they struck up against a communist picket, but the federates, thinking a whole regiment was at hand, took to their heels. And now they had but a short bit of the Rue d'Argenteuil to traverse and they would be safe in the Rue des Orties.

For four long hours that seemed like an eternity Jean's longing desire had been bent on that Rue des Orties with feverish impatience, and now they were there it appeared like a haven of safety. It was dark, silent, and deserted, as if there were no battle raging within a hundred leagues of it. The house, an old, narrow house without a concierge, was still as the grave.

"I have the keys in my pocket," murmured Maurice. "The big one opens the street door, the little one is the key of my room, way at the top of the house."

He succumbed and fainted dead away in Jean's arms, whose alarm and distress were extreme. They made him forget to close the outer door, and he had to grope his way up that strange, dark staircase, bearing his lifeless burden and observing the greatest caution not to stumble or make any noise that might arouse the sleeping inmates of the rooms. When he had gained the top he had to deposit the wounded man on the floor while he searched for the chamber door by striking matches, of which he fortunately had a supply in his pocket, and only when he had found and opened it did he return and raise him in his arms again. Entering, he laid him on the little iron bed that faced the window, which he threw open to its full extent in his great need of air and light. It was broad day; he dropped on his knees beside the bed, sobbing as if his heart would break, suddenly abandoned by all his strength as the fearful thought again smote him that he had slain his friend.

Minutes passed; he was hardly surprised when, raising his eyes, he saw Henriette standing by the bed. It was perfectly natural: her brother was dying, she had come. He had not even seen her enter the room; for all he knew she might have been standing there for hours. He sank into a chair and watched her with stupid eyes as she hovered about the bed, her heart wrung with mortal anguish at sight of her brother lying there senseless, in his blood-stained garments. Then his memory began to act again; he asked:

"Tell me, did you close the street door?"

She answered with an affirmative motion of the head, and as she came toward him, extending her two hands in her great need of sympathy and support, he added:

"You know it was I who killed him."

She did not understand; she did not believe him. He felt no flutter in the two little hands that rested confidingly in his own.

"It was I who killed him-yes, 'twas over yonder, behind a barricade, I did it. He was fighting on one side, I on the other-"

There began to be a fluttering of the little hands.

"We were like drunken men, none of us knew what he as about-it was I who killed him."

Then Henriette, shivering, pale as death, withdrew her hands, fixing on him a gaze that was full of horror. Father of Mercy, was the end of all things come! was her crushed and bleeding heart to know no peace for ever more! Ah, that Jean, of whom she had been thinking that very day, happy in the unshaped hope that perhaps she might see him once again! And it was he who had done that abominable thing; and yet he had saved Maurice, for was it not he who had brought him home through so many perils? She could not yield her hands to him now without a revolt of all her being, but she uttered a cry into which she threw the last hope of her tortured and distracted heart.

"Oh! I will save him; I *must* save him, now!"

She had acquired considerable experience in surgery during the long time she had been in attendance on the hospital at Remilly, and now she proceeded without delay to examine her brother's hurt, who remained unconscious while she was undressing him. But when she undid the rude bandage of Jean's invention, he stirred feebly and uttered a faint cry of pain, opening wide his eyes that were bright with fever. He recognized her at once and smiled.

"You here! Ah, how glad I am to see you once more before I die!"

She silenced him, speaking in a tone of cheerful confidence.

348

"Hush, don't talk of dying; I won't allow it! I mean that you shall live! There, be quiet, and let me see what is to be done."

However, when Henriette had examined the injured arm and the wound in the side, her face became clouded and a troubled look rose to her eyes. She installed herself as mistress in the room, searching until she found a little oil, tearing up old shirts for bandages, while Jean descended to the lower regions for a pitcher of water. He did not open his mouth, but looked on in silence as she washed and deftly dressed the wounds, incapable of aiding her, seemingly deprived of all power of action by her presence there. When she had concluded her task, however, noticing her alarmed expression, he proposed to her that he should go and secure a doctor, but she was in possession of all her clear intelligence. No, no; she would not have a chance-met doctor, of whom they knew nothing, who, perhaps, would betray her brother to the authorities. They must have a man they could depend on; they could afford to wait a few hours. Finally, when Jean said he must go and report for duty with his company, it was agreed that he should return as soon as he could get away, and try to bring a surgeon with him.

He delayed his departure, seemingly unable to make up his mind to leave that room, whose atmosphere was pervaded by the evil he had unintentionally done. The window, which had been closed for a moment, had been opened again, and from it the wounded man, lying on his bed, his head propped up by pillows, was looking out over the city, while the others, also, in the oppressive silence that had settled on the chamber, were gazing out into vacancy.

From that elevated point of the Butte des Moulins a good half of Paris lay stretched beneath their eyes in a vast panorama: first the central districts, from the Faubourg Saint-Honore to the Bastille, then the Seine in its entire course through the city, with the thickly-built, densely-populated regions of the left bank, an ocean of roofs, treetops, steeples, domes, and towers. The light was growing stronger, the abominable night, than which there have been few more terrible in history, was ended; but beneath the rosy sky, in the pure, clear light of the rising sun, the fires were blazing still. Before them lay the burning Tuileries, the d'Orsay barracks, the Palaces of the Council of State and the Legion of Honor, the flames from which were paled by the superior refulgence of the day-star. Even beyond the houses in the Rue de Lille and the Rue du Bac there must have been other structures burning, for clouds of smoke were visible rising from the carrefour of la Croix-Rouge, and, more distant still, from the Rue Vavin and the Rue Notre-Dame-des-Champs. Nearer at hand and to their right the fires in the Rue Saint-Honore were dying out, while to the left, at the Palais-Royal and the new Louvre, to which the torch had not been applied until near morning, the work of the incendiaries was apparently a failure. But what they were unable to account for at first was the dense volume of black smoke which, impelled by the west wind, came driving past their window. Fire had been set to the Ministry of Finance at three o'clock in the morning and ever since that time it had been smoldering, emitting no blaze, among the stacks and piles of documents that were contained in the low-ceiled, fire-proof vaults and chambers. And if the terrific impressions of the night were not there to preside at the awakening of the great city -the fear of total destruction, the Seine pouring its fiery waves past their doors, Paris kindling into flame from end to end-a feeling of gloom and despair, hung heavy over the quartiers that had been spared, with that dense, on-pouring smoke, whose dusky cloud was ever spreading. Presently the sun, which had risen bright and clear, was hid by it, and the golden sky was filled with the great funeral pall.

Maurice, who appeared to be delirious again, made a slow, sweeping gesture that embraced the entire horizon, murmuring:

"Is it all burning? Ah, how long it takes!"

Tears rose to Henriette's eyes, as if her burden of misery was made heavier for her by the share her brother had had in those deeds of horror. And Jean, who dared neither take her hand nor embrace his friend, left the room with the air of one crazed by grief.

"I will return soon. Au revoir!"

It was dark, however, nearly eight o'clock, before he was able to redeem his promise. Notwithstanding his great distress he was happy; his regiment had been transferred from the first to the second line and assigned the task of protecting the quartier, so that, bivouacking with his company in the Place du Carrousel, he hoped to get a chance to run in each evening to see how the wounded man was getting on. And he did not return alone; as luck would have it he had fallen in with the former surgeon of the 106th and had brought him along with him, having been unable to find another doctor, consoling himself with the reflection that the terrible, big man with the lion's mane was not such a bad sort of fellow after all.

When Bouroche, who knew nothing of the patient he was summoned with such insistence to attend and grumbled at having to climb so many stairs, learned that it was a Communist he had on his hands he commenced to storm.

"God's thunder, what do you take me for? Do you suppose I'm going to waste my time on those thieving, murdering, house-burning scoundrels? As for this particular bandit, his case is clear, and I'll take it upon me to see he is cured; yes, with a bullet in his head!"

But his anger subsided suddenly at sight of Henriette's pale face and her golden hair streaming in disorder over her black dress.

"He is my brother, doctor, and he was with you at Sedan."

He made no reply, but uncovered the injuries and examined them in silence; then, taking some phials from his pocket, he made a fresh dressing, explaining to the young woman how it was done. When he had finished he turned suddenly to the patient and asked in his loud, rough voice:

"Why did you take sides with those ruffians? What could cause you to be guilty of such an abomination?"

Maurice, with a feverish luster in his eyes, had been watching him since he entered the room, but no word had escaped his lips. He answered in a voice that was almost fierce, so eager was it:

"Because there is too much suffering in the world, too much wickedness, too much infamy!"

Bouroche's shrug of the shoulders seemed to indicate that he thought a young man was likely to make his mark who carried such ideas about in his head. He appeared to be about to say something further, but changed his mind and bowed himself out, simply adding:

"I will come in again."

To Henriette, on the landing, he said he would not venture to make any promises. The injury to the lung was serious; hemorrhage might set in and carry off the patient without a moment's warning. And when she re-entered the room she forced a smile to her lips, notwithstanding the sharp stab with which the doctor's words had pierced her heart, for had she not promised herself to save him? and could she permit him to be snatched from them now that they three were again united, with a prospect of a lifetime of affection and happiness before them? She had not left the room since morning, an old woman who lived on the landing having kindly offered to act as her messenger for the purchase of such things as she required. And she returned and resumed her place upon a chair at her brother's bedside.

But Maurice, in his febrile excitation, questioned Jean, insisting on knowing what had happened since the morning. The latter did not tell him everything, maintaining a discreet silence upon the furious rage which Paris, now it was delivered from its tyrants, was

350

manifesting toward the dying Commune. It was now Wednesday. For two interminable days succeeding the Sunday evening when the conflict first broke out the citizens had lived in their cellars, quaking with fear, and when they ventured out at last on Wednesday morning, the spectacle of bloodshed and devastation that met their eyes on every side, and more particularly the frightful ruin entailed by the conflagrations, aroused in their breasts feelings the bitterest and most vindictive. It was felt in every quarter that the punishment must be worthy of the crime. The houses in the suspected quarters were subjected to a rigorous search and men and women who were at all tainted with suspicion were led away in droves and shot without formality. At six o'clock of the evening of that day the army of the Versaillese was master of the half of Paris, following the line of the principal avenues from the park of Montsouris to the station of the Northern Railway, and the remainder of the braver members of the Commune, a mere handful, some twenty or so, had taken refuge in the *mairie* of the eleventh arrondissement, in the Boulevard Voltaire.

They were silent when he concluded his narration, and Maurice, his glance vaguely wandering over the city through the open window that let in the soft, warm air of evening, murmured:

"Well, the work goes on; Paris continues to burn!"

It was true: the flames were becoming visible again in the increasing darkness and the heavens were reddened once more with the ill-omened light. That afternoon the powder magazine at the Luxembourg had exploded with a frightful detonation, which gave rise to a report that the Pantheon had collapsed and sunk into the catacombs. All that day, moreover, the conflagrations of the night pursued their course unchecked; the Palace of the Council of State and the Tuileries were burning still, the Ministry of Finance continued to belch forth its billowing clouds of smoke. A dozen times Henriette was obliged to close the window against the shower of blackened, burning paper that the hot breath of the fire whirled upward into the sky, whence it descended to earth again in a fine rain of fragments; the streets of Paris were covered with them, and some were found in the fields of Normandy, thirty leagues away. And now it was not the western and southern districts alone which seemed devoted to destruction, the houses in the Rue Royale and those of the Croix-Rouge and the Rue Notre-Dame-des-Champs: the entire eastern portion of the city appeared to be in flames, the Hotel de Ville glowed on the horizon like a mighty furnace. And in that direction also, blazing like gigantic beacon-fires upon the mountain tops, were the Theatre-Lyrique, the *mairie* of the fourth arrondissement, and more than thirty houses in the adjacent streets, to say nothing of the theater of the Porte-Saint-Martin, further to the north, which illuminated the darkness of its locality as a stack of grain lights up the deserted, dusky fields at night. There is no doubt that in many cases the incendiaries were actuated by motives of personal revenge; perhaps, too, there were criminal records which the parties implicated had an object in destroying. It was no longer a question of self-defense with the Commune, of checking the advance of the victorious troops by fire; a delirium of destruction raged among its adherents: the Palace of Justice, the Hotel-Dieu and the cathedral of Notre-Dame escaped by the merest chance. They would destroy solely for the sake of destroying, would bury the effete, rotten humanity beneath the ruins of a world, in the hope that from the ashes might spring a new and innocent race that should realize the primitive legends of an earthly paradise. And all that night again did the sea of flame roll its waves over Paris.

"Ah; war, war, what a hateful thing it is!" said Henriette to herself, looking out on the sore-smitten city.

Was it not indeed the last act, the inevitable conclusion of the tragedy, the blood-madness for which the lost fields of Sedan and Metz were responsible, the epidemic of destruction born from the siege of Paris, the supreme struggle of a nation in peril of dissolution, in the midst of slaughter and universal ruin?

But Maurice, without taking his eyes from the fires that were raging in the distance, feebly, and with an effort, murmured:

"No, no; do not be unjust toward war. It is good; it has its appointed work to do-"

There were mingled hatred and remorse in the cry with which Jean interrupted him.

"Good God! When I see you lying there, and know it is through my fault- Do not say a word in defense of it; it is an accursed thing, is war!"

The wounded man smiled faintly.

"Oh, as for me, what matters it? There is many another in my condition. It may be that this blood-letting was necessary for us. War is life, which cannot exist without its sister, death."

And Maurice closed his eyes, exhausted by the effort it had cost him to utter those few words. Henriette signaled Jean not to continue the discussion. It angered her; all her being rose in protest against such suffering and waste of human life, notwithstanding the calm bravery of her frail woman's nature, with her clear, limpid eyes, in which lived again all the heroic spirit of the grandfather, the veteran of the Napoleonic wars.

Two days more, Thursday and Friday, passed, like their predecessors, amid scenes of slaughter and conflagration. The thunder of the artillery was incessant; the batteries of the army of Versailles on the heights of Montmartre roared against those that the federates had established at Belleville and Pare-Lachaise without a moment's respite, while the latter maintained a desultory fire on Paris. Shells had fallen in the Rue Richelieu and the Place Vendome. At evening on the 25th the entire left bank was in possession of the regular troops, but on the right bank the barricades in the Place Chateau d'Eau and the Place de la Bastille continued to hold out; they were veritable fortresses, from which proceeded an uninterrupted and most destructive fire. At twilight, while the last remaining members of the Commune were stealing off to make provision for their safety, Delescluze took his cane and walked leisurely away to the barricade that was thrown across the Boulevard Voltaire, where he died a hero's death. At daybreak on the following morning, the 26th, the Chateau d'Eau and Bastille positions were carried, and the Communists, now reduced to a handful of brave men who were resolved to sell their lives dearly, had only la Villette, Belleville, and Charonne left to them, And for two more days they remained and fought there with the fury of despair.

On Friday evening, as Jean was on his way from the Place du Carrousel to the Rue des Orties, he witnessed a summary execution in the Rue Richelieu that filled him with horror. For the last forty-eight hours two courts-martial had been sitting, one at the Luxembourg, the other at the Theatre du Chatelet; the prisoners convicted by the former were taken into the garden and shot, while those found guilty by the latter were dragged away to the Lobau barracks, where a platoon of soldiers that was kept there in constant attendance for the purpose mowed them down, almost at point-blank range. The scenes of slaughter there were most horrible: there were men and women who had been condemned to death on the flimsiest evidence: because they had a stain of powder on their hands, because their feet were shod with army shoes; there were innocent persons, the victims of private malice, who had been wrongfully denounced, shrieking forth their entreaties and explanations and finding no one to lend an ear to them; and all were driven pell-mell against a wall, facing the muzzles of the muskets, often so many poor wretches in the band at once that the bullets did not suffice for all and it became necessary to finish the wounded with the bayonet. From morning until night the place was streaming with blood; the tumbrils were kept busy bearing away the bodies of the dead. And throughout the length and breadth of the city, keeping pace with the revengeful clamors of the people, other executions were continually taking place, in front of barricades, against the walls in the deserted streets, on the steps of the public buildings. It was under such circumstances that Jean saw a woman and two men dragged by the residents of the quartier before the officer commanding the detachment that was guarding the Theatre Francais. The citizens showed themselves more bloodthirsty than the soldiery, and those among the newspapers that had resumed publication were howling for measures of extermination. A threatening crowd surrounded the prisoners and was particularly violent against the woman, in whom the excited bourgeois beheld one of those *petroleuses* who were the constant bugbear of terror-haunted imaginations, whom they accused of prowling by night, slinking along the darkened streets past the dwellings of the wealthy, to throw cans of lighted petroleum into unprotected cellars. This woman, was the cry, had been found bending over a coal-hole in the Rue Sainte-Anne. And notwithstanding her denials, accompanied by tears and supplications, she was hurled, together with the two men, to the bottom of the ditch in front of an abandoned barricade, and there, lying in the mud and slime, they were shot with as little pity as wolves caught in a trap. Some by-passers stopped and looked indifferently on the scene, among them a lady hanging on her husband's arm, while a baker's boy, who was carrying home a tart to someone in the neighborhood, whistled the refrain of a popular air.

As Jean, sick at heart, was hurrying along the street toward the house in the Rue des Orties, a sudden recollection flashed across his mind. Was not that Chouteau, the former member of his squad, whom he had seen, in the blouse of a respectable workman, watching the execution and testifying his approval of it in a loud-mouthed way? He was a proficient in his role of bandit, traitor, robber, and assassin! For a moment the corporal thought he would retrace his steps, denounce him, and send him to keep company with the other three. Ah, the sadness of the thought; the guilty ever escaping punishment, parading their unwhipped infamy in the bright light of day, while the innocent molder in the earth!

Henriette had come out upon the landing at the sound of footsteps coming up the stairs, where she welcomed Jean with a manner that indicated great alarm.

"'Sh! he has been extremely violent all day long. The major was here, I am in despair-"

Bouroche, in fact, had shaken his head ominously, saying he could promise nothing as yet. Nevertheless the patient might pull through, in spite of all the evil consequences he feared; he had youth on his side.

"Ah, here you are at last," Maurice said impatiently to Jean, as soon as he set eyes on him. "I have been waiting for you. What is going on -how do matters stand?" And supported by the pillows at his back, his face to the window which he had forced his sister to open for him, he pointed with his finger to the city, where, on the gathering darkness, the lambent flames were beginning to rise anew. "You see, it is breaking out again; Paris is burning. All Paris will burn this time!"

As soon as daylight began to fade, the distant quarters beyond the Seine had been lighted up by the burning of the Grenier d'Abondance. From time to time there was an outburst of flame, accompanied by a shower of sparks, from the smoking ruins of the Tuileries, as some wall or ceiling fell and set the smoldering timbers blazing afresh. Many houses, where the fire was supposed to be extinguished, flamed up anew; for the last three days, as soon as darkness descended on the city it seemed as if it were the signal for the conflagrations to break out again; as if the shades of night had breathed upon the still glowing embers, reanimating them, and scattering them to the four corners of the horizon. Ah, that city of the damned, that had harbored for a week within its bosom the demon of destruction, incarnadining the sky each evening as soon as twilight fell, illuminating with its infernal torches the nights of that week of slaughter! And when, that night, the docks at la Villette burned, the light they shed upon the huge city was so intense that it seemed to be on fire in every part at once, overwhelmed and drowned beneath the sea of flame.

"Ah, it is the end!" Maurice repeated. "Paris is doomed!"

He reiterated the words again and again with apparent relish, actuated by a feverish desire to hear the sound of his voice once more, after the dull lethargy that had kept him tongue-tied for three days. But the sound of stifled sobs causes him to turn his head.

"What, sister, you, brave little woman that you are! You weep because I am about to die-"

She interrupted him, protesting:

"But you are not going to die!"

"Yes, yes; it is better it should be so; it must be so. Ah, I shall be no great loss to anyone. Up to the time the war broke out I was a source of anxiety to you, I cost you dearly in heart and purse. All the folly and the madness I was guilty of, and which would have landed me, who knows where? in prison, in the gutter-"

Again she took the words from his mouth, exclaiming hotly:

"Hush! be silent!-you have atoned for all."

He reflected a moment. "Yes, perhaps I shall have atoned, when I am dead. Ah, Jean, old fellow, you didn't know what a service you were rendering us all when you gave me that bayonet thrust."

But the other protested, his eyes swimming with tears:

"Don't, I entreat you, say such things! do you wish to make me go and dash out my brains against a wall?"

Maurice pursued his train of thought, speaking in hurried, eager tones.

"Remember what you said to me the day after Sedan, that it was not such a bad thing, now and then, to receive a good drubbing. And you added that if a man had gangrene in his system, if he saw one of his limbs wasting from mortification, it would be better to take an ax and chop off that limb than to die from the contamination of the poison. I have many a time thought of those words since I have been here, without a friend, immured in this city of distress and madness. And I am the diseased limb, and it is you who have lopped it off-" He went on with increasing vehemence, regardless of the supplications of his terrified auditors, in a fervid tirade that abounded with symbols and striking images. It was the untainted, the reasoning, the substantial portion of France, the peasantry, the tillers of the soil, those who had always kept close contact with their mother Earth, that was suppressing the outbreak of the crazed, exasperated part, the part that had been vitiated by the Empire and led astray by vain illusions and empty dreams; and in the performance of its duty it had had to cut deep into the living flesh, without being fully aware of what it was doing. But the baptism of blood, French blood, was necessary; the abominable holocaust, the living sacrifice, in the midst of the purifying flames. Now they had mounted the steps of the Calvary and known their bitterest agony; the crucified nation had expiated its faults and would be born again. "Jean, old friend, you and those like you are strong in your simplicity and honesty. Go, take up the spade and the trowel, turn the sod in the abandoned field, rebuild the house! As for me, you did well to lop me off, since I was the ulcer that was eating away your strength!"

After that his language became more and more incoherent; he insisted on rising and going to sit by the window. "Paris burns, Paris burns; not a stone of it will be left standing. Ah! the fire that I invoked, it destroys, but it heals; yes, the work it does is good. Let me go down there; let me help to finish the work of humanity and liberty-"

Jean had the utmost difficulty in getting him back to bed, while Henriette tearfully recalled memories of their childhood, and entreated him, for the sake of the love they bore each other, to be calm. Over the immensity of Paris the fiery glow deepened and widened; the sea of flame seemed to be invading the remotest quarters of the horizon; the heavens were like the vaults of a colossal oven, heated to red heat. And athwart the red light of the conflagrations the dense black smoke-clouds from the Ministry of Finance, which had been burning three days and given forth no blaze, continued to pour in unbroken, slow procession.

The following, Saturday, morning brought with it a decided improvement in Maurice's condition: he was much calmer, the fever had subsided, and it afforded Jean inexpressible delight to behold a smile on Henriette's face once more, as the young woman fondly reverted to her cherished dream, a pact of reciprocal affection between the three of them, that should unite them in a future that might yet be one of happiness, under conditions that she did not care to formulate even to herself. Would destiny be merciful? Would it save them all from an eternal farewell by saving her brother? Her nights were spent in watching him; she never stirred outside that chamber, where her noiseless activity and gentle ministrations were like a never-ceasing caress. And Jean, that evening, while sitting with his friends, forgot his great sorrow in a delight that astonished him and made him tremble. The troops had carried Belleville and the Buttes-Chaumont that day; the only remaining point where there was any resistance now was the cemetery of Pere-Lachaise, which had been converted into a fortified camp. It seemed to him that the insurrection was ended; he even declared that the troops had ceased to shoot their prisoners, who were being collected in droves and sent on to Versailles. He told of one of those bands that he had seen that morning on the quai, made up of men of every class, from the most respectable to the lowest, and of women of all ages and conditions, wrinkled old bags and young girls, mere children, not yet out of their teens; pitiful aggregation of misery and revolt, driven like cattle by the soldiers along the street in the bright sunshine, and that the people of Versailles, so it was said, received with revilings and blows.

But Sunday was to Jean a day of terror. It rounded out and fitly ended that accursed week. With the triumphant rising of the sun on that bright, warm Sabbath morning he shudderingly heard the news that was the culmination of all preceding horrors. It was only at that late day that the public was informed of the murder of the hostages; the archbishop, the cure of the Madeleine and others, shot at la Roquette on Wednesday, the Dominicans of Arcueil coursed like hares on Thursday, more priests and gendarmes, to the number of forty-seven in all, massacred in cold blood in the Rue Haxo on Friday; and a furious cry went up for vengeance, the soldiers bunched the last prisoners they made and shot them in mass. All day long on that magnificent Sunday the volleys of musketry rang out in the courtyard of the Lobau barracks, that were filled with blood and smoke and the groans of the dying. At la Roquette two hundred and twenty-seven miserable wretches, gathered in here and there by the drag-net of the police, were collected in a huddle, and the soldiers fired volley after volley into the mass of human beings until there was no further sign of life. At Pere-Lachaise, which had been shelled continuously for four days and was finally carried by a hand-to-hand conflict among the graves, a hundred and forty-eight of the insurgents were drawn up in line before a wall, and when the firing ceased the stones were weeping great tears of blood; and three of them, despite their wounds, having succeeded in making their escape, they were retaken and despatched. Among the twelve thousand victims of the Commune, who shall say how many innocent people suffered for every malefactor who met his deserts! An order to stop the executions had been issued from Versailles, so it was said, but none the less the slaughter still went on; Thiers, while hailed as the savior of his country, was to bear the stigma of having been the Jack Ketch of Paris, and Marshal MacMahon, the vanquished of Froeschwiller, whose proclamation announcing the triumph of law and order was to be seen on every wall, was to receive the credit of the victory of Pere-Lachaise. And in the pleasant sunshine Paris, attired in holiday garb, appeared to be en fete; the reconquered streets were filled with an enormous crowd; men and women, glad to breathe the air of heaven once more, strolled leisurely from spot to spot to view the smoking ruins; mothers, holding their little children by the hand, stopped for a moment and listened with an air of interest to the deadened crash of musketry from the Lobau barracks.

When Jean ascended the dark staircase of the house in the Rue des Orties, in the gathering obscurity of that Sunday evening, his heart was oppressed by a chill sense of impending evil. He entered the room, and saw at once that the inevitable end was come; Maurice lay dead on the little bed; the hemorrhage predicted by Bouroche had done its work. The red light of the setting sun streamed through the open window and rested on the wall as if in a last farewell; two tapers were burning on a table beside the bed. And Henriette, alone with her dead, in her widow's weeds that she had not laid aside, was weeping silently.

At the noise of footsteps she raised her head, and shuddered on beholding Jean. He, in his wild despair, was about to hurry toward her and seize her hands, mingle his grief with hers in a sympathetic clasp, but he saw the little hands were trembling, he felt as by instinct the repulsion that pervaded all her being and was to part them for evermore. Was not all ended between them now? Maurice's grave would be there, a yawning chasm, to part them as long as they should live. And he could only fall to his knees by the bedside of his dead friend, sobbing softly. After the silence had lasted some moments, however, Henriette spoke:

"I had turned my back and was preparing a cup of bouillon, when he gave a cry. I hastened to his side, but had barely time to reach the bed before he expired, with my name upon his lips, and yours as well, amid an outgush of blood-"

Her Maurice, her twin brother, whom she might almost be said to have loved in the prenatal state, her other self, whom she had watched over and saved! sole object of her affection since at Bazeilles she had seen her poor Weiss set against a wall and shot to death! And now cruel war had done its worst by her, had crushed her bleeding heart; henceforth her way through life was to be a solitary one, widowed and forsaken as she was, with no one upon whom to bestow her love.

"Ah, bon sang!" cried Jean, amid his sobs, "behold my work! My poor little one, for whom I would have laid down my life, and whom I murdered, brute that I am! What is to become of us? Can you ever forgive me?"

At that moment their glances met, and they were stricken with consternation at what they read in each other's eyes. The past rose before them, the secluded chamber at Remilly, where they had spent so many melancholy yet happy days. His dream returned to him, that dream of which at first he had been barely conscious and which even at a later period could not be said to have assumed definite shape: life down there in the pleasant country by the Meuse, marriage, a little house, a little field to till whose produce should suffice for the needs of two people whose ideas were not extravagant. Now the dream was become an eager longing, a penetrating conviction that, with a wife as loving and industrious as she, existence would be a veritable earthly paradise. And she, the tranquillity of whose mind had never in those days been ruffled by thoughts of that nature, in the chaste and unconscious bestowal of her heart, now saw clearly and understood the true condition of her feelings. That marriage, of which she had not admitted to herself the possibility, had been, unknown to her, the object of her desire. The seed that had germinated had pushed its way in silence and in darkness; it was love, not sisterly affection, that she bore toward that young man whose company had at first been to her nothing more than a source of comfort and consolation. And that was what their eyes told each other, and the love thus openly expressed could have no other fruition than an eternal farewell. It needed but that frightful sacrifice, the rending of their heart-strings by that supreme parting, the prospect of their life's happiness wrecked amid all the other ruins, swept away by the crimson tide that ended their brother's life.

With a slow and painful effort Jean rose from his knees.

"Farewell!"

Henriette stood motionless in her place.

"Farewell!"

But Jean could not tear himself away thus. Advancing to the bedside he sorrowfully scanned the dead man's face, with its lofty forehead that seemed loftier still in death, its wasted features, its dull eyes, whence the wild look that had occasionally been seen there in life had vanished. He longed to give a parting kiss to his little one, as he had called him so many times, but dared not. It seemed to him that his hands were stained with his friend's blood; he shrank from the horror of the ordeal. Ah, what a death to die, amid the crashing ruins of a sinking world! On the last day, among the shattered fragments of the dying Commune, might not this last victim have been spared? He had gone from life, hungering for justice, possessed by the dream that haunted him, the sublime and unattainable conception of the destruction of the old society, of Paris chastened by fire, of the field dug up anew, that from the soil thus renewed and purified might spring the idyl of another golden age.

His heart overflowing with bitter anguish, Jean turned and looked out on Paris. The setting sun lay on the edge of the horizon, and its level rays bathed the city in a flood of vividly red light. The windows in thousands of houses flamed as if lighted by fierce fires within; the roofs glowed like beds of live coals; bits of gray wall and tall, sober-hued monuments flashed in the evening air with the sparkle of a brisk fire of brushwood. It was like the show-piece that is reserved for the conclusion of a fete, the huge bouquet of gold and crimson, as if Paris were burning like a forest of old oaks and soaring heavenward in a rutilant cloud of sparks and flame. The fires were burning still; volumes of reddish smoke continued to rise into the air; a confused murmur in the distance sounded on the ear, perhaps the last groans of the dying Communists at the Lobau barracks, or it may have been the happy laughter of women and children, ending their pleasant afternoon by dining in the open air at the doors of the wine-shops. And in the midst of all the splendor of that royal sunset, while a large part of Paris was crumbling away in ashes, from plundered houses and gutted palaces, from the torn-up streets, from the depths of all that ruin and suffering, came sounds of life.

Then Jean had a strange experience. It seemed to him that in the slowly fading daylight, above the roofs of that flaming city, he beheld the dawning of another day. And yet the situation might well be considered irretrievable. Destiny appeared to have pursued them with her utmost fury; the successive disasters they had sustained were such as no nation in history had ever known before; defeat treading on the heels of defeat, their provinces torn from them, an indemnity of milliards to be raised, a most horrible civil war that had been quenched in blood, their streets cumbered with ruins and unburied corpses, without money, their honor gone, and order to be re-established out of chaos! His share of the universal ruin was a heart lacerated by the loss of Maurice and Henriette, the prospect of a happy future swept away in the furious storm! And still, beyond the flames of that furnace whose fiery glow had not subsided yet, Hope, the eternal, sat enthroned in the limpid serenity of the tranquil heavens. It was the certain assurance of the resurrection of perennial nature, of imperishable humanity; the harvest that is promised to him who sows and waits; the tree throwing out a new and vigorous shoot to replace the rotten limb that has been lopped away, which was blighting the young leaves with its vitiated sap.

"Farewell!" Jean repeated with a sob.

357

"Farewell!" murmured Henriette, her bowed face hidden in her hands.

The neglected field was overgrown with brambles, the roof-tree of the ruined house lay on the ground; and Jean, bearing his heavy burden of affliction with humble resignation, went his way, his face set resolutely toward the future, toward the glorious and arduous task that lay before him and his countrymen, to create a new France.

www.ingramcontent.com/pod-product-compliance
Lightning Source LLC
Chambersburg PA
CBHW051204200326
41519CB00025B/6997